# REGULATION OF GENE EXPRESSION BY SMALL RNAs

# REGULATION OF GENE EXPRESSION BY SMALL RNAs

Edited by
## Rajesh K. Gaur and John J. Rossi

CRC Press
Taylor & Francis Group
Boca Raton London New York

CRC Press is an imprint of the
Taylor & Francis Group, an **informa** business

CRC Press
Taylor & Francis Group
6000 Broken Sound Parkway NW, Suite 300
Boca Raton, FL 33487-2742

**Library of Congress Cataloging-in-Publication Data**

Regulation of gene expression by small RNAs / editors, Rajesh K. Gaur and John J. Rossi.
    p. ; cm.
Includes bibliographical references and index.
ISBN-13: 978-0-8493-9169-9 (hardcover : alk. paper)
ISBN-10: 0-8493-9169-5 (hardcover : alk. paper)
    1. Small interfering RNA. I. Gaur, Rajesh K., 1961- II. Rossi, John J. III. Title.
    [DNLM: 1. MicroRNAs--genetics. 2. Gene Expression Regulation--genetics.
QU 58.7 R344 2009]

QP623.5.S63R44 2009
572.8'8--dc22

2008048939

Visit the Taylor & Francis Web site at
http://www.taylorandfrancis.com

and the CRC Press Web site at
http://www.crcpress.com

# Contents

# Preface

The discovery that gene expression can be controlled via Watson-Crick base pairing of short RNAs to complementary target RNAs (RNA interference) has significantly advanced our understanding of eukaryotic gene regulation and function. The ability of short RNA sequences to modulate gene expression has provided a powerful new tool to study gene function and is about to revolutionize the treatment of disease. Endogenous small RNAs have been found in various organisms, including humans, mouse, flies, worms, fungi, and bacteria. In higher eukaryotes microRNAs may regulate as much as 50% of gene expression. The biogenesis and processing of these regulatory RNAs is an area of intense research interest since they act as cellular rheostats, subtly modulating gene expression.

Targeting cellular genes by exogenous introduction of small interfering RNAs (siRNAs) takes advantage of the endogenous posttranscriptional gene silencing (PTGS) mechanism. The siRNAs can be transfected directly into cells wherein they enter the RNA induced silencing complex (RISC) directly. Alternatively, they can be generated within the cell via gene expression by the use of vectors containing Pol II or Pol III promoters for expression. These RNAi triggers can be expressed in the form of microRNAs or as short hairpins (shRNAs), which are processed into 21-22 nt RNAi triggers. Transcriptional gene silencing (TGS) involves small RNAs complementary to promoter regions that recruit components of the RNAi machinery as well as chromatin remodeling proteins to promoter regions, resulting in transcriptional silencing. TGS has been demonstrated in lower eukaryotes, plants, and most recently in mammalian cells. In addition to the small RNA mechanisms in higher organisms, there are numerous small RNAs in prokaryotic organisms that posttranscriptionally regulate gene expression by a variety of different mechanisms, including Watson-Crick base pairing.

The chapters in this volume cover a wide variety of small RNA regulatory pathways in organisms ranging from bacteria to humans. The breadth of these regulatory pathways and their functional importance in the host organisms are also covered within the volume. Aside from the biological aspects of small RNA mediated regulation of gene expression, techniques for utilizing small RNAs to study gene function or as therapeutic modalities are also discussed. We believe that this volume captures the essence of the breadth and excitement surrounding the newly discovered regulatory roles of small RNAs.

# Acknowledgments

We take this opportunity to acknowledge all the authors for their valuable contributions to the book and to the field of RNA in general. We are also grateful to our colleagues at City of Hope, Marieta Gencheva and Shikha Gaur, for their suggestions and advice throughout the course of this challenging project. Finally, we acknowledge the administrative support of Faith Osep.

# About the Editors

**Rajesh K. Gaur, Ph.D.**, received his doctorate in nucleic acids chemistry at the Institute of Genomics and Integrative Biology and Department of Chemistry, University of Delhi, working under the guidance of M. Atreyi and K.C. Gupta. He spent two years at the European Molecular Biology Laboratory (EMBL), Heidelberg, Germany, and Christian-Albrechts-Universität, Kiel, Germany, as an Alexander von Humboldt Fellow, working with Guido Krupp and Brian Sproat. He then moved to the United States in 1993, where he obtained postdoctoral training in RNA splicing in the laboratories of Tom Maniatis at Harvard University and Michael Green at the University of Massachusetts Medical School. He is currently an assistant professor in the Department of Molecular Biology, Beckman Research Institute of the City of Hope. His research interests include the regulation of alternative splicing in breast cancer and designing of novel RNA-based approaches to target disease state.

**John J. Rossi, Ph.D.**, received his doctorate in microbial genetics from the University of Connecticut at Storrs. For postdoctoral training, Dr. Rossi went to Brown University School of Medicine in Providence, Rhode Island, where he trained under Dr. Arthur Landy, studying the genomic structure, organization, and expression of two gene clusters encoding tRNA-tyrosine in *E. coli*. This research led to the first observation that a tRNA gene cluster was cotranscribed with and subsequently processed from an mRNA. In 1980 Dr. Rossi moved to the Department of Molecular Genetics at the City of Hope in Duarte, California. Dr. Rossi's laboratory began to develop and test the idea of utilizing catalytic RNAs or ribozymes for inhibition of HIV infection. This research program has led to two clinical trials in which ribozyme genes have been transduced into hematopoietic stem cells for autologous transplant in HIV-infected individuals. Work in the laboratory continues to focus upon enhancing the intracellular efficacy of ribozymes and RNA decoys via RNA trafficking and target colocalization approaches. At present a large percentage of the research effort of the lab is focused upon the biology and utilization of small interfering RNAs, or siRNA. This program has led to a first of its kind, a hematopoietic stem cell clinical trial using a triple gene therapy approach in AIDS/lymphoma patients.

# Contributors

**Manu Agarwal**
University of California
Riverside, California

**Hiroji Aiba**
Nagoya University
Nagoya, Japan

**Vladimir Benes**
European Molecular Biology
  Laboratory
Heidelberg, Germany

**Utpal Bhadra**
Centre for Cellular and Molecular
  Biology
Hyderabad, India

**Mirco Castoldi**
University of Heidelberg
Heidelberg, Germany

**Xuemei Chen**
University of California
Riverside, California

**Kwan-Ho Chung**
University of Michigan
Ann Arbor, Michigan

**Julien Curaba**
University of California
Riverside, California

**Alex S. Flynt**
Vanderbilt University
Nashville, Tennessee

**Yoichi R. Fujii**
Nagoya City University
Nagoya, Japan

**Rajesh K. Gaur**
Beckman Research Institute of the City
  of Hope
Duarte, California

**Adam Grundhoff**
Heinrich-Pette-Institute for
  Experimental Virology and
  Immunology
Hamburg, Germany

**Praveensingh B. Hajeri**
Centre for Cellular and Molecular
  Biology
Hyderabad, India

**Artemis G. Hatzigeorgiou**
Institute of Molecular Oncology
Biomedical Sciences Research Center
and
Computer and Information Sciences
University of Pennsylvania
Philadelphia, Pennsylvania

**Zoltan Havelda**
Agricultural Biotechnology Center
Gödöllő, Hungary

**Yusuke Hokii**
Iwate University
Morioka, Iwate,
Japan

**Sumedha D. Jayasena**
Amgen, Inc.
Thousand Oaks, California

**Daniel H. Kim**
Beckman Research Institute of the City
  of Hope
Duarte, California

**Shi-Lung Lin**
University of Southern California
Los Angeles, California

**Molly Megraw**
University of Pennsylvania
Philadelphia, Pennsylvania

**Kevin V. Morris**
The Scripps Research Institute
La Jolla, California

**Martina U. Muckenthaler**
University of Heidelberg
Heidelberg, Germany

**Lori A. Neely**
Exiqon
Woburn, Massachusetts

**Timothy W. Nilsen**
Case Western Reserve University
Cleveland, Ohio

**Mona J. Nolde**
Yale University
New Haven, Connecticut

**Manika Pal-Bhadra**
Indian Institute of Chemical
  Technology
Hyderabad, India

**Paresh D. Patel**
University of Michigan
Ann Arbor, Michigan

**James G. Patton**
Vanderbilt University
Nashville, Tennessee

**S.N.C.V.L. Pushpavalli**
Functional Genomics and Gene
  Silencing Group
Centre for Cellular and Molecular
  Biology
Hyderabad, India

**Linghua Qiu**
University of Massachusetts Medical
  School
Worcester, Massachusetts

**John J. Rossi**
Beckman Research Institute of the City
  of Hope
Duarte, California

**Pål Sætrom**
Norwegian University of Science and
  Technology
Trondheim, Norway

**Praveen Sethupathy**
University of Pennsylvania
Philadelphia, Pennsylvania

**Sunit Kumar Singh**
Centre for Cellular and Molecular
  Biology
Hyderabad, India

**Frank J. Slack**
Yale University
New Haven, Connecticut

**Ola Snøve**
Norwegian University of Science and
  Technology
Trondheim, Norway

**Elizabeth J. Thatcher**
Vanderbilt University
Nashville, Tennessee

**David L. Turner**
University of Michigan
Ann Arbor, Michigan

**Chisato Ushida**
Hirosaki University
Hirosaki, Aomori,
Japan

**Anne B. Vojtek**
University of Michigan
Ann Arbor, Michigan

**Zuoshang Xu**
University of Massachusetts Medical
  School
Worcester, Massachusetts

**Shao-Yao Ying**
University of Southern California
Los Angeles, California

**Yang Yu**
Case Western Reserve University
Cleveland, Ohio

# 1 MicroMining
## *Computational Approaches to microRNA Discovery*

*Adam Grundhoff*

## CONTENTS

## OVERVIEW

The recent past has seen the rapid identification of thousands of microRNAs (miRNAs) encoded by various metazoan organisms as well as some viruses, and it is very likely that many more still await discovery. Most of the hitherto-known miR-NAs have been identified via the cloning and sequencing of small RNAs. While very powerful, this approach is not without its limitations: especially those miRNAs that are of low abundance, or which are only expressed in certain cell types or only during brief periods of organismal development, or are easily missed in cloning-based screens. Thus, alternative means of miRNA discovery are needed.

Given that the signal that marks the miRNA precursor for the cellular processing machinery appears to be a relatively simple one (i.e., a hairpin structure), and considering the rapidly increasing availability of large-scale genomic sequencing data for many organisms, computational methods appear ideally suited for the comprehensive identification of hitherto-unknown miRNAs. This chapter discusses the general principles of computational miRNA identification methods, examines their advantages and disadvantages as compared to the cloning method, and takes a look at the various miRNA prediction algorithms that have been developed recently.

## 1.1   INTRODUCTION

miRNAs are small (~22 nt) RNA molecules that are able to regulate the expression of fully or partially complementary mRNA transcripts. As described in greater detail elsewhere in this book, they are initially transcribed as part of hairpin structures within much larger precursor transcripts (the so-called primary RNAs or pri-mi-RNAs). Following excision of the stem-loops by the RNase III–like enzyme Drosha, the isolated hairpins (called precursor miRNAs or pre-miRNAs) are exported to the cytoplasm and further processed by the Dicer complex to produce the mature, single-stranded miRNA molecule. Recent evidence suggests that plants and animals encode a multitude of miRNAs, many of which are evolutionarily conserved. As of this writing, it is still true that the majority of known miRNAs have been identified experimentally, that is, by cloning of small RNAs. However, this method has certain limitations, and alternative means for the prediction of novel miRNAs are therefore increasingly sought.

The observation that pre-miRNAs form characteristic stem-loops has spurred the development of a number of computational approaches designed to identify novel miRNA candidates based on the prediction and analysis of secondary structures. Given the already complete or near-complete sequencing of whole genomes from many species, such approaches hold great promise for identifying the full comple-ment of miRNAs encoded by a given organism. However, because the precise set of structural features that differentiate a pre-miRNA stem-loop from the large number of hairpins in the genome is not known, additional filters have to be employed to reduce the number of false-positive predictions, and experimental confirmation of the remaining candidates is required. In this chapter, I will compare the benefits and disadvantages of computational miRNA prediction methods in comparison to the cloning method, review principles of the existing miRNA prediction algorithms, discuss the general challenges and pitfalls of in silico miRNA identification, and provide an outlook of what might be expected from these approaches in the future. Finally, I will consider a special application of the miRNA prediction problem: the identification of miRNAs in viral genomes.

## 1.2   WHEN IS A SMALL RNA AN miRNA?

In order to devise approaches designed to identify miRNAs, be they experimental or computational, it is important to clearly define what an miRNA is. In a biologi-cal sense, such a definition is quite straightforward: an miRNA is simply a small, single-stranded regulatory RNA molecule that is generated from its precursor mol-ecules via successive processing by Drosha and Dicer. It is much more difficult, however, to define practicable criteria that are readily testable on an experimental or computational basis and that can unequivocally identify a candidate sequence as a genuine miRNA. Following the realization that miRNAs represent abundant mol-ecules expressed in a wide variety of organisms, a consortium of researchers agreed on a set of criteria that have to be fulfilled before a candidate can be called a bona fide miRNA.[1] According to these guidelines, it is necessary to provide evidence that (1) the candidate sequence is expressed as an appropriately sized RNA molecule in living cells and, furthermore, does not stem from random degradation (*Expression criteria*), and (2) that the maturation of the candidate involves processing by Drosha and Dicer (*Biogenesis criteria*). The expression criteria are preferentially satisfied by detection of a distinct band of approximately 22 nt on a Northern blot. Alternatively, the ability to detect the molecule in a library of cloned, size-selected RNAs is considered sufficient evidence, especially if the library contains high copy numbers of the particular candidate sequence.

To satisfy the biogenesis criteria, the guidelines by Ambros et al.[1] call for experimental proof of Dicer processing by demonstrating that increased levels of the precursor accumulate in cells with decreased Dicer expression. In contrast, experimental proof of Drosha processing is generally not required; instead, it is sufficient to show that the putative precursor transcript has the capacity to adopt a secondary structure that is likely to be amenable to Drosha processing. Of course, given the incomplete knowledge of the rules governing recognition of target mRNAs by Drosha, it is not known what exactly makes a given RNA structure amenable to Drosha processing, and (as will be discussed later) this complicates the computational prediction of miRNA candidates considerably. Based on the characteristics of known miRNA precursor structures, however, it is generally agreed that the minimal requirements are (1) the adopted structure is a hairpin that does not contain many or large internal bulges, and (2) the mature miRNA is to be found within the stem (not the loop) part of the hairpin.

Evolutionary conservation serves as a third biogenesis criterion: As miRNAs are often conserved in closely related (and sometimes even in distant) species, phylogenetic conservation of the miRNA sequence itself as well as its fold-back structure is considered strong evidence that the candidate sequence represents a genuine miRNA. An ideal miRNA candidate would meet all of the preceding criteria; however, it is generally considered sufficient to provide convincing evidence for at least one criterion out of the two categories. Indeed, because Dicer knockout cells are not readily available for most organisms, and effective knockdown of Dicer is technically challenging, positive experimental proof of Dicer processing is rarely shown.

## 1.3 ADVANTAGES AND DISADVANTAGES OF EXPERIMENTAL VERSUS COMPUTATIONAL miRNA IDENTIFICATION

The "traditional" approach to identifying miRNAs consists of cloning of small RNA moieties. Although several protocols for the efficient cloning of such molecules have been devised, they all rely on the common principle of ligating linkers to size-fractionated RNAs, followed by cDNA synthesis and typically PCR amplification. The obtained products are then either cloned (often after concatamerization to increase the information obtained in a single-sequence read) and sequenced, or subjected directly to massive parallel sequencing approaches ("deep sequencing"). According to the guidelines described earlier, these candidates are then further evaluated to ensure that the putative pre-miRNA sequence adopts an appropriate hairpin structure around the candidate. If this is the case, the candidate can generally be considered a bona fide miRNA, since the recovery of the clone from a small RNA library already satisfies the expression criterion (nevertheless, Northern blots are often performed to allow for proper quantification of the miRNA).

The cloning approach has been extremely successful, and although increasing numbers of miRNAs are being identified via computational means, the majority of confirmed miRNAs currently listed in the miRNA database (miRBase, http://microrna.sanger.ac.uk) still have been identified via this method. One of the great advantages of the cloning protocol is that it provides the precise sequence of the mature miRNA molecule. Therefore, in contrast to hybridization-based methods, even closely related miRNAs that differ in only one nucleotide position can be distinguished. Also, the currently available computational prediction tools generally only allow identification of miRNA precursors but do not reliably predict the location of Drosha and Dicer cleavage sites. In contrast, cloning identifies the precise 5′ and 3′ termini of the mature miRNA molecule.

As it appears that nucleotides 2 to 8 of the miRNA (the so-called *seed region*) are especially important for target recognition, knowledge of the precise ends (and particularly the 5′ terminus) is a distinct advantage if a computational prediction of target transcripts is to be performed. As might be expected, the frequency with which a given miRNA is cloned often is approximately equivalent to its abundance (although this frequency may also be affected by other factors; see the following text) and therefore provides a rough estimate of its expression levels. Thus, abundantly expressed miRNAs are usually readily identified. However, it can be challenging to achieve a saturated screen that also captures rare miRNAs. Furthermore, even if such miRNAs are contained within the library, one can never be entirely certain that enough clones have been sequenced to identify all of them.

In addition to these constraints, the scope of a cloning screen is also limited by its source material; naturally, only miRNAs that are expressed in the cells from which the RNA material was derived can be identified. Many miRNAs, however, are expressed in a tissue-dependent manner, or are only expressed at certain developmental stages. This limitation can be partially overcome for relatively simple organisms, where the RNA can be prepared from whole animals (e.g., mixed larvae stages and adults from worms or insects).

In organisms with higher complexity such as vertebrates, however, the situation is more difficult: RNA from different embryonic or adult tissues can be mixed, but the sensitivity of the screen will dramatically decrease with the complexity of the source material, and it is very unlikely that nonabundant miRNAs could be identified in such screens. While these problems could be theoretically solved by massive screening efforts, that is, performing separate screens with material prepared from every individual tissue at each developmental stage, the cloning approach also appears limited in a more fundamental way. Several observations suggest that some miRNAs are more readily cloned than others owing to intrinsic properties such as sequence composition, the presence of certain nucleotides at their termini, or posttranscriptional modifications such as methylation or RNA editing.[2–6]

Computational approaches to miRNA discovery are not subject to many of the limitations that apply to the cloning method. Certainly, one of the biggest advantages of computational miRNA identification is the universal scope of the analysis; as the prediction does not require experimental material, it can potentially discover all of the miRNAs encoded by a given organism, even those that are expressed only at very low levels, in rare cells, or during brief periods of development. However, this advantage is partially annulled by the insufficient precision of the presently available algorithms: as the programs (to varying degrees) produce large numbers of false-positive predictions, experimental verification is still a necessity. Northern blotting is frequently performed to investigate the expression of the computationally predicted candidates, or the predicted sequences are amplified from small RNA libraries. These procedures are not particularly compatible with high-throughput screening, and since many computational methods produce large numbers of candidates, only a small contingent of the predictions is usually subjected to experimental verification, whereas the majority remains untested. More importantly, the experimental valida-tion methods are subject to many of the same limitations that hamper the cloning approach. Thus, even if an experimental verification is attempted and fails, it is often impossible to decide whether the failure was due to a false-positive prediction, insuf-ficient sensitivity of the experimental detection method, or lack of expression in the tested tissue or cell line.

It is thus perhaps not surprising that the expression criterion has not been satisfied for most computationally predicted miRNA candidates. While some groups have attempted to reconcile these difficulties by developing expression analysis tools that are, for example, more sensitive or allow high-throughput screening, there is also tremendous effort to increase the reliability of computational prediction methods such that experimental confirmation is becoming less important.

## 1.4   COMPUTATIONAL PREDICTION OF miRNAs

A plethora of computational approaches aimed at the prediction of miRNAs have been devised, and although nearly all of them use the evaluation of features that are thought to be characteristic for miRNAs in order to identify novel candidates, they vary significantly in scope, complexity, and level of sophistication of the underlying algorithms. Some approaches strive to identify the totality of miRNAs encoded by a given organism, whereas others aim to identify only miRNAs that represent closely related ortho- or paralogs of those that are already known. Some programs investi-gate some of the largest genomes, those of mammals, whereas others consider only some of the smallest, those of viruses.

Despite these differences, most of the approaches function according to a com-mon scheme that might be abstracted as follows. First, a pool of input sequences (usually representing the complete genome of a given organism) is filtered in order to limit the number of candidates that have to be evaluated by downstream algorithms. I will refer to this process as *upstream filtering* in the following. The filtered pool is then subjected to a *structure prediction*. The obtained structures are then compared to those of known pre-miRNAs, and a *score calculation* is performed, depending on the degree of similarity. Finally, *experimental validation* is attempted, usually for a selection of the highest-scoring candidates.

There are considerable differences in the degree to which structural features are investigated during the scoring step; sometimes the filter might simply ensure that the candidate forms a hairpin structure, whereas in other cases it might investigate the candidate's structure down to the minutest detail. The level of sophistication, in large part, will depend on the design of the upstream filter and the efficiency with which this filter preselects a set of candidates enriched for genuine miRNAs. For example, phylogenetic conservation is the most widely used upstream filter (and at least presently, it is also appears to be the most efficient). Indeed, if the sequence of a known mature miRNA is perfectly conserved in a closely (or even distantly) related species, a relatively simple structural analysis that shows that the ability of the surrounding sequences to adopt a fold-back structure is conserved as well might suffice.

In contrast, an ab initio prediction method in which the upstream filter is minimal will require a much more detailed structural analysis during the downstream scoring step. Thus, a highly efficient upstream filter requires a less elaborate downstream structure evaluation, and vice versa. The cloning method might be considered a special case of this scheme in which the upstream filtering is based on an experimental procedure; since this method produces only little background, the subsequent structural investigation can be minimal.

All of the available computational approaches are subject to the production of false-positive (i.e., candidates that pass the filters but do not represent genuine miRNAs) and false-negative predictions (i.e., bona fide miRNAs that are rejected during the upstream filtering or the downstream scoring step). The ratio with which true-positive versus false-positive predictions are made will determine the algorithm's accuracy, while the ratio of true-positive versus false-negative predictions will determine its sensitivity. Such rates are frequently estimated in order to judge an algorithm's performance.

Estimating the rate of false-negative predictions is a relatively straightforward process. Often, only a limited number of the contingent of known miRNAs is used to establish the parameters of the filtering and scoring algorithms. The remaining miRNAs are then subjected to the prediction procedure, and the number of rejected versus retained miRNAs is determined. Alternatively, the full complement of miRNAs is repeatedly passed through the filters, and the method parameters are adjusted until an acceptable ratio between rejected and accepted miRNAs is achieved (what exactly an acceptable ratio is will greatly vary with the overall design and scope of the method).

The estimation of false-positive prediction rates is a more complicated matter: in order to measure such numbers with high reliability, one ideally would have a set of sequences that assuredly does not contain any miRNAs at all, or a set in which all of the genuine miRNAs are known beforehand. In theory, such a set can be created artificially from randomly generated sequences, or by shuffling naturally occurring ones, but since biological sequences are nonrandom, such a reference set would be hardly representative of the experimental sequence set. Alternatively, one might select genetic elements that have known functions and are thus unlikely to additionally represent miRNAs, but this would reduce the complexity of the reference set so drastically that the gained information would be close to meaningless.

In reality, the rate of false-positive predictions is often estimated on an experimental basis. For this purpose, a representative subset of the predictions (or all of them, if only relatively small numbers were predicted) is subjected to experimental verification, for example, by Northern blotting. The candidates for which positive expression data are obtained are then considered the minimal set of true positives within the experimentally investigated subset, and from their percentage the approximate frequencies of true and false positives among all of the predictions are deduced. Of course, given the limitations of experimental verification methods, this rate is only an estimate that represents the lower limit of true-positive predictions. Often, an allowance is made to account for undetectable, but nevertheless genuine, miRNAs, and this allowance is then used to calculate an upper bound. These numbers are often highly hypothetical, however, since they are usually not based on primary experimental or bioinformatic data.

### 1.4.1 GETTING STARTED: UPSTREAM FILTERING

Considering the various prediction methods, perhaps the highest level of diversity is found with regard to the upstream filtering methods. This first filter receives the input sequences (usually the complete genomic sequence of a given organism) and identifies the candidates that have a high probability of representing miRNAs. The filtering methods differ greatly in the degree by which they reduce the complexity of the initial sequence pool: some are minimally restrictive (e.g., filters that remove only repetitive sequences), whereas the discriminatory power of others might be greater than that of the downstream structural analysis and scoring step. Often, several filters are combined in order to achieve a more efficient preselection.

The largest group of upstream filters is based on phylogenetic conservation.[7–19] Once significant numbers of miRNAs had been cloned from different organisms, it became clear that many of them are conserved between closely related species, and some of them even in distant relatives. The simplest algorithms that exploit this fact are those that start with the sequences of already-known miRNAs, then search the genomes of closely or distantly related species for perfect or near-perfect matches.[9,15,20] Despite the fact that miRNAs are short, such matches are usually relatively rare even on a genomic scale, and thus the complexity of the sequence pool passed on to downstream analysis is comparably low. Of course, while such approaches are able to find orthologs of already-known miRNAs, they fail to identify novel miRNAs, even if these are conserved between different species. Therefore, the majority of approaches based on phylogenetic conservation filter for segments that are conserved between closely related species but do not require that their sequences be similar to known miRNAs. Such regions can be identified by using whole-genome alignments,[8,12,21–23] or by performing a BLAST alignment of a region of interest against the genome of a related organism.[10,16,17,19]

Whole-genome alignments have a higher probability of representing significant regions of homology, whereas a BLAST alignment can produce many spurious hits that have to be investigated separately.

On the other hand, conservation of an miRNA only requires sequence identity within the short region that encodes the mature molecule, and such regions might be missed in a global alignment if the flanking segments are not conserved, whereas they might score in a BLAST search. Of course, both types of alignments will potentially recover all different kinds of functional elements, not only those representing miRNAs. Therefore, additional filters are frequently used in order to exclude known elements that are unlikely to contain miRNAs. Some of the most prominent of these are protein-coding regions.

Since most miRNAs are located either in intergenic regions (approximately two-thirds of all miRNAs) or in the introns of known genes (the remaining third), but are rarely found within the exons of protein-coding transcripts, annotated exons are often excluded from the analysis. This also removes the bulk of highly conserved elements, thereby drastically reducing the complexity of the original sequence set. There are, however, a few miRNAs that are located within the 3′ or 5′ UTRs of protein-coding genes, or that even overlap with the coding regions themselves. Nevertheless, these miRNAs represent rare exceptions to the rule, and the increase in specificity gained by the removal of coding exons by far outweighs the disadvantage of losing such infrequent outliers.

Whereas many algorithms only require conservation of a putative miRNA sequence (i.e., conservation of an approximately 22 nt segment), other approaches search for more complex sequence conservation patterns that are typical for pre-miRNAs. By investigating a reference set of evolutionary conserved miRNAs, Lai and colleagues[12] observed that, whereas the segment encoding the mature miRNA molecule often is perfectly conserved, the sequences representing the other arm of the hairpin are subject to less (but nevertheless noticeable) evolutionary pressure, presumably because they only need to retain sufficient base paring with the miRNA-encoding arm to preserve a proper hairpin structure.

In contrast, the terminal loop part that separates the two segments is frequently only poorly conserved. If considering the whole pre-miRNA sequence, this often leads to a saddle-shaped conservation pattern, with the saddle nose being representative of the mature miRNA. By looking for the occurrence of such profiles in alignments between genomic alignments of *Drosophila melanogaster* and *Drosophila pseudoobscura*, Lai et al. could identify 48 novel candidates, 24 of which were experimentally confirmed.[12]

A similar but further refined method was used by Berezikov et al.[8] in order to identify vertebrate miRNAs. Using a phylogenetic shadowing approach, Berezikov and colleagues[8] could show that, in addition to the conservation patterns observed by Lai et al.,[12] the degree of conservation drops sharply within the sequence segments, which immediately flank the pre-miRNA hairpin. By scanning for regions with such conservation profiles in genomic human/mouse and human/rat alignments, Berezikov et al. identified nearly a thousand novel miRNA candidates, many of which are also conserved in other vertebrate genomes. Out of a representative set of 69 miRNAs, 16 could be verified by experimental means.

The potential existence of common sequence motifs located in sequences adjacent to pre-miRNA hairpins has also been investigated. Assuming that such elements could be important for the processing or transcriptional control of pri- and/or pre-miRNAs, Ohler et al.[17] extracted the 5′- or 3′-genomic segments (with sizes of 1 and 2 kbp, respectively) flanking known *Caenorhabditis elegans* pre-miRNAs, and searched these sequences for the occurrence of conserved sequence motifs. The authors identified a short (8 bp), highly conserved motif located approximately 200 bp upstream of many independently transcribed miRNAs (i.e., miRNAs that are not located in the introns of protein-coding genes). Indeed, incorporating a corresponding filter into their prediction algorithm improved its accuracy significantly. At present, it is unclear whether this sequence is involved in the transcription of pri-miRNAs (i.e., represents a transcription factor binding site) or is rather recognized by one of the factors involved in miRNA processing. The latter, however, appears unlikely since the motif appears not to be present upstream of miRNAs located in the introns of protein-coding transcripts (which are very likely processed by the same factors required for the maturation of independently transcribed miRNAs). The fact that another study did not find any conserved sequences in the vicinity of vertebrate miRNAs[8] also argues against a principal role for simple-sequence motifs in miRNA processing.

Not only conservation of the miRNAs themselves but also conservation of their target sites can be used to identify novel miRNA candidates. This approach appears especially promising in plants, since plant miRNAs exhibit perfect or near-perfect complementarity to their target transcripts. Accordingly, in order to discover novel miRNAs in *Arabidopsis thaliana*, Adai and colleagues[7] aligned all annotated transcripts to intergenic regions, reasoning that, in such alignments, miRNAs (or, strictly speaking, their reverse complement) contained within the intergenic set would reveal themselves as short segments that align with their target sites contained within the transcript set. Since this method is not based on interspecies conservation, it should be able to identify conserved as well as species-specific miRNAs. Because of the abundance of putative miRNA/miRNA target pairs, however, Adai et al. had to additionally filter their list of candidates for those that are conserved in rice in order to obtain a more specific data set.[7]

While the preceding approaches are based on sequence alignments alone, several approaches use simultaneous alignment of sequence as well as structure in order to identify conserved miRNA para- or orthologs. Since the structure information augments recognition of homologous pre-miRNAs, such algorithms can sometimes recognize pre-miRNAs that have diverged too far to be identified by primary sequence homology alone. The lower the requirement for primary sequence identity, the higher the probability that such an approach will be able to identify distant miRNA homologue. Legendre et al.[14] employed the ERPIN program[25] to capture primary sequence as well as secondary structure information of known pre-miRNAs in the form of profiles.

Using these profiles to perform homology searches in several animal genomes, the authors were able to detect several miRNA orthologs that could not be identified by conventional BLAST searches. Instead of ERPIN profiles, Wang et al.[18] used an algorithm called RNAforester[26] to generate local structure alignments. Nam and colleagues[27] developed a probabilistic colearning approach that uses an abstracted model to capture sequence and structure information, as well as information regarding the location of the mature miRNA within the pre-miRNA hairpin, and employed an algorithm trained with known miRNAs to screen four human chromosomes. Out of 23 candidates, 9 could be confirmed experimentally by demonstrating pre-miRNA accumulation upon Drosha knockdown. The sequence divergence between the 23 candidates and the training set was high enough so that the primary similarity was not readily detectable, indicating that the approach was indeed able to identify distant homologue.

Whereas the approaches described use conservation patterns to search specifically for miRNA candidates, a number of recent studies aimed at the universal identification of all functional DNA or RNA elements have also revealed miRNAs or their target sites. Lunter et al.[28] used a neutral indel model to search for functional DNA. This model assumes that insertions or deletions (indels) should be evenly (or neutrally) distributed across DNA that does not carry functional elements but should less frequently affect segments that contain such elements. Compared to methods based on primary sequence conservation, the indel approach has the advantage of being able to identify elements that are not primarily dependent on sequence conservation to retain their function, or require only the conservation of motifs that are too short to register in global alignments (a fitting definition for pre-miRNAs, where often only the sequence encoding the mature miRNA is highly conserved). By analyzing indel frequencies in human, mouse, and dog alignments, Lunter et al.[28] found that 2.6–3.3% of human euchromatin exhibit lower indel frequencies as would be expected under a neutral model. Since only 1.2% of the euchromatin are accounted for by annotated protein-coding genes, the rest is likely to represent other functional elements. Interestingly, while the identified sequences represent only approximately 3% of the genome, they contain almost 75% of all hitherto-known miRNAs. Although the authors noted that this sequence pool is also very likely to contain novel miRNAs, they did not attempt to identify any such candidates.

Whereas the indel method identifies functional elements in DNA, several other approaches use sequences corresponding to known mRNA transcripts or RNA structures as a starting point.[22,23,29,30] Xie et al.[23] performed a systematic analysis of human, mouse, rat, and dog genomes to generate a global catalog of common sequence motifs in promoters and 3′ UTRs. For this purpose, the 3′ UTRs of reference mRNAs were extracted from the RefSeq database, and a library of likely promoter sequences was built by selecting 4-kbp windows centered on the transcriptional start sites. By performing multispecies sequence alignments, the authors identified 174 candidate motifs in the promoter set, and 106 motifs in the 3′-UTR set. The promoter set contained many known sequence elements, especially transcription factor binding sites, whereas nearly all of the 3′-UTR motifs were unknown. However, Xie and colleagues noticed that the motifs were strongly biased toward a length of eight bases and had a propensity to begin with a "U."[23] Since miRNAs often begin with an "A," and the motif length corresponded well with the length of animal miRNA seed regions, Xie et al.[23] hypothesized that these motifs might represent signatures of miRNA binding sites. Indeed, searching among the known miRNAs revealed that almost one-half of the 8-mer 3′-UTR motifs were likely to represent such sites. By performing a genome-wide search for complementary matches to the 8-mer motifs, and subjecting the obtained sequences to secondary structure prediction and analysis, the authors were able to identify 129 novel miRNA candidates. Out of a representative set of 12 candidates, 6 could be confirmed experimentally by screening a small RNA library generated from pooled human tissues.

In a similar approach, Chan et al.[29] searched for conserved 3′-UTR motifs using an approach based on network-level conservation between the genomes of *Drosophila melanogaster* and *pseudoobscura*, or *Caenorhabditis elegans* and *briggsae*. They, too, observed that the resulting sequence pool was highly enriched for motifs that were similar to those observed by Xie et al.,[23] and identified 50 novel miRNA candidates in worms, although none of these were subjected to experimental validation attempts. Why do miRNA binding sites, but not the miRNAs themselves, register as conserved motifs in these screens? Since the input sequences consisted of spliced mRNAs, but nearly all miRNAs are located in intergenic regions or in the introns of protein-coding genes, the overwhelming majority of miRNA sequences had simply not been present in the analyzed datasets.

In another approach, Washietl and colleagues[22,30] developed a different method to detect functional RNA on a global scale, based on a search for conserved, thermodynamically stable structures. In this approach, the most highly conserved regions from genome-wide alignments of several vertebrate species were first identified. Since the authors were interested in noncoding RNAs, they then removed all annotated protein-coding exons from the data set (thus, in contrast to the previous approaches, this sequence pool retains the majority of miRNA-encoding regions). The remaining sequences were subsequently screened for functionally important RNA using a computational algorithm (called RNAz) that (1) scores conservation of structural features across multiple species and (2) determines a measure of the thermodynamic stability of the structures. The latter is calculated as a z-score and does not simply represent minimal free-folding energy (MFE) values.

Depending on their nucleotide composition and length, even random sequences can have quite low, and therefore seemingly significant, MFEs. This must be taken into account when calculating thermodynamic stability values, for example, by demonstrating that structures adopted by randomly shuffled control sequences have, on average, markedly higher MFEs. Indeed, Bonnet et al. had previously observed that pre-miRNAs score exceptionally high in this regard.[31] Thus, it is perhaps not surprising that the analysis by Washietl et al., among a multitude of other noncoding elements, detected not only known miRNAs but also many new candidates; 472 of these candidates overlapped with the nearly 1000 predictions from the phylogenetic shadowing approach by Berezikow et al.[8] While no attempt was made to confirm the candidates experimentally, several other studies further examined the produced candidate data set computationally, or employed the RNAz algorithm to search for novel miRNAs in other organism.[32–34]

All of the algorithms mentioned above use various aspects of phylogenetic conservation to search for novel miRNA candidates. Of course, while conservation represents a powerful filter, it will also efficiently eliminate species-specific miRNAs. Therefore, it is desirable to also devise procedures that are able to perform an ab initio prediction of potential miRNAs. As discussed later, in the special case of virus genomes, such an approach is even mandatory, since viral miRNAs are often not conserved even within closely related families. However, whereas the small size of viral genomes allows prediction methods that do not use any upstream filters at all, in more complex organisms the increased frequency of false-positive predictions poses a far greater problem. To reduce the rate of false positives, Sewer et al.[35] selectively focused on miRNAs that are found within close proximity of already-known miRNA clusters. For this purpose, the authors extended the limits of existing clusters by 10 kbp on each side, followed by systematic analysis of all hairpin structures found within such regions. Out of 89, 66, and 105 candidates that were identified in the human, rat, and mouse genome, respectively, experimental evidence of expression could be found in 20, 17, and 6 cases.

Whereas this method is limited to the detection of clustered miRNAs, Bentwich et al.[36] developed a global ab initio prediction and validation method to search for conserved as well as nonconserved miRNAs within the complete human genome. Their approach consisted of scanning the entire human genome for hairpin structures with a minimal filter that removed only hairpins overlapping with repeated or protein-coding sequences. These hairpins were then subjected to structure evaluation and scoring. As might be expected from the absence of additional filtering methods, a high number of hairpins (~434,000) exceeded a minimal score threshold. To cope with this large number, Bentwich et al. used a high-throughput microarray analysis to test a selection of 5300 candidates for evidence of expression. Out of 886 candidates that had survived this test, 359 were subjected to a confirmatory test. For this purpose, bead-bound oligos complementary to the sequences of interest were used to capture, amplify, and sequence specific clones from an expression library of small RNAs. A total of 89 experimentally verified novel miRNAs could be identified using this approach. Thus, a relatively elaborate downstream procedure had to be developed in order to balance the loss of specificity caused by the absence of stringent upstream filters.

Recently, Miranda et al.[37] reported a novel, pattern-based method to identify miRNA targets. Using a set of known miRNAs, the authors trained an algorithm to discover variable-length motifs (so called *patterns*) among the mature miRNA sequences. These patterns were represented by regular expressions (Regex), a method that is frequently used in bioinformatics to condense a complex pattern shared by multiple strings within a single, specially formatted character sequence. By searching mRNA sequences for matches to the total collection of such expressions, the authors were able to predict with a very high accuracy of almost 75% whether a given transcript was subject to miRNA regulation. Using patterns derived from precursors instead of mature miRNAs (approximately 190,000 patterns from a total of 530 nonredundant pre-miRNA sequences), Miranda et al. also adapted this method to predict pre-miRNA candidates. The algorithm positively identifies 93% of the miRNAs contained in the training set, and under conditions where the estimated error rate is 1%, it predicts the lower limit of the total number of miRNAs within the mouse and human genomes to be at an astounding 25,000. This by far exceeds previous estimates of approximately 1000 human miRNAs.[8] As of this writing, however, none of the novel predictions have been validated experimentally.

### 1.4.2 FOLLOWING THROUGH: STRUCTURE PREDICTION AND SCORING

Prediction of secondary structures and their scoring is the step that follows most of the filtering procedures described in the previous paragraph. The task of this step is to determine which structure is most likely to be adopted by the sequences contained within the filtered pool and then to calculate a score by comparing its similarity to the predicted structure of known pre-miRNAs. Given that the Drosha complex is likely to recognize its substrates primarily by structure, such methods alone should theoretically be able to discriminate genuine pre-miRNAs from other hairpin structures. However, a number of problems, some of which will be discussed in the following text, prevent efficient classification based exclusively on structural analysis.

Structure prediction is most frequently performed by submitting sequence segments of 80 to 500 bases to either the Mfold[38] or the RNAfold[39] algorithm. Both algorithms perform a structure prediction based on free-energy minimization, and, especially if larger sequences are analyzed, frequently return multiple alternative structures with very similar MFE values. To keep the computational burden on a manageable scale, miRNA prediction programs usually evaluate only the structure with the lowest MFE value. This poses the risk that genuine pre-miRNA hairpins might be missed, especially if larger sequence windows that also contain flanking sequences are analyzed. As the pre-miRNA hairpins themselves frequently exhibit low MFEs,[31] however, they are often prominently represented within the contingent of alternative structure predictions. Following secondary structure prediction, the obtained hairpins are evaluated for a number of criteria that are thought to be characteristic for genuine pre-miRNAs. The various scoring algorithms differ vastly in the number, nature, and weighting of the features that are considered. A detailed examination of all these methods would by far exceed the scope of this chapter; in the following text, I will therefore only discuss the general principles and fundamental problems.

The difficulties in selecting criteria that positively identify pre-miRNAs result in large part from the fact that the molecular basis of substrate recognition by the Drosha complex is very poorly understood. The selection of considered criteria is therefore almost exclusively based on statistical analysis of the pool of already known miRNAs and the predicted structure of their precursors. The problem, then, is to find the set of features that most efficiently discriminates bona fide pre-miRNAs from other structures. In bioinformatics, such problems are commonly encountered in machine-learning scenarios in which a training set (in this case, the pool of known pre-miRNA hairpins) is compared to a control set (e.g., a collection of randomly selected sequences) in order to define parameters that allow the classification of a test set (the pool of genomic sequences, or a filtered subselection thereof) into members being more similar to either the training or the control set. A popular algorithm choice for performing such tasks are the support vector machines (SVMs), which represent a set of related supervised learning methods. Several studies employed SVMs for miRNA classification,[33,35,40] whereas others used classifiers specifically developed for miRNA prediction purposes or, following statistical evaluation of known pre-miRNA precursors, even manually defined scoring formulas.

Certainly, a great deal of variability between the results of different miRNA prediction programs stems from the usage of different classifiers, but even using the same classifier would not guarantee identical (or even similar) results. While computational classifiers such as SVMs are powerful tools, they can neither define what a feature is nor can they generate the set of features that is evaluated during training. Instead, this has to be done manually. Since this selection is often somewhat arbitrary and depends on the preferences of the individual groups, classifiers have been fed with various combinations of different sequence and structural features, including (but not limited to) stem-loop length, length of the longest helix within the stem, terminal loop size, length distribution between hairpin arms, number of internal bulges, symmetry and size of internal bulges, position of bulges, sequence composition of bulges, average distance between internal loops, overall base composition of the hairpin, and frequency of certain bases within the stem or bulges. Depending on the combination of input features, the same classifier will return different results even when trained with the same sets of positive and negative controls.

Of course, the ability of any classifier to make an effective prediction will critically depend on the set of features with which it is initially trained, and the inaccuracy of the current algorithms suggests that the right set has not been found yet. Considering the list of features that have already been used, it appears possible that only more complex ones, such as variable-length motifs, will allow the efficient classification of miRNAs, as suggested by the recent study from Miranda et al.[37] It is also possible, however, that a better understanding of the molecular Drosha and/or Dicer processing steps will help to improve structure-based miRNA prediction. For example, recent studies have shown that miRNA maturation requires unstructured RNA sequences that flank the pre-miRNA, and that these regions are bound by DGCR8, a component of the Dicer complex,[41,42] which also positions the processing centre of Drosha within the stem of the pre-miRNA. Incorporating such findings into computational algorithms might help to both improve the prediction of novel pre-miRNAs as well as the prediction of the precise Drosha cleavage sites (and therefore of one of the ends of the mature miRNA).

### 1.4.3  WRAPPING IT UP: EXPERIMENTAL VALIDATION

While Northern blotting remains the gold standard for experimental miRNA verification, it is not suited to the analysis of the typically large numbers of candidates produced by computational prediction tools. For this and other reasons (namely, the often unsatisfactory sensitivity of Northern blots), alternative methods for miRNA validation have been sought (for a more in-depth reviews of confirmatory methods, see Aravin and Tuschl[43] and Bentwich[44]). For example, some approaches employ linker-based PCR amplification as a confirmatory test. The method uses linker-ligated cDNAs generated from small RNA material; these cDNAs are then probed by PCR amplification using one linker-specific primer and one primer that is specific to the miRNA candidate of interest. By sequencing the amplificates, one of the miRNA's termini can be identified, and the invariability of this terminus also excludes the possibility that the amplified cDNA originated from random degradation products. Due to the amplification step, the method has the advantage of being much more sensitive than Northern blotting. Although it is also less laborious, it is nevertheless not feasible to screen hundreds or even thousands of candidates using this method. The same is true for RNase protection or primer extension assays. So far, the only high-throughput methods that have been used to test the expression of large miRNA numbers are microarray- or bead-based hybridization protocols.

Present microarray layouts can accommodate tens of thousands of sequences and are therefore especially suited for downstream validation of computationally predicted candidates. Small RNAs can be efficiently labeled with fluorophores, either directly or during cDNA synthesis, and microarray hybridization of such material can be used to accurately monitor miRNA expression (Reference 45, reviewed in References 43 and 44). While vast numbers of candidates can be tested simultaneously in this manner, microarrays cannot rule out the possibility that a positive signal results from random degradation products or from the cross-hybridization of closely related miRNAs of similar sequence (although the latter is also true for Northern blotting). Nevertheless, microarray screening can be enormously useful to identify the most promising among a large list of computationally predicted candidates.[36,46] These select candidates can then be subjected to either Northern blotting[46] or other, more sensitive, tests[36] for final validation.

## 1.5  VIRAL miRNAs

Given the importance of miRNAs in cellular gene expression, it would be surprising if viruses would not exploit the pathway for their own purposes. Indeed, a number of viral miRNAs have been identified by the cloning method[40,47–52] as well as by computational approaches.[40,46,53–56] For several reasons, computational prediction represents an especially attractive method for viral miRNA discovery. First, if productive viral replication results in the destruction of the host cell (which frequently is the case), small RNA libraries cloned from such cell populations will be highly enriched for degradation products, which makes it more difficult to achieve a saturated screen (of course, there are also many viruses for which infected tissues are difficult to acquire, and for which no in vitro infection system exists; this, however, will also make the experimental confirmation of predicted candidates impossible). Second, because of their small genome size, the number of predicted miRNA candidates in viruses is substantially lower as compared to more complex organisms. On the downside, miRNAs appear to be highly specific for individual viruses and are frequently not conserved even between the members of a given virus family, likely reflecting the fact that viruses are highly adapted organisms. There are a few examples in which miRNAs are conserved between very closely related viruses,[52–54] but for most viruses such close relatives either do not exist or remain yet to be discovered. Thus, conservation-based filters are generally ill suited to predict virally encoded miRNAs, and ab initio prediction methods are needed instead.

If viruses with genomes of just a few thousand bases are investigated, one may be able to employ relatively low-stringency ab initio prediction methods without the use of any upstream filters and, despite the relatively high rate of false positives produced by such methods, nevertheless end up with just a few candidates that can be easily subjected to Northern blot validation.[53] The situation is different, however, for large viruses such as herpesviruses (which have DNA genomes up to ~200 kbp in size). Here, either more stringent prediction algorithms or high-throughput screening methods are required to weed out false positives. Higher stringency, of course, also increases the risk that genuine miRNAs are missed. A recent study not only used a SVM-based classifier to predict miRNAs in several herpesviruses but also performed independent cloning experiments from virally infected cells.[40] The algorithm predicted some of the cloned miRNAs but missed several others that were cloned in the same and/or were identified in other studies.[46,49,50] In another study, the authors used constraints for GC-content and complexity to filter their predictions,[56] but these filters also removed a genuine miRNA of antiapoptotic function identified by another group.[55] The use of microarrays represents an excellent alternative to the use of stringent prediction methods, since the number of predictions from large viral genomes, even under minimal filtering conditions, will still be small enough to easily fit on an array. Starting by applying a refined, low-stringency ab initio prediction algorithm originally developed for small viral genomes[53] to the genomes of herpesviruses, we have used microarrays to select a subset of candidates that were then subjected to Northern blot confirmation.[46] By doing so, we could identify several miRNAs that had been missed by other approaches using more stringent computational algorithms or the traditional cloning method.[40,49–52]

## 1.6 CONCLUSIONS

In recent years, miRNAs have emerged as central regulators of cellular gene expression. Knowledge of the full repertoire of miRNAs encoded by a given organism is a necessary (but certainly not sufficient) prerequisite to understand the highly complex network underlying miRNA-mediated gene regulation. While it is still difficult to give an estimate of the total numbers of miRNAs, recent studies indicate that there may be thousands or even tens of thousands of mammalian miRNAs, and the rapid identification of such large numbers will require highly efficient computational tools. Massive efforts to improve the accuracy of existing miRNA prediction algorithms are being made, and novel approaches are developed on an almost daily basis. However, to what degree these algorithms will become independent from experimental verification remains unknown. In the best possible case, tomorrow's prediction methods will be able to globally identify miRNAs with such high confidence that validation by experimental means will be superfluous. At worst, these tools will remain more or less restricted to their present role, that is, they will continue to deliver candidate sequences that are to be tested experimentally. The real-world scenario will very likely be somewhere between the two extremes. Whatever the outcome, there can be little doubt that, without computational methods, the identification of all the miRNAs encoded by complex organisms would be impossible.

## REFERENCES

1. Ambros, V., Bartel, B., Bartel, D. P., Burge, C. B., Carrington, J. C., Chen, X., Dreyfuss, G., Eddy, S. R., Griffiths-Jones, S., Marshall, M., Matzke, M., Ruvkun, G., and Tuschl, T., A uniform system for microRNA annotation, *RNA* 9 (3), 277–279, 2003.
2. Elbashir, S. M., Lendeckel, W., and Tuschl, T., RNA interference is mediated by 21- and 22-nucleotide RNAs, *Genes Dev* 15 (2), 188–200, 2001.
3. Ambros, V., Lee, R. C., Lavanway, A., Williams, P. T., and Jewell, D., MicroRNAs and other tiny endogenous RNAs in *C. elegans*, *Curr Biol* 13 (10), 807–818, 2003.
4. Luciano, D. J., Mirsky, H., Vendetti, N. J., and Maas, S., RNA editing of a miRNA precursor, *RNA* 10 (8), 1174–1177, 2004.

5. Yang, W., Chendrimada, T. P., Wang, Q., Higuchi, M., Seeburg, P. H., Shiekhattar, R., and Nishikura, K., Modulation of microRNA processing and expression through RNA editing by ADAR deaminases, *Nat Struct Mol Biol* 13 (1), 13–21, 2006.
6. Kim, V. N. and Nam, J. W., Genomics of microRNA, *Trends Genet* 22 (3), 165–173, 2006.
7. Adai, A., Johnson, C., Mlotshwa, S., Archer-Evans, S., Manocha, V., Vance, V., and Sundaresan, V., Computational prediction of miRNAs in *Arabidopsis* thaliana, *Genome Res* 15 (1), 78–91, 2005.
8. Berezikov, E., Guryev, V., van de Belt, J., Wienholds, E., Plasterk, R. H., and Cuppen, E., Phylogenetic shadowing and computational identification of human microRNA genes, *Cell* 120 (1), 21–24, 2005.
9. Dezulian, T., Remmert, M., Palatnik, J. F., Weigel, D., and Huson, D. H., Identification of plant microRNA homologs, *Bioinformatics* 22 (3), 359–360, 2006.
10. Grad, Y., Aach, J., Hayes, G. D., Reinhart, B. J., Church, G. M., Ruvkun, G., and Kim, J., Computational and experimental identification of C. elegans microRNAs, *Mol Cell* 11 (5), 1253–1263, 2003.
11. Jones-Rhoades, M. W. and Bartel, D. P., Computational identification of plant microRNAs and their targets, including a stress-induced miRNA, *Mol Cell* 14 (6), 787–799, 2004.
12. Lai, E. C., Tomancak, P., Williams, R. W., and Rubin, G. M., Computational identification of *Drosophila* microRNA genes, *Genome Biol* 4 (7), R42, 2003.
13. Lee, R. C. and Ambros, V., An extensive class of small RNAs in *Caenorhabditis elegans*, *Science* 294 (5543), 862–864, 2001.
14. Legendre, M., Lambert, A., and Gautheret, D., Profile-based detection of microRNA precursors in animal genomes, *Bioinformatics* 21 (7), 841–845, 2005.
15. Li, Y., Li, W., and Jin, Y. X., Computational identification of novel family members of microRNA genes in *Arabidopsis thaliana* and *Oryza sativa*, *Acta Biochim Biophys Sin (Shanghai)* 37 (2), 75–87, 2005.
16. Lim, L. P., Lau, N. C., Weinstein, E. G., Abdelhakim, A., Yekta, S., Rhoades, M. W., Burge, C. B., and Bartel, D. P., The microRNAs of *Caenorhabditis elegans*, *Genes Dev* 17 (8), 991–1008, 2003.
17. Ohler, U., Yekta, S., Lim, L. P., Bartel, D. P., and Burge, C. B., Patterns of flanking sequence conservation and a characteristic upstream motif for microRNA gene identification, *RNA* 10 (9), 1309–1322, 2004.
18. Wang, X., Zhang, J., Li, F., Gu, J., He, T., Zhang, X., and Li, Y., MicroRNA identification based on sequence and structure alignment, *Bioinformatics* 21 (18), 3610–3614, 2005.
19. Wang, X. J., Reyes, J. L., Chua, N. H., and Gaasterland, T., Prediction and identification of *Arabidopsis thaliana* microRNAs and their mRNA targets, *Genome Biol* 5 (9), R65, 2004.
20. Weber, M. J., New human and mouse microRNA genes found by homology search, *FEBS J* 272 (1), 59–73, 2005.
21. Lim, L. P., Glasner, M. E., Yekta, S., Burge, C. B., and Bartel, D. P., Vertebrate microRNA genes, *Science* 299 (5612), 1540, 2003.
22. Washietl, S., Hofacker, I. L., Lukasser, M., Huttenhofer, A., and Stadler, P. F., Mapping of conserved RNA secondary structures predicts thousands of functional noncoding RNAs in the human genome, *Nat Biotechnol* 23 (11), 1383–1390, 2005.
23. Xie, X., Lu, J., Kulbokas, E. J., Golub, T. R., Mootha, V., Lindblad-Toh, K., Lander, E. S., and Kellis, M., Systematic discovery of regulatory motifs in human promoters and 3′ UTRs by comparison of several mammals, *Nature* 434 (7031), 338–345, 2005.
24. Boffelli, D., McAuliffe, J., Ovcharenko, D., Lewis, K. D., Ovcharenko, I., Pachter, L., and Rubin, E. M., Phylogenetic shadowing of primate sequences to find functional regions of the human genome, *Science* 299 (5611), 1391–1394, 2003.
25. Gautheret, D. and Lambert, A., Direct RNA motif definition and identification from multiple sequence alignments using secondary structure profiles, *J Mol Biol* 313 (5), 1003–1011, 2001.

26. Hochsmann, M., Toller, T., Giegerich, R., and Kurtz, S., Local similarity in RNA secondary structures, in *Proceedings of the IEEE Computational Systems Bioinformatics Conference*, Stanford, CA, 2003, 159–168.

27. Nam, J. W., Shin, K. R., Han, J., Lee, Y., Kim, V. N., and Zhang, B. T., Human microRNA prediction through a probabilistic co-learning model of sequence and structure, *Nucleic Acids Res* 33 (11), 3570–3581, 2005.

28. Lunter, G., Ponting, C. P., and Hein, J., Genome-wide identification of human functional DNA using a neutral indel model, *PLoS Comput Biol* 2 (1), e5, 2006.

29. Chan, C. S., Elemento, O., and Tavazoie, S., Revealing posttranscriptional regulatory elements through network-level conservation, *PLoS Comput Biol* 1 (7), e69, 2005.

30. Washietl, S., Hofacker, I. L., and Stadler, P. F., Fast and reliable prediction of noncoding RNAs, *Proc Natl Acad Sci USA* 102 (7), 2454–2459, 2005.

31. Bonnet, E., Wuyts, J., Rouze, P., and Van de Peer, Y., Evidence that microRNA precursors, unlike other non-coding RNAs, have lower folding free energies than random sequences, *Bioinformatics* 20 (17), 2911–2917, 2004.

32. Hsu, P. W., Huang, H. D., Hsu, S. D., Lin, L. Z., Tsou, A. P., Tseng, C. P., Stadler, P. F., Washietl, S., and Hofacker, I. L., miRNAMap: Genomic maps of microRNA genes and their target genes in mammalian genomes, *Nucleic Acids Res* 34 (Database issue), D135–139, 2006.

33. Hertel, J. and Stadler, P. F., Hairpins in a Haystack: Recognizing microRNA precursors in comparative genomics data, *Bioinformatics* 22 (14), e197–202, 2006.

34. Missal, K., Zhu, X., Rose, D., Deng, W., Skogerbo, G., Chen, R., and Stadler, P. F., Prediction of structured non-coding RNAs in the genomes of the nematodes *Caenorhabditis elegans* and *Caenorhabditis briggsae*, *J Exp Zoolog B Mol Dev Evol* 306 (4), 379–392, 2006.

35. Sewer, A., Paul, N., Landgraf, P., Aravin, A., Pfeffer, S., Brownstein, M. J., Tuschl, T., van Nimwegen, E., and Zavolan, M., Identification of clustered microRNAs using an ab initio prediction method, *BMC Bioinformatics* 6, 267, 2005.

36. Bentwich, I., Avniel, A., Karov, Y., Aharonov, R., Gilad, S., Barad, O., Barzilai, A., Einat, P., Einav, U., Meiri, E., Sharon, E., Spector, Y., and Bentwich, Z., Identification of hundreds of conserved and nonconserved human microRNAs, *Nat Genet* 37 (7), 766–770, 2005.

37. Miranda, K. C., Huynh, T., Tay, Y., Ang, Y. S., Tam, W. L., Thomson, A. M., Lim, B., and Rigoutsos, I., A pattern-based method for the identification of MicroRNA binding sites and their corresponding heteroduplexes, *Cell* 126 (6), 1203–1217, 2006.

38. Mathews, D. H., Sabina, J., Zuker, M., and Turner, D. H., Expanded sequence dependence of thermodynamic parameters improves prediction of RNA secondary structure, *J Mol Biol* 288 (5), 911–940, 1999.

39. Hofacker, I. L., Fontana, W., Stadler, P. F., Bonhoeffer, S., Tacker, P., and Schuster, P., Fast folding and comparison of RNA secondary structures, *Monatshefte f. Chemie* 125, 167–188, 1994.

40. Pfeffer, S., Sewer, A., Lagos-Quintana, M., Sheridan, R., Sander, C., Grasser, F. A., van Dyk, L. F., Ho, C. K., Shuman, S., Chien, M., Russo, J. J., Ju, J., Randall, G., Lindenbach, B. D., Rice, C. M., Simon, V., Ho, D. D., Zavolan, M., and Tuschl, T., Identification of microRNAs of the herpesvirus family, *Nat Methods* 2 (4), 269–276, 2005.

41. Han, J., Lee, Y., Yeom, K. H., Nam, J. W., Heo, I., Rhee, J. K., Sohn, S. Y., Cho, Y., Zhang, B. T., and Kim, V. N., Molecular basis for the recognition of primary microRNAs by the Drosha-DGCR8 complex, *Cell* 125 (5), 887–901, 2006.

42. Zeng, Y. and Cullen, B. R., Efficient processing of primary microRNA hairpins by Drosha requires flanking nonstructured RNA sequences, *J Biol Chem* 280 (30), 27595–27603, 2005.

43. Aravin, A. and Tuschl, T., Identification and characterization of small RNAs involved in RNA silencing, *FEBS Lett* 579 (26), 5830–5840, 2005.

44. Bentwich, I., Prediction and validation of microRNAs and their targets, *FEBS Lett* 579 (26), 5904–5910, 2005.
45. Barad, O., Meiri, E., Avniel, A., Aharonov, R., Barzilai, A., Bentwich, I., Einav, U., Gilad, S., Hurban, P., Karov, Y., Lobenhofer, E. K., Sharon, E., Shiboleth, Y. M., Shtutman, M., Bentwich, Z., and Einat, P., MicroRNA expression detected by oligonucleotide microarrays: System establishment and expression profiling in human tissues, *Genome Res* 14 (12), 2486–2494, 2004.
46. Grundhoff, A., Sullivan, C. S., and Ganem, D., A combined computational and microarray-based approach identifies novel microRNAs encoded by human gamma-herpesviruses, *RNA* 12 (5), 733–750, 2006.
47. Burnside, J., Bernberg, E., Anderson, A., Lu, C., Meyers, B. C., Green, P. J., Jain, N., Isaacs, G., and Morgan, R. W., Marek's disease virus encodes MicroRNAs that map to meq and the latency-associated transcript, *J Virol* 80 (17), 8778–8786, 2006.
48. Dunn, W., Trang, P., Zhong, Q., Yang, E., van Belle, C., and Liu, F., Human cytomegalovirus expresses novel microRNAs during productive viral infection, *Cell Microbiol* 7 (11), 1684–1695, 2005.
49. Samols, M. A., Hu, J., Skalsky, R. L., and Renne, R., Cloning and identification of a microRNA cluster within the latency-associated region of Kaposi's sarcoma-associated herpesvirus, *J Virol* 79 (14), 9301–9305, 2005.
50. Cai, X., Lu, S., Zhang, Z., Gonzalez, C. M., Damania, B., and Cullen, B. R., Kaposi's sarcoma-associated herpesvirus expresses an array of viral microRNAs in latently infected cells, *Proc Natl Acad Sci USA* 102 (15), 5570–5575, 2005.
51. Pfeffer, S., Zavolan, M., Grasser, F. A., Chien, M., Russo, J. J., Ju, J., John, B., Enright, A. J., Marks, D., Sander, C., and Tuschl, T., Identification of virus-encoded microRNAs, *Science* 304 (5671), 734–736, 2004.
52. Cai, X., Schafer, A., Lu, S., Bilello, J. P., Desrosiers, R. C., Edwards, R., Raab-Traub, N., and Cullen, B. R., Epstein–Barr virus microRNAs are evolutionarily conserved and differentially expressed, *PLoS Pathog* 2 (3), e23, 2006.
53. Sullivan, C. S., Grundhoff, A. T., Tevethia, S., Pipas, J. M., and Ganem, D., SV40-encoded microRNAs regulate viral gene expression and reduce susceptibility to cytotoxic T cells, *Nature* 435 (7042), 682–686, 2005.
54. Grey, F., Antoniewicz, A., Allen, E., Saugstad, J., McShea, A., Carrington, J. C., and Nelson, J., Identification and characterization of human cytomegalovirus-encoded microRNAs, *J Virol* 79 (18), 12095–12099, 2005.
55. Gupta, A., Gartner, J. J., Sethupathy, P., Hatzigeorgiou, A. G., and Fraser, N. W., Anti-apoptotic function of a microRNA encoded by the HSV-1 latency-associated transcript, *Nature* 442 (7098), 82–85, 2006.
56. Cui, C., Griffiths, A., Li, G., Silva, L. M., Kramer, M. F., Gaasterland, T., Wang, X. J., and Coen, D. M., Prediction and identification of herpes simplex virus 1-encoded microRNAs, *J Virol* 80 (11), 5499–5508, 2006.

# 2 Animal microRNA Gene Prediction

*Ola Snøve and Pål Sætrom*

## CONTENTS

## OVERVIEW

MicroRNAs (miRNA) are small non-protein-coding RNAs with important roles in animal development and disease. While some miRNAs are abundantly expressed and can be cloned easily, others show a temporal and tissue-specific expression pattern that makes them harder to identify. This creates the need for computational methods to predict miRNA genes. This chapter presents the basis for such algorithms and provides a guide to both conservation-based and nonconservation-based miRNA prediction algorithms.

## 2.1   INTRODUCTION

MicroRNAs were discovered only 6 years ago [1–3], but the research field has moved with impressive determination and progress. Before the realization that miRNAs constituted an abundant class of non-protein-coding RNAs, only *lin-4* and *let-7* were known. The founding members of the miRNA class were called small temporal RNAs due to their expression and function during specific stages of larval development in *Caenorhabditis elegans* [4–7]. Unlike *lin-4*, the *let-7* miRNA has orthologs in many species, including vertebrates [8]. This prompted the search for other RNAs and laid the foundation for the initial discovery of the class as a whole.

MicroRNAs are either transcribed from independent polymerase II promoters [9] or excised from the introns or exons of protein-coding genes [10]. At least one example exists to show that an miRNA may stem from a transcript that could also function as an mRNA [11]. One interesting feature of miRNAs is that they often appear in clusters throughout the genome [12,13]. One study showed that miRNAs separated by up to 50 kb were coexpressed [14]. Little is known about the regulation of miRNAs, but recent evidence suggests that posttranscriptional mechanisms are important determinants for miRNA expression [15,16].

The relatively long primary miRNA transcripts (pri-miRNA) depend on at least four different protein complexes to mature into the single-stranded RNAs that guide targeting (reviewed in References 17 and 18): First is the Microprocessor complex, which comprise the ribonuclease III Drosha and DGCR8, processes the pri-miRNAs into miRNA precursors (pre-miRNA) [19–22]. Second is a carrier protein called Exportin-5 which participates with Ran GTPase in pre-miRNA transport from the nucleus and release into the cytoplasm [23–25]. Third is the Dicer, which is another ribonuclease III, processes pre-miRNAs into the mature duplexes that are unwound and incorporated into the RNA-induced silencing complex (RISC) in a process that involves the Tar RNA-binding protein (TRBP) [26–28]. Finally, the RISC can cleave [29,30], degrade [31,32], or suppress translation [4,5] of mRNAs that have some degree of sequence complementarity to the miRNA component of RISC.

Based on the analysis of conserved sites in genes' 3′ untranslated regions (UTR), miRNAs appear to have the potential to regulate thousands of genes, maybe as much as one-third of the human transcriptome [33–35]. Overexpression of miR-1 and miR-124 confirmed this potential, as each miRNA downregulated about a hundred messages as measured by microarrays—a result that suggests that degradation may be more important than previously recognized, as translational suppression should not be observable by microarrays [36]. As with RNA interference (RNAi; reviewed in References 37–39) off-target effects [40,41], the reverse complement of nucleotides 2–7 on miRNAs' 5′ end were the most significant sequence element in miRNA-regulated genes. Many miRNA targets do not require such a seed match [42–44], however, which is why the seed requirement by itself is not enough to construct a good miRNA target prediction algorithm [45].

Approaches for miRNA gene discovery include both computational and experimental methods (reviewed in Reference 46). While forward genetics approaches were used to identify the first miRNA examples, these methods are generally not well suited for miRNA discovery for at least two reasons. First, even if a mutation in the short miRNAs can be introduced, it may not change the phenotype, as targeting is flexible with respect to the degree of target complementarity [42]. Second, several miRNAs may collaborate to silence targets to varying degree [47], which means that some miRNAs can be redundant [48]. Cloning approaches have been much more successful in identifying new miRNAs. For some time, mapping of ~22 nucleotide clones from size-fractioned cDNA libraries to genomic hairpin regions have been the preferred approach for miRNA discovery. Deep sequencing has lately been used to identify a new type of abundant RNAs in mammalian testes [49–51]. The technology's ability to generate tens of thousands of reads in a single run could overcome the problem with rare, low, or condition-dependent miRNA expression, as sequencing from a range of samples may be feasible.

Despite the advances in cloning techniques, computational approaches are still needed. First, their sequence composition makes some miRNAs difficult to clone. Second, editing mechanisms may alter the sequence from the genomic version [52–54]. Third, mapping of clones to a genomic hairpin is not trivial, and the hairpin still needs to possess the characteristics that would make it a likely miRNA. Close to 500 human miRNAs are currently known [55], but recent estimates suggest that as many as 1000 miRNAs may exist [56], which means that miRNA discovery is still important.

This chapter will outline the general principles for miRNA gene prediction, review some of the algorithms that are available to researchers, and identify the challenges that we believe are most important for the future of computational miRNA discovery.

## 2.2 MICRORNA CHARACTERISTICS

The miRNA definition has been and should still be debated. The online clearing-house for newly discovered miRNAs is miRbase [55,57], which assigns unique names to miRNA candidates that have been validated according to criteria for expression and biogenesis [58]. Specifically, the original criteria demands evidence for expression by the detection of a distinct ~22 nucleotide RNA transcript, typically by Northern blotting or from a cDNA library. Furthermore, this mature transcript should map to a region with the potential to fold into a hairpin structure—prefer-ably one that is conserved in other species—and the precursor should accumulate in Dicer-deficient cells [58]. It may be difficult, however, to separate miRNAs from similar small non-protein-coding RNAs [59,60], which is why some have argued that the transcript should also have several other characteristics that, among other things, make the miRNA candidate a likely or validated Drosha substrate [46,61]. One should be aware that any classifier built from examples in miRbase depends on the database's accuracy, and computational approaches will enforce errone-ous annotations by looking for similar molecules. As miRbase already contain entries with some characteristics that make them different from the first miRNAs that were identified [61], stronger evidence requirements should be considered for miRNA annotation.

Even though most review articles describe mature miRNAs as ~22 nucleotides, one should not forget that there is a range of sizes. Actually, the mature human miRNAs in the current version of miRbase (release 8.2; 462 entries) have a median length of 23 nucleotides, but range from 18 through 26 nucleotides. Similarly, human miRNA precursors come in different lengths, but most are between 50 and 80 nucle-otides with a median of 60.

Mature miRNAs stem from hairpin transcripts whose characteristics determine how efficiently they can be processed. Consequently, information about the critical factors—be it sequence or structure elements—that are required for efficient biogen-esis can be used to separate true from false miRNA candidates. In principle, some characteristics could exist to support efficient transition from one state to the next for all the steps of miRNA biogenesis—that is, for pri-miRNA to pre-miRNA process-ing by the Microprocessor complex, nuclear export of pre-miRNAs by Exportin-5, pre-miRNA processing into miRNA*:miRNA duplexes by Dicer, and unwinding of the duplex and efficient incorporation of mature miRNAs into RISC by TRBP and other proteins.

We have previously shown that known mammalian miRNAs possess several characteristics that are conserved between different miRNA families [62]. Both sequence and structure features are common among known miRNAs. For example, the GC-content of pri-miRNAs is markedly higher outside of the pre-miRNA region than in other hairpins. Also, some bases are over- and underrepresented in certain positions in the pre-miRNA, such as uracil being common in positions 1, 6, 9, and 18, and cytosine uncommon in position 20, counted from the precursor 5' end. The regions corresponding to positions 3–8 and 14–21 are often stable, whereas internal loops and bulges often disrupt positions 1 and 10–12. These observations are consistent with other accounts in the literature [63–65]. It is important to relate these features to the biogenesis and function of miRNAs, as sequence and structure characteristics could be important determinants for appropriate processing.

Several groups have shown that the initial miRNA biogenesis step, the processing of pri-miRNAs into pre-miRNAs by the Microprocessor complex, is important for efficient progression through the remaining processing steps. Interestingly, we found that both sequence and structure are conserved in the 5' and 3' flanking regions of the pre-miRNA. First, as in the pre-miRNA, there are some position-dependent nucleotide conservation patterns; for example, cytosine is more common in nucleotides 19 to 23 upstream of the 5'-processing site, whereas adenine is less common in nucleotides 2 to 11 downstream of the 3'-processing site. Second, the double-stranded hairpin structure is conserved in the 13 nucleotides upstream of the Microprocessor site that defines the 5' end of pre-miRNAs, but more distal nucleotides tend to be unpaired or base-pair with other nucleotides in the same flanking region [62]. Others have also seen a dependency on the sequences flanking the pre-miRNA for Microprocessor processing [22,66,67]. A recent paper confirmed the importance of this region, as the Microprocessor complex seems to depend on a downstream distance of about 11 nucleotides from the first unpaired 3' nucleotide to create the correct precursor (see Figure 2.1) [68]. Including the two nucleotides that form the 3' overhang, this distance corresponds exactly to the one found in the conservation analysis. The authors claim that the terminal loop is unessential, which is in contrast to previous evidence suggesting that the terminal loop must be relatively large [67].

If the loop turns out to be important, it seems to matter only for the Microprocessor recognition step. Several reports have shown that short hairpin RNA constructs—synthetic molecules that are similar to Microprocessor products and that trigger stable RNAi—with loop sizes ranging from one through nine nucleotides function almost equally well [69–72]. Thus, since these shRNA constructs were not designed to go through Microprocessor processing, a small terminal loop does not seem to compromise Exportin-5 function. One study of the ability of a mir-30 precursor to migrate from the nucleus to the cytoplasm shows that Exportin-5 likely recognizes most other parts of the pre-miRNA [73]. Proper Microprocessor processing seems to be important, as 5' overhangs are inhibitory for transport, suggesting that the two-nucleotide 3' overhang of pre-miRNAs is pivotal for cytoplasmic export. The study

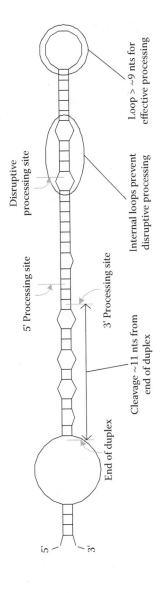

**FIGURE 2.1** Structural features implicated in Microprocessor processing of pri-miRNAs. The figure shows the hsa-mir-23a hairpin structure as predicted by RNAfold [77]. The current pri-miRNA-processing model predicts that the Microprocessor complex recognizes double-stranded structures flanked by single-stranded regions, such as large internal or terminal loops, and processes the duplex about 11 nucleotides from the single-stranded structure [68]. In addition to the correct miRNA-processing sites, this can result in disruptive processing where the Microprocessor cleaves the pri-miRNA about 11 nucleotides from the terminal loop, but other features likely ensure that the Microprocessor favors the correct processing site. A large terminal loop is one likely feature, as a small terminal loop can give reduced pre-miRNA processing [67] and increased disruptive processing [68]. V. Narry Kim and colleagues also speculated that internal loops close to a potential processing site may prevent disruptive Microprocessor processing [68], but the general abundance of internal loops close to the correct processing site seems to contradict this theory [62].

also confirmed that the stem must be more than about 16 nucleotides, which means that hairpin structures without the potential to encode a mature miRNA of close to 20 nucleotides are suboptimal substrates for Exportin-5.

The next step of miRNA biogenesis is that of Dicer processing, which results in a cut approximately two helical turns, or about 20 nucleotides, away from the double-stranded stem's 5' end. Again, the precursor's two-nucleotide 3' overhangs have been shown to be crucial both for precursor recognition and correct processing [74]. The product, often an imperfect duplex with two-nucleotide 3' overhangs at both ends, is then unwound and incorporated into RISC. The duplex's thermodynamic stability profile plays a role in the preferential incorporation of one of the strands [63]. Even though it is not a perfect correlation, and small differences between the ends may render the process somewhat stochastic, the strand with a lower 5' stability is more likely to be incorporated into RISC [75]. Another protein may bind to the other end and thereby determine the orientation of the duplex before RISC incorporation, but so far this has only been demonstrated in *Drosophila* [75].

MicroRNA gene prediction algorithms can take advantage of all the aforementioned features to predict a novel candidate or confirm that an expressed hairpin is a true miRNA. As we will see in the next section, many of the features have indeed been used in currently available miRNA gene prediction algorithms. Note that sequence or structure conservation between species is a powerful filter that can significantly reduce an algorithm's number of false-positive predictions. While several algorithms use this step, however, such filters can be applied to the output of any prediction algorithm. Thus, where possible, algorithms should generally be compared without the use of these filters.

Several waves of algorithm development have occurred. The first approaches were pragmatic in terms of the biological end product, which was to identify as many new miRNAs as possible. Sequence similarity search algorithms can take advantage of the sequence of known miRNAs to search for paralogs in the same species and orthologs in related species. For example, the let-7 family consists of at least 11 highly conserved human members. Currently, 12 paralogs are known in mouse, and 10 in rat. As time went by, developers of miRNA prediction algorithms started to pay more attention to the fine-tuning of parameters to optimize various performance measures. This was a natural development, as it became harder to identify new miRNAs with time. The coming sections will present the underlying logic of the pioneering algorithms and the three approaches to miRNA gene prediction that are currently used: prediction of homologs to known miRNAs, prediction based on conservation patterns between species, and de novo prediction based on characteristic miRNA features.

## 2.3 ALGORITHMS THAT PIONEERED COMPUTATIONAL miRNA PREDICTION

Three pioneering algorithms came out about 2 years after the initial discovery of the miRNA gene class. Altogether, 106 novel miRNAs in worms, flies, and humans were validated from candidates identified with the first versions of these algorithms [76].

Arguably the most prominent of the pioneer algorithms, MiRscan, has been used to predict miRNAs in both nematodes and vertebrates. In both cases, the secondary structure properties of 110-nucleotide sequence windows were evaluated for hairpin potential using RNAfold [77]. In the nematode predictions, conservation between *Caenorhabditis elegans* and *Caenorhabditis briggsae* were used to reduce false-positive predictions [78]. The authors suggest that more genomes should be used to increase the performance, as conservation between three vertebrates, namely *Homo sapiens, M. musculus,* and *Fubu rubribes,* increased the performance of the same algorithm despite the larger vertebrate genomes [65]. This may, however, be because the human miRNAs that were known at the time were highly conserved. Using more genomes will reduce false positives but will restrict the predictions to those miRNAs that are well conserved.

MiRscan slides a 21-nucleotide window along the 110-nucleotide hairpin and scores each window according to a weighted set of properties ranging from distance from the loop to 3′- and 5′-stem conservation. These weights were optimized to capture the characteristics of the miRNAs that were known. A revision of MiRscan came when the same group discovered sequence conservation patterns and a motif in the region flanking the stem [79]. The difference between the two versions is simply that features from the flanking region were included among the others to better discriminate between miRNAs and random hairpins. In their first publication, the authors point out that the base pairing within the miRNA contributes most to the algorithm's performance [78].

Another algorithm, miRseeker, was proposed following analysis of a reference set that consisted of 24 miRNAs in *Drosophila* [80]. The algorithm demanded only moderate conservation, as only two species were used to obtain conserved candidates. One hundred nucleotide regions conserved between *Drosophila melanogaster* and *D. pseudoobscura* were ranked according to their secondary structure as determined with Mfold [81]. Finally, Boolean filters were applied to ensure that the conservation pattern of miRNA candidates were similar to those in the reference set—that is, conserved nucleotides are more prevalent in the active strand than in the loop or nonactive strand (miRNA*-strand).

The third algorithm in the first wave of development also predicted miRNAs in *C. elegans* [82]. Hairpin candidates were extracted from the genome using an algorithm called Srnaloop that looks for small reverse complementary sequence windows within a certain distance from each other. Several filter steps, including one for GC-content, another for folding energy, and one for structure similarity to known miRNAs, were applied before three sets were created based on conservation to other species. The miRNA candidates in the first set consisted of those conserved between *C. elegans, D. melanogaster,* and *C. briggsae;* the second of those conserved between *C. elegans, D. melanogaster,* and *H. sapiens;* and the third of *C. elegans* homologs to known miRNAs from *D. melanogaster, M. musculus,* and *H. sapiens.*

From a bioinformatics perspective, only miRscan remains relevant, as the Boolean filter approach of the other algorithms does not allow for tradeoffs between different miRNA features; either a sequence meets all criteria or it does not. This will prevent the methods from taking full advantage of all the information that is available on known miRNAs, as adding more Boolean filters will increase an algorithm's specificity but decrease its sensitivity.

## 2.4 ALGORITHMS THAT IDENTIFY HOMOLOGOUS miRNAs

Together with the initial and subsequent small RNA cloning studies, the pioneering miRNA gene prediction algorithms identified several hundred miRNAs in mammals, fish, worms, and flies (reviewed in Reference 83). Many of the miRNAs were orthologs, and some were paralogs; some of these were identified by homology searches. Consequently, developing algorithms that find homologous miRNA genes became an active research topic. The simplest method to search for genes that are homologous to known miRNAs is to identify sequences that share the miRNAs' primary sequence—for example, by using standard tools such as BLAST [84] or BLAT [85]. Finding a sequence that is similar to a known miRNA does not, however, guarantee that the new sequence also is an miRNA. As a minimum, the sequence must also fold into an miRNA's characteristic hairpin structure. Consequently, using standard sequence homology search tools to look for miRNA genes requires additional secondary structure filters to ensure that the hairpin structure is conserved [86,87].

Several algorithms have integrated the search for sequences with homologous sequence and structure to known miRNAs and have shown superior performance to the simpler-sequence homology tools [88,89]. Legendre et al. described an approach that used primary and secondary structure profiles to identify homologous miRNA precursors [88]. From a multiple alignment of animal miRNAs, they built 18 different profiles, each representing miRNAs with similar sequences and structures, and showed that these profiles could make more sensitive and specific predictions than a pure sequence-based method. Instead of building profiles for similar miRNAs, miRAlign scores miRNA candidates based on their total sequence and structure similarity to known miRNAs [89]. The similarity score also considers the putative mature miRNA's position in the hairpin. miRAlign uses BLAST to identify miRNA candidates, and although their specificities are comparable, miRAlign's predictions are more sensitive than the profile search by Legendre and colleagues.

## 2.5 ALGORITHMS THAT USE CONSERVATION CHARACTERISTICS TO IDENTIFY miRNAs

Detecting a hairpin that has sequence and structure similar to a known miRNA is a good indication that the hairpin is an miRNA. Indeed, sequence and structure similarity to a known miRNA is an accepted criterion for annotating new miRNA orthologs [58]. Sequence and structure conservation can, however, also be used to find miRNAs with no homology to a known miRNA. Sequence and structure conservation in genome multiple alignments is a general method for finding non-protein-coding RNAs (ncRNAs; reviewed in Reference 90), and programs such as RNAz have rediscovered many of the known miRNAs [91]. Many classes of ncRNAs exist (reviewed in Reference 92), but these general programs do not attempt to annotate their predictions. The RNAmicro program attempts to alleviate this problem by specifically detecting miRNA genes among the ncRNAs predicted by RNAz [93]. In addition to structural characteristics, RNAmicro bases its predictions on the characteristic pattern that the mature miRNA sequence is more conserved than other parts of the miRNA hairpin [78,80].

An miRNA gene-finding approach that depends solely on conservation was published 2 years after the pioneering algorithms. Phylogenetic shadowing is a technique for comparison of sequences from closely related species that allows short alignments to determine the conservation for each nucleotide [94]. To determine the conservation details for the regions surrounding known miRNAs, Plasterk and co-workers sequenced 700 bp around 122 miRNAs in 10 different primates to analyze the conservation pattern of miRNAs [95]. As was already known, mutations in the loop are much more frequent than mutations in the stem, especially in the functional arm. They used this information to search for similar conservation profiles in whole-genome alignments for human and mouse on the one hand and human and rat on the other hand. The Randfold algorithm [96] was used to ensure that these typically conserved regions could indeed fold into hairpins that could be miRNA candidates. With their algorithm, they found 83% of the human miRNAs that were known at the time, but validated only 16 of the 69 experimentally tested candidates.

An interesting and indirect way of exploiting conservation is to search for target motifs in 3' UTRs and then relate these to conserved hairpins in the genome. Xie et al. found 106 motifs in human 3' UTRs that are highly conserved in rat, mouse, and dog UTRs [35]. Almost half of these approximately 8-mer motifs corresponded to the reverse complement of the 5' end of known miRNA genes, which confirms the results of others showing that the 5' end is more conserved than the 3' end [33–36]. When the authors searched for stretches of nucleotides within hairpin structures that, when reverse complemented, matched the conserved motifs, they predicted 129 new miRNAs. Six of 12 candidates that were chosen for experimental validation turned out to be expressed.

The premise for all miRNA prediction algorithms that depend on sequence or structure conservation is that all miRNAs are well conserved. Since miRNAs were discovered following the realization that *let-7* was widely conserved, it was only natural to include some kind of conservation in the first prediction algorithms. One should remember, however, that the first miRNA to be discovered was *lin-4*—necessary for normal transition from the first to the second larval stage in worms [7], yet probably dispensable for other species as it is an example of a nonconserved miRNA. A large group of miRNAs that appear to have emerged in primates was also recently discovered [56]. This further questions the validity of conservation filters—at least the most stringent ones. Since conservation filters can be applied to all algorithms as a way of lowering the false-positive rate if nonconserved candidates can be sacrificed, it makes more sense to develop algorithms that perform well without using a conservation filter by default.

## 2.6   DE NOVO PREDICTION ALGORITHMS

A common theme for the algorithms that do not rely on conservation is that they put more emphasis on the recognition step and less emphasis on the filters, including the requirements for conservation in other species. One should note, however, that some of the algorithms that were described in the previous sections could easily omit the conservation requirement of the initial step. In particular, this is the case for MiRscan, which has a powerful recognition step at its core [78]. Many of the algorithms that will be described in the following text use machine learning techniques to take maximum advantage of the information contained in the sequences and structures of experimentally validated miRNAs. Needless to say, as time goes by, more information becomes available, and it is therefore not a surprise that more recent algorithms perform better than some of the pioneering algorithms.

Microarrays that detect miRNA expression enable researchers to submit more candidates to an initial step that can precede cloning and sequencing. One group tested about 5300 human hairpin candidates for expression, detected expression from 359 hairpins, and confirmed 89 of these to be novel miRNAs by sequencing [56]. The algorithm that detected the candidates consisted of two steps. First, the entire human genome was folded into potential hairpins using the Vienna package [77]. Second, the PalGrade algorithm scored each candidate according to several criteria that separates known miRNAs from negative controls: minimum thresholds were assigned to 28 features that were ranked according to their stringency. The final score depends on the number of features that are satisfied—if a hairpin does not pass the first feature's threshold, it receives a zero score; if a hairpin passes the five least stringent thresholds but misses the sixth, it receives a score of 5/28 no matter how many of the remaining features it satisfies. The predictions were then separated in conserved and nonconserved candidates. Importantly, 53 of the 89 validated miRNAs were not conserved beyond primates.

Many of the aforementioned algorithms have tried to assign weights to features that were assumed to be important. For example, Lim et al. [65,78] use log-odds scores to find the weights that best separate true miRNAs from random genomic hairpins. Even though the approach by Bentwich et al. [56] is less founded in statistical methods, their aim is the same, namely to find a way to rank the features according to their importance. Another way to find such weights is to use a machine-learning algorithm to optimize classifiers that separate between true and false examples. Usually, machine-learning algorithms will provide better classifiers than more ad hoc approaches. The trade-off is that many machine-learning algorithms are black-box classifiers, meaning that it can be difficult to comprehend their biological meaning.

## 2.7  BLACK-BOX APPROACHES

Typically, a two-way classification problem, such as separating miRNA genes from other types of genomic hairpins, starts by generating a training set of positive and negative examples. It is important that the information is balanced and that all families are well represented. Hence, it is also important that any significant bias is removed from the training set. For example, as previously mentioned, the human *let-7* family currently has 11 members that are very similar. If all 11 members are included in a training set, some machine-learning algorithms may produce less reliable classifiers than if this bias is removed.

When the training set has been constructed, algorithms are trained to separate between true and false examples based on input that, in principle, may be anything from raw sequences to user-generated feature vectors. The trained classifiers output a score, and a cut-off score must be chosen so that the candidates that receive a score higher than the threshold are perceived to be miRNAs, whereas the others are discarded on grounds of being too dissimilar to known miRNAs, as measured by the particular classifier. This does not mean they cannot be miRNAs; just that they are different from most known miRNAs. Note that, as one increases the threshold, an increasing number of known miRNAs will be missed by the algorithm, which illustrates the importance of an appropriate cut-off score.

Of course, higher cut-off scores mean higher probabilities for predicted candidates being true miRNA genes. However, if you opt for a threshold that is too high, you will lose many miRNA genes as well. The trade-off between sensitivity and specificity is important and must be studied when two or more classifiers are compared. Receiver operating characteristics (ROC) curves show this relationship. The area under the curve says something about a classifier's overall performance, but certain regions of the ROC curve may be more interesting, depending on the application. For example, if the validation protocol can only handle a small number of candidates, it may not be that important to pick up all miRNAs, and so the high-specificity region is more important than the high-sensitivity region. Also, one should keep in mind that the sensitivity and specificity measures depend on the data set characteristics, that is, the number of positive and negative examples in the set and how similar they are. Thus, sensitivities and specificities may not be directly comparable across data sets.

Various machine-learning algorithms have been used to predict miRNA genes. ProMir uses Hidden Markov Models (HMM) to predict miRNA candidates in a sequence [97]. The algorithm folds the sequence into a probable hairpin structure using programs in the Vienna package [77]. All HMMs work by estimating the hidden parameters based on the observable parameters [98]. Here, the observable parameters correspond to the sequence and structure properties of the hairpins. ProMir scans each position of the duplex in an miRNA candidate and assigns one of four states that depend on the structural relationship to the opposite base—it is paired, unpaired, deleted, or inserted. Furthermore, each state is further classified based on the base identity. For example, pairs can be G:C, A:U, G:U, or identical pairs with the order reversed. The hidden states of the Markov model correspond to whether the hairpin is an miRNA or not. The authors estimated ProMir's performance using fivefold cross validation. Even though these measures of performance may not be comparable across data sets, ProMir's sensitivity of 73% and specificity of 96% is an indication of its performance. The algorithm is publicly available and accessible through a Web interface [99].

Instead of trying to discover the features directly, as in the HMM approach, the features can be predetermined based on evidence in the literature. Then, machine learning's job is to find a classifier based on these features—an optimal weighting of sequence and structure characteristics—that captures the differences between known miRNAs and background hairpin samples. Support vector machines (SVM) were popularized in the late nineties and are ideally suited to such tasks [100].

Previous uses of SVMs include predictions of promoter regions [101], splicing sites [102], translation initiation sites [103], proteins [104], and siRNA efficacy [105,106]. Several groups have also used SVMs to predict miRNAs. For example, Xue et al. presented a triplet-SVM classifier based on the secondary structure [107]. The nucleotides of hairpins whose secondary structure did not contain multiple loops were annotated as paired or unpaired. With each position either paired or unpaired, three adjacent positions can have eight encodings based on its structure (2^3). When the identity of the middle nucleotide is encoded as well, that brings the total number of possibilities to 32 (4 * 2^3) triplets. The SVM was then trained to separate between pre-miRNAs and other hairpin structures. Another approach computed a number of features, including hairpin length, distance from the hairpin loop, number of base pairs, and more—40 features in total—and used an SVM to separate between miRNAs and other hairpins [108]. To improve the prediction rate, they take advantage of the clustering property of many known miRNAs. Consequently, the authors focus especially on the regions close to known miRNAs and limit the output to those candidates that are conserved between species.

One of the main problems when setting up a machine-learning experiment to create an miRNA gene predictor is to assemble a good training set. The positive set should consist of all known miRNAs, and while some measures should be taken to account for many examples within some families—the let-7 family is particularly big—the positive set can be expected to do a decent job at representing known types of miRNAs. The problem, however, is to construct a training set with representative negative examples. We recently used a two-step approach to create a classifier that predicts miRNA genes [61]. First, a classifier was trained to separate true Microprocessor processing sites—that is, the end of pre-miRNAs—from the other positions of known pri-miRNAs. If the highest-scoring position for a given hairpin candidate is used to indicate whether or not that structure is a Microprocessor substrate, it was a decent miRNA predictor by itself. However, when the output was used to find good negative samples, and the algorithm retrained on the new training set, it produced a far better classifier.

## 2.8  OUTLOOK

A comparison of the various algorithms' performance would naturally follow the presentation of the various miRNA gene prediction algorithms in the previous sections. Figure 2.2 shows a comparison of the performance of various machine-learning-based algorithms. This comparison is based on the algorithms' estimated sensitivities and specificities reported in the respective papers, but as all the five publications used different data sets, the results may not be directly comparable. To partially solve this problem, we tested the algorithms' ability to predict the 130 new human miRNAs in miRBase 8.1. Importantly, these miRNAs were, to the best of our knowledge, not part of the algorithms' training sets, and should therefore be a fair test-set for all the algorithms. As Figure 2.3 shows, the four algorithms we were able to test have reduced performance on this test set—the sensitivities of all algorithms are about 20 percentage points less on the 8.1 test set. A straightforward interpretation of this result is that all algorithms have overfitted their training sets

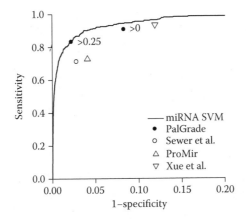

**FIGURE 2.2** Performance comparison of the five current de novo miRNA gene prediction algorithms. The plot shows the reported specificities and sensitivities for miRNA SVM [61], PalGrade [56], Sewer et al. [108], ProMir [97], and Xue et al. [107].

and therefore have problems recognizing these miRNAs, possibly because some of these new sequences have slightly different characteristics than the previously known miRNAs. Indeed, they do differ, as these sequences were verified as miRNAs only based on a single sequence clone. Interestingly, the sequences that have additional evidence for being miRNAs more resemble the previously known miRNAs than the single clone sequences [61]. Whether these sequences are a subspecies of lowly expressed miRNAs or RNAs from a different pathway remains to be shown.

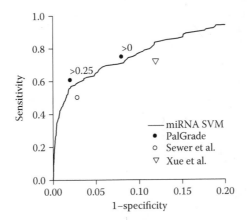

**FIGURE 2.3** All de novo algorithms have reduced sensitivity when predicting the 130 new human miRNAs in miRBase 8.1. The reported sensitivities are based on the same cut-offs as in Figure 2.2. The sensitivities for Sewer et al. [108] and Xue et al. [107] are based on the 123 and 126 miRNAs that the algorithms scored, whereas the sensitivities for miRNA SVM and PalGrade are based on all 130 miRNAs. ProMir was not included because of problems with using the ProMir Web server for multiple predictions [99].

We can roughly divide the different methods discussed in this review into three groups: (1) methods that find miRNAs that are homologous to already known miRNAs, (2) de novo methods that find novel unconserved miRNAs, and (3) methods that use sequence and structure conservation to make more specific predictions. These three different approaches to miRNA gene prediction all have their distinct uses: we see at least five different scenarios where these approaches are useful. First, one needs to identify known miRNA genes in newly sequenced animal genomes, for which the homology-based methods are ideal. Second, when someone identifies a new miRNA in one species, homology-based methods can verify the miRNA's presence or absence in all relevant genomes. Third, de novo methods can routinely annotate newly sequenced genomes for potentially unconserved miRNAs. Fourth, conservation features can improve de novo predictions by identifying miRNAs that are conserved between related species, such as, for example, primate-specific miRNAs. Fifth, de novo methods can determine whether a newly sequenced small RNA shares the characteristics of known miRNAs and therefore is a likely miRNA candidate. An increasing number of mammalian and animal genomes have been sequenced since the *C. elegans* genome was published [109], and ongoing large-scale sequencing projects such as the mammalian genome project (http://www.broad.mit.edu/mammals/) will ensure that this figure continues to increase.

Homology-based methods are routinely used to find miRNAs in newly sequenced genomes, but we believe that de novo methods and conservation analyses also will be routinely used to annotate newly sequenced genomes, as these methods are invaluable for detecting species or lineage-specific miRNAs. Routine use of de novo miRNA gene prediction methods to annotate genomes is, however, still not a reality. Current algorithms have one major problem: they predict a large number of candidate miRNAs. This is partly related to the large number of genomic sequences that can form miRNA-like hairpins—the human genome contains several million hairpin candidates [56]—as even a method that is highly specific and only has 0.01% false-positive predictions will return 1000 false positives from 10 million hairpin candidates. Current de novo methods report false positive rates ranging from 3% (Reference 108) to 12% (Reference 107). In other words, miRNA gene prediction algorithms still have a great potential for improvement, and better methods to address secondary structure prediction, miRNA transcription, miRNA processing, and database quality can be first steps toward improved algorithms.

Secondary structure prediction is a major aspect of miRNA gene prediction, as the prediction algorithms derive several of the features associated with miRNA genes from the hairpin's predicted secondary structure (see Table 2.1). Secondary structure prediction is, however, notoriously difficult [110], and secondary structure prediction algorithms, in many cases, even get the relatively simple structure of a miRNA precursor wrong [64]. Methods that give more reliable secondary structure predictions should therefore lead to improved miRNA gene predictions. This is especially true for the hairpin loop and the sequences flanking the miRNA stem, as these seem to be related to miRNA processing [67,68,111] and were the regions the secondary structure prediction algorithms failed to predict [64].

## TABLE 2.1
## Features Used by Different miRNA Gene Prediction Algorithms

| | Feature | Reference |
|---|---|---|
| Hairpin structure characteristics | Hairpin stability | 56,80,82,89,93,108 |
| | Hairpin stem and loop length | 56,93,108 |
| | Mature sequence distance from loop | 78,89,108 |
| | Bulge symmetry | 78 |
| | Bulge size | 56, 108 |
| | Distance between internal loops | 108 |
| | Multiloops? | 82 |
| | Base pairing in precursor | 56,78,80,108 |
| | Base pairing in flanking region | 78 |
| | Local sequence/structure triplets | 107 |
| | Longest stem features | 108 |
| | Longest near-perfect stem features | 108 |
| | miRNA window features | 108 |
| Hairpin sequence characteristics | Initial pentamer | 78 |
| | Precursor GC content | 56,82 |
| | Stem nucleotide content | 108 |
| | Stem base-pair types | 108 |
| | Sequence repetitiveness | 56 |
| | Internal repeats | 56 |
| Conservation characteristics | Mature sequence conservation | 78,80,93,95 |
| | Loop divergence | 80,95 |
| | Precursor flanking region conservation | 95 |
| | Secondary structure conservation | 82 |
| | Distant upstream conservation | 79 |
| | Distant downstream conservation | 79 |
| Distal characteristics | Distant upstream sequence motifs | 79 |

Most current miRNA gene prediction algorithms disregard transcription potential and focus only on the characteristics of the hairpin candidate itself. Intergenic miRNAs in *C. elegans*, however, do have a characteristic upstream motif, which MiRscan used to improve its predictions [79]. Although it is unclear whether mammalian miRNAs have a similar pattern [79,95], better understanding of miRNA transcription can lead to computational models of miRNA promoters. By using such computational models to predict whether an miRNA candidate will be transcribed, microRNA gene prediction algorithms can strengthen their predictions.

MicroRNAs are defined by their unique processing pathway, and the four distinct steps—Microprocessor processing of the pri-miRNA, nuclear export, Dicer processing of the pre-miRNA, and RISC incorporation—separate miRNAs from other small RNAs. Consequently, miRNA gene prediction essentially amounts to predicting whether a given DNA sequence, if transcribed, can be a substrate in the four processing steps and then predicting the processing products. Current understanding of miRNA processing is lacking, however, and although complete knowledge of all processing steps is important, more knowledge about the Microprocessor processing step is essential. This is both because, as we have mentioned, current literature contains inconsistencies and conflicting evidence regarding Microprocessor processing, and because the Microprocessor complex recognizes and processes the initial transcript and its processing site defines the mature miRNA.

Recent discoveries of other types of non-protein-coding RNAs, such as the Piwi-associated RNAs (piRNAs) [49,112,113] have emphasized that rigorous annotation is necessary to avoid misclassified miRNAs. Previous versions of the miRNA registry have contained other RNAs, such as tRNAs, erroneously annotated as miRNAs, and despite the curators' ongoing effort, the current version of the registry likely contain erroneously annotated miRNAs as well. As others also have pointed out [46], we saw that some newly annotated miRNAs were markedly different from previous entries in the database, which may suggest that these RNAs either represent a new family of miRNAs or that they may belong to a different class of RNA genes altogether [61]. As most miRNA gene prediction algorithms build on the currently known miRNAs, RNAs wrongly annotated to be miRNAs will lead to poorer predictions. This is especially true for homology-based approaches, as these algorithms will look for homologous RNAs and falsely predict these to be miRNAs as well, but classifier-based approaches are also vulnerable. An ideal algorithm would identify wrongly annotated RNAs in its training set as noise and ignore such sequences, and even though some algorithms tolerate noise [114], there is always a risk that the classifier tries to accommodate the noise and as a result becomes suboptimal.

We believe that more stringent miRNA annotations would alleviate some of these problems. Although current annotation practice requires both expression and biogenesis evidence before an miRNA candidate can be accepted as a true miRNA [58], in practice, detecting a single clone of a ~22-nt RNA that maps to a genomic hairpin is enough to declare the RNA as an miRNA [115]. Knowing that the human genome contains around 10 million miRNA-like hairpins [56,61], however, we can expect that a 22-mer randomly extracted from the genome has a high probability of coming from one of these hairpins.

Is it possible that, in a large sequencing study that identifies a thousand small RNAs, at least one of the RNAs maps to a hairpin by chance and is not a true miRNA? Assuming that the probability that a single ~22-nt sequence maps to a genomic hairpin without being an miRNA is 0.001, which is a conservative estimate given the large number of genomic hairpins, the probability that none of a thousand small RNAs will be falsely annotated as an miRNA is $0.999^{1000} = 0.368$. Assuming that the probability of a chance mapping is 0.01, the probability of no falsely annotated miRNAs is $0.99^{1000} = 4.32 \cdot 10^{-5}$. Current high-throughput sequencing methods can effectively sequence thousands of small RNAs [116]. Thus, one should expect that some of the RNAs from a large-scale sequencing effort will map to genomic hairpins by chance.

For this reason, we believe that, as was initially suggested by Berezikov and colleagues [46], one should remain skeptical to sequences that were annotated as miRNAs, based only on a single clone that maps to a genomic hairpin. Instead of the current practice, where miRNA biogenesis evidence equals a genomic hairpin sequence, we suggest that one should clearly separate sequences with verified miRNA biogenesis from short RNAs that map to genomic hairpins. Consequently, unless one can show evidence of differential expression in Dicer- and Microprocessor-deficient cells, or otherwise show that the miRNA candidate is a Microprocessor and Dicer substrate, the RNA should be clearly annotated as an miRNA candidate.

## 2.9  CONCLUSIONS

In the end, identifying an miRNA's targets is more interesting than identifying the miRNA itself. Thus, one can argue that once one has identified a RISC-dependent target of a small RNA that maps to a genomic hairpin, the RNA should be annotated as an miRNA. We would, however, caution that, even though miRNAs and siRNAs are the only distinct RNAs currently known to be incorporated into RISC, the latter may also associate with other small RNAs that have their own distinct processing pathways. By defining miRNA genes based on their processing pathway, we clearly separate miRNAs from other small RNAs that can be incorporated into the RISC complex.

## ACKNOWLEDGMENTS

The authors received support from the bioinformatics platform in the Norwegian Functional Genomics program (FUGE) and the Norwegian Research Council's Leiv Eriksson program. We thank A. Sewer and R. Aharonov for kindly providing miRNA gene predictions.

## REFERENCES

1. Lagos-Quintana, M., Rauhut, R., Lendeckel, W., and Tuschl, T. (2001). Identification of novel genes coding for small expressed RNAs. *Science 294*, 853–858.
2. Lau, N.C., Lim, L.P., Weinstein, E.G., and Bartel, D.P. (2001). An abundant class of tiny RNAs with probable regulatory roles in *Caenorhabditis elegans*. *Science 294*, 858–862.
3. Lee, R.C., and Ambros, V. (2001). An extensive class of small RNAs in *Caenorhabditis elegans*. *Science 294*, 862–864.
4. Olsen, P.H., and Ambros, V. (1999). The lin-4 regulatory RNA controls developmental timing in *Caenorhabditis elegans* by blocking LIN-14 protein synthesis after the initiation of translation. *Dev Biol 216*, 671–680.
5. Reinhart, B.J., Slack, F.J., Basson, M., Pasquinelli, A.E., Bettinger, J.C., Rougvie, A.E., Horvitz, H.R., and Ruvkun, G. (2000). The 21-nucleotide let-7 RNA regulates developmental timing in *Caenorhabditis elegans*. *Nature 403*, 901–906.
6. Wightman, B., Ha, I., and Ruvkun, G. (1993). Posttranscriptional regulation of the heterochronic gene lin-14 by lin-4 mediates temporal pattern formation in *C. elegans*. *Cell 75*, 855–862.
7. Lee, R.C., Feinbaum, R.L., and Ambros, V. (1993). The *C. elegans* heterochronic gene lin-4 encodes small RNAs with antisense complementarity to lin-14. *Cell 75*, 843–854.

8.  Pasquinelli, A.E., Reinhart, B.J., Slack, F., Martindale, M.Q., Kuroda, M.I., Maller, B., Hayward, D.C., Ball, E.E., Degnan, B., Muller, P., et al. (2000). Conservation of the sequence and temporal expression of let-7 heterochronic regulatory RNA. *Nature 408*, 86–89.

9.  Lee, Y., Kim, M., Han, J., Yeom, K.H., Lee, S., Baek, S.H., and Kim, V.N. (2004). MicroRNA genes are transcribed by RNA polymerase II. *EMBO J 23*, 4051–4060.

10. Lagos-Quintana, M., Rauhut, R., Meyer, J., Borkhardt, A., and Tuschl, T. (2003). New microRNAs from mouse and human. *RNA 9*, 175–179.

11. Cai, X., Hagedorn, C.H., and Cullen, B.R. (2004). Human microRNAs are processed from capped, polyadenylated transcripts that can also function as mRNAs. *RNA 10*, 1957–1966.

12. Altuvia, Y., Landgraf, P., Lithwick, G., Elefant, N., Pfeffer, S., Aravin, A., Brownstein, M.J., Tuschl, T., and Margalit, H. (2005). Clustering and conservation patterns of human microRNAs. *Nucleic Acids Res 33*, 2697–2706.

13. Lagos-Quintana, M., Rauhut, R., Yalcin, A., Meyer, J., Lendeckel, W., and Tuschl, T. (2002). Identification of tissue-specific microRNAs from mouse. *Curr Biol 12*, 735–739.

14. Baskerville, S., and Bartel, D.P. (2005). Microarray profiling of microRNAs reveals frequent coexpression with neighboring miRNAs and host genes. *RNA 11*, 241–247.

15. Obernosterer, G., Leuschner, P.J., Alenius, M., and Martinez, J. (2006). Post-transcriptional regulation of microRNA expression. *RNA 12*, 1161–1167.

16. Thomson, J.M., Newman, M., Parker, J.S., Morin-Kensicki, E.M., Wright, T., and Hammond, S.M. (2006). Extensive post-transcriptional regulation of microRNAs and its implications for cancer. *Genes Dev 20*, 2202–2207.

17. Kim, V.N. (2005). MicroRNA biogenesis: Coordinated cropping and dicing. *Nat Rev Mol Cell Biol 6*, 376–385.

18. Murchison, E.P., and Hannon, G.J. (2004). miRNAs on the move: miRNA biogenesis and the RNAi machinery. *Curr Opin Cell Biol 16*, 223–229.

19. Denli, A.M., Tops, B.B., Plasterk, R.H., Ketting, R.F., and Hannon, G.J. (2004). Processing of primary microRNAs by the Microprocessor complex. *Nature 432*, 231–235.

20. Gregory, R.I., Yan, K.P., Amuthan, G., Chendrimada, T., Doratotaj, B., Cooch, N., and Shiekhattar, R. (2004). The Microprocessor complex mediates the genesis of micro-RNAs. *Nature 432*, 235–240.

21. Han, J., Lee, Y., Yeom, K.H., Kim, Y.K., Jin, H., and Kim, V.N. (2004). The Drosha-DGCR8 complex in primary microRNA processing. *Genes Dev 18*, 3016–3027.

22. Lee, Y., Ahn, C., Han, J., Choi, H., Kim, J., Yim, J., Lee, J., Provost, P., Radmark, O., Kim, S., et al. (2003). The nuclear RNase III Drosha initiates microRNA processing. *Nature 425*, 415–419.

23. Bohnsack, M.T., Czaplinski, K., and Gorlich, D. (2004). Exportin 5 is a RanGTP-dependent dsRNA-binding protein that mediates nuclear export of pre-miRNAs. *RNA 10*, 185–191.

24. Lund, E., Guttinger, S., Calado, A., Dahlberg, J.E., and Kutay, U. (2004). Nuclear export of microRNA precursors. *Science 303*, 95–98.

25. Yi, R., Qin, Y., Macara, I.G., and Cullen, B.R. (2003). Exportin-5 mediates the nuclear export of pre-microRNAs and short hairpin RNAs. *Genes Dev 17*, 3011–3016.

26. Bernstein, E., Caudy, A.A., Hammond, S.M., and Hannon, G.J. (2001). Role for a bidentate ribonuclease in the initiation step of RNA interference. *Nature 409*, 363–366.

27. Chendrimada, T.P., Gregory, R.I., Kumaraswamy, E., Norman, J., Cooch, N., Nishikura, K., and Shiekhattar, R. (2005). TRBP recruits the Dicer complex to Ago2 for microRNA processing and gene silencing. *Nature 436*, 740–744.

28. Hutvagner, G., McLachlan, J., Pasquinelli, A.E., Balint, E., Tuschl, T., and Zamore, P.D. (2001). A cellular function for the RNA-interference enzyme Dicer in the maturation of the let-7 small temporal RNA. *Science 293*, 834–838.

29. Yekta, S., Shih, I.H., and Bartel, D.P. (2004). MicroRNA-directed cleavage of HOXB8 mRNA. *Science 304*, 594–596.

30. Zamore, P.D., Tuschl, T., Sharp, P.A., and Bartel, D.P. (2000). RNAi: Double-stranded RNA directs the ATP-dependent cleavage of mRNA at 21 to 23 nucleotide intervals. *Cell 101*, 25–33.

31. Bagga, S., Bracht, J., Hunter, S., Massirer, K., Holtz, J., Eachus, R., and Pasquinelli, A.E. (2005). Regulation by let-7 and lin-4 miRNAs results in target mRNA degradation. *Cell 122*, 553–563.

32. Wu, L., Fan, J., and Belasco, J.G. (2006). MicroRNAs direct rapid deadenylation of mRNA. *Proc Natl Acad Sci USA 103*, 4034–4039.

33. Lewis, B.P., Burge, C.B., and Bartel, D.P. (2005). Conserved seed pairing, often flanked by adenosines, indicates that thousands of human genes are microRNA targets. *Cell 120*, 15–20.

34. Lewis, B.P., Shih, I.H., Jones-Rhoades, M.W., Bartel, D.P., and Burge, C.B. (2003). Prediction of mammalian microRNA targets. *Cell 115*, 787–798.

35. Xie, X., Lu, J., Kulbokas, E.J., Golub, T.R., Mootha, V., Lindblad-Toh, K., Lander, E.S., and Kellis, M. (2005). Systematic discovery of regulatory motifs in human promoters and 3′ UTRs by comparison of several mammals. *Nature 434*, 338–345.

36. Lim, L.P., Lau, N.C., Garrett-Engele, P., Grimson, A., Schelter, J.M., Castle, J., Bartel, D.P., Linsley, P.S., and Johnson, J.M. (2005). Microarray analysis shows that some microRNAs downregulate large numbers of target mRNAs. *Nature 433*, 769–773.

37. Dykxhoorn, D.M., Novina, C.D., and Sharp, P.A. (2003). Killing the messenger: short RNAs that silence gene expression. *Nat Rev Mol Cell Biol 4*, 457–467.

38. Hannon, G.J. (2002). RNA interference. *Nature 418*, 244–251.

39. Hannon, G.J., and Rossi, J.J. (2004). Unlocking the potential of the human genome with RNA interference. *Nature 431*, 371–378.

40. Birmingham, A., Anderson, E.M., Reynolds, A., Ilsley-Tyree, D., Leake, D., Fedorov, Y., Baskerville, S., Maksimova, E., Robinson, K., Karpilow, J., et al. (2006). 3′ UTR seed matches, but not overall identity, are associated with RNAi off-targets. *Nat Methods 3*, 199–204.

41. Jackson, A.L., Burchard, J., Schelter, J., Chau, B.N., Cleary, M., Lim, L., and Linsley, P.S. (2006). Widespread siRNA "off-target" transcript silencing mediated by seed region sequence complementarity. *RNA 12*, 1179–1187.

42. Brennecke, J., Stark, A., Russell, R.B., and Cohen, S.M. (2005). Principles of microRNA-target recognition. *PLoS Biol 3*, e85.

43. Sethupathy, P., Corda, B., and Hatzigeorgiou, A.G. (2006). TarBase: A comprehensive database of experimentally supported animal microRNA targets. *RNA 12*, 192–197.

44. Vella, M.C., Choi, E.Y., Lin, S.Y., Reinert, K., and Slack, F.J. (2004). The *C. elegans* microRNA let-7 binds to imperfect let-7 complementary sites from the lin-41 3′ UTR. *Genes Dev 18*, 132–137.

45. Didiano, D., and Hobert, O. (2006). Perfect seed pairing is not a generally reliable predictor for miRNA-target interactions. *Nat Struct Mol Biol 13*, 849–851.

46. Berezikov, E., Cuppen, E., and Plasterk, R.H. (2006). Approaches to microRNA discovery. *Nat Genet 38 Suppl*, S2–7.

47. Bartel, D.P., and Chen, C.Z. (2004). Micromanagers of gene expression: The potentially widespread influence of metazoan microRNAs. *Nat Rev Genet 5*, 396–400.

48. Abbott, A.L., Alvarez-Saavedra, E., Miska, E.A., Lau, N.C., Bartel, D.P., Horvitz, H.R., and Ambros, V. (2005). The let-7 MicroRNA family members mir-48, mir-84, and mir-241 function together to regulate developmental timing in *Caenorhabditis elegans*. *Dev Cell 9*, 403–414.

49. Aravin, A., Gaidatzis, D., Pfeffer, S., Lagos-Quintana, M., Landgraf, P., Iovino, N., Morris, P., Brownstein, M.J., Kuramochi-Miyagawa, S., Nakano, T., et al. (2006). A novel class of small RNAs bind to MILI protein in mouse testes. *Nature 442*, 203–207.
50. Girard, A., Sachidanandam, R., Hannon, G.J., and Carmell, M.A. (2006). A germline-specific class of small RNAs binds mammalian Piwi proteins. *Nature 442*, 199–202.
51. Lau, N.C., Seto, A.G., Kim, J., Kuramochi-Miyagawa, S., Nakano, T., Bartel, D.P., and Kingston, R.E. (2006). Characterization of the piRNA complex from rat testes. *Science 313*, 363–367.
52. Blow, M.J., Grocock, R.J., van Dongen, S., Enright, A.J., Dicks, E., Futreal, P.A., Wooster, R., and Stratton, M.R. (2006). RNA editing of human microRNAs. *Genome Biol 7*, R27.
53. Luciano, D.J., Mirsky, H., Vendetti, N.J., and Maas, S. (2004). RNA editing of a miRNA precursor. *RNA 10*, 1174–1177.
54. Yang, W., Chendrimada, T.P., Wang, Q., Higuchi, M., Seeburg, P.H., Shiekhattar, R., and Nishikura, K. (2006). Modulation of microRNA processing and expression through RNA editing by ADAR deaminases. *Nat Struct Mol Biol 13*, 13–21.
55. Griffiths-Jones, S., Grocock, R.J., van Dongen, S., Bateman, A., and Enright, A.J. (2006). miRBase: microRNA sequences, targets and gene nomenclature. *Nucleic Acids Res 34*, D140–144.
56. Bentwich, I., Avniel, A., Karov, Y., Aharonov, R., Gilad, S., Barad, O., Barzilai, A., Einat, P., Einav, U., Meiri, E., et al. (2005). Identification of hundreds of conserved and nonconserved human microRNAs. *Nat Genet 37*, 766–770.
57. Griffiths-Jones, S. (2004). The microRNA Registry. *Nucleic Acids Res 32*, D109–111.
58. Ambros, V., Bartel, B., Bartel, D.P., Burge, C.B., Carrington, J.C., Chen, X., Dreyfuss, G., Eddy, S.R., Griffiths-Jones, S., Marshall, M., et al. (2003). A uniform system for microRNA annotation. *RNA 9*, 277–279.
59. Griffiths-Jones, S., Bateman, A., Marshall, M., Khanna, A., and Eddy, S.R. (2003). Rfam: An RNA family database. *Nucleic Acids Res 31*, 439–441.
60. Griffiths-Jones, S., Moxon, S., Marshall, M., Khanna, A., Eddy, S.R., and Bateman, A. (2005). Rfam: Annotating non-coding RNAs in complete genomes. *Nucleic Acids Res 33*, D121–124.
61. Helvik, S.A., Snove, O., Jr., and Saetrom, P. (2007). Reliable prediction of Drosha processing sites improves microRNA gene prediction. *Bioinformatics 23*, 142–149.
62. Saetrom, P., Snove, O., Nedland, M., Grunfeld, T.B., Lin, Y., Bass, M.B., and Canon, J.R. (2006). Conserved MicroRNA characteristics in mammals. *Oligonucleotides 16*, 115–144.
63. Khvorova, A., Reynolds, A., and Jayasena, S.D. (2003). Functional siRNAs and miRNAs exhibit strand bias. *Cell 115*, 209–216.
64. Krol, J., Sobczak, K., Wilczynska, U., Drath, M., Jasinska, A., Kaczynska, D., and Krzyzosiak, W.J. (2004). Structural features of MicroRNA (miRNA) precursors and their relevance to miRNA biogenesis and small interfering RNA/Short hairpin RNA design. *J Biol Chem 279*, 42230–42239.
65. Lim, L.P., Glasner, M.E., Yekta, S., Burge, C.B., and Bartel, D.P. (2003). Vertebrate microRNA genes. *Science 299*, 1540.
66. Chen, C.Z., Li, L., Lodish, H.F., and Bartel, D.P. (2004). MicroRNAs modulate hematopoietic lineage differentiation. *Science 303*, 83–86.
67. Zeng, Y., Yi, R., and Cullen, B.R. (2005). Recognition and cleavage of primary microRNA precursors by the nuclear processing enzyme Drosha. *EMBO J 24*, 138–148.
68. Han, J.J., Lee, Y., Yeom, K.H., Nam, J.W., Heo, I., Rhee, J.K., Sohn, S.Y., Cho, Y.J., Zhang, B.T., and Kim, V.N. (2006). Molecular basis for the recognition of primary microRNAs by the Drosha-DGCR8 complex. *Cell 125*, 887–901.

69. Brummelkamp, T.R., Bernards, R., and Agami, R. (2002). A system for stable expression of short interfering RNAs in mammalian cells. *Science 296*, 550–553.

70. McManus, M.T., Petersen, C.P., Haines, B.B., Chen, J., and Sharp, P.A. (2002). Gene silencing using micro-RNA designed hairpins. *RNA 8*, 842–850.

71. Sui, G., Soohoo, C., Affar el, B., Gay, F., Shi, Y., Forrester, W.C., and Shi, Y. (2002). A DNA vector-based RNAi technology to suppress gene expression in mammalian cells. *Proc Natl Acad Sci USA 99*, 5515–5520.

72. Yu, J.Y., DeRuiter, S.L., and Turner, D.L. (2002). RNA interference by expression of short-interfering RNAs and hairpin RNAs in mammalian cells. *Proc Natl Acad Sci USA 99*, 6047–6052.

73. Zeng, Y., and Cullen, B.R. (2004). Structural requirements for pre-microRNA binding and nuclear export by Exportin 5. *Nucleic Acids Res 32*, 4776–4785.

74. Vermeulen, A., Behlen, L., Reynolds, A., Wolfson, A., Marshall, W.S., Karpilow, J., and Khvorova, A. (2005). The contributions of dsRNA structure to Dicer specificity and efficiency. *RNA 11*, 674–682.

75. Schwarz, D.S., Hutvagner, G., Du, T., Xu, Z., Aronin, N., and Zamore, P.D. (2003). Asymmetry in the assembly of the RNAi enzyme complex. *Cell 115*, 199–208.

76. Bentwich, I. (2005). Prediction and validation of microRNAs and their targets. *FEBS Lett 579*, 5904–5910.

77. Hofacker, I.L. (2003). Vienna RNA secondary structure server. *Nucleic Acids Res 31*, 3429–3431.

78. Lim, L.P., Lau, N.C., Weinstein, E.G., Abdelhakim, A., Yekta, S., Rhoades, M.W., Burge, C.B., and Bartel, D.P. (2003). The microRNAs of *Caenorhabditis elegans*. *Genes Dev 17*, 991–1008.

79. Ohler, U., Yekta, S., Lim, L.P., Bartel, D.P., and Burge, C.B. (2004). Patterns of flanking sequence conservation and a characteristic upstream motif for microRNA gene identification. *RNA 10*, 1309–1322.

80. Lai, E.C., Tomancak, P., Williams, R.W., and Rubin, G.M. (2003). Computational identification of Drosophila microRNA genes. *Genome Biol 4*, R42.

81. Zuker, M. (2003). Mfold web server for nucleic acid folding and hybridization prediction. *Nucleic Acids Res 31*, 3406–3415.

82. Grad, Y., Aach, J., Hayes, G.D., Reinhart, B.J., Church, G.M., Ruvkun, G., and Kim, J. (2003). Computational and experimental identification of *C. elegans* microRNAs. *Mol Cell 11*, 1253–1263.

83. Bartel, D.P. (2004). MicroRNAs: Genomics, biogenesis, mechanism, and function. *Cell 116*, 281–297.

84. Altschul, S.F., Gish, W., Miller, W., Myers, E.W., and Lipman, D.J. (1990). Basic local alignment search tool. *J Mol Biol 215*, 403–410.

85. Kent, W.J. (2002). BLAT—the BLAST-like alignment tool. *Genome Res 12*, 656–664.

86. Weber, M.J. (2005). New human and mouse microRNA genes found by homology search. *FEBS J 272*, 59–73.

87. Hertel, J., Lindemeyer, M., Missal, K., Fried, C., Tanzer, A., Flamm, C., Hofacker, I.L., and Stadler, P.F. (2006). The expansion of the metazoan microRNA repertoire. *BMC Genomics 7*, 25.

88. Legendre, M., Lambert, A., and Gautheret, D. (2005). Profile-based detection of microRNA precursors in animal genomes. *Bioinformatics 21*, 841–845.

89. Wang, X., Zhang, J., Li, F., Gu, J., He, T., Zhang, X., and Li, Y. (2005). MicroRNA identification based on sequence and structure alignment. *Bioinformatics 21*, 3610–3614.

90. Eddy, S.R. (2002). Computational genomics of noncoding RNA genes. *Cell 109*, 137–140.

91. Washietl, S., Hofacker, I.L., and Stadler, P.F. (2005). Fast and reliable prediction of noncoding RNAs. *Proc Natl Acad Sci USA 102*, 2454–2459.

92. Eddy, S.R. (2001). Non-coding RNA genes and the modern RNA world. *Nat Rev Genet* 2, 919–929.
93. Hertel, J., and Stadler, P.F. (2006). Hairpins in a Haystack: Recognizing microRNA precursors in comparative genomics data. *Bioinformatics 22*, e197–202.
94. Boffelli, D., McAuliffe, J., Ovcharenko, D., Lewis, K.D., Ovcharenko, I., Pachter, L., and Rubin, E.M. (2003). Phylogenetic shadowing of primate sequences to find functional regions of the human genome. *Science 299*, 1391–1394.
95. Berezikov, E., Guryev, V., van de Belt, J., Wienholds, E., Plasterk, R.H., and Cuppen, E. (2005). Phylogenetic shadowing and computational identification of human microRNA genes. *Cell 120*, 21–24.
96. Bonnet, E., Wuyts, J., Rouze, P., and Van de Peer, Y. (2004). Evidence that microRNA precursors, unlike other non-coding RNAs, have lower folding free energies than random sequences. *Bioinformatics 20*, 2911–2917.
97. Nam, J.W., Shin, K.R., Han, J., Lee, Y., Kim, V.N., and Zhang, B.T. (2005). Human microRNA prediction through a probabilistic co-learning model of sequence and structure. *Nucleic Acids Res 33*, 3570–3581.
98. Eddy, S.R. (2004). What is a hidden Markov model? *Nat Biotechnol 22*, 1315–1316.
99. Nam, J.W., Kim, J., Kim, S.K., and Zhang, B.T. (2006). ProMiR II: A web server for the probabilistic prediction of clustered, nonclustered, conserved and nonconserved microRNAs. *Nucleic Acids Res 34*, W455–458.
100. Yang, Z.R. (2004). Biological applications of support vector machines. *Brief Bioinform 5*, 328–338.
101. Gordon, L., Chervonenkis, A.Y., Gammerman, A.J., Shahmuradov, I.A., and Solovyev, V.V. (2003). Sequence alignment kernel for recognition of promoter regions. *Bioinformatics 19*, 1964–1971.
102. Zhang, X.H., Heller, K.A., Hefter, I., Leslie, C.S., and Chasin, L.A. (2003). Sequence information for the splicing of human pre-mRNA identified by support vector machine classification. *Genome Res 13*, 2637–2650.
103. Zien, A., Ratsch, G., Mika, S., Scholkopf, B., Lengauer, T., and Muller, K.R. (2000). Engineering support vector machine kernels that recognize translation initiation sites. *Bioinformatics 16*, 799–807.
104. Leslie, C.S., Eskin, E., Cohen, A., Weston, J., and Noble, W.S. (2004). Mismatch string kernels for discriminative protein classification. *Bioinformatics 20*, 467–476.
105. Saetrom, P. (2004). Predicting the efficacy of short oligonucleotides in antisense and RNAi experiments with boosted genetic programming. *Bioinformatics 20*, 3055–3063.
106. Teramoto, R., Aoki, M., Kimura, T., and Kanaoka, M. (2005). Prediction of siRNA functionality using generalized string kernel and support vector machine. *FEBS Lett 579*, 2878–2882.
107. Xue, C., Li, F., He, T., Liu, G.P., Li, Y., and Zhang, X. (2005). Classification of real and pseudo microRNA precursors using local structure-sequence features and support vector machine. *BMC Bioinformatics 6*, 310.
108. Sewer, A., Paul, N., Landgraf, P., Aravin, A., Pfeffer, S., Brownstein, M.J., Tuschl, T., van Nimwegen, E., and Zavolan, M. (2005). Identification of clustered microRNAs using an ab initio prediction method. *BMC Bioinformatics 6*, 267.
109. *C. elegans* Sequencing Consortium. (1998). Genome sequence of the nematode *C. elegans*: A platform for investigating biology. *Science 282*, 2012–2018.
110. Eddy, S.R. (2004). How do RNA folding algorithms work? *Nat Biotechnol 22*, 1457–1458.
111. Zeng, Y., and Cullen, B.R. (2005). Efficient processing of primary microRNA hairpins by Drosha requires flanking nonstructured RNA sequences. *J Biol Chem 280*, 27595–27603.

112. Girard, A., Sachidanandam, R., Hannon, G.J., and Carmell, M.A. (2006). A germline-specific class of small RNAs binds mammalian Piwi proteins. *Nature 442*, 199–202.
113. Lau, N.C., Seto, A.G., Kim, J., Kuramochi-Miyagawa, S., Nakano, T., Bartel, D.P., and Kingston, R.E. (2006). Characterization of the piRNA Complex from Rat Testes. *Science 313*, 363–367.
114. Müller, K.R., Mika, S., Rätsch, G., and Tsuda, K. (2001). An introduction to kernel-based learning algorithms. *IEEE Trans Neural Networks 12*, 181–201.
115. Cummins, J.M., He, Y., Leary, R.J., Pagliarini, R., Diaz, L.A., Jr., Sjoblom, T., Barad, O., Bentwich, Z., Szafranska, A.E., Labourier, E., et al. (2006). The colorectal micro-RNAome. *Proc Natl Acad Sci USA 103*, 3687–3692.
116. Margulies, M., Egholm, M., Altman, W.E., Attiya, S., Bader, J.S., Bemben, L.A., Berka, J., Braverman, M.S., Chen, Y.J., Chen, Z., et al. (2005). Genome sequencing in micro-fabricated high-density picolitre reactors. *Nature 437*, 376–380.

# 3 A Suite of Resources for the Study of microRNA Ontology and Function

*Praveen Sethupathy, Molly Megraw, and Artemis G. Hatzigeorgiou*

## CONTENTS

## OVERVIEW

MicroRNAs (miRNAs) are ~22-nt small RNA molecules involved in posttranscriptional gene silencing in organisms ranging from plant to human (Bartel, 2004). They have been shown to bind, with perfect complementarity in plants, and usually imperfect complementarity in animals, to specific sites along the 3' UTR of various cytoplasmic messenger RNAs. Since 2001, when miRNAs were appreciated as abundant and widespread regulators (Lagos-Quintana et al., 2001; Lee et al., 2001; Lau et al., 2001), they have been implicated in fat metabolism, development, differentiation, apoptosis, antiviral defense, insulin secretion, and many other diverse biological processes (Kloosterman et al., 2005). There is also strong evidence that the misexpression of miRNAs abets the development of many diseases such as various forms of cancer (Calin and Croce, 2006; Esquela-Kerscher and Slack, 2006). Since it has been predicted that miRNAs target at least 30% of human genes (Lewis et al., 2005; Rajewsky, 2006), it is likely that they underlie an even larger set of disease processes than is currently appreciated. In this chapter we present a suite of resources we have developed to facilitate the study of miRNA ontology and function.

## 3.1 INTRODUCTION

### 3.1.1 BIOGENESIS AND GENOMIC ARRANGEMENT OF miRNAs

An miRNA is processed from a longer transcript, referred to as the primary transcript (pri-miRNA). Pri-miRNAs can be located within the introns of protein-coding genes, outside protein-coding genes entirely (intergenic), or, more rarely, in a coding exon, UTR region, or exon of a noncoding transcript. There is some evidence that the majority of pri-miRNAs located within the introns of protein-coding genes are processed from the pre-mRNAs of these "host genes" (Baskerville and Bartel, 2005; Kim and Kim, 2007). The very few pri-miRNAs that have been experimentally characterized in animals have lengths up to approximately 4 kilobases (kb) (Bracht et al., 2004; Cai et al., 2004; Lee et al., 2004). Furthermore, these pri-miRNAs often contain several miRNAs (Tanzer and Stadler, 2004). These cotranscribed miRNAs are referred to as a *cluster.* In cases where pri-miRNAs are unknown, miRNAs located close to each other on the genome are thought to be likely members of a cluster. The maximum length of an miRNA cluster is still unknown. Evidence from microarray expression data has suggested that mammalian pri-miRNAs may be up to 50 kb in length (Baskerville and Bartel, 2005). After a pri-miRNA is transcribed, precursor miRNA (pre-miRNA) hairpin structures are cleaved from the transcript and exported to the cytoplasm where the final "mature" miRNAs of 21–24 nt are produced (Kim and Nam, 2006; Seitz and Zamore, 2006; Du and Zamore, 2005).

### 3.1.2 miRNA TARGET SITES

MicroRNAs have been shown to guide the RNA-induced silencing complex (RISC) to specific target sites predominantly within the 3' UTR of mRNAs in order to

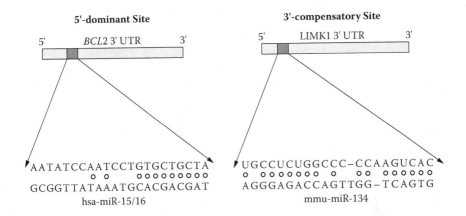

**FIGURE 3.1**   Two categories of microRNA target sites.

induce immediate cleavage, localization to P-bodies, or translational repression (Kloosterman and Plasterk, 2006). To bind to the mRNA, an miRNA may use only a part of its sequence, predominantly its 5′ end. Based on the nucleotides of the miRNA that bind to the mRNA, miRNA target sites can be classified into two categories: (1) 5′ dominant, and (2) 3′ compensatory (Figure 3.1). The "seed" region is defined as the consecutive stretch of seven nucleotides starting from either the first or the second nucleotide at the 5′ end of an miRNA. The 5′-dominant sites have perfect base pairing to at least the seed portion of the 5′ end of the miRNA. The 3′-compensatory sites have extensive base pairing to the 3′ end of the miRNA to compensate for imperfect or a shorter stretch of base pairing to the seed portion of the miRNA (Stark et al., 2005).

## 3.2   EXPERIMENTALLY SUPPORTED miRNA TARGET SITES

### 3.2.1   EXPERIMENTAL METHODS FOR miRNA: TARGET GENE INTERACTION VERIFICATION

Several different experimental techniques have been utilized to verify miRNA targets, ranging from small-scale forward genetics experiments to large-scale microarray studies. Here, we briefly review two of the most widely used methods.

#### 3.2.1.1   Reporter-Silencing Assay

The experimental strategy consists of cloning predicted target sites (as single or multiple copies) into the 3′ UTR of a reporter construct, such as Renilla Luciferase. Because target sites are necessary and sufficient to confer miRNA-dependent translational repression (Kiriakidou et al., 2004; Lewis et al., 2003), the placement of predicted target sites for specific miRNAs in the 3′ UTR of a reporter construct, followed by transfections in cells expressing the miRNAs that recognize the target sites, should lead to a decrease of the reporter protein levels if the target site is functional. If there is a decrease in reporter activity, then this assay is often repeated with a mutation or set of mutations in the putative target site. If this negative control experiment recovers reporter activity, then it is accepted that the wild-type target site induces translational repression (Kiriakidou et al., 2004; Lewis et al., 2003).

### 3.2.1.2 miRNA Inhibition

More recent strategies involve the use of chemically modified oligomers, such as antagomirs, to bind miRNAs and sequester them away from mRNAs (Krutzfeldt et al., 2005). To test a particular predicted miRNA:target interaction, target protein levels can be measured and compared in the absence and presence of the miRNA antagomir. For examples of such studies, we refer the reader to Care et al., 2007; Scherr et al., 2007; and Krutzfeldt et al., 2007.

### 3.2.2  TarBase: A Database of Experimentally Supported miRNA Target Sites

TarBase provides an up-to-date, manually curated collection of all experimentally tested miRNA target sites in eight different species (Sethupathy et al., 2006a). The total number of target sites recorded in TarBase for human/mouse, fruit fly, worm, and zebra fish exceeds 750. Furthermore, TarBase describes each supported target site by the miRNA that binds it, the gene in which it occurs, the location within the 3′ UTR where it occurs, the nature of the experiments that were conducted to validate it, and the sufficiency of the site to induce translational repression and/or cleavage (Figure 3.2). Such a comprehensive description of each target site will be useful for focused bioinformatic and experimental studies to further understand the features of miRNA targeting, the mechanisms of miRNA-based translational repression and/or cleavage, and the roles of miRNAs in various biological networks. TarBase can be accessed at http://www.diana.cslab.ece.ntua.gr./tarbase.

## 3.3  miRNA TARGET PREDICTION

### 3.3.1  Existing Methodologies

Currently, there exists more than a dozen target prediction programs (Sethupathy et al. 2006b; Rajewsky 2006). In the following text we briefly describe a few of the most widely used programs for the prediction of human miRNA targets. These programs essentially perform two major steps: In the first, they identify potential miRNA-binding sites according to specific base-pairing rules. The actual rules vary from program to program. Some programs apply very specific base-pairing rules that were observed in a few experimentally supported cases (Diana-microT—Kiriakidou et al., 2004; EMBL program—Stark et al., 2003). Others focus more on the thermodynamic energy of the base pairing (miRanda—John et al., 2004). Still others choose the simplest rules that still allow them to observe a strong statistical signal (TargetScanS—Lewis et al., 2005; PicTar—Krek et al., 2005; reviewed in Rajewsky, 2006).

In the second step, the programs implement cross-species conservation requirements. Again, the actual implementation varies slightly from program to program. TargetScanS and PicTar both require conservation between at least five species for the portion of the target site that binds to the miRNA seed. However, they define

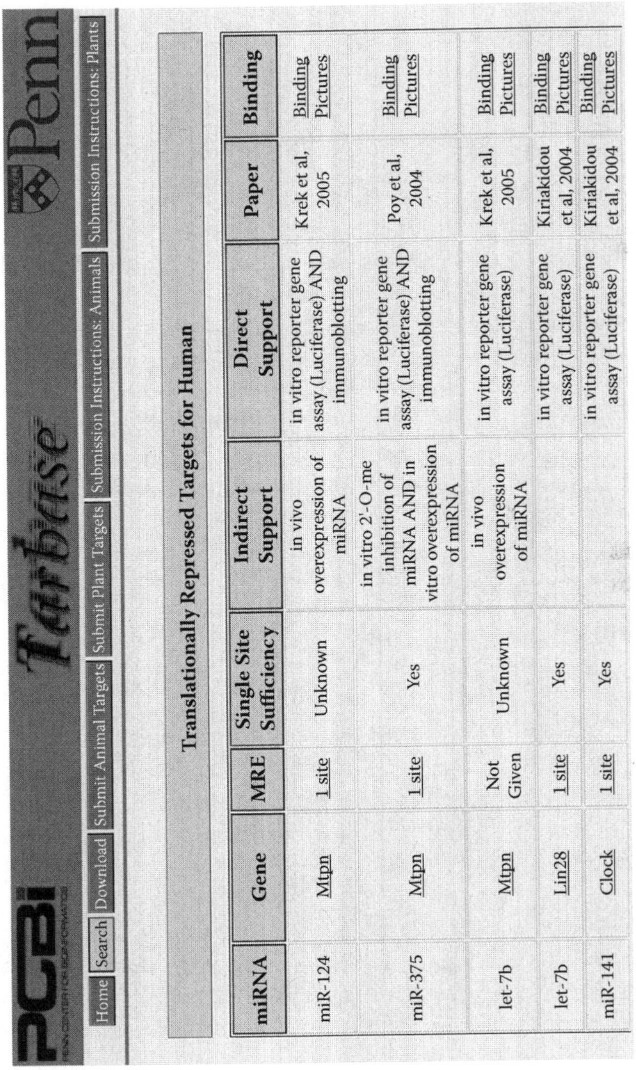

**Translationally Repressed Targets for Human**

| miRNA | Gene | MRE | Single Site Sufficiency | Indirect Support | Direct Support | Paper | Binding |
|---|---|---|---|---|---|---|---|
| miR-124 | Mtpn | 1 site | Unknown | in vivo overexpression of miRNA | in vitro reporter gene assay (Luciferase) AND immunoblotting | Krek et al, 2005 | Binding Pictures |
| miR-375 | Mtpn | 1 site | Yes | in vitro 2'-O-me inhibition of miRNA AND in vitro overexpression of miRNA | in vitro reporter gene assay (Luciferase) AND immunoblotting | Poy et al, 2004 | Binding Pictures |
| let-7b | Mtpn | Not Given | Unknown | in vivo overexpression of miRNA | in vitro reporter gene assay (Luciferase) | Krek et al, 2005 | Binding Pictures |
| let-7b | Lin28 | 1 site | Yes | | in vitro reporter gene assay (Luciferase) | Kiriakidou et al, 2004 | Binding Pictures |
| miR-141 | Clock | 1 site | Yes | | in vitro reporter gene assay (Luciferase) | Kiriakidou et al, 2004 | Binding Pictures |

**FIGURE 3.2** A snapshot of a TarBase results page.

conservation slightly differently. TargetScanS requires that a seed match occur at exactly corresponding positions in a cross-species UTR alignment, whereas PicTar requires only that the seed match occur at overlapping positions in a cross-species UTR alignment. Diana-microT, on the other hand, defines conservation as an entire target site occurring with at least 90% identity at exactly corresponding positions in a cross-species UTR alignment. A diagrammatic summary of a few of the current methodologies is presented in Figure 3.3. An extended description of miRNA target prediction programs and the statistical significance of their predictions can be found in Rajewsky, 2006.

### 3.3.2   COMPARATIVE ANALYSIS OF EXISTING METHODOLOGIES

An extensive evaluation of five mammalian target prediction programs is described in Sethupathy et al., 2006b. The performance of these individual programs, as well as various combinations of the programs, was measured by calculating their (1) prediction sensitivity on experimentally supported targets from TarBase, and (2) total number of predictions. The results revealed a large disparity in program performance. Although it is not possible to suggest one program as consistently superior to all of the rest, suggestions were made as to which program combinations to use in order to maximize both sensitivity and specificity (Sethupathy et al., 2006b). Target predictions of various program combinations were stored in internal databases and made accessible through a Web site, *TargetCombo* (http://www.diana.pcbi.upenn. edu/cgi-bin/TargetCombo.cgi), through which users can query search for miRNA targets by miRNA or gene.

## 3.4   ANALYSIS OF miRNA GENOMIC ORGANIZATION

### 3.4.1   EXISTING TOOLS

The discovery of novel miRNAs, the characterization of their biogenesis, and the identification of their functions are areas of research that have been highly interdisciplinary in nature, bringing together a number of experimental and computational groups. There are several existing tools and resources that provide updated data regarding each of these areas of research.

#### 3.4.1.1   *miRBase*

Sanger Institute's *miRBase* serves as the central database for experimentally supported mature miRNA sequences. For each supported miRNA, *miRBase* provides the genomic coordinates of the predicted precursor sequence, the nucleotide composition of both the precursor and mature miRNA sequences, and predicted targets of the mature miRNA according to miRanda—a target prediction tool published in 2004 (John et al., 2004). *miRBase* can be accessed at http://microrna.sanger.ac.uk/.

#### 3.4.1.2   *miRNAMap* and Argonaute

The recently published miRNAMap provides genomic coordinates, nucleotide composition, and experimentally supported targets for each miRNA in four mammalian

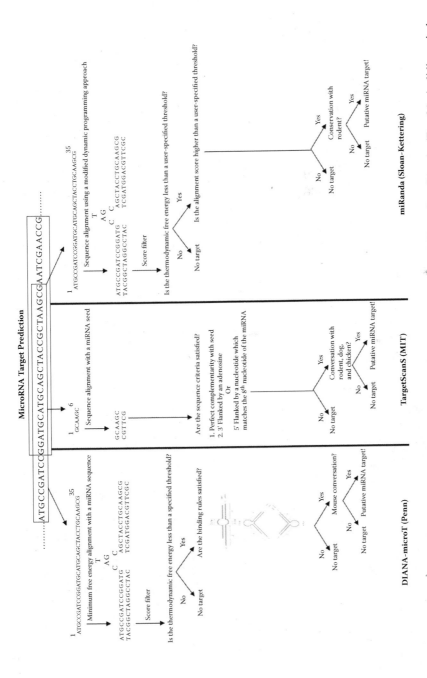

**FIGURE 3.3** Target prediction methodology for three widely used programs (Diana-microT, TargetScanS, and miRanda). A sliding window across the 3′ UTR of messenger is processed in different ways according to each program.

genomes (Hsu et al., 2006). It also reports computationally predicted miRNAs and their predicted targets according to miRanda. Finally, it provides crosslinks to other biological databases, such as the UCSC Genome Browser, in order to provide tissue expression and cross-species sequence conservation data for each supported and predicted miRNA. The Argonaute (Shahi et al., 2006), published simultaneously with miRNA-Map (Hsu et al., 2006), provides much of the same information with perhaps a larger miRNA tissue expression data set collected from various miRNA expression studies. *miRNAMap* can be accessed at http://mirnamap.mbc.nctu.edu.tw/, and Argonaute can be accessed at http://www.ma.uni-heidelberg.de/apps/zmf/argonaute/interface/.

### 3.4.2   miRGen: A Database for the Study of miRNA Genomic Organization and Function

The aforementioned resources have been highly useful as centralized sources of basic miRNA genomic and target information. However, the current array of tools is not sufficient to conduct systematic analyses of the relationship between miRNA genomic organization and miRNA function. Here we discuss miRGen, a highly integrated resource that contains three connected interfaces: miRNA genomic context, miRNA clusters, and miRNA targets.

miRGen (Megraw et al., 2007) is an integrated database of (1) positional relationships between miRNAs and genomic annotation sets, and (2) miRNA targets according to combinations of widely used target prediction programs, as well as experimentally supported targets from TarBase. A major goal of the database is the study of the relationship between miRNA genomic organization and miRNA function. miRGen's three connected interfaces—miRNA genomic context, miRNA clusters, and miRNA targets—provide a user-friendly context for performing rapid up-to-date investigations into miRNA genomic organization, cotranscription, and targeting. miRGen can be freely accessed at http://www.diana.pcbi.upenn.edu/miRGen.

### 3.4.3   Performing Analyses with miRGen

Investigations with miRGen are made possible by three integrated and user-friendly interfaces. The Genomics interface allows the user to explore where whole-genome collections of miRNAs are located with respect to UCSC Genome Browser annotation sets such as Known Genes, Refseq Genes, Genscan predicted genes, CpG islands, and pseudogenes. These miRNAs are connected through the *Targets* interface to their experimentally supported target genes from TarBase, as well as computationally predicted target genes from optimized intersections and unions of several widely used mammalian target prediction programs. Finally, the *Clusters* interface provides predicted miRNA clusters at any given inter-miRNA distance and specific functional information on the targets of miRNAs within each cluster.

#### 3.4.3.1   Analysis of miRNA Genomic Organization with miRGen

The miRGen Genomics interface provides direct queries for miRNAs that are located only in introns, exons, overlapping exon boundaries, or in UTRs for several popular gene sets including UCSC Known Genes, Refseq Genes, and Genscan genes. It also provides direct queries for miRNAs located in other genomic entities such as pseudogenes and CpG islands. If predicted or experimentally supported targets are available for the selected organism, an miRNA Targets interface query link is provided for each miRNA. Within all selected subsets, links to the UCSC genome browser for each miRNA and gene also allow a graphical view of any individual entity of interest, shown in Figure 3.4.

### 3.4.3.2    Analysis of miRNA Function with miRGen

The miRGen Targets interface allows the user to search unions and intersections of predicted targets from the most up-to-date versions of the programs PicTar, TargetScanS, miRanda, and Diana-microT in humans, the mouse, rat, worm, and fruit fly. The user can search either for targets of a particular miRNA or miRNAs that target a particular gene. Since computational programs for target prediction use different gene id systems (Ensembl gene id, Refseq id, Gene symbol, etc.) to represent the predicted targets genes, miRGen performs gene id integration so that users can search for and compare target predictions of several programs.

### 3.4.3.3    Analysis of miRNA Clusters with miRGen

It is now well established that miRNAs are arranged in the genome as polycistronic units. In other words, miRNAs are not usually transcribed independently but along with other neighboring miRNAs. The Clusters interface of miRGen allows the user to define a cluster distance—the maximum distance between any two miRNAs considered to be in the same cluster—and view the resulting clusters of miRNAs on each chromosome. miRGen gives careful consideration to the issue of how clusters should be defined in relationship to protein-coding genes. The Clusters interface provides a set of advanced options that allow the user to choose whether all miRNAs within the same gene are automatically defined to constitute a separate cluster, whether a cluster should be split by a gene that is located between miRNA members of the cluster, and which (if any) available gene sets for the chosen organism should be used for these tasks.

## 3.5    CONCLUSIONS

miRNAs have emerged as widespread and abundant regulators of eukaryotic gene expression. Due to their involvement in a wide range of biological processes, they are likely to underlie an array of genetic disorders. The rapid rate of discovery in the miRNA field has necessitated the development of bioinformatic resources and tools for analyzing miRNA genomics and function. In this chapter, we have discussed the utility of the published databases, TarBase, TargetCombo, and miRGen, which facilitate the study of miRNA genomic organization and molecular function. Specifically, TarBase records a comprehensive set of experimentally supported miRNA target sites, TargetCombo provides precompiled lists of miRNA target predictions from combinations of various existing methodologies, and miRGen relates the genomic context of miRNAs to their function. As the miRNA field continues to expand, these and other databases will be valuable resources for the research community

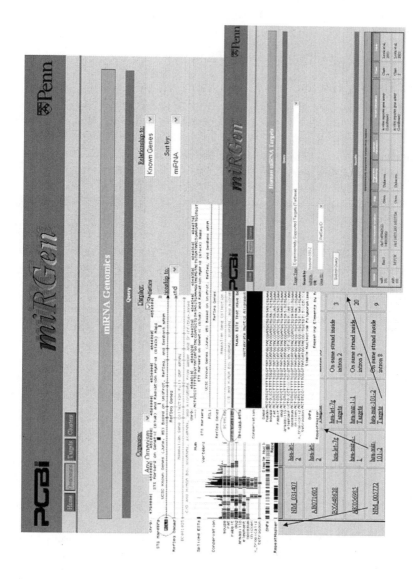

**FIGURE 3.4**  The miRGen genomics interface links to the UCSC genome browser and to the miRGen Targets interface.

and can contribute to the creative application of synergistic experimental–informatic approaches to miRNA research.

## REFERENCES

Bartel, D.P. MicroRNAs: Genomics, biogenesis, mechanism, and function. *Cell* **116**, 281–297 (2004).

Baskerville, S. and Bartel, D.P. Microarray profiling of microRNAs reveals frequent coexpression with neighboring miRNAs and host genes. *RNA* **11**, 241–247 (2005).

Bracht, J., Hunter, S., Eachus, R., Weeks, P., and Pasquinelli, A.E. Trans-splicing and polyadenylation of let-7 microRNA primary transcripts. *RNA* **10**, 1586–1594 (2004).

Cai, X., Hagedorn, C.H., and Cullen, B.R. Human microRNAs are processed from capped, polyadenylated transcripts that can also function as mRNAs. *RNA* **10**, 1957–1966 (2004).

Calin, G.A. and Croce, C.M. MicroRNA signatures in human cancers. *Nat Rev Cancer* **6**, 857: 866 (2006).

Care, A. Catalucci, D., Felicetti, F., Bonci, D., Addario, A., Gallo, P., Bang, M.L., Segnalini, P., Gu, Y., Dalton, N.D. et al. MicroRNA-133 controls cardiac hypertrophy. *Nat Med* **13**, 613–618 (2007).

Du, T. and Zamore, P.D. microPrimer: The biogenesis and function of microRNA. *Development* **132**, 4645–4652 (2005).

Esquela-Kerscher, A. and Slack, F.J. Oncomirs—microRNAs with a role in cancer. *Nat Rev Cancer* **6**, 259–269 (2006).

Hsu, P.W., Huang, H.D., Hsu, S.D., Lin, L.Z., Tsou, A.P., Tseng, C.P., Stadler, P.F., Washietl, S., and Hofacker, I.L. miRNAMap: Genomic maps of microRNA genes and their target genes in mammalian genomes. *Nucleic Acids Res* **34**, D135–139 (2006).

John, B., Enright, A.J., Aravin, A., Tuschl, T., Sander, C., and Marks, D.S. Human MicroRNA targets. *PLoS Biol* **2**, e363 (2004).

Kim, V.N. and Nam, J.W. Genomics of microRNA. *Trends Genet* **22**, 165–173 (2006).

Kim, Y.K. and Kim, V.N. Processing of intronic microRNAs. *EMBO J* **26**, 775–783 (2007).

Kiriakidou, M. et al. A combined computational-experimental approach predicts human microRNA targets. *Genes Dev* **18**, 1165–1178 (2004).

Kloosterman, W.P. and Plasterk, R.H. The diverse functions of microRNAs in animal development and disease. *Dev Cell* **11**, 441–450 (2006).

Krek, A. et al. Combinatorial microRNA target predictions. *Nat Genet* **37**, 495–500 (2005).

Krutzfeldt, J., Rajewsky, N., Braich, R., Rajeev, K.G., Tuschl, T., Manoharan, M., and Stoffel, M. Silencing of microRNAs in vivo with "antagomirs." *Nature* **438**, 685–689 (2005).

Krutzfeldt, J., Kuwajima, S., Braich, R., Rajeev, K.G., Pena, J., Tuschl, T., Manoharan, M., and Stoeffel, M. Specificity, duplex degradation and subcellular localization of antagomirs. *Nucleic Acids Res* **35**, 2885–2892 (2007).

Lagos-Quintana, M., Rauhut, R., Lendeckel, W., and Tuschl, T. Identification of novel genes coding for small expressed RNAs. *Science* **294**, 853–858 (2001).

Lau, N.C., Lim, L.P., Weinstein, E.G., and Bartel, D.P. An abundant class of tiny RNAs with probably regulatory roles in *Caenorhabditis elegans*. *Science* **294**, 858–862 (2001).

Lee, R.C. and Ambros, V. An extensive class of small RNAs in *Caenorhabditis elegans*. *Science* **294**, 862–864 (2001).

Lee, Y., Kim, M., Han, J., Yeom, K.H., Lee, S., Baek, S.H., and Kim, V.N. MicroRNA genes are transcribed by RNA polymerase II. *EMBO J* **23**, 4051–4060 (2004).

Lewis, B.P., Shih, I., Jones-Rhoades, M.W., Bartel, D.P., and Burge, C.B. Prediction of mammalian microRNA targets. *Cell* **115**, 787–798 (2003).

Lewis, B.P., Burge, C.B., and Bartel, D.P. Conserved seed pairing, often flanked by adenosines, indicates that thousands of human genes are microRNA targets. *Cell* **120**, 15–20 (2005).

Megraw, M., Sethupathy, P., Corda, B., and Hatzigeorgiou, A.G. miRGen: a database for the study of animal microRNA genomic organization and function, *Nucleic Acids Res* **35**, D149–155 (2007).

Rajewsky, N. microRNA target predictions in animals. *Nat Genet* **38**, Suppl., S8–13 (2006).

Scherr, M., Venturini, L., Battmer, K., Schaller-Schoenitz, M., Schaefer, D., Dallmann, I., Ganser, A., and Eder, M. Lentivirus-mediated antagomir expression for specific inhibition of miRNA function. *Nucleic Acids Res* **35**, e149 (2007).

Seitz, H. and Zamore, P.D. Rethinking the microprocessor. *Cell* **125**, 827–829 (2006).

Sethupathy, P., Corda, B., and Hatzigeorgiou, A.G. TarBase: A comprehensive database of experimentally supported animal microRNA targets. *RNA* **12**, 192–197 (2006a).

Sethupathy, P., Megraw, M., and Hatzigeorgiou, A.G. A guide through present computational approaches for the identification of mammalian microRNA targets. *Nat Methods* **3**: 881–886 (2006b).

Shahi, P., Loukianiouk, S., Bohne-Lang, A., Kenzelmann, M., Kuffer, S., Maertens, S., Eils, R., Grone, H.J., Gretz, N., and Brors, B. Argonaute—a database for gene regulation by mammalian microRNAs. *Nucleic Acids Res* **34**, D115–118 (2006).

Stark, A., Brennecke, J., Russell, R.B., and Cohen, S.M. Identification of Drosophila microRNA targets. *PLoS Biol* **1**, E60 (2003).

Stark, A., Brennecke, J., Bushati, N., Russell, R.B., and Cohen, S.M. Animal microRNAs confer robustness to gene expression and have a significant impact on 3′ UTR evolution. *Cell* **123**, 1133–1146 (2005).

Tanzer, A. and Stadler, P.F. Molecular evolution of a microRNA cluster. *J Mol Biol* **339**, 327–335 (2004).

# 4 Regulation of Translation and mRNA Stability by Hfq-Binding Small RNAs in *Escherichia coli*

*Hiroji Aiba*

## CONTENTS

## OVERVIEW

Recent studies have established that a major class of small RNAs (sRNAs) bind to an RNA chaperon Hfq and act by imperfect base pairing to regulate the translation and stability of target mRNAs under specific physiological conditions in *Escherichia coli*. I will discuss here how the posttranscriptional regulation of gene expression mediated by Hfq-binding sRNAs has been discovered and what we have learned about their physiological roles and their mechanisms of action by focusing on two stress-induced sRNAs: SgrS and RyhB. These sRNAs form a specific ribonucleo-protein complex with RNase E through Hfq, resulting in translation inhibition and RNase E-dependent degradation of target mRNAs. Translation inhibition is the primary event for gene silencing. Degradation of both target mRNAs and sRNAs occurs simultaneously. The crucial base pairs for the action of SgrS are confined to the 6-nt region, overlapping the Shine–Dalgarno sequence of the target mRNA. Hfq accelerates the rate of duplex formation between SgrS and the target mRNA. Membrane localization of target mRNA contributes to efficient SgrS action by com-peting with ribosome loading.

## 4.1   INTRODUCTION

Small noncoding RNAs are widely involved in the regulation of diverse cellular functions in all organisms. The existence of small RNAs (sRNAs) and their involve-ment in the regulation of cellular functions were first recognized in *Escherichia coli*. However, regulation of gene expression mediated by sRNAs had been thought to be the exception for a long time because only a few chromosomally encoded regula-tory RNAs were known. In the last 10 years, an increasing number of sRNAs have been newly identified in *E. coli* and also in other organisms. Although the functions of many of them remain to be elucidated, an emerging feature is that a major class of *E. coli* sRNAs binds to an RNA chaperon Hfq and acts by imperfect base pair-ing to regulate the translation and stability of target mRNAs under specific physi-ological conditions. I will discuss here how the posttranscriptional regulation of gene expression mediated by Hfq-binding sRNAs has been discovered, and what we have learned about their physiological roles and their mechanisms of action by focusing on the most recent studies. In the last several years, many reviews have dealt with bacterial sRNAs, including Hfq-binding sRNAs, from different points of view [1–8]. sRNAs belonging to other categories such as *cis*-encoded base-pairing RNAs and those involved in modulation of the activity of a protein are covered in some of these previous reviews.

## 4.2   THE BEGINNING

The idea that an RNA molecule can exert its regulatory role through base pairing was already described in the historic paper by Jacob and Monod, in which the regulatory gene such as *lacI* was postulated to encode an RNA repressor that can interact spe-cifically with an operator DNA or RNA [9]. In a little while, the regulatory gene was shown to produce a protein rather than an RNA. Thus, transcriptional control medi-ated by regulatory proteins became a major focus of attention, and people almost forgot the idea of RNA regulator. The concept of antisense RNA regulator had come true in 1981 when an sRNA was shown to control the replication of ColE1 plasmid [10]. Tomizawa and colleagues discovered that the *cis*-encoded RNAI interferes with the processing of the RNA primer required for plasmid replication through RNA–RNA base pairing, resulting in a decrease in the frequency of DNA replication and therefore lowering plasmid copy number. RNAI is the first small regulatory RNA to be discovered in any system. Following this pioneering work, a number of examples had been found in which small antisense RNAs control the replication or stability of extrachromosomal elements, typically by inhibiting translation of the target RNAs [7]. These antisense RNAs act by base-pairing with RNAs that are encoded in *cis* and thus have perfect complementarity to the target RNAs.

The sRNAs expressed from the *E. coli* genome were first identified by direct biochemical analyses of in vivo $^{32}$P-labeled total RNA [11–13]. Later studies had revealed that these RNAs have a variety of biological functions. They are the 4.5S RNA, a component of the protein secretion machinery; the 10Sb RNA, the catalytic component of RNase P; the 10Sa RNA (now known as tmRNA), involved in translational quality control; and the 6S RNA, involved in modulation of RNA polymerase activity [14]. Spot 42 RNA, whose function as an antisense regulator has been worked out recently [15], was also among them.

## 4.3   HFQ-BINDING sRNAs AND TARGET mRNAs

A summary of the features of Hfq-binding sRNAs characterized to some extent is given in Table 4.1. Predicted base-pairing interactions between these sRNAs and

## TABLE 4.1
## Hfq-Binding Small RNAs in *E. coli*

| Name (Alternative) | Size (nt) | Stress/Condition | Regulator | Targets | Reference |
|---|---|---|---|---|---|
| MicF | 93 | High osmolarity | EnvZ/OmpR | *ompF* | 31–35 |
| | | Oxidative stress | SoxR/S | | |
| | | High temperature | ? | | |
| DicF | 56 | ? | ? | *ftsZ* | 37 |
| DsrA | 85 | Low temperature | ? | *hns/rpoS* | 38–42 |
| RprA | 105 | ? | RcsC/B | *rpoS* | 43,44 |
| OxyS | 109 | Oxidative stress | OxyR | *fhlA/rpoS* | 22,23,45–47 |
| GcvB | 204 | Glycine | GcvA | *oppA/dppA* | 48 |
| RyhB (SraI) | 90 | Iron limitation | Fur | *sodB/sdh* | 58,59 |
| Spot42 | 109 | Rich carbon source | CRP/cAMP | *galK* | 15,24 |
| SgrS (RyaA) | 220 | G6P accumulation | SgrR | *ptsG* | 30,60,61,64–66,68,69 |
| MicC (IS063) | 108 | Reciprocal to MicF | ? | *ompC* | 71 |
| MicA (SraD) | 153 | Stationary | $\sigma^E$ | *ompA* | 29,72 |
| RybB | 180 | Stationary | $\sigma^E$ | *ompC/ompW* | 6 |
| RseX | 91 | ? | ? | *ompC/ompA* | 73 |
| OmrA (RygA/SraE) | 88 | High osmolarity | EnvZ/OmpR | *ompT/citA/ fecA/fepA* | 75 |
| OmrB (RygB) | 82 | High osmolarity | EnvZ/OmpR | *ompT/citA/fecA* | 75 |

their target mRNAs are shown in Figure 4.1. Some of these base pairings have been supported by biochemical and mutational studies, while others remain to be proved. Studies on representative RNAs have illuminated several features of the physiological roles and the mechanisms of action of Hfq-binding sRNAs. For example, most of these RNAs are transcribed from their own single genes and range from about 50 to 220 nucleotides in length. Their transcription is induced in response to specific physiological (stress) conditions, depending on specific transcription factors. They act on *trans*-encoded target mRNAs along with Hfq through limited regions of base pairing, changing the translation and stability of the mRNAs. Studies on two newly identified sRNAs, SgrS and RyhB, in the last few years have uncovered previously unrecognized features of the mechanisms of action of Hfq-binding sRNAs.

The properties of Hfq and its roles in RNA functions are described in a recent review [16]. I will touch on them briefly here. Hfq was originally identified as a host factor required for the in vitro replication of the RNA phage Qβ in *E. coli* [17]. The pleiotropic phenotypes of an *hfq* mutant suggested that Hfq is also deeply involved in cellular physiology [18]. Indeed, Hfq was shown to affect the translation and stability of several mRNAs [19–21]. The involvement of Hfq in sRNA functions was first recognized during the study on the mechanism of *rpoS* regulation by OxyS [22,23]. Subsequently, Hfq was shown to bind to, and be necessary for the regulatory activity of, a number of chromosomally encoded antisense sRNAs. It is highly possible that most Hfq-binding sRNAs act as antisense RNAs on specific target mRNAs. Hfq also binds to the target mRNAs of the sRNAs. It binds AU-rich single-stranded RNA, with a preference for binding next to a stem-loop region. Hfq is a heat-stable protein with subunits of about 11 kDa and forms a doughnut-shaped hexameric structure. It is a homologue of the Sm and Sm-like proteins involved in splicing and mRNA degradation in eukaryotic cells [23,24]. The structural and mutational studies indicate that RNA binds around the internal cavity on one side of the hexameric ring [25–27].

Hfq has been demonstrated to stimulate the base pairing between a given sRNA and its target mRNA in vitro in several cases [23,24,28–30]. It may promote the base pairing by changing the RNA structures acting as an RNA chaperon, thereby allowing accessibility of two complementary RNAs. It could also facilitate the base pairing by binding to a given sRNA and its target mRNA simultaneously, thereby increasing the local concentrations of two RNAs. Recently, it has been demonstrated that Hfq stimulates the base pairing by increasing the rate of duplex formation in vitro [30]. Another role of Hfq is just to stabilize sRNAs, thereby supporting appropriate expression levels of these molecules.

## 4.4 DISCOVERY OF MicF

The first chromosomally encoded antisense regulatory RNA to be characterized is MicF (mRNA-interfering complementary RNA for *ompF*), which was discovered fortuitously during a study on expression of *ompF* encoding a major outer membrane porin OmpF. Mizuno et al. [31] observed that a multicopy plasmid carrying a genomic DNA region located upstream of the *ompC* encoding another porin OmpC markedly reduced the levels of OmpF protein and *ompF* mRNA. They showed that an sRNA called MicF, transcribed divergently from the *ompC* promoter, is responsible for the

```
5'-ACCAUGAGGGUAAUAAAUAAUGAUGAAGC-3'    ompF
   ****************** * *** *
3'-ACUUACUGCCAUUAUUUAUUUCAAUUACU-5'    MicF


5'-GGCGACAGGCACAAAUCGGAGAGAAACUAUG-3'    ftsZ
   **** ****        ***** ******
3'-CCGCCGUCUCGUCAGUGCCUCAUUUUGACUA-5'    DicF


5'-UUUCCUGGAAGAACAAAAUG-(24)-GGACAACAAGGGUUGU-3'    fhlA
   ******* *  *  ** *        ***--*********
3'-CUAGGACCUCUAGGCGUUUU-(50)-ACUGAAGUUCCCAAUU-5'    OxyS


5'-AAAUUAAUAAUAAAGGAGAGUAGCAAUGUCAUU-3'    sodB
   ** ***** *  *** ** * *
3'-AUUCAUUAUGACCUUCGUUACACUCGUUACAG-5'    RyhB


5'-GCGAAUCCGGA-GUGUAA---GAAAUGAGUCUGAAAGAAAAAACAC-3'    galK
   ********  ****    ****  ******** * * ****
3'-GGUUUAGGCUAAUGCACUUCAUUUU--CCAGACUUUCUAUCUUGUA-5'    Spot42


5'-AAAUAAAGGCAUAUAACAGAGGGUUAAUAACAUG-3'    ompC
   ****************
3'-GUUAUUUCCGUAUAUUG-5'                      MicC


5'-UUGGAUGAUAACGAG-GCGCAAAAAAUG-3'    ompA
   ************ ****
3'-CUACUACUAUUGUUUACGCGCAGAAAG-5'  2  MicA


5'-AAAGCACCCAUACUCAGGAGCACUCUCAAUUAUGUUUA-3'    ptsG
   ***** ********  ****  * *  ** **
3'-AAAUGUGGUUAUGAGUCAGUGUGU-ACUACGUCCGUUC-5'    SgrS


5'-AAGUUUGAGAUUACUACAAUGAGCGAAGCACUUAAAAU-3'    hns
    *  *    * *  *   * * **   ** ***
3'---GCCCUACUUUGAACGAAUUCGUUCUUCGUGAAUUUUU-5'  DsrA


5'-GCAUUUUGAAAUUCGUUACAAGGGGAAAUCCGU-(80)-UAGGAGCCACCUUAUG-3'    rpoS
   **  ************* ********
3'-GUGAAUUUUUUAAGCAAUGUGGUCCUUUAGACU-5'                          DsrA
```

**FIGURE 4.1**    Base pairing of Hfq-binding sRNAs and target mRNAs. The predicted base pairings between Hfq-binding sRNAs and target mRNAs are shown. The sequence of each sRNA is shown in red, oriented 3′ to 5′. The target mRNA sequence is shown in black, oriented 5′ to 3′. The Shine–Dalgarno sequences and the AUG initiation codons of the target mRNAs are underlined.

downregulation of *ompF*. A portion of MicF was shown to be partially complementary to the translation initiation region of *ompF*. These observations led the authors to propose a model in which MicF RNA downregulates *ompF* expression by inhibiting the translation through RNA–RNA base pairing.

Later studies showed that the synthesis of 93-nucleotide (nt) MicF is induced in response to temperature increase, high osmolarity, and ethanol, resulting in significant decreases in the levels of *ompF* mRNA and of newly synthesized OmpF [32]. Thus, the chromosomally derived MicF RNA was proved to be able to repress OmpF expression in response to environmental stresses. MicF is also induced in response to oxidative stress through the SoxR/S regulator [33]. In addition, the two-component EnvZ/OmpR is responsible for the MicF induction at high osmolarity [34]. Although the biological significance of *ompF* downregulation by MicF in response to a variety of stresses remains to be studied, one role would be to prevent the entry of toxic molecules. In vitro study established that MicF RNA binds to a fragment of *ompF* mRNA containing the predicted base-pairing region, resulting in a stable RNA–RNA duplex [35]. More recently, MicF was shown to coimmunoprecipitate with Hfq, though an Hfq requirement for MicF action has not been reported yet [36].

## 4.5  MORE SMALL RNAs BEFORE SYSTEMATIC SEARCHES

MicF had been the only antisense RNA of *E. coli* until DicF, encoded by a cryptic prophage, was also discovered serendipitously during the study on the expression of *dicB* encoding a cell division inhibitor protein in 1992 [37]. The multicopy plasmid carrying the *dicF* region was shown to inhibit cell division. The 53-nt DicF RNA is partially complementary to the translation initiation region of *ftsZ* encoding the major cell division protein, suggesting that it inhibits the *ftsZ* translation through base pairing. No further studies on DicF have been reported except that it coimmunoprecipitates with Hfq [36]. Thus, little is known about the biological significance and the mechanism of *ftsZ* downregulation by DicF.

Subsequently, several chromosomally encoded sRNAs (DsrA, RprA, OxyS, and GcvB) were discovered again by chance during the study of expression of particular genes. DsrA was discovered based on the observation of a phenotype conferred by multicopy plasmids [38]. RcsA is a transcriptional activator for genes responsible for the synthesis of polysaccharide capsule in *E. coli*. The transcription of *rcsA* is repressed normally at a low level by the bacterial histonelike protein H-NS encoded by *hns*. The Gottesman group found that multicopy plasmids carrying a DNA segment downstream from *rcsA* increase transcription of a number of H-NS-repressed genes, including *rcsA*. This region, *dsr*, was shown to encode an 85-nt sRNA that presumably counteracts the H-NS action. DsrA was shown to be partially complementary to the translation initiation region of *hns*, and it was proposed that DsrA acts as an antisense RNA to inhibit the *hns* translation [39]. In addition, DsrA was shown to be induced at a low temperature where RpoS, stationary-phase sigma factor, also accumulates to high levels [40]. How low temperature induces DsrA expression is not known.

Subsequent studies suggested that the 5′ leader region of *rpoS* mRNA can fold itself to mask the ribosome-binding site, and DsrA RNA can pair with the upper part of this *rpoS* mRNA structure, freeing the translation initiation region and allowing an increased *rpoS* translation [39,41]. Mutational studies confirmed that the proposed base pairing is indeed necessary for DsrA function [39,41]. In addition, it was demonstrated that Hfq is required for DsrA action in vivo, and DsrA binds to Hfq in vitro [42]. The physiological role of an increased expression of *rpoS* by DsrA would be to protect the cell from cold temperatures by regulating expression of genes involved in synthesis of osmoprotectant trehalose. Subsequently, another sRNA called RprA was identified as a multicopy suppressor of *dsrA* mutants [43,44]. RprA stimulates the *rpoS* translation by interacting with the same region of the *rpoS* leader as DsrA. The synthesis of RprA is regulated by the RcsC/YojN/RcsB phosphorelay system, which is involved also in the regulation of capsule synthesis [44]. Thus, two different sRNAs positively regulate translation of a common mRNA through RNA–RNA base pairing. Regulation of *rpoS* translation by DsrA and RprA is one of the few examples of positive regulation of a gene by a small antisense RNA in *E. coli*.

OxyS was discovered by Northern analysis during the study of expression of OxyR, a major transcription factor responsible for the synthesis of oxidative stress-inducible proteins [45]. The Storz group found that a 109-nt sRNA denoted that OxyS is induced in response to oxidative stress in an OxyR-dependent manner. The *oxyS* gene is located just upstream of and transcribed divergently from the *oxyR* gene. Remarkably, OxyS was shown to affect expression of more than 40 genes. The *fhlA* (transcriptional activator of formate metabolism) and *rpoS* are two genes about which the regulation by OxyS was examined in some detail. OxyS is partially complementary to the ribosome-binding site of *fhlA* mRNA [46]. Both mutational and biochemical analyses established that OxyS inhibits the binding of ribosome to *fhlA* mRNA, and therefore, its translation through base pairing [46]. Subsequent study revealed that the OxyS-*fhlA* mRNA interaction occurs also at a second site within the coding sequence of the *fhlA*, and base pairing at two sites contributes to the translational repression [47]. OxyS also inhibits translation of *rpoS*, but the mode of OxyS action on the *rpoS* mRNA remains obscure because no significant sequence complementary to OxyS is recognized in the *rpoS* leader region [22]. Hfq requirement for OxyS action in vivo and OxyS binding to Hfq both in vivo and in vitro were demonstrated [22,23]. In addition, Hfq was shown to increase the OxyS interaction with *fhlA* and *rpoS* mRNAs in vitro, though not dramatically [23]. It remains to be studied how the downregulation of *fhlA* and *rpoS* by OxyS is integrated in the oxidative stress response of the cell.

GcvB RNA was found during the study on expression of *gcvA* encoding a transcription factor of the glycine cleavage operon [48]. It is transcribed divergently from the *gcvA* in response to the presence of glycine in a GcvA-dependent manner and downregulates the expression of OppA and DppA, components of two major peptide transport systems. Although GcvB is likely to act by a base-pairing mechanism, this has not been proved yet.

## 4.6    GENOME-WIDE SEARCHES OF sRNAs

As mentioned earlier, six sRNAs (MicF, DicF, DsrA, RprA, OxyS, and GcvB) had been recognized to act as antisense RNAs before systematic searches for sRNAs were carried out in 2001. The properties of these RNAs allowed prediction of new sRNAs from the genome sequences of *E. coli* and other organisms. Thus, computational searches for sRNAs were carried out by several groups focusing on intergenic regions of the *E. coli* genome by using different characteristics of known sRNAs such as sequence conservation with relatives of *E. coli*, conservation of secondary structure, and existence of a promoter and a terminator [49–53]. Expression of the predicted RNAs was experimentally tested by microarrays and by Northern analysis in some cases. Thus, a number of sRNAs were newly identified. Remarkably, many of them were demonstrated to interact with Hfq [49]. More recently, RNAs bound to Hfq were isolated by coimmunoprecipitation with Hfq, allowing the identification of additional new RNAs [36]. The cloning-based RNA approach was also used to search for sRNAs [54,55]. All together more than 70 sRNAs, including known ones, have been detected in *E. coli,* and more than 30 RNAs appear to bind to Hfq [2,56]. The functions of many of them remain to be studied.

## 4.7   FUNCTION OF Sᴘᴏᴛ42 RNA

Spot42 was identified as one of the stable and abundant RNAs more than three decades ago [13]. Subsequent studies showed that this RNA is made from the *spf* gene, and its transcription is negatively regulated by CRP-cAMP [57]. However, the function of Spot42 RNA and the physiological relevance of the *spf* regulation by CRP-cAMP remained unknown. The Valentin–Hansen group noticed by BLAST search that Spot42 RNA has a significant complementarity to the translation initiation region of *galK* encoding galactokinase [15]. This prompted them to test the possible involvement of Spot 42 in the expression of *galK*, the third gene in the *gal* operon. They discovered that Spot42 RNA indeed downregulates *galK* expression without affecting expression of proximal *galE* and *galT* genes. In vitro studies demonstrated that Spot 42 RNA specifically interacts with the ribosome-binding site of *galK,* thereby blocking ribosome binding. Thus, Spot 42 RNA was shown to act as an antisense RNA that mediates discoordinate expression of the *gal* operon [15]. Subsequent studies demonstrated that Hfq binds to Spot42 RNA and stimulates the base pairing between Spot42 and the target RNA corresponding to the translation initiation region of *galK* in vitro [24]. The *galETKM* operon encodes four enzymes involved in galactose metabolism.

Although the four proteins are translated from a polycistronic mRNA, the relative synthesis of UDP-galactose epimerase (GalE), galactose-1 Phosphate uridyltransferase (GalT), and galactokinase (GalK) varies discoordinately, depending on metabolic conditions. When cAMP levels are low, for example, owing to the presence of glucose, the levels of GalK, but not GalE, decrease, resulting in an increase in the GalE-to-GalK ratio. However, the mechanism by which cAMP affects GalK synthesis specifically was again unknown. The discovery of the function of Spot42 RNA as an antisense RNA of *galK* had explained the mechanism by which the discoordinated expression of *gal* operon is achieved and the physiological relevance of the *spf* regulation by CRP-cAMP. The transcription of *spf* is negatively regulated by CRP-cAMP. Therefore, the Spot42 level increases when cells are growing on glucose, resulting in inhibition of *galK* translation without perturbing *galE* translation. The biological rationale for the *gal* regulation by Spot42 would be that GalK is needed only in the presence of galactose, while GalE is always required because it has a second role in synthesis of UDP-galactose, a building block for the cell wall.

## 4.8   RyhB AND IRON METABOLISM

The 90-nt RyhB RNA (SraI), one of the Hfq-binding sRNAs identified by genome-wide searches, has also led to the solution of a regulatory mystery concerning iron metabolism [58]. A tight regulation of iron concentration is important for the cell because it is an essential nutrient, but abundant iron can be toxic to the cell. Fur protein is a central regulator of iron metabolism in the cell, acting as a transcriptional repressor for various genes encoding iron uptake proteins when iron is abundant. It binds to specific sites called *Fur boxes* located in the promoter region of target genes in an $Fe^{2+}$-dependent manner to inhibit the binding of RNA polymerase to the promoter of these genes. When iron becomes scarce in the cell, Fur is inactivated, resulting in derepression of target genes. Expression of some genes encoding proteins for iron binding and storage depends on the repressor Fur protein. Among them are the *sdhCDAB* operon encoding succinate dehydrogenase and the *sodB* encoding superoxide dismutase. Thus, transcription of these genes is stimulated when iron is plentiful, while it is repressed when iron is scarce. In other words, Fur and $Fe^{2+}$ positively regulate these genes, but the mechanism underlying this regulation was unknown.

It was noticed that the *ryhB* promoter contains the Fur-binding sites, and RyhB is complementary to the translation initiation region of *sdhD*, the second gene of the *sdhCDAB* operon [58]. This raised an interesting possibility that Fur positively regulates the *sdh* operon through the negative regulation of RyhB expression. Then, the wild-type but not *hfq* mutant cells showed poor growth on media containing succinate as a sole carbon source when RyhB was overproduced on a multicopy plasmid, suggesting that RyhB is involved in the downregulation of the *sdh* operon in an Hfq-dependent manner. Indeed, RyhB RNA was found to be dramatically induced in response to iron limitation, resulting in a marked decrease of *sdh* mRNA as well as messages of five other genes known to be positively regulated by Fur. Sequences within RyhB are complementary to regions within each of the target genes. Thus, it was proposed that RyhB downregulates the target mRNAs through RNA–RNA base pairing. The biological relevance of the downregulation of the target mRNAs by RyhB is rather clear because it makes sense to stop the synthesis of iron-storage proteins when iron is limited.

## 4.9   COUPLED DEGRADATION OF sRNAs AND mRNA TARGETS

Further studies on the downregulation mediated by RyhB have illuminated intriguing features of the sRNA action [59]. First, RyhB induced under iron limitation causes rapid degradation of target mRNAs in an RNase E-dependent manner. The rapid degradation of target mRNAs was thought to be an additional mode of sRNA function. However, it should be noted that reduction of target mRNA level was already observed in the downregulation of *ompF* by MicF [31,32]. More recent studies indicate that destabilization of target mRNAs is deeply linked with translational inhibition and is probably an intrinsic feature of sRNA functions [60]. Second, RyhB itself and several other sRNAs are also destabilized in the presence but not absence of ongoing transcription. On the basis of these findings, it has been proposed that the turnover of sRNAs is coupled to and dependent on pairing with target mRNAs, and that sRNAs act stoichiometrically rather than catalytically. The coupled degradation of sRNAs and their target mRNAs provides an elegant way by which cells escape from sRNA actions when the stress signals are gone. Third, degradation of both mRNA targets and RyhB is dependent on RNase E, the major endoribonuclease of *E. coli*, and slowed in degradosome mutants. This issue is discussed further later.

## 4.10   S_GR_S AND GLUCOSE PHOSPHATE STRESS

SgrS (RyhA) is among Hfq-binding novel RNAs that were detected by a genome-wide search [36]. Recent study established that SgrS downregulates the expression of *ptsG* encoding the glucose transporter at posttranscriptional levels [61]. Regulation of *ptsG* by SgrS is becoming an excellent system to understand mechanisms by which Hfq-binding sRNAs regulate their target mRNAs. How the involvement of SgrS in the *ptsG* regulation was discovered is quite interesting. In *E. coli*, external glucose is transported into the cells coupled with phosphorylation by the phosphoenolpyruvate: sugar phosphotransferase system (PTS) and metabolized through the glycolytic pathway. The PTS consists of two common cytoplasmic proteins, enzyme I and HPr, and a series of sugar-specific enzyme II complexes. IICB^Glc encoded by *ptsG* is the membrane component of glucose-specific EII. The transport of glucose by PTS into the cells inhibits the *lac* operon expression by preventing the entry of lactose as an inducer [62,63]. Thus, glucose repression is dependent on the glucose PTS.

During mutational study on glucose repression, the *pgi* or *pfkA* mutation was found to eliminate the inhibitory effect of glucose on the *lac* operon through a marked reduction of IICB^Glc expression and destabilization of the *ptsG* mRNA in the presence of glucose [64]. The *pgi* and *pfkA* encode phosphoglucose isomerase and phosphofructokinase of glycolytic pathway, respectively. The rapid degradation of *ptsG* mRNA no longer occurred when RNase E was inactivated, indicating that RNase E is responsible for the destabilization of *ptsG* mRNA [64]. Then, evidence indicated that accumulation of glucose-6-phosphate or α-methylglucoside-6-phosphate somehow causes the RNase E-mediated destabilization of *ptsG* mRNA [65]. The physiological relevance of the posttranscriptional regulation of *ptsG* would be to avoid too much accumulation of glucose phosphate that could be deleterious to the cell. However, the mechanism underlying this posttranscriptional regulation had been unknown for a while.

A key observation to understanding the mechanism of *ptsG* downregulation in response to glucose phosphate stress was due to Vanderpool and Gotteman, who found that overproduction of an uncharacterized Hfq-binding sRNA, RyaA (SgrS), inhibited cell growth on glucose [61]. This raised the possibility that SgrS is involved in the downregulation of *ptsG*. Indeed, they demonstrated that the synthesis of SgrS RNA is induced in response to glucose phosphate stress and noted that a region of about 30 nt within SgrS RNA is partially complementary to the translation initiation region of *ptsG* mRNA. These observations led them to propose that SgrS acts on *ptsG* mRNA, presumably through base pairing. They also found that another protein, SgrR, encoded divergently from SgrS, is required for the induction of SgrS under the stress condition. The observation that the destabilization of *ptsG* mRNA under stress condition no longer occurs in cells lacking an RNA chaperon Hfq indicates that SgrS acts on *ptsG* mRNA in an Hfq-dependent manner [66].

## 4.11   RIBONUCLEOPROTEIN COMPLEX CONSISTING OF AN sRNA, Hfq, AND RNase E

As mentioned earlier, RNase E is responsible for the rapid degradation of *ptsG* mRNA mediated by SgrS. RNAse E forms, through its C-terminal scaffold region, a multiprotein complex called *RNA degradosome* containing PNPase, RhlB RNA helicase, and a glycolytic enzyme enolase [67]. Interestingly, the rapid degradation of *ptsG* mRNA in response to glucose phosphate stress no longer occurred when the C-terminal scaffold region of RNase E was removed [68]. Thus, the C-terminal scaffold region of RNase E is required for the rapid degradation of *ptsG* mRNA under the stress condition. An important finding was that Hfq associates with RNase E through its C-terminal scaffold region [69]. More importantly, SgrS RNA also associates with RNase E through Hfq [69].

In other words, RNAse E forms a ribonucleoprotein complex with SgrS through Hfq. Interestingly, the RNase E-based ribonucleoprotein complex containing SgrS and Hfq is distinct from the conventional RNA degradosome. The physical association of SgrS with RNase E nicely explains how the functional cooperation of SgrS/Hfq and RNase E is achieved. Furthermore, RyhB RNA induced by iron limitation was also shown to associate with RNase E through Hfq [69]. Thus, formation of a specific ribonucleoprotein complex with RNase E and Hfq could be a universal feature of Hfq-binding sRNAs. In addition, the rapid degradation of *ptsG* mRNA in response to glucose phosphate stress was prevented when enolase was depleted [68] because of failure of SgrS induction, although how enolase is involved in SgrS induction is not known (Morita and Aiba, unpublished result).

## 4.12   IMPLICATION OF MEMBRANE LOCALIZATION OF TARGET mRNAs

An unexpected finding concerning the *ptsG* downregulation by SgrS was that localization of target mRNA at the membrane facilitates the action of SgrS [66,70]. IICB$^{Glc}$ consists of the N-terminal IIC domain containing eight transmembrane segments and the C-terminal IIB domain. During the study on the downregulation of *ptsG* mRNA mediated by SgrS, it was found that the region corresponding to the first two transmembrane stretches of *ptsG* is required for the rapid degradation of *ptsG* mRNA. When the transmembrane region was replaced with that of another protein, the mRNA was still destabilized by SgrS, suggesting that the membrane-targeting property is important for the destabilization of mRNA mediated by SgrS. In addition, the cytoplasmic target mRNA is destabilized by SgrS when its translation is reduced by mutations.

On the basis of these observations, the following model has been proposed. The cotranslational membrane insertion of nascent peptide of IICB$^{Glc}$ brings the *ptsG* mRNA near the membrane. This membrane localization of *ptsG* mRNA may allow SgrS RNA, by competing with ribosome entry for the subsequent rounds of translation, to base-pair with the *ptsG* mRNA, resulting in the RNase E-dependent degradation of the *ptsG* mRNA. This suggests that sRNA action on target mRNAs could be modulated by translational status (efficiency of ribosome loading) in some cases. It is interesting to see whether this mechanism is operating also in other membrane-associated mRNAs regulated by sRNAs.

## 4.13   TRANSLATIONAL REPRESSION IS THE PRIMARY EVENT FOR GENE SILENCING

The regulatory outcomes of SgrS and RyhB RNAs are RNase E-dependent degradation of target mRNAs and translational repression. A recent study has demonstrated that translational repression is primarily responsible for gene silencing because the downregulation of proteins by two sRNAs occurs in the absence of RNase E-dependent rapid degradation of mRNA [60]. This conclusion was supported by experiments using strains in which the truncated RNase E lacking the C-terminal scaffold region was expressed. Effect of SgrS on biosynthesis of IICB$^{Glc}$ was examined by pulse-labeling with $^{35}$S-methionine and immunoprecipitation following the induction of SgrS. The synthesis of $^{35}$S-IICB$^{Glc}$ was markedly reduced in wild-type cells, but not in cells lacking SgrS under the stress condition, indicating that the stress-induced SgrS efficiently inhibits the translation of *ptsG* mRNA. Importantly, a marked reduction of $^{35}$S-IICB$^{Glc}$ was observed in cells expressing the truncated RNase E in which the destabilization of the *ptsG* mRNA no longer occurred even in the presence of SgrS. In addition, efficient translational inhibition was also observed when RNase E is thermally inactivated.

These results imply that SgrS downregulates the IICB$^{Glc}$ expression primarily by directly inhibiting its translation, and rapid degradation of *ptsG* mRNA is not required for *ptsG* downregulation. RyhB was also shown to inhibit the translation of *sodB* mRNA in the absence of RNase E-dependent rapid degradation of the mRNA. The rapid destabilization of target mRNAs would make gene silencing irreversible and ensure an effective turnover of mRNAs by eliminating translationally inactive mRNAs.

## 4.14   BASE-PAIRING REQUIREMENT AND ROLE OF H$_{FQ}$ IN S$_{GRS}$-*PTSG* SYSTEM

There is no doubt that Hfq-binding sRNAs act on target mRNAs through base pairing. Indeed, the requirement of base pairing for sRNA action has been supported in part by mutational studies in several cases such as OxyS-*fhlA* [46], DrsA-*rpoS* [40], RyhB-*sodB* [28], MicC-*ompC* [71], and MicA-*ompA* [29,72]. Typically, particular mutations either in a given sRNA or its target, which disrupt the predicted pairing, were shown to decrease the regulation. The compensatory mutations restored the regulation in some cases. However, to what extent is the predicted pairing required and which base pairs are important for the regulation have not been studied in any of these cases. Recently, a systematic mutational study concerning the predicted base-pairing requirement for sRNA action was carried out in the SgrS-*ptsG* system [30]. Interestingly, mutations either in the *ptsG* or SgrS that affected the SgrS action were confined to the 6-nt region overlapping the *ptsG* SD sequence. In particular, two single mutations that disrupt a G-C base pair within this short sequence completely eliminated SgrS function, and compensatory mutations restored it. In addition, the duplex formation in vitro was eliminated by a single mutation disrupting the C-G base pair within the critical region and restored by the compensatory mutation. Furthermore, the sequences of SgrS corresponding to the 6-nt region are perfectly conserved, while the overall sequences of SgrS are rather variable among organisms [61]. Thus, the 6 base pairs overlapping the *ptsG* SD sequence among the predicted 23 base pairs are important for SgrS action. The requirement for base pairing between SgrS and *ptsG* at a short sequence overlapping the *ptsG* SD sequence is consistent with the view that SgrS competes with and prevents the binding of 16S rRNA of the ribosome resulting in translation inhibition. If the six base pairs around the *ptsG* SD sequence are sufficient for SgrS action, this raises an interesting question of how SgrS can specify an mRNA among all other mRNA by using only six base pairs. Although most Hfq-binding sRNAs with known targets apparently are able to base-pair to a region in the vicinity of or overlapping the translation initiation region, it remains to be studied whether base pairing at a short sequence is sufficient for sRNA action in general.

Another important question is the role of Hfq in SgrS action. It has been demonstrated that Hfq dramatically enhances the duplex formation between *ptsG* RNA and SgrS by increasing its rate [30]. This finding nicely explains why Hfq is required for SgrS function. The duplex formation between the *ptsG* mRNA and SgrS proceeds slowly without Hfq in vitro. If this reflects the in vivo situation, the slow duplex formation would not be sufficient for SgrS to prevent the ribosome loading. The rapid association of SgrS with the *ptsG* mRNA in the presence of Hfq would efficiently prevent ribosome binding to the *ptsG* mRNA. Enhancement of base pairing between sRNAs and target mRNAs by Hfq has been shown in several cases such as OxyS-*fhlA* [23], Spot42-*galK* [24], and MicA-*ompA* [29] interactions. It is interesting to see whether Hfq also stimulates the rate of duplex formation in general. It remains to be studied whether it also affects the binding constant between two RNAs.

## 4.15  sRNAs AND OUTER MEMBRANE PROTEINS

Recent studies have shown that increasing numbers of Hfq-binding sRNAs are widely involved in the regulation of synthesis of outer membrane proteins. This topic has been reviewed very recently [6]. Therefore, I will summarize it briefly here. The second sRNA to be demonstrated as a regulator of the expression of outer membrane proteins, after the discovery of MicF, is MicC, which was originally denoted as IS063 in a computational screen [71]. MicC was shown to be complementary to the region around the ribosome-binding site of *ompC*. Both in vivo and in vitro experiments demonstrated that MicC negatively regulates OmpC expression by base-pairing with *ompC* mRNA, depending on Hfq. Similarly, MicA (SraD) has been found to be complementary to the translation initiation region of *ompA* [29,72]. Overexpression of MicA reduced the expression of OmpA protein and *ompA* mRNA in an Hfq-dependent fashion. MicA binds to the *ompA* mRNA leader, and Hfq stimulates the binding between two RNAs in vitro.

Two more Hfq-binding sRNAs, RseX and RybB, have been shown to be involved in downregulation of major outer membrane proteins. The 80-nt RybB RNA, identified in a genome-wide search, was found to be synthesized in a $\sigma^E$-dependent fashion and to downregulate the *ompC* and *ompW* mRNAs [6]. The 91-nt RseX RNA was identified as a multicopy suppressor of extracytoplasmic stress protease and shown to downregulate the *ompA* and *ompC* through base-pairing in an Hfq-dependent manner [73]. Furthermore, overexpression of $\sigma^E$ was shown to decrease the expression of major outer membrane proteins, including OmpF, OmpC, OmpA, and OmpX depending on Hfq, suggesting the existence of more $\sigma^E$-regulated Hfq-binding sRNAs [74]. Outer membrane proteins other than major porins have been shown to be regulated by Hfq-binding sRNAs, OmrA and OmrB, that were also identified by computational searches, although the mode of action of OmrA and OmrB on their targets is not clear at this moment [75].

It is quite intriguing why the synthesis of many outer membrane proteins is regulated at posttranscriptional levels by sRNAs. Several possible reasons for this have been proposed [6]. First, this regulation may allow the cell to balance porin levels and respond to the environment quickly. Second, posttranscriptional control may be necessary, in addition to transcriptional regulation, for efficient regulation of the levels of proteins made from abundant mRNAs such as porin message. Third, translocation across the inner membrane of outer membrane proteins may render the mRNAs susceptible to sRNA regulation, as demonstrated in the *ptsG* mRNA regulation by SgrS.

## 4.16 CONCLUSIONS AND PERSPECTIVES

Studies on gene silencing mediated by Hfq-binding sRNAs, in particular by SgrS and RyhB RNAs, in the last few years have clearly illuminated the regulatory cascade in which sRNAs are integrated and the mode of their action (Figures 4.2 and 4.3). It has been confirmed that sRNAs are induced in response to specific metabolic stresses depending on specific transcription factors and that sRNAs act by base-pairing with the ribosome-binding site of target mRNAs to downregulate their expression. More importantly, studies on SgrS and RyhB RNAs uncovered several novel features of the mechanisms of action of Hfq-binding sRNAs. Both SgrS and RyhB RNAs act to destabilize target mRNAs in an RNase E-dependent manner. They form a specific ribonucleoprotein complex with RNase E through Hfq to cause the translation repression and rapid degradation of target mRNAs. Translational inhibition rather than mRNA destruction is the primary event for gene silencing. In addition, it has been shown that RyhB and also other sRNAs are degraded coupled with the degradation of target mRNAs. In the case of SgrS, membrane localization of target mRNA facilitates the action of SgrS presumably by affecting competition between

| Stress | Accumulation of G6P | Iron limitation |
|---|---|---|
| Sensor/Regulator | SgrR | Fur |
| Small RNA | SgrS | RyhB |
| Silencing complex | SgrS/Hfq/RNase E | RyhB/Hfq/RNase E |
| Base-pairing with target mRNA | *ptsG* mRNA (membrane localization) | *sodB* mRNA |
| Regulatory outcomes | Translation inhibition of target mRNAs RNase E-dependent degradation of target mRNAs and small RNAs | |

**FIGURE 4.2**  Regulatory cascades of gene silencing mediated by Hfq-binding sRNAs. The regulatory pathways in which Hfq-binding sRNAs are involved in downregulation of the target mRNAs are shown focusing on SgrS and RyhB. A specific stress condition (accumulation of glucose-6-phosphate or iron limitation) leads to the synthesis of a specific sRNA (SgrS or RyhB) by affecting the activity of a transcription factor (SgrR or Fur). The sRNA forms a ribonucleoprotein complex with Hfq and RNase E and then acts on the target mRNA (*ptsG* mRNA or *sodB* mRNA) through base pairing. The base pairing occurs in competition with ribosome entry because the base-pairing region is around the ribosome-binding site. Thus, any event that slows ribosome loading, for example, membrane localization of the target mRNA as shown in the case of SgrS, may favor the base pairing. The base pairing and recruitment of Hfq/RNase E on the target mRNA causes inhibition of the translation and RNase E-dependent rapid degradation of the target mRNA. Evidence shows that the translational inhibition occurs efficiently without the recruitment of RNase E, indicating that the translational inhibition is the primary event for gene silencing.

the sRNA and ribosomes. Hfq stimulates base pairing between SgrS and the target mRNA by accelerating the rate of duplex formation.

However, many important questions remain to be studied concerning the mechanisms of action and the physiological roles of Hfq-binding sRNAs. First, whether many of the features of SgrS and RyhB hold true for other sRNAs must be tested experimentally. Second, while it is clear that sRNAs along with Hfq act to inhibit the translation of target mRNAs, the roles of base pairing and Hfq in gene silencing are less clear. Thus, one can ask whether the base pairing itself is responsible for the translational inhibition or Hfq acts directly to inhibit translation. If the former is true, the major role of Hfq is to stimulate the base pairing. If the latter is the case, the role of sRNAs would be to guide or recruit Hfq around the ribosome-binding site. Third, we have to learn more about the mechanisms and roles of RNase E-dependent degradation of both target mRNAs and sRNAs. Fourth, it is largely unknown how a specific stress modulates the activity of the cognate transcription factor to induce transcription of a given sRNA. Fifth, we do not know the functions and targets of

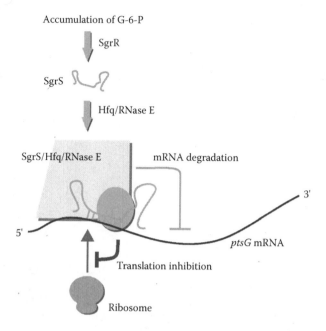

**FIGURE 4.3**   *Color version of this figure follows page 238.* Model for gene silencing by SgrS. SgrS induced in response to the accumulation of glucose-6-phosphate forms a ribonucleoprotein complex with Hfq (blue) and RNase E (yellow) and acts on the ribosome-binding site of *ptsG* mRNA through imperfect base pairing. The base pairing and recruitment of Hfq/RNase E on the target mRNA lead to translation inhibition and RNase E-dependent rapid degradation of the target mRNA.

many physically identified Hfq-binding sRNAs. In addition, we do not know how many more Hfq-binding sRNAs are expressed in *E. coli*. Finally, bacterial sRNAs seem to resemble eukaryotic sRNAs such as miRNAs and siRNAs. Both RNAs act on target mRNAs through base pairing by forming ribonucleoprotein complexes. The major regulatory outcomes of both RNAs are translational inhibition and mRNA destabilization. However, more studies on both systems are necessary to compare the mechanisms of their action in detail.

## ACKNOWLEDGMENT

I thank Teppei Morita for comments on this review and for preparing figures.

## REFERENCES

1. Gottesman, S., Stealth regulation: Biological circuits with small RNA switches. *Genes Dev*, 2002. 16(22): 2829–2842.
2. Gottesman, S., The small RNA regulators of *Escherichia coli*: Roles and mechanisms. *Annu Rev Microbiol*, 2004.
3. Gottesman, S., Micros for microbes: Non-coding regulatory RNAs in bacteria. *Trends Genet*, 2005. 21(7): 399–404.

4. Storz, G., S. Altuvia, and K. M. Wassarman, An abundance of RNA regulators. *Annu Rev Biochem*, 2005. 74: 199–217.

5. Storz, G., J. A. Opdyke, and A. Zhang, Controlling mRNA stability and translation with small, noncoding RNAs. *Curr Opin Microbiol*, 2004. 7(2): 140–144.

6. Guillier, M., S. Gottesman, and G. Storz, Modulating the outer membrane with small RNAs. *Genes Dev*, 2006. 20(17): 2338–2348.

7. Wagner, E. G., S. Altuvia, and P. Romby, Antisense RNAs in bacteria and their genetic elements. *Adv Genet*, 2002. 46: 361–398.

8. Storz, G. and S. Gottesman, Versatile roles of small RNA regulators in bacteria. Third edition, *The RNA World*, ed. R. F. Gesteland, T. R. Cech, and J. F. Atkins. 2006, New York: Cold Spring Harbor Laboratory Press. 567–594.

9. Jacob, F. and J. Monod, Genetic regulatory mechanisms in the synthesis of proteins. *J Mol Biol*, 1961. 3: 318–356.

10. Tomizawa, J. et al., Inhibition of ColE1 RNA primer formation by a plasmid-specified small RNA. *Proc Natl Acad Sci USA*, 1981. 78(3): 1421–1425.

11. Hindley, J., Fractionation of 32P-labelled ribonucleic acids on polyacrylamide gels and their characterization by fingerprinting. *J Mol Biol*, 1967. 30(1): 125–136.

12. Griffin, B. E., Separation of 32P-labelled ribonucleic acid components. The use of polyethylenimine-cellulose (TLC) as a second dimension in separating oligoribonucleotides of "4.5 S" and 5 S from *E. coli. FEBS Lett*, 1971. 15(3): 165–168.

13. Ikemura, T. and J. E. Dahlberg, Small ribonucleic acids of *Escherichia coli*. I. Characterization by polyacrylamide gel electrophoresis and fingerprint analysis. *J Biol Chem*, 1973. 248(14): 5024–5032.

14. Wassarman, K. M., A. Zhang, and G. Storz, Small RNAs in *Escherichia coli. Trends Microbiol*, 1999. 7(1): 37–45.

15. Moller, T. et al., Spot 42 RNA mediates discoordinate expression of the *E. coli* galactose operon. *Genes Dev*, 2002. 16(13): 1696–1706.

16. Valentin-Hansen, P., M. Eriksen, and C. Udesen, The bacterial Sm-like protein Hfq: A key player in RNA transactions. *Mol Microbiol*, 2004. 51(6): 1525–1533.

17. Franze de Fernandez, M. T., L. Eoyang, and J. T. August, Factor fraction required for the synthesis of bacteriophage Qbeta-RNA. *Nature*, 1968. 219(154): 588–590.

18. Tsui, H. C., H. C. Leung, and M. E. Winkler, Characterization of broadly pleiotropic phenotypes caused by an hfq insertion mutation in *Escherichia coli*. K-12. *Mol Microbiol*, 1994. 13(1): 35–49.

19. Muffler, A., D. Fischer, and R. Hengge-Aronis, The RNA-binding protein HF-I, known as a host factor for phage Qbeta RNA replication, is essential for rpoS translation in *Escherichia coli. Genes Dev*, 1996. 10(9): 1143–1151.

20. Brown, L. and T. Elliott, Mutations that increase expression of the rpoS gene and decrease its dependence on hfq function in *Salmonella typhimurium. J Bacteriol*, 1997. 179(3): 656–662.

21. Vytvytska, O. et al., Hfq (HF1) stimulates ompA mRNA decay by interfering with ribosome binding. *Genes Dev*, 2000. 14(9): 1109–1118.

22. Zhang, A. et al., The OxyS regulatory RNA represses rpoS translation and binds the Hfq (HF-I) protein. *EMBO J*, 1998. 17(20): 6061–6068.

23. Zhang, A. et al., The Sm-like Hfq protein increases OxyS RNA interaction with target mRNAs. *Mol Cell*, 2002. 9(1): 11–22.

24. Moller, T. et al., Hfq: A bacterial Sm-like protein that mediates RNA-RNA interaction. *Mol Cell*, 2002. 9(1): 23–30.

25. Sauter, C., J. Basquin, and D. Suck, Sm-like proteins in Eubacteria: The crystal structure of the Hfq protein from *Escherichia coli. Nucleic Acids Res*, 2003. 31(14): 4091–4098.

26. Schumacher, M. A. et al., Structures of the pleiotropic translational regulator Hfq and an Hfq-RNA complex: A bacterial Sm-like protein. *EMBO J*, 2002. 21(13): 3546–3556.

27. Mikulecky, P. J. et al., *Escherichia coli* Hfq has distinct interaction surfaces for DsrA, rpoS and poly(A) RNAs. *Nat Struct Mol Biol*, 2004. 11(12): 1206–1214.

28. Geissmann, T. A. and D. Touati, Hfq, a new chaperoning role: Binding to messenger RNA determines access for small RNA regulator. *EMBO J*, 2004. 23(2): 396–405.

29. Rasmussen, A. A. et al., Regulation of ompA mRNA stability: The role of a small regulatory RNA in growth phase-dependent control. *Mol Microbiol*, 2005. 58(5): 1421–1429.

30. Kawamoto, H. et al., Base-pairing requirement for RNA silencing by a bacterial small RNA and acceleration of duplex formation by Hfq. *Mol Microbiol*, 2006. 61(4): 1013–1022.

31. Mizuno, T., Regulation of gene expression by a small RNA transcript (micRNA). *Tanpakushitsu Kakusan Koso*, 1984. 29(11): 908–913.

32. Andersen, J. et al., The function of micF RNA. micF RNA is a major factor in the thermal regulation of OmpF protein in *Escherichia coli. J Biol Chem*, 1989. 264(30): 17961–17970.

33. Chou, J. H., J. T. Greenberg, and B. Demple, Posttranscriptional repression of *Escherichia coli* OmpF protein in response to redox stress: Positive control of the micF antisense RNA by the soxRS locus. *J Bacteriol*, 1993. 175(4): 1026–1031.

34. Coyer, J. et al., micF RNA in ompB mutants of *Escherichia coli*: Different pathways regulate micF RNA levels in response to osmolarity and temperature change. *J Bacteriol*, 1990. 172(8): 4143–4150.

35. Andersen, J. and N. Delihas, micF RNA binds to the 5' end of ompF mRNA and to a protein from *Escherichia coli. Biochemistry*, 1990. 29(39): 9249–9256.

36. Zhang, A. et al., Global analysis of small RNA and mRNA targets of Hfq. *Mol Microbiol*, 2003. 50(4): 1111–1124.

37. Tetart, F. and J. P. Bouche, Regulation of the expression of the cell-cycle gene ftsZ by DicF antisense RNA. Division does not require a fixed number of FtsZ molecules. *Mol Microbiol*, 1992. 6(5): 615–620.

38. Sledjeski, D. and S. Gottesman, A small RNA acts as an antisilencer of the H-NS-silenced rcsA gene of *Escherichia coli. Proc Natl Acad Sci USA*, 1995. 92(6): 2003–2007.

39. Lease, R. A., M. E. Cusick, and M. Belfort, Riboregulation in *Escherichia coli*: DsrA RNA acts by RNA:RNA interactions at multiple loci. *Proc Natl Acad Sci USA*, 1998. 95(21): 12456–12461.

40. Sledjeski, D. D., A. Gupta, and S. Gottesman, The small RNA, DsrA, is essential for the low temperature expression of RpoS during exponential growth in *Escherichia coli. EMBO J*, 1996. 15(15): 3993–4000.

41. Majdalani, N. et al., DsrA RNA regulates translation of RpoS message by an anti-anti-sense mechanism, independent of its action as an antisilencer of transcription. *Proc Natl Acad Sci USA*, 1998. 95(21): 12462–12467.

42. Sledjeski, D. D., C. Whitman, and A. Zhang, Hfq is necessary for regulation by the untranslated RNA DsrA. *J Bacteriol*, 2001. 183(6): 1997–2005.

43. Majdalani, N. et al., Regulation of RpoS by a novel small RNA: The characterization of RprA. *Mol Microbiol*, 2001. 39(5): 1382–1394.

44. Majdalani, N., D. Hernandez, and S. Gottesman, Regulation and mode of action of the second small RNA activator of RpoS translation, RprA. *Mol Microbiol*, 2002. 46(3): 813–826.

45. Altuvia, S. et al., A small, stable RNA induced by oxidative stress: Role as a pleiotropic regulator and antimutator. *Cell*, 1997. 90(1): 43–53.

46. Altuvia, S. et al., The *Escherichia coli* OxyS regulatory RNA represses fhlA translation by blocking ribosome binding. *EMBO J*, 1998. 17(20): 6069–6075.

47. Argaman, L. and S. Altuvia, fhlA repression by OxyS RNA: Kissing complex formation at two sites results in a stable antisense-target RNA complex. *J Mol Biol*, 2000. 300(5): 1101–1112.

48. Urbanowski, M. L., L. T. Stauffer, and G. V. Stauffer, The gcvB gene encodes a small untranslated RNA involved in expression of the dipeptide and oligopeptide transport systems in *Escherichia coli*. *Mol Microbiol*, 2000. 37(4): 856–868.

49. Wassarman, K. M. et al., Identification of novel small RNAs using comparative genomics and microarrays. *Genes Dev*, 2001. 15(13): 1637–1651.

50. Rivas, E. et al., Computational identification of noncoding RNAs in *E. coli* by comparative genomics. *Curr Biol*, 2001. 11(17): 1369–1373.

51. Argaman, L. et al., Novel small RNA-encoding genes in the intergenic regions of *Escherichia coli*. *Curr Biol*, 2001. 11(12): 941–950.

52. Chen, S. et al., A bioinformatics based approach to discover small RNA genes in the *Escherichia coli* genome. *Biosystems*, 2002. 65(2–3): 157–177.

53. Carter, R. J., I. Dubchak, and S. R. Holbrook, A computational approach to identify genes for functional RNAs in genomic sequences. *Nucleic Acids Res*, 2001. 29(19): 3928–3938.

54. Kawano, M. et al., Detection of low-level promoter activity within open reading frame sequences of *Escherichia coli*. *Nucleic Acids Res*, 2005. 33(19): 6268–6276.

55. Vogel, J. et al., RNomics in *Escherichia coli* detects new sRNA species and indicates parallel transcriptional output in bacteria. *Nucleic Acids Res*, 2003. 31(22): 6435–6443.

56. Vogel, J. and C. M. Sharma, How to find small non-coding RNAs in bacteria. *Biol Chem*, 2005. 386(12): 1219–1238.

57. Polayes, D. A. et al., Cyclic AMP-cyclic AMP receptor protein as a repressor of transcription of the spf gene of *Escherichia coli*. *J Bacteriol*, 1988. 170(7): 3110–3114.

58. Masse, E. and S. Gottesman, A small RNA regulates the expression of genes involved in iron metabolism in *Escherichia coli*. *Proc Natl Acad Sci USA*, 2002. 99(7): 4620–4625.

59. Masse, E., F. E. Escorcia, and S. Gottesman, Coupled degradation of a small regulatory RNA and its mRNA targets in *Escherichia coli*. *Genes Dev*, 2003. 17(19): 2374–2383.

60. Morita, T., Y. Mochizuki, and H. Aiba, Translational repression is sufficient for gene silencing by bacterial small noncoding RNAs in the absence of mRNA destruction. *Proc Natl Acad Sci USA*, 2006. 103(13): 4858–4863.

61. Vanderpool, C. K. and S. Gottesman, Involvement of a novel transcriptional activator and small RNA in post-transcriptional regulation of the glucose phosphoenolpyruvate phosphotransferase system. *Mol Microbiol*, 2004. 54(4): 1076–1089.

62. Inada, T., K. Kimata, and H. Aiba, Mechanism responsible for glucose-lactose diauxie in *Escherichia coli*: Challenge to the cAMP model. *Genes Cells*, 1996. 1(3): 293–301.

63. Kimata, K. et al., cAMP receptor protein-cAMP plays a crucial role in glucose-lactose diauxie by activating the major glucose transporter gene in *Escherichia coli*. *Proc Natl Acad Sci USA*, 1997. 94(24): 12914–12919.

64. Kimata, K. et al., Expression of the glucose transporter gene, ptsG, is regulated at the mRNA degradation step in response to glycolytic flux in *Escherichia coli*. *EMBO J*, 2001. 20(13): 3587–3595.

65. Morita, T. et al., Accumulation of glucose 6-phosphate or fructose 6-phosphate is responsible for destabilization of glucose transporter mRNA in *Escherichia coli*. *J Biol Chem*, 2003. 278(18): 15608–15614.

66. Kawamoto, H. et al., Implication of membrane localization of target mRNA in the action of a small RNA: Mechanism of post-transcriptional regulation of glucose transporter in *Escherichia coli*. *Genes Dev*, 2005. 19(3): 328–338.

67. Carpousis, A. J., The *Escherichia coli* RNA degradosome: Structure, function and relationship in other ribonucleolytic multienzyme complexes. *Biochem Soc Trans*, 2002. 30(2): 150–155.

68. Morita, T. et al., Enolase in the RNA degradosome plays a crucial role in the rapid decay of glucose transporter mRNA in the response to phosphosugar stress in *Escherichia coli*. *Mol Microbiol*, 2004. 54(4): 1063–1075.

69. Morita, T., K. Maki, and H. Aiba, RNase E-based ribonucleoprotein complexes: Mechanical basis of mRNA destabilization mediated by bacterial noncoding RNAs. *Genes Dev*, 2005. 19(18): 2176–2186.
70. Vanderpool, C. K. and S. Gottesman, Noncoding RNAs at the membrane. *Nat Struct Mol Biol*, 2005. 12(4): 285–286.
71. Chen, S. et al., MicC, a second small-RNA regulator of Omp protein expression in *Escherichia coli. J Bacteriol*, 2004. 186(20): 6689–6697.
72. Udekwu, K. I. et al., Hfq-dependent regulation of OmpA synthesis is mediated by an antisense RNA. *Genes Dev*, 2005. 19(19): 2355–2366.
73. Douchin, V., C. Bohn, and P. Bouloc, Down-regulation of porins by a small RNA bypasses the essentiality of the regulated intramembrane proteolysis protease RseP in *Escherichia coli. J Biol Chem*, 2006. 281(18): 12253–12259.
74. Rhodius, V. A. et al., Conserved and variable functions of the sigmaE stress response in related genomes. *PLoS Biol*, 2006. 4(1): e2.
75. Guillier, M. and S. Gottesman, Remodelling of the *Escherichia coli* outer membrane by two small regulatory RNAs. *Mol Microbiol*, 2006. 59(1): 231–247.

# 5 Mechanisms by which microRNAs Regulate Gene Expression in Animal Cells

*Yang Yu and Timothy W. Nilsen*

## CONTENTS

## 5.1 INTRODUCTION

MicroRNAs comprise a large family of regulatory molecules found in all multicellular organisms. As detailed in other chapters in this book, these small RNAs have been demonstrated to have important functions in a wide variety of biological processes and have been implicated in many diseases including cancer.

Despite rapid and impressive progress in characterizing miRNAs and in elucidating their physiological roles, the mechanisms by which they regulate gene expression are still largely mysterious. It is clear that the major role of miRNAs is to downregulate the protein production from targeted mRNAs posttranscriptionally. It is also clear that miRNAs do not function as naked RNAs but as components of ribonucleoprotein particles known as miRNPs, or miRISC; an essential constituent of these particles is a member of the Argonaute (AGO) protein family. In this chapter we review our current understanding of how miRNPs recognize their targets and how they repress those targets. The discussion will be limited to animal cells because plant miRNAs are thought to work primarily via directed cleavage of their target mRNAs, a situation only rarely observed in animal cells (reviewed in Reference 1). Furthermore, other interesting activities of miRNAs (e.g., stimulation of viral replication[2] and potential activities in the nucleus[3,4]) will not be discussed because mechanisms of these activities are even more poorly understood than those of repression. Finally, the reader is referred to several recent reviews[5-7] of mechanistic aspects of miRNA function for different perspectives and additional references.

## 5.2 miRNA-MEDIATED REGULATION: TARGET RECOGNITION

The first examples of mRNAs targeted by miRNAs came from studies in *Caenorhabditis elegans*, where it was found that the 3′ UTRs of regulated mRNAs contained multiple copies of sequences with imperfect complementarily to the miRNAs.[8-10] Mutational analyses established that these sites were necessary for miRNA-mediated regulation and, accordingly, that target recognition was accomplished at least in part by base-pairing interactions.[11]

Subsequent studies, both informatic and experimental, in a variety of systems have examined the degree of base complementarily that is required to confer regulation and, by inference, to promote the binding of an miRNP to its recognition site.[12-20] Cumulatively, these analyses have revealed that a remarkably low (6–7 bp) extent of base pairing is sufficient to elicit regulation.[13,15,21,22] This limited pairing must occur between the 5′ end of the miRNA (bases 2–7 or 2–8) and the target mRNA. The critical region of the miRNA is known as the seed region, and "seed-pairing" forms the basis for popular target prediction algorithms.[13,14,16,17,19] Although the physical basis for the importance of seed-pairing is not yet fully understood, crystal structures of an archeal Argonaute protein in complex with RNA indicate that this region of the miRNA may be highly constrained for presentation to the target.[23]

The fact that miRNP–target mRNA interactions can be mediated by specific, short base-pairing regions suggested the possibility that miRNAs could potentially regulate hundreds of distinct mRNAs. Indeed, informatic analyses coupled with experimental evidence indicate that this is the case, and it is now estimated that over a third of all mRNAs are subject to miRNA regulation.[22] Despite the power of seed-pairing as a predictor of miRNA targets, there remain several open questions. In this regard, there are now several examples of biologically relevant miRNA–target mRNA interactions where the seed-pairing rule does not hold (i.e., let-7:lin-41, and lys6:cog-1). For example, wobble pairing and mismatches within the seed region are sometimes tolerated.[24-27] At present, it is not clear if these examples represent rare exceptions to the rule or are more common. In either case, it will be important to determine what features allow such sites to be recognized so that target prediction can become more accurate.

In addition, it will be important to determine the fraction of seed-pairing sites that are actually functional. One extensive study indicated that most sites that contained seed-pairing were functional by the criteria of reporter assays.[22] These analyses therefore indicated that features other than primary sequence—for example, secondary structure, protein binding, or pairing with the 3′ end of the miRNA—were often not important in determining whether a site would be recognized.[13,22] Nevertheless, a recent study has demonstrated that secondary structure can have profound effects on the ability of miRNAs to recognize perfect seed matches.[28] Clearly, more studies are required before the principles of miRNA–target mRNA interactions are fully understood.

It will also be important to elucidate the mechanism whereby miRNAs encounter and engage their mRNA targets. At present, it is assumed that miRNPs freely diffuse in the cytoplasm, and the recognition occurs passively via random collision. However, given the elaborate machinery that is involved in loading miRNAs into their protein companions,[29–32] it would not be surprising if miRNA engagement with target mRNAs was an active process. Thus, it will be important to determine where in the cell and when during the lifetime of an mRNA recognition takes place.

Finally, it is not at all clear what determines the "strength" of a specific miRNA–mRNA interaction in terms of magnitude of regulation. As noted earlier, the first miRNA-targeted mRNAs identified contained multiple imperfectly complementary sites to the miRNAs in their 3′ UTRs. Subsequently, several other examples have been described where multiple target sites are present. A simple hypothesis would be that multiple sites would be necessary to confer regulation either by ensuring that any one site would be occupied by an miRNA or through additive or cooperative interactions between miRNAs bound to a single UTR. However, this hypothesis has not been borne out experimentally since there are now clear examples where a single miRNA recognition site can confer substantive recognition,[18,33] and additional examples where multiple sites confer only modest regulation.[13,22] The biochemical basis of these phenomena is currently unknown. It is possible that different sites bind miRNAs more strongly or weakly. It is also possible that the specific conformation of the miRNP bound to different sites allows the recruitment of ancillary factors that could aid or inhibit the ability of the miRNP to regulate expression. Elucidation of molecular details that determine the "strength" or magnitude of regulation by specific miRNAs is an important area of current research.

In summary, miRNAs and their protein partners play widespread and important roles in nearly every aspect of biology. Although significant progress has been made in understanding some of these roles, many fundamental questions regarding principles of target recognition and the different outcomes of such recognition remain unanswered.

## 5.3   MECHANISMS OF miRNA-MEDIATED REGULATION

### 5.3.1   Effects of Translation

In the preceding section, our current understanding of target recognition by miRNAs was discussed. Once an mRNA target is engaged by an miRNA, how is regulation actually achieved? Again, the first hints of mechanism came from studies in *C. elegans*. It was known that the protein products of specific miRNA-targeted genes were essentially absent.[34,35] Surprisingly, however, there was no comparable decrease in the mRNAs encoding those proteins. These observations indicated that the miRNAs were exerting their effect posttranscriptionally at the level of translation of the targeted mRNAs. One well-understood form of translational regulation is at the level of initiation of protein synthesis, a situation in which regulated mRNAs are prevented from engaging with ribosomes. Remarkably, however, the miRNA-regulated mRNAs were found to be associated with polysomes; polysomes form when multiple ribosomes are translating an mRNA.[34,35] Two possible explanations were put forward to explain these phenomena. First, the ribosomes associated with the miRNA-regulated mRNA might not be actively translating; if this was the case, the miRNA would have to arrest translational elongation after the mRNA was fully loaded with ribosomes, perhaps near the termination codon. The alternative explanation was that the miRNA in some way induced the degradation of protein as it was being synthesized—cotranslational degradation.[35] Two recent studies have provided support for the latter mechanism of miRNA action. In one, an miRNA-regulated mRNA was shown to be exclusively in polysomes, and by several criteria it was engaged with translating ribosomes.[36] In the second study, a reporter mRNA carrying a 3′ UTR responsive to an miRNA was constructed so that the protein synthesized bore an amino-terminal epitope tag.[37] Importantly, the reporter was shown to be downregulated in an miRNA-dependent fashion, and there was no difference in the polysome association of the miRNA-responsive mRNA and a control mRNA.[37] To determine if the miRNA-responsive mRNA was synthesizing protein, immunoprecipitation of nascent polypeptides was performed. Strikingly, while the control mRNA was efficiently precipitated via the N-terminal tag, no miRNA-regulated mRNA could be recovered. The simplest explanation for these results is that the polypeptides synthesized from the miRNA-regulated mRNA were rapidly degraded as they were synthesized.[37] Substantive evidence exists for the process of cotranslational proteolysis in other systems (e.g., see Reference 38), but definitive proof that this occurs in response to miRNAs awaits the identification and inhibition of the relevant proteases. Although the studies just discussed provide strong evidence that miRNA-mediated regulation can occur after initiation of translation at the level of protein turnover, many studies have shown that other pathways of regulation exist.

Specifically, several studies have found that miRNAs can inhibit protein synthesis at the level of initiation of translation. In two cases, one studying a reporter mRNA and one studying an endogenous mRNA, miRNAs were shown to prevent, to a large degree, ribosome association of the targeted mRNAs.[39,40] In a third study, the effect of miRNAs on cap-dependent or internal ribosome entry site (IRES)-driven translation was examined.[41] Because cap-dependent translation was inhibited while IRES-dependent translation was not, it was deduced that miRNAs affected translation at the level of cap recognition (i.e., initiation). Consistent with this interpretation, it has been recently shown that Ago proteins that function in miRNA-mediated regulation have intrinsic cap-binding activity.[42] These findings suggest that Ago proteins, upon interacting with their target, could compete with eIF4E and thus prevent cap recognition, thereby inhibiting initiation.[42] In vitro analysis in *Drosophila* embryo extracts also suggests that inhibition of initiation of protein synthesis can occur in a cap-dependent manner.[43]

Finally, an alternative mechanism of initiation inhibition has been proposed in which Ago recruits the initiation factor eIF6.[44] eIF6 acts to prevent association of the large ribosomal subunit with the small subunit at the initiation codon. In contrast to inhibition of cap recognition, inhibition via recruitment of eIF6 would not be expected to be dependent on the cap.[44]

In addition to these studies, another effect on translation has been observed. Using a reporter construct responsive to an miRNA, it was found that the mRNA was polysome associated, but several lines of evidence suggested that the miRNA-dependent inhibition of protein synthesis occurred because of premature ribosome drop off (premature termination of translation). Furthermore, in this case, both cap-dependent and IRES-driven translation were equally susceptible to miRNA-mediated inhibition.[45]

At present, there is no easy way to rationalize the different observations made in different laboratories that have studied similar but distinct mRNAs. For example, the two mRNAs analyzed that showed inhibition after initiation of protein synthesis were responsive to the same miRNA (let-7) as the mRNA that was inhibited at the level of initiation. The only notable difference between the three mRNAs lies in the target sites; in the two mRNAs whose translation was inhibited post initiation, the let-7 binding sites were in the context of whole 3′ UTRs, while in the mRNA inhibited at the level of initiation, the target sites were designed. How the target sites differ in terms of promoting distinct mechanisms is currently unclear, although, as discussed, specific conformations could dictate the association of specific cofactors.

The reasons for the differences in susceptibility of IRES-driven translation to miRNA repression are equally unclear. Specifically, two studies that used the same reporter and the same miRNA came to opposite conclusions. In these cases, a possible explanation is that one study examined the behavior of transfected RNAs, whereas the other examined the behavior of mRNAs expressed from transfected plasmids. It will be of interest to determine how an mRNA that is transcribed in the nucleus differs from the same RNA introduced via transfection and how those differences impact miRNA function.

In summary, it seems clear that miRNA-mediated repression of translation can occur at several different levels. However, only a few examples have been studied in any detail, and it will be important to define the molecular features that dictate the specific mechanism of inhibition for specific miRNA–mRNA target pairs, as well as what fraction of total regulation occurs by each pathway.

### 5.3.2   EFFECTS ON mRNA STABILITY

In addition to direct effects on translation and/or protein accumulation, it is now evident that miRNAs can exert effects on other aspects of mRNA metabolism, including enhancing mRNA turnover. The first indication that miRNAs could promote mRNA degradation came from analyses in which an miRNAs was transfected into a cell that did not normally express that miRNA.[15] Following transfection, microarray profiling of mRNA expression revealed that many mRNAs were downregulated. Strikingly, the downregulated mRNAs were highly enriched for seed matches to the transfected miRNA. Similar results were obtained with two different miRNAs, and subsequent analyses established a strong correlation between low levels of mRNA expression and high levels of miRNA expression in specific tissues.

These studies were soon followed by the demonstration that knockdown of the miRNA-specific Ago protein in *Drosophila* lead to upregulation of a number of mRNAs that again were strongly enriched for miRNA target sites.[46,47] In addition, it was demonstrated that certain miRNA-regulated mRNAs in *C. elegans* were regulated at the level of mRNA abundance.[48] Finally, it was shown that the half-life of certain miRNA-targeted mRNAs was greatly accelerated.[46,49] Key questions are, how do miRNAs affect mRNA stability, and what fraction of miRNA targets are affected in this way? Unfortunately, there is no easy way to answer the second question. There are numerous examples of regulated mRNAs where abundance does not change, but there is, at present, no high-throughput method analogous to microarrays that can identify mRNAs regulated strictly at the level of translation. With regard to mechanisms of destabilization, insight has now been obtained from several systems.

In particular, studies in both human and *Drosophila* cells as well as zebra fish embryos have shown that miRNAs can promote rapid deadenylation of targeted mRNAs.[46,50,51] In *Drosophila*, knockdown experiments have demonstrated that the CCR4–NOT complex is the responsible deadenylase, but the relevant deadenylases in other organisms have not yet been identified. Because deadenylation followed by decapping is the pathway by which most mRNAs are degraded in eukaryotic cells, miRNA-dependent accelerated deadenylation could explain why certain targeted mRNAs are reduced in abundance.

Nevertheless, several significant questions remain open. First, does deadenylation occur on all targeted mRNAs? In studies so far, it would appear that the answer may be yes, but relatively few examples have been analyzed. It will be of particular interest to determine if polysome-associated miRNA-targeted mRNAs are deadenylated.

A second major question is what determines whether a deadenylated mRNA is subject to enhanced degradation. In the *Drosophila* studies, several reporter and endogenous mRNA-targeted mRNAs were analyzed, and all appeared to be completely deadenylated. However, one mRNA remained stable, while the others were destabilized to different degrees. Elucidation of the specific features of these mRNAs that lead to distinct fates will be illuminating not only in terms of miRNA function but also in terms of the process of miRNA decay in general. Third, it is unclear how the deadenylation machinery is recruited to mRNA-targeted mRNAs. In this regard, it has been shown in *Drosophila* that a protein known as GW182 (see the following text) is required for deadenylation,[46] but it has not been established whether it communicates directly with the deadenylase complex or whether other factors are involved.

In summary, it is now clear that miRNAs can exert their regulatory function by accelerating the turnover of targeted mRNAs. However, the detailed features of specific mRNPs that are susceptible to this type of regulation remain to be determined.

### 5.3.3 THE ROLE OF P-BODIES

For several years, it has been known that a variety of enzymes involved in eukaryotic mRNA decay colocalize in the cytoplasm in foci called *processing bodies* (P-bodies), also known as P-granules or GW bodies. Extensive studies in budding yeast suggest that these foci serve many functions, including decapping and subsequent exonucleolytic destruction of mRNAs (reviewed in Reference 52). Interestingly, they may also serve as storage sites for untranslated mRNAs that are not destined for decay but may return to the translating pool.[52,53]

The finding that Argonaute proteins showed the same pattern of localization as known P-body components suggested that these foci might somehow be involved in miRNA function. This idea was considerably strengthened by the demonstration that the P-body constituent GW182 and related proteins were required for miRNA-mediated regulation in a number of systems.[39,46,54–59] Furthermore, several studies revealed that certain miRNA-regulated mRNAs were localized to P-bodies in an miRNA- and/or GW182-dependent manner. Collectively, these data provide the basis for a mechanistic framework in which to view miRNA function.

In this regard, miRNA-targeted mRNAs could be directly recruited to P-bodes where they could either be sequestered from the translational apparatus (P-bodes lack ribosomes and most translation factors) or degraded by the mRNA turnover machinery. These distinct fates could explain why some regulated mRNAs display accelerated decay while others do not (see earlier text). However, what properties distinguish these mRNAs have not yet been determined. Direct recruitment could also explain examples where inhibition of initiation of translation have been observed,[39,40] that is, targeted mRNAs would not encounter ribosomes. Alternatively, if miRNAs inhibit initiation of translation by interfering with the initiation process itself (e.g., cap recognition), targeted mRNAs might be recruited to P-bodes as a consequence of this inhibition.

While the P-body hypothesis is attractive and explains much of the current data, it cannot explain examples of regulation where regulated mRNAs are found to be in translating polysomes. Furthermore, recent studies have shown that miRNA-mediated regulation is not compromised when several P-body constituents were knocked down by RNAi.[60,61] The conclusions of these analyses were that P-bodes were a consequence, not the cause, of miRNA-mediated regulation. The role of P-bodes in the miRNA pathway is further complicated by the observation that only a very small fraction of total cellular Argonaute proteins (<2%) are localized to visible foci.[62]

In summary, the function of P-bodes in miRNA-mediated regulation is currently puzzling. In part, this confusion results from the fact that they are defined visually and not biochemically; to date, the foci have not been purified. While there exist clear examples of localization of miRNA-regulated mRNAs to these foci, it has not yet been shown that localization is required to achieve regulation. Finally, although several components of P-bodes have been shown to be involved in miRNA-mediated regulation—most notably GW182, decapping enzymes, and the translation regulator Rck/p54—it is not clear whether their localization plays a role in their activity.

## 5.4   CONCLUSIONS

From the foregoing discussion of mechanisms of miRNA-mediated regulation, it is apparent that there is no single mechanism of action but rather multiple means by which miRNAs can exert their effects. These include inhibition of translation (at different stages), and enhanced decay, and sequestration. In hindsight, it may not be surprising that there is no one universal mechanism of action, even though common factors (e.g., Argonaute proteins and GW182) are involved. For many years, it has been known that the same (or similar) factors can have markedly different effects on specific mRNAs. Particularly striking examples include AUF1/hnRNP D,[63] which can act to either destabilize or stabilize certain mRNAs, and the Puf family of proteins, which can promote deadenylation and decay or inhibit translation.[64] While the mechanistic basis for these diverse effects is not fully understood, it is reasonable to hypothesize that they result from the interplay of common factors with distinct factors bound to the 3′ UTRs of different mRNAs.

A similar scenario could apply to miRNA-mediated regulation where the action of common factors (Argonaute proteins and GW182) is dictated by the presence of mRNP-specific factors. Unfortunately, our understating of mRNP-binding proteins and their influence on stability and/or translation of specific mRNAs is still in its infancy. A major challenge will be defining the protein composition of individual mRNPs and how that constellation of proteins ultimately dictates both the magnitude and mechanism of regulation.

Finally, it should be noted that nearly all studies of mechanism of regulation have been conducted in transformed tissue culture cells. It is likely, if not certain, that the spectrum of mRNP-binding proteins differs substantially, depending on cell type, state of differentiation, or growth status. This raises the interesting possibility that individual miRNA-targeted mRNAs might be regulated by different mechanisms and to different extents in specific tissues or at distinct stages of development.

## REFERENCES

1. Vaucheret, H., Post-transcriptional small RNA pathways in plants: Mechanisms and regulations, *Genes Dev.* 20 (7), 759–771, 2006.
2. Jopling, C. L., Yi, M., Lancaster, A. M., Lemon, S. M., and Sarnow, P., Modulation of Hepatitis C virus RNA abundance by a liver-specific microRNA, *Science* 309 (5740), 1577–1581, 2005.
3. Hwang, H. -W., Wentzel, E. A., and Mendell, J. T., A hexanucleotide element directs microRNA nuclear import, *Science* 315 (5808), 97–100, 2007.
4. Ritland Politz, J. C., Zhang, F., and Pederson, T., MicroRNA-206 colocalizes with ribosome-rich regions in both the nucleolus and cytoplasm of rat myogenic cells, *PNAS* 103 (50), 18957–18962, 2006.
5. Pillai, R. S., Bhattacharyya, S. N., and Filipowicz, W., Repression of protein synthesis by miRNAs: How many mechanisms?, *Trends Cell Biol.* 17 (3), 118–126, 2007.
6. Jackson, R. J. and Standart, N., How do microRNAs regulate gene expression?, *Sci. STKE* 2007 (367), re1-, 2007.
7. Nilsen, T. W., Mechanisms of microRNA-mediated gene regulation in animal cells, *Trends Genet.* 23 (5), 243–249, 2007.
8. Wightman, B., Burglin, T. R., Gatto, J., Arasu, P., and Ruvkun, G., Negative regulatory sequences in the lin-14 3′-untranslated region are necessary to generate a temporal switch during *Caenorhabditis elegans* development, *Genes Dev.* 5 (10), 1813–1824, 1991.
9. Wightman, B., Ha, I., and Ruvkun, G., Posttranscriptional regulation of the heterochronic gene lin-14 by lin-4 mediates temporal pattern formation in *C. elegans*, *Cell* 75 (5), 855–862, 1993.
10. Lee, R. C., Feinbaum, R. L., and Ambros, V., The *C. elegans* heterochronic gene lin-4 encodes small RNAs with antisense complementarity to lin-14, *Cell* 75 (5), 843–854, 1993.
11. Ha, I., Wightman, B., and Ruvkun, G., A bulged lin-4/lin-14 RNA duplex is sufficient for *Caenorhabditis elegans* lin-14 temporal gradient formation, *Genes Dev.* 10 (23), 3041–3050, 1996.
12. Brennecke, J., Stark, A., Russell, R. B., and Cohen, S. M., Principles of microRNA-target recognition, *PLOS Biol.* 3 (3), e85–e85, 2005.
13. Stark, A., Brennecke, J., Bushati, N., Russell, R. B., and Cohen, S. M., Animal microRNAs confer robustness to gene expression and have a significant impact on 3′ UTR evolution, *Cell* 123 (6), 1133–1146, 2005.

14. Stark, A., Brennecke, J., Russell, R. B., and Cohen, S. M., Identification of *Drosophila* microRNA targets, *PLoS Biol.* 1 (3), E60, 2003.
15. Lim, L. P., Lau, N. C., Garrett-Engele, P., Grimson, A., Schelter, J. M., Castle, J., Bartel, D. P., Linsley, P. S., and Johnson, J. M., Microarray analysis shows that some microRNAs downregulate large numbers of target mRNAs, *Nature* 433 (7027), 769–773, 2005.
16. Lewis, B. P., Burge, C. B., and Bartel, D. P., Conserved seed pairing, often flanked by adenosines, indicates that thousands of human genes are microRNA targets, *Cell* 120 (1), 15–20, 2005.
17. Lewis, B. P., Shih, I.-H., Jones-Rhoades, M. W., Bartel, D. P., and Burge, C. B., Prediction of mammalian microRNA targets, *Cell* 115 (7), 787–798, 2003.
18. Kiriakidou, M., Nelson, P. T., Kouranov, A., Fitziev, P., Bouyioukos, C., Mourelatos, Z., and Hatzigeorgiou, A., A combined computational-experimental approach predicts human microRNA targets, *Genes Dev.* 18 (10), 1165–1178, 2004.
19. Krek, A., Grun, D., Poy, M. N., Wolf, R., Rosenberg, L., Epstein, E. J., MacMenamin, P., da Piedade, I., Gunsalus, K. C., Stoffel, M., and Rajewsky, N., Combinatorial microRNA target predictions, *Nat. Genet.* 37 (5), 495–500, 2005.
20. Chen, K. and Rajewsky, N., Natural selection on human microRNA binding sites inferred from SNP data, *Nat. Genet.* 38 (12), 1452–1456, 2006.
21. Doench, J. G. and Sharp, P. A., Specificity of microRNA target selection in translational repression, *Genes Dev.* 18 (5), 504–511, 2004.
22. Farh, K. K.-H., Grimson, A., Jan, C., Lewis, B. P., Johnston, W. K., Lim, L. P., Burge, C. B., and Bartel, D. P., The widespread impact of mammalian microRNAs on mRNA repression and evolution, *Science* 310 (5755), 1817–1821, 2005.
23. Ma, J.-B., Yuan, Y.-R., Meister, G., Pei, Y., Tuschl, T., and Patel, D. J., Structural basis for 5′-end-specific recognition of guide RNA by the *A. fulgidus* Piwi protein, *Nature* 434 (7033), 666–670, 2005.
24. Reinhart, B. J., Slack, F. J., Basson, M., Pasquinelli, A. E., Bettinger, J. C., Rougvie, A. E., Horvitz, H. R., and Ruvkun, G., The 21-nucleotide let-7 RNA regulates developmental timing in *Caenorhabditis elegans*, *Nature* 403 (6772), 901–906, 2000.
25. Vella, M. C., Choi, E.-Y., Lin, S.-Y., Reinert, K., and Slack, F. J., The *C. elegans* microRNA let-7 binds to imperfect let-7 complementary sites from the lin-41 3′ UTR, *Genes Dev.* 18 (2), 132–137, 2004.
26. Vella, M. C., Reinert, K., and Slack, F. J., Architecture of a validated microRNA:target interaction, *Chem. Biol.* 11 (12), 1619–1623, 2004.
27. Didiano, D. and Hobert, O., Perfect seed pairing is not a generally reliable predictor for miRNA-target interactions, *Nat. Struct. Mol. Biol.* 13, 849–851, 2006.
28. Long, D., Lee, R., Williams, P., Chan, C. Y., Ambros, V., and Ding, Y., Potent effect of target structure on microRNA function, *Nat. Struct. Mol. Biol.* 14 (4), 287–294, 2007.
29. Hutvagner, G. and Zamore, P. D., A microRNA in a multiple-turnover RNAi enzyme complex, *Science* 297 (5589), 2056–2060, 2002.
30. Schwarz, D. S., Hutvagner, G., Du, T., Xu, Z., Aronin, N., and Zamore, P. D., Asymmetry in the assembly of the RNAi enzyme complex, *Cell* 115 (2), 199–208, 2003.
31. Jiang, F., Ye, X., Liu, X., Fincher, L., McKearin, D., and Liu, Q., Dicer-1 and R3D1-L catalyze microRNA maturation in *Drosophila*, *Genes Dev.* 19 (14), 1674–1679, 2005.
32. Gregory, R. I., Chendrimada, T. P., Cooch, N., and Shiekhattar, R., Human RISC couples microRNA biogenesis and posttranscriptional gene silencing, *Cell* 123 (4), 631–640, 2005.
33. Nelson, P. T., Hatzigeorgiou, A. G., and Mourelatos, Z., miRNP:mRNA association in polyribosomes in a human neuronal cell line, *RNA* 10 (3), 387–394, 2004.
34. Olsen, P. H. and Ambros, V., The lin-4 regulatory RNA controls developmental timing in *Caenorhabditis elegans* by blocking LIN-14 protein synthesis after the initiation of translation, *Devel. Biol.* 216 (2), 671–680, 1999.

35. Seggerson, K., Tang, L., and Moss, E. G., Two genetic circuits repress the *Caenorhabditis elegans* heterochronic gene lin-28 after translation initiation, *Devel. Biol.* 243 (2), 215–225, 2002.

36. Maroney, P. A., Yu, Y., Fisher, J., and Nilsen, T. W., Evidence that microRNAs are associated with translating messenger RNAs in human cells, *Nat. Struct. Mol. Biol.* 13 (12), 1102–1107, 2006.

37. Nottrott, S., Simard, M. J., and Richter, J. D., Human let-7a miRNA blocks protein production on actively translating polyribosomes, *Nat. Struct. Mol. Biol.* 13 (12), 1108–1114, 2006.

38. Turner, G. C. and Varshavsky, A., Detecting and measuring cotranslational protein degradation in vivo, *Science* 289 (5487), 2117–2120, 2000.

39. Pillai, R. S., Bhattacharyya, S. N., Artus, C. G., Zoller, T., Cougot, N., Basyuk, E., Bertrand, E., and Filipowicz, W., Inhibition of translational initiation by let-7 microRNA in human cells, *Science* 309 (5740), 1573–1576, 2005.

40. Bhattacharyya, S. N., Habermacher, R., Martine, U., Closs, E. I., and Filipowicz, W., Relief of microRNA-mediated translational repression in human cells subjected to stress, *Cell* 125 (6), 1111–1124, 2006.

41. Humphreys, D. T., Westman, B. J., Martin, D. I. K., and Preiss, T., MicroRNAs control translation initiation by inhibiting eukaryotic initiation factor 4E/cap and poly(A) tail function, *PNAS* 102 (47), 16961–16966, 2005.

42. Kiriakidou, M., Tan, G. S., Lamprinaki, S., De Planell-Saguer, M., Nelson, P. T., and Mourelatos, Z., An mRNA m7G cap binding-like motif within human Ago2 represses translation, *Cell* 129 (6), 1141–1151, 2007.

43. Thermann, R. and Hentze, M. W., *Drosophila* miR2 induces pseudo-polysomes and inhibits translation initiation, *Nature* 447 (7146), 875–878, 2007.

44. Chendrimada, T. P., Finn, K. J., Ji, X., Baillat, D., Gregory, R. I., Liebhaber, S. A., Pasquinelli, A. E., and Shiekhattar, R., MicroRNA silencing through RISC recruitment of eIF6, *Nature* 447 (7146), 823–828, 2007.

45. Petersen, C. P., Bordeleau, M.-E., Pelletier, J., and Sharp, P. A., Short RNAs repress translation after initiation in mammalian cells, *Mol. Cell* 21 (4), 533–542, 2006.

46. Behm-Ansmant, I., Rehwinkel, J., Doerks, T., Stark, A., Bork, P., and Izaurralde, E., mRNA degradation by miRNAs and GW182 requires both CCR4:NOT deadenylase and DCP1:DCP2 decapping complexes, *Genes Dev.* 20 (14), 1885–1898, 2006.

47. Rehwinkel, J., Natalin, P., Stark, A., Brennecke, J., Cohen, S. M., and Izaurralde, E., Genome-wide analysis of mRNAs regulated by Drosha and Argonaute proteins in *Drosophila melanogaster*, *Mol. Cell. Biol.* 26 (8), 2965–2975, 2006.

48. Bagga, S., Bracht, J., Hunter, S., Massirer, K., Holtz, J., Eachus, R., and Pasquinelli, A. E., Regulation by let-7 and lin-4 miRNAs results in target mRNA degradation, *Cell* 122 (4), 553–563, 2005.

49. Wu, L. and Belasco, J. G., Micro-RNA regulation of the mammalian lin-28 gene during neuronal differentiation of embryonal carcinoma cells, *Mol. Cell. Biol.* 25 (21), 9198–9208, 2005.

50. Wu, L., Fan, J., and Belasco, J. G., MicroRNAs direct rapid deadenylation of mRNA, *PNAS* 103 (11), 4034–4039, 2006.

51. Giraldez, A. J., Mishima, Y., Rihel, J., Grocock, R. J., Van Dongen, S., Inoue, K., Enright, A. J., and Schier, A. F., Zebrafish MiR-430 promotes deadenylation and clearance of maternal mRNAs, *Science* 312 (5770), 75–79, 2006.

52. Parker, R. and Sheth, U., P bodies and the control of mRNA translation and degradation, *Mol. Cell* 25 (5), 635–646, 2007.

53. Brengues, M., Teixeira, D., and Parker, R., Movement of Eukaryotic mRNAs between polysomes and cytoplasmic processing bodies, *Science* 310 (5747), 486–489, 2005.

54. Ding, L., Spencer, A., Morita, K., and Han, M., The developmental timing regulator AIN-1 interacts with miRISCs and may target the Argonaute protein ALG-1 to cytoplasmic P bodies in *C. elegans*, *Mol. Cell* 19 (4), 437–447, 2005.

55. Rehwinkel, J. A. N., Behm-Ansmant, I., Gatfield, D., and Izaurralde, E., A crucial role for GW182 and the DCP1:DCP2 decapping complex in miRNA-mediated gene silencing, *RNA* 11 (11), 1640–1647, 2005.

56. Liu, J., Rivas, F. V., Wohlschlegel, J., Yates, J. R., III, Parker, R., and Hannon, G. J., A role for the P-body component, GW182, in microRNA function, *Nat. Cell Biol.* 7 (12), 1261–1266, 2005.

57. Liu, J., Valencia-Sanchez, M. A., Hannon, G. J., and Parker, R., MicroRNA-dependent localization of targeted mRNAs to mammalian P-bodies, *Nat. Cell Biol.* 7 (7), 719–723, 2005.

58. Sen, G. L. and Blau, H. M., Argonaute 2/RISC resides in sites of mammalian mRNA decay known as cytoplasmic bodies, *Nat. Cell Biol.* 7, 633–636, 2005.

59. Jakymiw, A., Pauley, K. M., Li, S., Ikeda, K., Lian, S., Eystathioy, T., Satoh, M., Fritzler, M. J., and Chan, E. K. L., The role of GW/P-bodies in RNA processing and silencing, *J. Cell Sci.* 120 (8), 1317–1323, 2007.

60. Chu, C. Y. and Rana, T. M., Translation repression in human cells by microRNA-induced gene silencing requires RCK/p54, *PLoS Biol.* 4 (7), e210, 2006.

61. Eulalio, A., Behm-Ansmant, I., and Izaurralde, E., P bodies: At the crossroads of post-transcriptional pathways, *Nat. Reviews Mol. Cell Biol.* 8 (1), 9–22, 2007.

62. Leung, A. K. L., Calabrese, J. M., and Sharp, P. A., Quantitative analysis of Argonaute protein reveals microRNA-dependent localization to stress granules, *PNAS* 103 (48), 18125–18130, 2006.

63. Wilson, G. M., Lu, J., Sutphen, K., Sun, Y., Huynh, Y., and Brewer, G., Regulation of A + U-rich element-directed mRNA turnover involving reversible phosphorylation of AUF1, *J. Biol. Chem.* 278 (35), 33029–33038, 2003.

64. Wickens, M., Bernstein, D. S., Kimble, J., and Parker, R., A PUF family portrait: 3′ UTR regulation as a way of life, *Trends Genet.* 18 (3), 150–157, 2002.

# 6 The microRNAs of Caenorhabditis elegans

## Mona J. Nolde and Frank J. Slack

## CONTENTS

## OVERVIEW

MicroRNAs (miRNAs) are abundant, noncoding RNAs that regulate gene expression in diverse organisms. miRNAs control biological events, including developmental timing, cell proliferation, apoptosis, stress resistance, and metabolism. Much of what is known about miRNA function and biogenesis comes from studies done in *Caenorhabditis elegans*. The *C. elegans* genome encodes hundreds of miRNAs, many of which are highly conserved in vertebrates. Despite their widespread occurrence, the majority of miRNA targets remain undetermined.

The temporal regulators, *lin-4* and *let-7*, were the first miRNAs to be discovered, over 10 years ago, and represent two of the best-characterized miRNAs in *C. elegans*. *lin-4* and *let-7* act as genetic switches to control developmental timing events in many different tissues. They negatively regulate translation by binding to distinct sequences in the 3′ untranslated regions of their respective targets, *lin-14* and *lin-41*. Together, *lin-4* and *let-7* have set the precedence for miRNA function and processing in *C. elegans* and beyond. Two other miRNAs, *lsy-6* and *mir-273*, regulate key spatial determinants to establish left or right asymmetry in neural development. The assigned roles for miRNAs in biological processes in *C. elegans* are quickly expanding, and because of the high degree of miRNA conservation across phylogeny, it is likely that *C. elegans* miRNA mechanisms will give insight into the functions of miRNAs in other organisms.

## 6.1 INTRODUCTION

MicroRNAs (miRNAs) encompass a recently discovered class of endogenous noncoding regulatory RNAs of ~21–22 nt in length. They are widely distributed across phyla and are thought to regulate genes involved in a range of biological functions during development, including cell proliferation, apoptosis, stress resistance, and metabolism [1]. Much of what is known about miRNA function and processing comes from studies done in *C. elegans*. A variety of molecular and bioinformatics approaches have located many new miRNA-encoding genes in the *C. elegans* genome, putting the validated number of miRNAs at 114 [2–12]. It has been proposed that miRNAs make up at least 1% of both the *C. elegans* and human genomes [7]. However, this number will most certainly grow as additional predicted miRNAs are validated. The miRNAs of *C. elegans* are grouped into 48 families, based on sequence similarity. Among them, 46 families are conserved in *Caenorhabditis briggsae*, a closely related nematode species, and 22 families are conserved in humans [7]. Thus, the high amount of miRNA conservation between vertebrate and invertebrate species makes studies of *C. elegans* miRNAs relevant to higher-order organisms.

## 6.2 miRNA BIOGENESIS AND PROCESSING

In *C. elegans*, much of what is known about miRNA biogenesis and maturation has come from studies involving a second class of small regulatory RNAs called *small interfering RNAs* (siRNAs). siRNAs are involved in RNA interference (RNAi) in animals [13,14] and posttranscriptional gene silencing (PTGS) in plants [15,16]. While miRNAs are endogenously encoded in introns, exons, and coding regions of an animal genome, siRNAs are introduced by transposons or viruses, or are supplied as exogenous synthetic double-stranded RNA (dsRNA) sequences [1,17].

The processing of mature miRNAs in animals involves a multistep cleavage process whereby a primary miRNA transcript (pri-miRNA) is processed by a complex containing Drosha and Pasha proteins, collectively referred to as the *Microprocessor* [18], to form an ~70-nt precursor miRNA (pre-miRNA) hairpin in the nucleus [19,20]. The pre-miRNA is then transported to the cytoplasm, where it is further cleaved by the Dicer RNase III enzyme to release the mature single-stranded ~21–22-nt miRNA from one strand of the hairpin [19,21] (Figure 6.1). In a similar manner, Dicer cleaves a long dsRNA to the siRNAs that initiate RNAi [15]. Following Dicer processing, the miRNA is incorporated into an miRNP complex that resembles the RNA-induced silencing complex (RISC) [22] of the siRNA pathway. The miRNA, in the miRISC, is used as a sequence-specific guide to target mRNAs for translational inhibition or degradation [23,24] (Figure 6.1). The RNA-cleaving ability of RISC is thought to rely on members of the Argonaute family of proteins [25,26], which represent molecular regulators involved in general miRNA and siRNA processing in many different organisms [27].

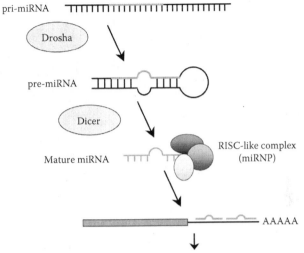

pri-miRNA

Drosha

pre-miRNA

Dicer

Mature miRNA

RISC-like complex
(miRNP)

AAAAA

Blocked translation or mRNA degradation

**FIGURE 6.1** miRNA processing and target recognition. The pri-miRNA transcript is cleaved by the Drosha enzyme to release the pre-miRNA, which forms a hairpin. The mature miRNA is cut out of one side of the pre-miRNA by the action of the Dicer enzyme. The miRNA, in the miRNP RISC-like complex, binds to complementary sequences in the 3′ UTR of target mRNAs, leading to blocked translation or mRNA degradation.

## 6.3 ASSIGNED miRNA DEVELOPMENTAL ROLES IN *C. ELEGANS* AND BEYOND

The roles for miRNAs are quickly expanding to diverse biological processes. The best-characterized roles for miRNAs during *C. elegans* development include developmental timing, neuronal spatial patterning, and aging.

### 6.3.1 TEMPORAL REGULATION BY THE *LIN-4* AND *LET-7* miRNA FAMILIES

The temporal regulators, *let-7* and *lin-4*, are the pioneering members of the large family of miRNAs and were first identified through genetic analysis to play a role in the *C. elegans* heterochronic pathway [28–30]. During *C. elegans* development, hypodermal specification is under the control of a cascade of dynamically regulated gene products. It has been hypothesized that the transitions between each of the four *C. elegans* postembryonic developmental stages are controlled by a specific set of miRNAs working in concert to downregulate a number of target genes [31]. *lin-4* and *let-7* were originally classified as small temporal RNAs (stRNAs) for their roles in regulating developmental timing and have since been grouped into the much larger, general class of miRNAs.

#### 6.3.1.1 *lin-4*

*lin-4* encodes an abundant 22-nt transcript that is first detected at the late L1 stage and is continuously expressed throughout development [32,33]. Loss-of-function (lf)

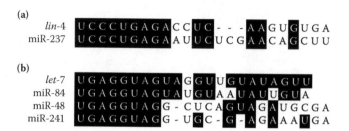

**FIGURE 6.2**   The *lin-4* and *let-7* family members of *C. elegans.* (A) The *lin-4* family: *lin-4* and *mir-237*. (B) The *let-7* family: *let-7, mir-48, mir-84,* and *mir-241*. Identical nucleotides are highlighted in black.

mutations in the *lin-4* miRNA result in a reiteration of L1 cell fates, while over-expression of *lin-4* allows cells to precociously adopt L2 and L3 fates [30]. The *lin-4* family member, *mir-237* (Figure 6.2A), is also temporally expressed during *C. elegans* development [7] with initial expression seen at the L2 stage, one stage later than *lin-4* [33]. However, the function of *mir-237* remains unknown.

The onset of *lin-4* expression coincides with a drop in LIN-14 protein levels [34] (Figure 6.3A). Gain-of-function (gf) mutations in *lin-14* phenocopy *lin-4(lf)* mutants and reiterate L1 stage cell fates [29,35]. The *lin-14* 3′ UTR contains seven *lin-4* complementary elements (LCE) that direct *lin-4* binding and the downregulation of the *lin-14* mRNA [34]. These data lead to the conclusion that in early larval development, *lin-4* promotes the L2-to-L3 transition by regulating the translation of *lin-14* (Figure 6.3B) in a posttranscriptional manner [35–37]. It is postulated that this regulation occurs after the initiation of translation, based on the observation that *lin-4* and *lin-14* are both associated with polyribosomes [37]. Genetic analysis revealed *lin-28* as a second *lin-4* target [36]. The 3′ UTR of *lin-28* contains a single LCE sequence that is able to direct the stage-specific downregulation of a reporter gene at the L2-to-L3 transition (Figure 6.3A) in a similar manner to *lin-14* [36].

### 6.3.1.2   *let-7*

*let-7* is first expressed in the L3 stage [28,38,39] and acts to promote adult fates in the hypodermis [40] by directly repressing genes that promote larval fates (Figure 6.3B). *let-7* is highly conserved in both invertebrate and vertebrate organisms [40], and multiple *let-7* homologues have been found in humans [7]. There are four *let-7* family members encoded by the *C. elegans* genome, including *let-7, mir-48, mir-84,* and *mir-241* [7] (Figure 6.2B). Of the *let-7* family members, *mir-84* is expressed at the earliest stage of development, in the early L1 stage [33,39]. *mir-48* and *mir-241* are thought to share upstream regulatory elements and, as such, are both expressed in a parallel manner from the L2 stage onward [33]. Despite their differential onset of expression, all four *let-7* family members have peak expression in the L4 stage [33].

*let-7 lf* mutant animals display a retarded developmental phenotype, where the lateral hypodermal seam cells fail to secrete alae at the young adult stage and instead undergo an extra molt [28]. *let-7* mutants also display an almost completely penetrant

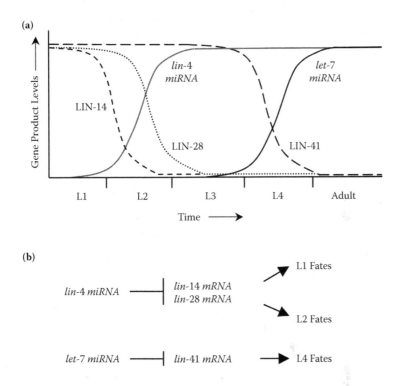

**FIGURE 6.3** *lin-4* and *let-7* act as temporal control switches during *C. elegans* development. (A) *lin-4* negatively regulates *lin-14* and *lin-28* translation at the L1/L2 and L2/L3 transitions, respectively, to allow L3 stage fates to progress. *let-7* negatively regulates expression of LIN-41 protein during the L4 stage, to allow cells to adopt adult fates. (B) *lin-14* promotes L1 fates, *lin-28* promotes L2 fates, and *lin-41* promotes L4 fates.

adult lethal phenotype owing to hemorrhaging through the vulva, the egg-laying organ [28]. The remaining *let-7* family members, *mir-48*, *mir-84*, and *mir-241*, act redundantly to control hypodermal cell fates at the L2 stage [39]. A triple mutant in miR-48, miR-84, and miR-241 displays a reiteration of L2 cell divisions, leading to extra seam cells at the L3 stage [39].

*let-7* has been shown in vivo to bind directly to the *lin-41* 3′ UTR at specific regulatory sequences called *let-7* complementary sites (LCS) [41,42], and thus, *let-7* negatively regulates translation of the *lin-41* message at the L4 stage in the hypodermis [40] (Figure 6.3A). *lin-41* promotes fates specific to the fourth larval stage [40,43,44] (Figure 6.3B), and its downregulation allows for the progression to the final adult stage of development [41].

A second message, *hbl-1*, is a common target of all four *let-7* family members, but in different tissues at different stages. *let-7* and *lin-4* miRNAs are required for regulation of *hbl-1* in the nervous system at the adult stage [43,44], while *mir-48*, *mir-84*, and *mir-241* are redundantly required for downregulation of *hbl-1* in the hypodermis during the L2 to L3 transition [39]. Similar to *lin-41*, the *hbl-1* 3′ UTR contains sequences that resemble complementary sites for the *let-7* and *lin-4* family of miRNAs [43,44], which suggests that a combination of the *let-7* and *lin-4* family members bind directly to the *hbl-1* 3′ UTR to mediate its downregulation.

Based on the presence of potential LCS existing in the 3′ UTRs of many of the genes involved in the heterochronic pathway, it has been suggested that *let-7* may act as a global regulator of developmental timing in *C. elegans*. In addition to *lin-41* and *hbl-1*, LCS are found in the 3′ UTRs of *lin-14, lin-28, daf-12,* and *lin-42* messages [28,45].

*let-7* is one of the most highly conserved genes with nearly 100% sequence identity across metazoan phylogeny, with orthologs found in mouse, human, *Drosophila,* and zebra fish [40]. The temporal expression patterns of *let-7* are also highly conserved in other invertebrates and vertebrates [40,46]. For example, in *Drosophila, let-7* RNA appears at the end of the third larval instar, just prior to metamorphosis [40], which correlates with the larval-to-adult transition in *C. elegans*. Although vertebrates do not go through larval stages, *let-7* expression is also temporally regulated and is consistently seen at later developmental stages [46]. In addition, *lin-41* orthologs retain *let-7* complementary binding sites in their 3′ UTRs [40,46], suggesting that the mechanism of *lin-41* regulation by *let-7* is also conserved.

## 6.3.2   LIFE SPAN REGULATION

The timing genes, *lin-4* and its target *lin-14,* have been shown to play reciprocal roles in the regulation of ageing in *C. elegans* [47], thus mirroring the molecular relationship that they share in developmental timing (Figure 6.4A). Hypomorphic alleles of *lin-4* cause animals to have a shortened life span compared to wild-type animals, while *lin-14* mutants display a long-lived phenotype [47]. *lin-4* acts through

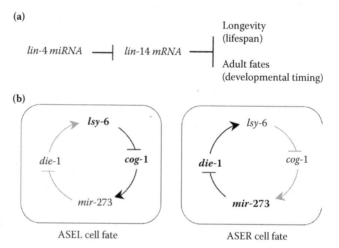

**FIGURE 6.4**   miRNAs control aging and spatial patterning. (A) *lin-4* acts in an opposing manner to *lin-14* to direct life span regulation, similar to the relationship seen for these two factors in developmental timing. (B) *lsy-6* negatively regulates the *cog-1* mRNA to promote ASEL cell fates, while *mir-273* negatively regulates the *die-1* mRNA to promote ASER cell fates. All four of these gene products act in a double-negative feedback loop to collectively promote ASE neuronal asymmetry.

the insulin signaling pathway to modulate aging and represents the first instance of an miRNA involved in life span regulation [47].

### 6.3.3  MIRNAS AND CANCER

Several recent advances have linked miRNAs to cancer and, in fact, many miRNAs are present within loci that are designated as fragile regions for various cancers [33,48]. In *C. elegans*, the *let-7* family member, *mir-84*, is temporally expressed in vulval precursor cells and negatively regulates the *let-60*/RAS oncogene during vulva development [49]. Reduction in *let-60* is also able to partially suppress *let-7(lf)* mutant phenotypes [45], implicating *let-7* and its family members in important regulatory roles in cancer.

### 6.3.4  SPATIAL PATTERNING THROUGH MIRNA ACTION

Two additional *C. elegans* miRNAs, *lsy-6* and *mir-273,* have been assigned a role in left/right spatial patterning of the adult ASE gustatory neurons [50–52]. *lsy-6* and *mir-273* act to promote asymmetric chemoreceptor gene expression and have reciprocal expression patterns in the ASE left (ASEL) and ASE right (ASER) neurons, respectively [50,52]. *lsy-6* restricts the expression of the *cog-1* mRNA to the ASER neuron by binding to a single complementary site in the *cog-1* 3′ UTR, thus promoting ASEL cell fates [52] (Figure 6.4B). *mir-273* promotes ASER cell fates by negatively regulating *die-1* mRNA expression in the ASER neuron [50] (Figure 6.4B). The *lsy-6* and *die-1* miRNAs indirectly regulate their own expression through the subsequent regulation of *cog-1* and *die-1,* leading to a double-negative feedback loop [51]. It is still to be determined if the autoregulation displayed by these two miRNAs represents a general feature of *C. elegans* miRNAs.

## 6.4  REGULATION OF MIRNA EXPRESSION AND FUNCTION

*C. elegans* is a powerful model system for uncovering the mechanisms by which miRNAs are regulated in vivo. The multiple processing steps in miRNA maturation provide many points at which miRNAs themselves might be regulated. Mutations in the *C. elegans* homologues of Drosha, *drsh-1,* and Pasha, *pash-1,* lead to the misregulation of an *let-7*-mediated reporter gene, suggesting that *drsh-1* and *pash-1* are required for wild-type *let-7* function [18]. DCR-1, the *C. elegans* Dicer homologue, physically interacts with the Argonaute homologues, ALG-1 and ALG-2 [53]. *dcr-1* and *alg-1/alg-2* are required for *lin-4* and *let-7* mature miRNA expression [21,23,27]. Mutations in *dcr-1, alg-1,* or *alg-2* phenocopy *lf* mutations in *let-7* [21,23,27]. While *let-7* represents a single miRNA example, it is likely that the dependence of *let-7* on *drsh-1, pash-1, dcr-1,* and *alg-1/alg-2* for processing and function is a common feature to many, if not all, *C. elegans* miRNAs.

## 6.5  MECHANISMS OF MIRNA-MEDIATED GENE REGULATION

The explosion of miRNAs as a novel gene family represents a new paradigm for gene regulation. However, the exact mechanism by which miRNAs act to regulate their messenger RNA (mRNA) targets is still unclear. miRNAs are thought to regulate gene expression at the posttranscriptional level by binding to imperfect complementary sequences in the 3′ UTRs of their target mRNAs, causing translational inhibition and, possibly, mRNA decay [54]. It has been proposed that, as a rule in animals, if the miRNA pairs with perfect complementarity to its target mRNA, then degradation of the mRNA occurs [17]. On the other hand, if the miRNA:mRNA binding site includes imperfectly paired nucleotides, then translational inhibition occurs [17]. However, in the case of *let-7* and *lin-4*, imperfect mRNA binding can facilitate target mRNA degradation for the *let-7* target, *lin-41,* and the *lin-4* targets, *lin-14* and *lin-28* [54]. It is hypothesized, based on previous data, that translational repression of the target mRNA precedes cleavage. Further investigation is required to determine if miRNA-directed mRNA degradation represents a mechanism that is common to other *C. elegans* miRNAs.

A model has been proposed whereby AIN-1 directs the miRISC, containing the miRNA:mRNA duplex, to cytoplasmic processing bodies (P bodies), where the mRNAs are degraded, possibly by a 5′ to 3′ exonuclease mechanism [55].

The abundance and apparent functional diversity of these small RNAs indicates their importance as both global and gene-specific regulators. However, the potential range of miRNA gene targets is only just beginning to be discovered and, to date, only a few miRNA targets have been validated. Because of the variability of miRNA:mRNA complementarity, it is difficult to predict miRNA targets accurately. Based on computational models, it is predicted that at least 10% of *C. elegans* genes are directly regulated by miRNAs [56].

## 6.6  CONCLUSIONS

The field of posttranscriptional regulation has rapidly expanded in the past few years, and the relatively recent discovery of numerous conserved miRNAs further emphasizes the importance of 3′-UTR-mediated regulation in the control of normal cellular processes. Computational models predict that ~30% of human genes are regulated by miRNAs [57]. This prediction, along with the fact that many of the miRNAs in *C. elegans* are highly conserved in humans, suggests that what is elucidated about miRNA function in *C. elegans* can be directly applied to humans. On account of their documented roles as sensitive regulators of gene expression and their association with multiple cancers, miRNAs represent an untapped resource for gene-specific therapies in the treatment of disease.

## REFERENCES

1. Ambros, V. (2003). MicroRNA pathways in flies and worms: Growth, death, fat, stress, and timing. *Cell 113*, 673–676.
2. Ambros, V., Lee, R.C., Lavanway, A., Williams, P.T., and Jewell, D. (2003). MicroRNAs and other tiny endogenous RNAs in *C. elegans. Curr Biol 13*, 807–818.
3. Grad, Y., Aach, J., Hayes, G.D., Reinhart, B.J., Church, G.M., Ruvkun, G., and Kim, J. (2003). Computational and experimental identification of *C. elegans* microRNAs. *Mol Cell 11*, 1253–1263.

4. Lagos-Quintana, M., Rauhut, R., Lendeckel, W., and Tuschl, T. (2001). Identification of novel genes coding for small expressed RNAs. *Science 294*, 853–858.
5. Lau, N.C., Lim, L.P., Weinstein, E.G., and Bartel, D.P. (2001). An abundant class of tiny RNAs with probable regulatory roles in *Caenorhabditis elegans*. *Science 294*, 858–862.
6. Lee, R.C. and Ambros, V. (2001). An extensive class of small RNAs in *Caenorhabditis elegans*. *Science 294*, 862–864.
7. Lim, L.P., Lau, N.C., Weinstein, E.G., Abdelhakim, A., Yekta, S., Rhoades, M.W., Burge, C.B., and Bartel, D.P. (2003). The microRNAs of *Caenorhabditis elegans*. *Genes Dev 17*, 991–1008.
8. Ambros, V., Bartel, B., Bartel, D.P., Burge, C.B., Carrington, J.C., Chen, X., Dreyfuss, G., Eddy, S.R., Griffiths-Jones, S., Marshall, M., Matzke, M., Ruvkun, G., and Tuschl, T. (2003). A uniform system for microRNA annotation. *RNA 9*, 277–279.
9. Griffiths-Jones, S. (2004). The microRNA registry. *Nucleic Acids Res 32*, D109–111.
10. Griffiths-Jones, S. (2006). miRBase: The microRNA sequence database. *Methods Mol Biol 342*, 129–138.
11. Griffiths-Jones, S., Grocock, R.J., van Dongen, S., Bateman, A., and Enright, A.J. (2006). miRBase: microRNA sequences, targets and gene nomenclature. *Nucleic Acids Res 34*, D140–144.
12. Griffiths-Jones, S., Saini, H.K., van Dongen, S., and Enright, A.J. (2008). miRBase: Tools for microRNA genomics. *Nucleic Acids Res 36*, D154–158.
13. Fire, A., Xu, S., Montgomery, M.K., Kostas, S.A., Driver, S.E., and Mello, C.C. (1998). Potent and specific genetic interference by double-stranded RNA in *Caenorhabditis elegans*. *Nature 391*, 806–811.
14. Timmons, L. and Fire, A. (1998). Specific interference by ingested dsRNA. *Nature 395*, 854.
15. Bernstein, E., Caudy, A.A., Hammond, S.M., and Hannon, G.J. (2001). Role for a bidentate ribonuclease in the initiation step of RNA interference. *Nature 409*, 363–366.
16. Hamilton, A.J. and Baulcombe, D.C. (1999). A species of small antisense RNA in post-transcriptional gene silencing in plants. *Science 286*, 950–952.
17. Bartel, D.P. (2004). MicroRNAs: Genomics, biogenesis, mechanism, and function. *Cell 116*, 281–297.
18. Denli, A.M., Tops, B.B., Plasterk, R.H., Ketting, R.F., and Hannon, G.J. (2004). Processing of primary microRNAs by the Microprocessor complex. *Nature 432*, 231–235.
19. He, L. and Hannon, G.J. (2004). MicroRNAs: small RNAs with a big role in gene regulation. *Nat Rev Genet 5*, 522–531.
20. Lee, Y., Ahn, C., Han, J., Choi, H., Kim, J., Yim, J., Lee, J., Provost, P., Radmark, O., Kim, S., and Kim, V.N. (2003). The nuclear RNase III Drosha initiates microRNA processing. *Nature 425*, 415–419.
21. Hutvagner, G., McLachlan, J., Pasquinelli, A.E., Balint, E., Tuschl, T., and Zamore, P.D. (2001). A cellular function for the RNA-interference enzyme Dicer in the maturation of the let-7 small temporal RNA. *Science 293*, 834–838.
22. Hammond, S.M., Bernstein, E., Beach, D., and Hannon, G.J. (2000). An RNA-directed nuclease mediates post-transcriptional gene silencing in *Drosophila* cells. *Nature 404*, 293–296.
23. Ketting, R.F., Fischer, S.E., Bernstein, E., Sijen, T., Hannon, G.J., and Plasterk, R.H. (2001). Dicer functions in RNA interference and in synthesis of small RNA involved in developmental timing in *C. elegans*. *Genes Dev 15*, 2654–2659.
24. Schwarz, D.S. and Zamore, P.D. (2002). Why do miRNAs live in the miRNP? *Genes Dev 16*, 1025–1031.

25. Song, J.J., Smith, S.K., Hannon, G.J., and Joshua-Tor, L. (2004). Crystal structure of Argonaute and its implications for RISC slicer activity. *Science 305*, 1434–1437.

26. Vaucheret, H., Vazquez, F., Crete, P., and Bartel, D.P. (2004). The action of ARGONAUTE1 in the miRNA pathway and its regulation by the miRNA pathway are crucial for plant development. *Genes Dev 18*, 1187–1197.

27. Grishok, A., Pasquinelli, A.E., Conte, D., Li, N., Parrish, S., Ha, I., Baillie, D.L., Fire, A., Ruvkun, G., and Mello, C.C. (2001). Genes and mechanisms related to RNA interference regulate expression of the small temporal RNAs that control *C. elegans* developmental timing. *Cell 106*, 23–34.

28. Reinhart, B.J., Slack, F.J., Basson, M., Pasquinelli, A.E., Bettinger, J.C., Rougvie, A.E., Horvitz, H.R., and Ruvkun, G. (2000). The 21-nucleotide let-7 RNA regulates developmental timing in *Caenorhabditis elegans*. *Nature 403*, 901–906.

29. Ambros, V. and Horvitz, H.R. (1984). Heterochronic mutants of the nematode *Caenorhabditis elegans*. *Science 226*, 409–416.

30. Chalfie, M., Horvitz, H.R., and Sulston, J.E. (1981). Mutations that lead to reiterations in the cell lineages of *C. elegans*. *Cell 24*, 59–69.

31. Slack, F. and Ruvkun, G. (1997). Temporal pattern formation by heterochronic genes. *Annu Rev Genet 31*, 611–634.

32. Feinbaum. R. and Ambros, V. (1999). The timing of lin-4 RNA accumulation controls the timing of postembryonic developmental events in *Caenorhabditis elegans*. *Dev Biol 210*, 87–95.

33. Esquela-Kerscher, A., Johnson, S.M., Bai, L., Saito, K., Partridge, J., Reinert, K.L., and Slack, F.J. (2005). Post-embryonic expression of *C. elegans* microRNAs belonging to the *lin-4* and *let-7* families in the hypodermis and the reproductive system. *Dev Dyn 234*, 868–877.

34. Wightman, B., Ha, I., and Ruvkun, G. (1993). Posttranscriptional regulation of the heterochronic gene lin-14 by lin-4 mediates temporal pattern formation in *C. elegans*. *Cell 75*, 855–862.

35. Ambros, V. and Horvitz, H.R. (1987). The lin-14 locus of *Caenorhabditis elegans* controls the time of expression of specific postembryonic developmental events. *Genes Dev 1*, 398–414.

36. Moss, E.G., Lee, R.C., and Ambros, V. (1997). The cold shock domain protein LIN-28 controls developmental timing in *C. elegans* and is regulated by the lin-4 RNA. *Cell 88*, 637–646.

37. Olsen, P.H. and Ambros, V. (1999). The lin-4 regulatory RNA controls developmental timing in *Caenorhabditis elegans* by blocking LIN-14 protein synthesis after the initiation of translation. *Dev Biol 216*, 671–680.

38. Johnson, S.M., Lin, S.Y., and Slack, F.J. (2003). The time of appearance of the *C. elegans* let-7 microRNA is transcriptionally controlled utilizing a temporal regulatory element in its promoter. *Dev Biol 259*, 364–379.

39. Abbott, A.L., Alvarez-Saavedra, E., Miska, E.A., Lau, N.C., Bartel, D.P., Horvitz, H.R., and Ambros, V. (2005). The let-7 MicroRNA family members mir-48, mir-84, and mir-241 function together to regulate developmental timing in *Caenorhabditis elegans*. *Dev Cell 9*, 403–414.

40. Pasquinelli, A.E., Reinhart, B.J., Slack, F., Martindale, M.Q., Kuroda, M.I., Maller, B., Hayward, D.C., Ball, E.E., Degnan, B., Muller, P., Spring, J., Srinivasan, A., Fishman, M., Finnerty, J., Corbo, J., Levine, M., Leahy, P., Davidson, E., and Ruvkun, G. (2000). Conservation of the sequence and temporal expression of let-7 heterochronic regulatory RNA. *Nature 408*, 86–89.

41. Vella, M.C., Choi, E.Y., Lin, S.Y., Reinert, K., and Slack, F.J. (2004). The *C. elegans* microRNA let-7 binds to imperfect let-7 complementary sites from the lin-41 3′ UTR. *Genes Dev 18*, 132–137.

42. Vella, M.C., Reinert, K., and Slack, F.J. (2004). Architecture of a validated microRNA: Target interaction. *Chem Biol 11*, 1619–1623.
43. Abrahante, J.E., Daul, A.L., Li, M., Volk, M.L., Tennessen, J.M., Miller, E.A., and Rougvie, A.E. (2003). The *Caenorhabditis elegans* hunchback-like gene lin-57/hbl-1 controls developmental time and is regulated by microRNAs. *Dev Cell 4*, 625–637.
44. Lin, S.Y., Johnson, S.M., Abraham, M., Vella, M.C., Pasquinelli, A., Gamberi, C., Gottlieb, E., and Slack, F.J. (2003). The *C. elegans* hunchback homolog, hbl-1, controls temporal patterning and is a probable microRNA target. *Dev Cell 4*, 639–650.
45. Grosshans, H., Johnson, T., Reinert, K.L., Gerstein, M., and Slack, F.J. (2005). The temporal patterning microRNA let-7 regulates several transcription factors at the larval to adult transition in *C. elegans*. *Dev Cell 8*, 321–330.
46. Schulman, B.R., Esquela-Kerscher, A., and Slack, F.J. (2005). Reciprocal expression of lin-41 and the microRNAs let-7 and mir-125 during mouse embryogenesis. *Dev Dyn 234*, 1046–1054.
47. Boehm, M. and Slack, F. (2005). A developmental timing microRNA and its target regulate life span in *C. elegans*. *Science 310*, 1954–1957.
48. Calin, G.A., Sevignani, C., Dumitru, C.D., Hyslop, T., Noch, E., Yendamuri, S., Shimizu, M., Rattan, S., Bullrich, F., Negrini, M., and Croce, C.M. (2004). Human microRNA genes are frequently located at fragile sites and genomic regions involved in cancers. *Proc Natl Acad Sci USA 101*, 2999–3004.
49. Johnson, S.M., Grosshans, H., Shingara, J., Byrom, M., Jarvis, R., Cheng, A., Labourier, E., Reinert, K.L., Brown, D., and Slack, F.J. (2005). RAS is regulated by the let-7 microRNA family. *Cell 120*, 635–647.
50. Chang, S., Johnston, R.J., Jr., Frokjaer-Jensen, C., Lockery, S., and Hobert, O. (2004). MicroRNAs act sequentially and asymmetrically to control chemosensory laterality in the nematode. *Nature 430*, 785–789.
51. Johnston, R.J., Jr., Chang, S., Etchberger, J.F., Ortiz, C.O., and Hobert, O. (2005). MicroRNAs acting in a double-negative feedback loop to control a neuronal cell fate decision. *Proc Natl Acad Sci USA 102*, 12449–12454.
52. Johnston, R.J. and Hobert, O. (2003). A microRNA controlling left/right neuronal asymmetry in *Caenorhabditis elegans*. *Nature 426*, 845–849.
53. Duchaine, T.F., Wohlschlegel, J.A., Kennedy, S., Bei, Y., Conte, D., Jr., Pang, K., Brownell, D.R., Harding, S., Mitani, S., Ruvkun, G., Yates, J.R., 3rd, and Mello, C.C. (2006). Functional proteomics reveals the biochemical niche of *C. elegans* DCR-1 in multiple small-RNA-mediated pathways. *Cell 124*, 343–354.
54. Bagga, S., Bracht, J., Hunter, S., Massirer, K., Holtz, J., Eachus, R., and Pasquinelli, A.E. (2005). Regulation by let-7 and lin-4 miRNAs results in target mRNA degradation. *Cell 122*, 553–563.
55. Ding, L., Spencer, A., Morita, K., and Han, M. (2005). The developmental timing regulator AIN-1 interacts with miRISCs and may target the argonaute protein ALG-1 to cytoplasmic P bodies in *C. elegans*. *Mol Cell 19*, 437–447.
56. Lall, S., Grun, D., Krek, A., Chen, K., Wang, Y.L., Dewey, C.N., Sood, P., Colombo, T., Bray, N., Macmenamin, P., Kao, H.L., Gunsalus, K.C., Pachter, L., Piano, F., and Rajewsky, N. (2006). A genome-wide map of conserved microRNA targets in *C. elegans*. *Curr Biol 16*, 460–471.
57. Lewis, B.P., Burge, C.B., and Bartel, D.P. (2005). Conserved seed pairing, often flanked by adenosines, indicates that thousands of human genes are microRNA targets. *Cell 120*, 15–20.

# 7 Isolation and Characterization of Small RNAs in *Caenorhabditis elegans*

*Chisato Ushida and Yusuke Hokii*

## CONTENTS

## OVERVIEW

Various RNomics and transcriptome studies have established that there are more small RNAs than had been previously thought. As the characteristics and function of individual RNA are apparent, the significance of small RNAs in the complex genetic network has been recognized. However, there are only certain RNAs whose physiology and molecular mechanism of function are clear. *Caenorhabditis elegans* is a good model to study gene products of unknown function. In this chapter, we introduce how *C. elegans* small RNAs are isolated and what features they have.

## 7.1  INTRODUCTION

Powerful cDNA sequencing approaches and computational analyses of genome sequences have revealed that there are more non-protein-coding RNAs (npcRNAs, ncRNAs) than had been thought previously both in prokaryotes and eukaryotes [1–8]. These RNAs have a wide variety of functions, including translation, splicing, structural scaffold, translational inhibition, mRNA degradation, transcriptional regulation, RNA editing, modification, processing, and protein secretion. There are now 607 RNA families registered in the Rfam database (version 8.1, October 2007)

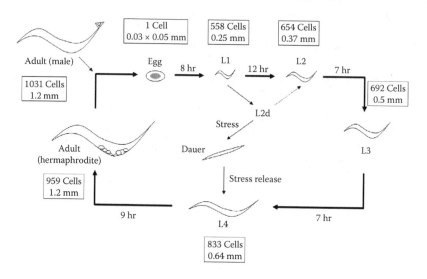

**FIGURE 7.1**   Life cycle of *C. elegans*: *C. elegans* has two sexes—hermaphrodite and male. Hermaphrodites can self-fertilize or cross with males. After fertilization, it takes 8 h for embryogenesis and 35 h for postembryonic development through four larval stages (L1: 12 h; L2: 7 h; L3: 7 h; L4: 9 h) at 25°C. The life span of a worm is about 3 weeks. When growth conditions are insufficient for reproduction, for example, starvation or temperature downshifts, the L1 stage larva enters into the "dauer" stage instead of proceeding to the L2 stage. Release from stress results in recovery of the worm from dauer to L4.

[9], but this is only the tip of the iceberg. How many ncRNAs exist in cells? What are their functions, and how do they work? Obtaining answers to these questions about ncRNAs is one of the important issues of the postgenomic era to understand the complex genetic system of life.

  *C. elegans* is a good model for studying multicellular organisms. The generation time is as short as about 3 days, and one worm produces 300–350 progeny in a short period (Figure 7.1). The anatomy is simple, 959 cells in a hermaphrodite and 1031 cells in a male, but *C. elegans* has tissues and organs in common with other animals. The complete cell lineage from the fertilized egg stage to the adult stage has been determined. *C. elegans* has five pairs of autosomes and either a pair of sex chromosomes (XX, hermaphrodites) or a single sex chromosome (X, in a male) [10–14]. The entire genome sequence was determined in 1998, and was the first example of the complete genome sequencing in multicellular organisms [15]. The encompassing protein gene expression profiles have been examined by microarray analysis, whole mount *in situ* hybridization, and the EST project (NextDB, http://nematode.lab.nig. ac.jp/db2/index.php, http://nematode.lab.nig.ac.jp/dbest/keysrch.html). Systematic RNA interference (RNAi) experiments have revealed which protein genes are necessary for worms. These data have been deposited in databases that are linked to one another [16]. The aforementioned features of *C. elegans* are useful for understanding the functions and mechanisms of novel ncRNAs.

  In this chapter, recent studies on identification and characterization of *C. elegans* small ncRNAs, the first step in functional genomics of ncRNAs, are described.

## 7.2    NcRNAs SMALLER THAN 50 NT IN LENGTH: MiRNAs, siRNAs, TNcRNAs, AND 21U-RNAs

There have been two breakthroughs in the field of RNA study as a result of *C. elegans* investigations. One is the discovery of small temporal RNAs (stRNAs) of 22-nt length, *lin-4* RNA and *let-7* RNA, which were identified by genetic approaches to find factors involved in the regulation of developmental timing [17,18]. The other is the discovery of RNAi, the phenomenon of gene silencing induced by exogenous double-stranded RNAs (dsRNAs) [19]. The introduced dsRNAs are cut into short fragments of about 22 nt in length by the RNase III-type nuclease Dicer. These fragments, designated small interfering RNAs (siRNAs), guide the degradation of targeted mRNAs, which have the complementary sequence to siRNAs [20–22]. These findings led to the discovery of a large family of functional RNAs, microRNAs (miRNAs) [23–25]. Details of the function and mechanism of RNAi, siRNAs, and miRNAs are discussed in Chapters 2 and 6 of this book.

The siRNAs and miRNAs have many common characteristics. For example, both are 20–30 nt in length, both are processed by Dicer from larger precursors, both are present in various organisms, and both function in downregulating the gene expression by base-pairing with target mRNAs [5,26]. Although siRNAs were first found to be derived from exogenously delivered dsRNAs, endogenous siRNAs were subsequently detected. A clear difference between siRNAs and miRNAs is in their biogenesis: while siRNAs are produced from long dsRNAs, miRNAs are processed from hairpin-structured precursors of 60–80 nt in length. A guideline for the classification of small RNAs as miRNA has been provided by Ambros et al. [27]. There are five criteria (criteria A to E in Reference 26) to distinguish miRNAs from other small RNAs. Two criteria, criteria A and B, are *expression criteria*, and the other three, criteria C, D, and E, are *biogenesis criteria*. The miRNAs require evidence of both expressional and biogenesis criteria. These are summarized as follows:

*Criteria A:* Expression of small RNA of about 22 nt is detected by a hybridization technique such as Northern blotting.

*Criteria B:* cDNA of about 22 nt in length, the sequence of which precisely corresponds to the genomic sequence, is made from size-fractionated RNA.

*Criteria C:* A potential fold-back precursor structure that contains the miRNA sequence within one arm of the hairpin is predicted. The precursor hairpin structure must have the lowest level of free energy among the candidate structures predicted by a conventional RNA-folding computer program such as mfold. The hairpin must include at least 16 base pairings involving the first 22 nt of the miRNA. No large internal loops or bulges should be included (Figure 7.2).

*Criteria D:* The miRNA sequence and its predicted precursor secondary structure are phylogenetically conserved.

*Criteria E:* Accumulation of the precursor is detected in organisms with reduced Dicer function.

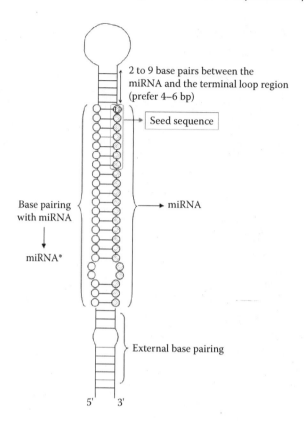

**FIGURE 7.2**  A secondary structure model of *C. elegans* precursor miRNA (pre-miRNA). Pre-miRNA is a 60–80-nt hairpin-structured RNA, which encodes miRNA (the coding nucleotides are indicated as hatched circles) at either 5′ or 3′ arm of the hairpin. miRNA* (the coding nucleotides are indicated as white circles) is a less stable RNA that is derived from the opposite arm. Small symmetric rather than asymmetric internal loops and bulges are included in the stem. There are 2–9 base pairings between the miRNA and the terminal loop region. Base pairings between the miRNA and miRNA* are extended toward the 5′ and 3′ terminal region of the pre-miRNA (indicated as "External base pairing"). The 5′ half of miRNA is more conserved with that of *C. briggsae* ortholog than the 3′ half. The first five nucleotides of the miRNA, which are included in the seed sequence, show sequence biases such that the first nucleotide prefers uridine but the nucleotides at positions 2–4 do not [24,29].

An miRNA that fulfills criteria A, D, and E is an ideal one. However, fulfillment of a combination of criteria A and D, A and C, B and D, or D and E is also sufficient for classification as an miRNA. See Reference 26 for details.

*C. elegans* miRNAs have been identified using cDNA cloning or bioinformatics approaches [24,25,28–30]. The first two reports were published in 2001 by Lau's group [24] and Ambros' group [25]. They independently identified 54 and 15 novel miRNAs, respectively, 11 of which overlapped each other. Two years later, both groups identified an additional 23 and 29 new miRNAs, 13 of which were identical, by scaled-up analyses. Together with *lin-4* and *let-7* RNAs, 99 miRNA species, of which expressions were validated by Northern blotting or reverse transcription polymerase chain reaction (RT-PCR), have so far been identified. Lim and colleagues developed a computer program, MiRscan, that searches for miRNA gene candidates from the genome sequence (Figure 7.2) [29]. Grad et al. predicted 214 miRNAs on the basis of sequence conservation and structural similarity to known miRNAs of diverse eukaryotic organisms [28]. Among the candidates, 40 were conserved in *Drosophila melanogaster* and *Homo sapiens*. By calculating the ratio of experimentally verified miRNAs among candidates and the ratio of already-known miRNAs in the candidates, they estimated that 100–300 miRNAs are present in *C. elegans*.

Presently, 137 miRNAs of *C. elegans* are registered in an miRNA database, miRBase (release 10.1, December 2007, http://microrna.sanger.ac.uk/sequences/index.shtml, [31]). Recent studies have shed light on the biological function of each miRNA as shown in Chapter 6 of this book.

In the course of miRNA identification, Ambros and colleagues isolated 33 novel small RNAs of 19–21 nt in length that could not be classified into miRNAs [30]. Their expressions were detected by Northern blotting as signals of about 20–22 nt in length. However, the potential fold-back precursor structure, which is one of the criteria for classification as miRNAs [27], could not be predicted for these RNAs. Because of this, the newly found 33 RNAs were classified into a novel RNA family designated as tiny noncoding RNAs (tncRNAs). Among these, 18 tncRNAs exhibited distinct temporal expression during development. Interestingly, 5 tncRNAs had antisense sequences to parts of longer non-protein-coding EST sequences. In Dicer mutant worms, 6 tncRNAs were absent and 7 tncRNAs were accumulated as their precursors. These findings suggested that some tncRNAs share a part of the biogenesis pathway with siRNAs and miRNAs. The expressions of the remaining 3 tncRNAs did not change in the Dicer mutant worms [32].

Besides novel miRNAs and tncRNAs, Ambros and coworkers also identified 1799 endogenous siRNAs (endo-siRNAs) derived from 1085 protein-coding genes [32]. Two endo-siRNAs, Ct1182 and Ct1189, were less abundant or absent in mutants of *rde-2*, *rde-3*, *rde-4*, *mut-7*, or *mut-14*, which are RNAi defective. However, the behavior of Ct1182 and Ct1189 was not always identical to that of exogenous siRNAs (exo-siRNAs) in these mutants. Exo-siRNAs accumulate in *mut-14* mutants, but the amounts of Ct1182 and Ct1189 are decreased. The effects of *eri-1* and *rrf-3* mutations, which enhance RNAi, also work in the opposite way between these endo-siRNAs and exo-siRNAs; the amounts of Ct1189 and Ct1182 are decreased, while the amounts of exo-siRNAs are increased in these mutants. These findings suggest that the pathways for biogenesis and function of endo- and exo-siRNAs are common in several respects but are not completely identical [32]. Bartel and colleagues also isolated novel small RNAs other than miRNAs that had the corresponding sequence to the intergenic regions of the genome. They suggested that these RNAs are heterochromatic siRNAs [29]. In the recent study by Ruby et al., at least 0.7% of about 400,000 sequence reads of small RNA libraries were predicted to be endogenous siRNAs [33].

The high-throughput sequencing of small RNA libraries represented a novel class of *C. elegans* small ncRNAs, designated 21-U RNAs. These are 21 nt in length and have a uridine at their 5′ ends [33]. The genes were mostly mapped to three regions of chromosome IV; mainly at 4.5–7.0 M and 13.5–17.2 M, and a few at 9–9.7 M. Two conserved sequence motifs exist at the upstream of many 21-U RNA coding regions: a large motif (34 bp), which has a core consensus sequence 5′-CTGTTTCA-3′, and a small motif, which has a core consensus sequence 5′-YRNT-3′ (Y, C, or T; R, A, or G; N, any of A, C, G, or T; "T" at the 3′ termini corresponds to the 5′ terminal "U" of the 21-U RNA). Interestingly, these upstream sequences were more conserved between *C. elegans* and *C. briggsae* than the corresponding 21-U RNA sequences between them. These are expected to have roles in the biogenesis of 21-U RNAs. It is an attractive hypothesis that the transcription of 21-U RNA loci itself has an important biological function [33].

The use of two-dimensional gel electrophoresis (2D-PAGE) is an efficient way to fractionate small RNAs. The *C. elegans* small RNA fraction, which contained RNAs smaller than 50 nt in length, was separated into approximately 130 spots by 2D-PAGE, and 11 novel small ncRNA candidates were isolated from 23 randomly chosen spots [34]. A novel ncRNA candidate was found from a spot of 2D-PAGE specific for embryonic small RNAs. The spot patterns of mixed-stage worm RNAs and embryonic RNAs revealed by 2D-PAGE were significantly different, suggesting that the expression profiles of small RNAs change during development. Although the sequence analysis of all detected spots is still far from saturation, one of the low-abundant miRNAs, miR-252, was detected by using this method (Ushida, C., unpublished data).

## 7.3  snoRNAs AND OTHER SMALL RNAs OF ABOUT 50–500 NT IN LENGTH

Small nucleolar RNAs (snoRNAs) make up one of the large families of ncRNAs as well as miRNAs. They are classified into two groups according to their structures: C/D snoRNAs and H/ACA snoRNAs (Figures 7.3A and 7.3B). C/D snoRNAs have two conserved sequence motifs, 5′-UGAUGA-3′ (box C) and 5′-CUGA-3′ (box D), near the terminal stem of the RNA. Similar sequence motifs, designated C′ box and D′ box, respectively, often exist in the central region of the RNA. They guide the 2′-*O*-methylation of rRNA nucleotides. As shown in Figure 7.3A, the upstream region of box D or D′ of C/D snoRNA makes base pairings of 10–21 bp with the target region of rRNA. The modification site of rRNA is involved in the base pairings, where five nucleotides upstream of either box D or D′ are precisely positioned (indicated as "Me" in Figure 7.3A) [35,36]. H/ACA snoRNAs also have two conserved sequence motifs, 5′-ANANNA-3′ (box H) and 5′-ACA-3′ (box ACA), each positioned at the 3′ side of the "hairpin-hinge-hairpin-tail" secondary structure (Figure 7.3B). Boxes H and ACA are positioned at the 3′ side of each hairpin. There is an internal loop in one or both hairpins, a pseudouridylation pocket. This region of H/ACA snoRNA forms a bipartite base pairing, 3–10 bp at the 5′ side of the loop and 4–10 bp at the 3′ side of the loop, with the flanking region of the target uridine of rRNA [35,37]. The total number of base pairings is more than 9 bp. The distance between the conserved box motifs of H/ACA snoRNA, box H or box ACA, and the nucleotide that is complementary to the rRNA residue next to the target uridine is 14–16 nt (Figure 7.3B).

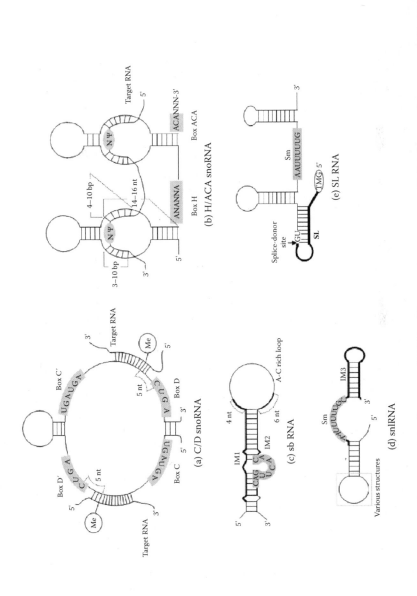

**FIGURE 7.3**   Schematic secondary structure of *C. elegans* small RNAs. The secondary structure with the conserved sequence motifs of the modification-guide box C/D RNA (a), the modification-guide box H/ACA RNA (b), sbRNA (c), snlRNA (d), and SL RNA (e) is shown. See text for details. The region indicated by the dotted square in (d) shows variety in length and structure among five snlRNAs listed in Table 7.1.

Small Cajal body-specific RNAs (scaRNAs) have a structure similar to that of C/D snoRNAs or H/ACA snoRNAs. They function in 2'-O-methylation or pseudouridylation of spliceosomal U small nuclear RNAs (snRNAs). Unlike snoRNAs, they have a conserved sequence motif, 5'-AGAG-3' (CAB box), necessary for their localization at the Cajal body [38]. There are several databases covering snoRNAs and scaRNAs [39].

About 100 nucleotides in yeast rRNAs are 2'-O-methylated or pseudouridylated [36,37,40]. The number of modifications increases up to about 200 in mammalian rRNAs, suggesting that there are corresponding numbers of snoRNA species that guide the modifications. More than 100 RNAs of this class are expected to be found in *C. elegans*. Until recently, only two *C. elegans* snoRNA species (U15 and U18 snoRNAs) were registered in the database (Rfam, as of the beginning of 2005). A computational search of C/D snoRNAs according to the experimentally determined 2'-O-methylated nucleotide positions on rRNAs revealed 16 additional C/D snoRNA genes [41]. The first *C. elegans* H/ACA snoRNAs were identified by cDNA sequence analysis of size-fractionated RNAs [42]. The following large-scale RNomics studies revealed about 200 genes encoding ncRNAs of this class [43,44].

Table 7.1 shows a summary of *C. elegans* ncRNAs of 50–500 nt in length identified by cDNA cloning studies [42–44]. Among 198 genes, 75 encode C/D-box RNAs and 53 encode H/ACA-box RNAs. For 58 C/D-box RNAs and 47 H/ACA-box RNAs, the guiding targets were predicted in rRNAs and U snRNAs. Interestingly, 4 C/D-box RNAs, Ce62 (CeN98), Ce63 (CeN27), Ce94 RNAs, and Ce246 (CeN70) RNAs, had the potential to guide 2'-O-methylation of tRNAs as well as rRNAs or U6 snRNA [44]. These are the first candidates of eukaryotic RNAs that guide modifications of tRNAs. Five C/D-box RNAs were U3 snoRNA homologues, and one H/ACA-box RNA was a *C. elegans* U17/snR30 RNA homologue. The nucleolar localizations of these RNAs as well as of several other modification-guiding snoRNAs were confirmed.[44a] No target sequences on rRNAs, U snRNAs, or tRNAs have been predicted for 11 box C/D RNAs and 15 box H/ACA RNAs. It is expected that these orphan C/D-box and H/ACA-box RNAs may have novel functions as in the case of MBII-52, which is a brain-specific box C/D snoRNA and probably guides 2'-O-methylation of the A-to-I editing site of 5-HT2C serotonin receptor pre-mRNA [45–47].

The remaining 70 genes were classified as follows: 23 for spliceosomal U snRNA genes, 14 for spliced leader RNA (SL RNA) genes, 4 for SRP RNA genes, 1 for Y RNA gene, 1 for RNase P RNA gene, and 27 for novel types of small ncRNAs. Among the 27 novel RNAs, 6 and 5 RNAs belong to the stem-bulge RNA (sbRNA) families and the snRNA-like RNA (snlRNA) families, respectively, which have been defined by Deng et al. [43]. The sbRNAs have two conserved motifs, designated IM1 and IM2, which are located at the 5' and 3' termini, respectively. The two motifs make an intramolecular base pairing with a bulge of the conserved sequence 5'-AACUU-3' in the 3' stem region (Figure 7.3C). The loop region between IM1 and IM2 is rich in nucleotides A and C. The snlRNAs have a motif, IM3, including the conserved sequence 5'-AARUUUUGGA-3', which corresponds to the Sm protein-binding site in spliceosomal snRNAs (Figure 7.3D). The IM3 motif was also found in U3 snoRNAs, two probable modification-guiding RNAs (Ce96/CeN25-2 and Ce135/

**TABLE 7.1**

**C. elegans Small RNAs Isolated by cDNA Sequencing Method**

| Serial No. | RNA Nomenclature[a] | | | Classification | Function[bk] |
|---|---|---|---|---|---|
| | Zemann, A. et al. | Deng, W. et al. | Other | | |
| 1 | Ce18 | — | — | C/D | Cm2407/26S |
| | | | | | Um292/26S |
| | | | | | Um 3136/26S |
| 2 | Ce39 | CeN121 | — | C/D | Gm2343/26S |
| | | | | | Um2654/26S |
| 3 | Ce55 | CeN61 | — | C/D | Gm 2927/26S |
| | | | | | Um 553/26S |
| 4 | Ce59 | — | — | C/D | Um14/5.8S |
| | | | | | Am931/26S |
| 5 | Ce61 | CeN118 | — | C/D | Am1185/26S |
| | | | | | Gm2498/26S |
| 6 | Ce62 | CeN98 | — | C/D | Gm20/tRNA$^{Ile}$ |
| | | | | | Cm55/U6 |
| 7 | Ce63 | CeN27 | — | C/D | Cm982/18S |
| | | | | | Am1060/26S |
| | | | | | Am31/tRNA$^{Ser}$ |
| 8 | Ce67 | — | — | C/D | Am1389/18S |
| | | | | | Gm336/18S |
| 9 | Ce75 | CeN63 | — | C/D | Gm2731/26S |
| | | | | | Am1205/26S |
| | | | | | Am730/18S |
| 10 | Ce81 | CeN117 | — | C/D | Um2417/26S |
| | | | | | Gm1805/26S |
| | | | | | Am300/26S |
| 11 | Ce83 | CeN57 | — | C/D | Am 2359/26S |
| 12 | Ce86 | CeN35 | — | C/D | Um560/18S |
| 13 | Ce94 | — | — | C/D | Gm1668/26S |
| | | | | | Gm46/tRNA$^{Asn}$ |
| 14 | Ce96[c] | CeN25-2 | — | C/D | Um2529/26S |
| 15 | Ce97 | — | — | C/D | Cm57/U6 |
| 16 | Ce98 | — | — | C/D | Cm1196/18S |
| | | | | | Cm367/18S |
| 17 | Ce100 | — | — | C/D | Um203/26S |
| | | | | | Cm2728 /26S |
| | | | | | Gm2731/26S |
| | | | | | Am1222/26S |

*Continued*

## TABLE 7.1 (*Continued*)
## *C. elegans* Small RNAs Isolated by cDNA Sequencing Method

| Serial No. | RNA Nomenclature[a] Zemann, A. et al. | Deng, W. et al. | Other | Classification | Function[bk] |
|---|---|---|---|---|---|
| 18 | Ce104.2 | — | — | C/D | Um538/26S |
| 19 | Ce110[d] | CeN42 | — | C/D | Gm1259/26S |
|   |   |   |   |   | Cm1010/18S |
| 20 | Ce118 | CeN24 | — | C/D | Gm272/26S |
|   |   |   |   |   | Am601/18S |
| 21 | Ce135[c] | CeN25-1 | — | C/D | Gm1067/26S |
|   |   |   |   |   | Gm 35/U4 |
|   |   |   |   |   | Cm187/18S |
| 22 | Ce138a | CeN50-1 | — | C/D | Cm2440/26S |
|   |   |   |   |   | Am2448/26S |
| 23 | Ce138b | CeN50-2 | — | C/D | Cm2440/26S |
|   |   |   |   |   | Am2448/26S |
| 24 | Ce139 | CeN109 | — | C/D | Am862/26S |
| 25 | Ce151 | CeN129 | — | C/D | Cm 3060/26S |
|   |   |   |   |   | Gm1373/18S |
|   |   |   |   |   | Gm1117/26S |
| 26 | Ce160 | CeN30 | — | C/D | Am3023/26S |
|   |   |   |   |   | Gm2027/26S |
| 27 | Ce161 | CeN108 | — | C/D | Am423/18S |
| 28 | Ce167 | — | — | C/D | Am28/18S |
|   |   |   |   |   | Um72/18S |
|   |   |   |   |   | Um1058/26S |
| 29 | Ce169 | — | — | C/D | Gm 1236/18S |
| 30 | Ce171 | CeN40 | — | C/D | Um68/U1 |
| 31 | Ce173.1 | CeN128[f] | — | C/D | Gm396/26S |
|   |   |   |   |   | Am96/5.8S |
| 32 | Ce173.2 |   | — | C/D | Am90/18S |
|   |   |   |   |   | Am2287/26S |
|   |   |   |   |   | Am46/U5 |
| 33 | Ce173.3 |   | — | C/D | Gm 2720/26S |
|   |   |   |   |   | Gm39/U5 |
| 34 | Ce177 | CeN33 | — | C/D | Am 75/18S |
|   |   |   |   |   | Cm400/18S |
| 35 | Ce209 | CeN124 | — | C/D | Cm111/U2 |
|   |   |   |   |   | Um2172/26S |
| 36 | Ce211 | CeN5 | — | C/D (U18) | Am678/26S |
| 37 | Ce223 | CeN60 | — | C/D | Am 1514/26S |

## TABLE 7.1 (*Continued*)
## *C. elegans* Small RNAs Isolated by cDNA Sequencing Method

| Serial No. | RNA Nomenclature[a] | | | Classification | Function[bk] |
|---|---|---|---|---|---|
| | Zemann, A. et al. | Deng, W. et al. | Other | | |
| 38 | Ce230 | CeN15 | — | C/D | Cm1502/26S |
| | | | | | Um3441/16S |
| 39 | Ce234.1 | CeN47[g] | — | C/D | Am 65/U6 |
| 40 | Ce234.2 | | — | C/D | Gm268/18S |
| | | | | | Cm209/26S |
| 41 | Ce238 | — | — | C/D | Um1566/26S |
| | | | | | Gm 1022/18S |
| 42 | Ce240 | — | — | C/D | Um 800/18S |
| | | | | | Um1688/18S |
| 43 | Ce243 | CeN111 | — | C/D | Am872/26S |
| | | | | | Gm963/26S |
| 44 | Ce245 | CeN28 | — | C/D | Am774/26S |
| | | | | | Cm2468/26S |
| 45 | Ce246 [e] | CeN70 | — | C/D | Gm860/26S |
| | | | | | Um17/tRNA[Ile] |
| | | | | | Um304/26S |
| 46 | Ce251 | — | — | C/D | Gm1879/26S |
| | | | | | Gm 2309/26S |
| | | | | | Am42/U6 |
| 47 | Ce252 | CeN122 | — | C/D | Am2429/26S |
| | | | | | Um247/18S |
| | | | | | Um668/18S |
| | | | | | Um310/18S |
| 48 | Ce254a | CeN23-1 | — | C/D | Cm2391/26S |
| | | | | | Am2384/26S |
| 49 | Ce254b | CeN23-2 | — | C/D | Cm2391/26S |
| | | | | | Am2384/26S |
| 50 | Ce271 | CeN44 | — | C/D | Um1298/18S |
| | | | | | Gm66/U5 |
| | | | | | Gm1313/26S |
| 51 | Ce282 | CeN52 | — | C/D | Cm138/5.8S |
| | | | | | Um1223/18S |
| | | | | | Um3163/26S |
| 52 | Ce285 | CeN106 | — | C/D | Um927/26S |
| | | | | | Cm154/U2 |
| | | | | | Um239/18S |

*Continued*

**TABLE 7.1 (*Continued*)**

**C. elegans Small RNAs Isolated by cDNA Sequencing Method**

| Serial No. | RNA Nomenclature[a] | | | Classification | Function[bk] |
|---|---|---|---|---|---|
| | Zemann, A. et al. | Deng, W. et al. | Other | | |
| 53 | Ce297 | CeN17 | — | C/D (U15) | Am2384/26S |
| 54 | Ce298 | — | — | C/D | Am48/U6 |
| | | | | | Um 1263/26S |
| 55 | Ce304 | CeN114 | — | C/D | Am395/26S |
| | | | | | Um1981/26S |
| 56 | Ce325 | CeN120 | — | C/D | Am3159/26S |
| | | | | | Am448/18S |
| 57 | Ce354 | CeN14 | — | C/D | Gm1985/26S |
| | | | | | Um2841/26S |
| 58 | Ce372 | CeN123 | — | C/D | Cm3071/26S |
| 59 | Ce1a | CeN26-4 | — | U3 | Pre-rRNA processing |
| 60 | Ce1b | CeN26-5 | — | U3 | Pre-rRNA processing |
| 61 | Ce1c | CeN26-1 | — | U3 | Pre-rRNA processing |
| 62 | Ce5a | CeN26-2 | — | U3 | Pre-rRNA processing |
| 63 | Ce5b | CeN26-3 | — | U3 | Pre-rRNA processing |
| 64 | — | CeN69 | CeR-19 | C/D | Um308/26S |
| | | | | | Cm642/18S[h] |
| 65 | Ce239 | CeN119 | — | C/D | Target unknown |
| 66 | — | CeN13 | — | C/D | Target unknown |
| 67 | — | CeN22 | — | C/D | Target unknown |
| 68 | — | CeN53 | — | C/D | Target unknown |
| 69 | — | CeN54 | — | C/D | Target unknown |
| 70 | — | CeN56 | CeR-5 | C/D | Target unknown |
| 71 | — | CeN62 | — | C/D | Target unknown |
| 72 | — | CeN65 | — | C/D | Target unknown |
| 73 | — | CeN89 | — | C/D | Target unknown |
| 74 | — | CeN103 | — | C/D | Target unknown |
| 75 | — | CeN113 | — | C/D | Target unknown |
| 76 | comCe6 | CeN87 | CeR-3 | H/ACA | P2:Ψ3035/26S[i] |
| 77 | comCe7 | CeN78 | — | H/ACA | P1:Ψ2237/26S |
| | | | | | P2:Ψ2484/26S |
| 78 | comCe8 | CeN95 | — | H/ACA | P1:Ψ18/U6 |
| | | | | | P2:Ψ1411/18S |
| 79 | comCe9 | CeN88 | — | H/ACA | P1:Ψ18/U6 |
| | | | | | P2:Ψ1411/18S |
| 80 | comCe12 | CeN110 | — | H/ACA | P2:Ψ2977/26S |

## TABLE 7.1 (*Continued*)
## *C. elegans* Small RNAs Isolated by cDNA Sequencing Method

| Serial No. | RNA Nomenclature[a] | | | Classification | Function[bk] |
|---|---|---|---|---|---|
| | Zemann, A. et al. | Deng, W. et al. | Other | | |
| 81 | comCe14 | — | CeR-7 | H/ACA | P1:Ψ1058/26S |
| | | | | | P2:Ψ3217/26S |
| | | | | | P2:Ψ1718/18S |
| 82 | comCe22 | CeN104 | — | H/ACA | P2:Ψ69/U6 |
| 83 | Ce23 | CeN55 | — | H/ACA | P1:Ψ6/U4 |
| | | | | | P2:Ψ2519/26S |
| 84 | Ce25 | CeN48 | — | H/ACA | P1:Ψ2401/26S |
| | | | | | P1:Ψ127/18S |
| 85 | Ce36 | CeN46 | — | H/ACA | P1:Ψ1082/18S |
| | | | | | P2:Ψ344/18S |
| | | | | | P2:Ψ683/26S |
| 86 | Ce57 | CeN39 | — | H/ACA | P1:Ψ2992/26S |
| 87 | Ce58 | CeN86 | — | H/ACA | P1:Ψ499/18S |
| 88 | Ce80 | — | — | H/ACA | P1:Ψ2237/26S |
| | | | | | P2:Ψ1237/18S |
| 89 | Ce104.1 | — | — | H/ACA | P1:Ψ29/18S |
| | | | | | P2:Ψ3427/26S |
| 90 | Ce105 | CeN66 | — | H/ACA | P1:Ψ1435/18S |
| | | | | | P2:Ψ1361/26S |
| 91 | Ce141 | CeN38 | — | H/ACA | P1:Ψ1198/26S |
| | | | | | P2:Ψ230/18S |
| 92 | Ce149 | CeN125 | — | H/ACA | P2:Ψ940/26S |
| 93 | Ce162 | CeN43 | — | H/ACA | P1:Ψ1996/26S |
| | | | | | P2:Ψ1667/18S |
| 94 | Ce166 | CeN126 | — | H/ACA | P1:Ψ3220/26S |
| | | | | | P2:Ψ2828/26S |
| 95 | Ce170 | — | — | H/ACA | P2:Ψ1156/18S |
| 96 | Ce175 | CeN93 | — | H/ACA | P1:Ψ2417/26S |
| | | | | | P2:Ψ2483/26S |
| 97 | Ce176 | CeN36-1 | — | H/ACA | P1:Ψ3056/26S |
| 98 | BlCe176 | CeN36-2 | — | H/ACA | P1:Ψ3056/26S |
| 99 | Ce192 | CeN100 | CeR-4 | H/ACA | P2:Ψ2558/26S |
| 100 | Ce210 | CeN96 | — | H/ACA (U17/ snR30) | Pre-rRNA processing |

*Continued*

**TABLE 7.1** (*Continued*)

**C. elegans Small RNAs Isolated by cDNA Sequencing Method**

| Serial No. | RNA Nomenclature[a] | | | Classification | Function[bk] |
|---|---|---|---|---|---|
| | Zemann, A. et al. | Deng, W. et al. | Other | | |
| 101 | Ce222 | CeN101 | CeR-7 | H/ACA | P1:Ψ580/18S |
| | | | | | P2:Ψ1266/18S |
| 102 | Ce236 | CeN90 | CeR-6 | H/ACA | P2:Ψ2977/26S |
| 103 | Ce244 | CeN83 | — | H/ACA | P1:Ψ2369/26S |
| 104 | Ce248 | CeN84 | — | H/ACA | P1:Ψ2294/26S |
| 105 | Ce273 | CeN85 | — | H/ACA | P1:Ψ2367/26S |
| | | | | | P2:Ψ1422/26S |
| | | | | | P2:Ψ1392/18S |
| 106 | Ce280 | CeN82 | CeR-8 | H/ACA | P1:Ψ2098/26S |
| | | | | | P2:Ψ3035/26S |
| 107 | Ce283 | CeN79 | — | H/ACA | P1:Ψ1523/18S |
| | | | | | P2:Ψ1156/18S |
| 108 | Ce286 | — | — | H/ACA | P1:Ψ26/U6 |
| | | | | | P2:Ψ22/ 5.8S |
| 109 | Ce309 | CeN41 | CeR-9 | H/ACA | P2:Ψ513/18S |
| 110 | Ce323 | CeN81 | — | H/ACA | P1:Ψ2361/26S |
| | | | | | P2:Ψ1443/18S |
| | | | | | P2:Ψ55/5S |
| 111 | Ce352 | — | — | H/ACA | P1:Ψ2294/ 26S |
| | | | | | P2:Ψ1671/ 26S |
| 112 | Ce356 | — | — | H/ACA | P1:Ψ2367/26S |
| | | | | | P2:Ψ1357/18S |
| 113 | Ce375 | CeN94 | — | H/ACA | P2:Ψ69/U6 |
| 114 | Ce27 | CeN49 | — | H/ACA | Target unknown |
| 115 | Ce48 | CeN58 | — | H/ACA | Target unknown |
| 116 | Ce87 | CeN51 | — | H/ACA | Target unknown |
| 117 | Ce182 | CeN99 | — | H/ACA | Target unknown |
| 118 | Ce345 | CeN92 | — | H/ACA | Target unknown |
| 119 | — | CeN67 | — | H/ACA | Target unknown |
| 120 | — | CeN68 | — | H/ACA | Target unknown |
| 121 | — | CeN80 | — | H/ACA | Target unknown |
| 122 | — | CeN45 | — | H/ACA | Target unknown |
| 123 | — | CeN59 | — | H/ACA | Target unknown |
| 124 | — | CeN91 | — | H/ACA | Target unknown |
| 125 | — | CeN97 | — | H/ACA | Target unknown |

**TABLE 7.1 (*Continued*)**
**C. elegans Small RNAs Isolated by cDNA Sequencing Method**

| Serial No. | RNA Nomenclature[a] | | | Classification | Function[bk] |
|---|---|---|---|---|---|
| | Zemann, A. et al. | Deng, W. et al. | Other | | |
| 126 | — | CeN102 | — | H/ACA | Target unknown |
| 127 | — | CeN105 | — | H/ACA | Target unknown |
| 128 | — | CeN127 | — | H/ACA | Target unknown |
| 129 | Ce11a | CeN1-1 | — | U1-8 | *cis*-splicing |
| 130 | Ce11b | CeN1-6 | — | U1-9 | *cis*-splicing |
| 131 | Ce11c | — | — | U1-1 | *cis*-splicing |
| 132 | Ce11d | CeN1-5 | — | U1-4 | *cis*-splicing |
| 133 | Ce11e | — | — | U1-7 | *cis*-splicing |
| 134 | Ce11f | CeN1-2 | — | U1 | *cis*-splicing |
| 135 | Ce11g | CeN1-7 | — | U1 | *cis*-splicing |
| 136 | Ce11h | CeN1-4 | — | U1 | *cis*-splicing |
| 137 | — | CeN1-3 | — | U1 | *cis*-splicing |
| 138 | Ce310a | — | — | U2-8 | *cis*- and *trans*-splicing |
| 139 | Ce310b | CeN18 | — | U2-5 | *cis*- and *trans*-splicing |
| 140 | Ce310c | — | — | U2-9 | *cis*- and *trans*-splicing |
| 141 | Ce310d | — | — | U2-14 | *cis*- and *trans*-splicing |
| 142 | Ce307a | CeN2-1 | — | U4-3 | *cis*- and *trans*-splicing |
| 143 | Ce307b | CeN2-2 | — | U4 | *cis*- and *trans*-splicing |
| 144 | Ce13 | CeN3-1 | — | U5 | *cis*- and *trans*-splicing |
| 145 | Ce117 | CeN3-3 | — | U5-1 | *cis*- and *trans*-splicing |
| 146 | Ce119a | — | — | U5-5 | *cis*- and *trans*-splicing |
| 147 | Ce119b | CeN3-2 | — | U5-7 | *cis*- and *trans*-splicing |
| 148 | Ce119c | CeN3-4 | — | U5 | *cis*- and *trans*-splicing |
| 149 | — | CeN3-5 | — | U5 | *cis*- and *trans*-splicing |
| 150 | — | CeN3-6 | — | U5 | *cis*- and *trans*-splicing |
| 151 | Ce79 | CeN4 | — | U6-2 | *cis*- and *trans*-splicing |
| 152 | Ce20 | CeN8-2 | — | SL8 | *trans*-splicing |
| 153 | Ce74 | CeN116 | — | SL1 | *trans*-splicing |
| 154 | Ce103a | CeN16-1 | — | SL2-2 | *trans*-splicing |
| 155 | Ce103b | CeN16-4 | — | SL2-4 | *trans*-splicing |
| 156 | Ce133 | CeN16-2 | — | SL2-1 | *trans*-splicing |
| 157 | Ce224 | — | — | SL10 | *trans*-splicing |
| 158 | Ce226 | CeN7 | — | SL4 | *trans*-splicing |
| 159 | — | CeN6 | — | SL3-2 | *trans*-splicing |
| 160 | — | CeN8-1 | — | SL9 | *trans*-splicing |

*Continued*

**TABLE 7.1 (*Continued*)**

**C. *elegans* Small RNAs Isolated by cDNA Sequencing Method**

| Serial No. | RNA Nomenclature[a] | | | Classification | Function[bk] |
|---|---|---|---|---|---|
| | Zemann, A. et al. | Deng, W. et al. | Other | | |
| 161 | — | CeN11 | — | SL11 | *trans*-splicing |
| 162 | — | CeN12 | — | SL3-1 | *trans*-splicing |
| 163 | — | CeN16-3 | — | SL2-3 | *trans*-splicing |
| 164 | — | CeN19 | — | SL13 | *trans*-splicing |
| 165 | — | CeN20 | — | SL6-1 | *trans*-splicing |
| 166 | Ce121a | CeN107-3 | — | SRP-C | Protein secretion |
| 167 | Ce121b | CeN107-2 | — | SRP-D | Protein secretion |
| 168 | Ce121c | CeN107-1 | — | SRP-A | Protein secretion |
| 169 | — | CeN107-4 | — | SRP | Protein secretion |
| 170 | — | CeN9 | — | yrn-1 scRNA | Unknown |
| 171 | — | CeN10 | — | RNase P | Pre-tRNA processing |
| 172 | — | CeN25-3 | Sm Y-5 | snl/Sm Y | Unknown |
| 173 | — | CeN25-4 | Sm Y-11 | snl/Sm Y | Unknown |
| 174 | — | CeN31 | Sm Y-8 | snl/Sm Y | Unknown |
| 175 | — | CeN32 | Sm Y-1 | snl/Sm Y | Unknown |
| 176 | — | CeN112 | Sm Y-10 | snl/Sm Y | Unknown |
| 177 | — | CeN115 | — | snl | Unknown |
| 178 | — | CeN73-1 | — | sb | Unknown |
| 179 | — | CeN73-2 | — | sb | Unknown |
| 180 | — | CeN75 | — | sb | Unknown |
| 181 | — | CeN76 | — | sb | Unknown |
| 182 | — | CeN77 | — | sb | Unknown |
| 183 | Ce9a | CeN21-1 | CeR-2a[j] | Unclassified | Unknown |
| 184 | Ce9b | CeN21-2 | CeR-2b[j] | Unclassified | Unknown |
| 185 | Ce26 | CeN34 | — | Unclassified | Unknown |
| 186 | Ce106 | — | — | Unclassified | Unknown |
| 187 | Ce129[e] | CeN74-2 | — | Unclassified | Unknown |
| 188 | Ce142 | CeN71 | — | Unclassified | Unknown |
| 189 | Ce148 | CeN37 | — | Unclassified | Unknown |
| 190 | Ce150 | — | — | Unclassified | Unknown |
| 191 | Ce178[e] | CeN72 | — | Unclassified | Unknown |
| 192 | Ce220[e] | CeN74-1 | — | Unclassified | Unknown |
| 193 | Ce296 | — | — | Unclassified | Unknown |
| 194 | Ce342 | — | — | Unclassified | Unknown |
| 195 | Ce347 | CeN29 | — | Unclassified | Unknown |

**TABLE 7.1 (*Continued*)**
***C. elegans* Small RNAs Isolated by cDNA Sequencing Method**

| Serial No. | Zemann, A. et al. | Deng, W. et al. | Other | Classification | Function[bk] |
|---|---|---|---|---|---|
| | RNA Nomenclature[a] | | | | |
| 196 | Ce376 | — | — | Unclassified | Unknown |
| 197 | Ce377 | — | — | Unclassified | Unknown |
| 198 | — | CeN64 | — | Unclassified | Unknown |

[a] RNA nomenclatures in each study of Zemann, A. et al. (Reference 44), Deng, W. et al. (Reference 43), Wachi, M. et al. (Reference 42), and MacMorris, M. et al. (Reference 48).

[b] For C/D and H/ACA RNAs, the modification sites predicted by Zemann, A. et al. (Reference 44) are indicated.

[c] Ce96 and Ce135 are classified as snlRNAs in Reference 43.

[d] Ce110 is classified as an H/ACA RNA in Reference 43.

[e] Ce129, Ce178, Ce220, and Ce246 are classified as sbRNAs in Reference 43.

[f] Ce173.1, Ce173.2, and Ce173.3 are described as one H/ACA RNA (CeN128) in Reference 43.

[g] Ce234.1 and Ce234.2 are described as one C/D RNA (CeN47) in Reference 43.

[h] Reference 42.

[i] "P1" and "P2" indicate the pseudouridine pockets at upstream of H box and ACA box, respectively.

[j] Ushida, C. et al., unpublished data.

[k] The latest RNomics study by Huang, Z.-P. et al. (Reference 63) presented the putative targets of several C/D or H/ACA snoRNAs of which targets were unidentified formerly: Ce239/CeN119, CeN13, CeN54, CeN65, and CeN89 probably guide 2′-*O*-methylation of A2317 of 26S rRNA, G668 of 26S rRNA, C2300 of 26S rRNA, A90 of 18S rRNA, and C42 of U2 snRNA, respectively. Ce48/CeN58, Ce87/CeN51, CeN80, CeN59, and CeN105 probably guide pseudouridylation of U1573 of 26S rRNA, U2836 of 26S rRNA, U1523 of 18S rRNA, U1152 of 18S rRNA, and U45 of U5 snRNA, respectively. CeN69/CeR-19 RNA was predicted as a C/D snoRNA that guides 2′-*O*-methylation of G12 of 18S rRNA, which was inconsistent with the prediction by Wachi, M. et al. (Reference 42). Huang, Z.-P. et al. also isolated 5 C/D-type and 10 H/ACA-type novel RNAs by cDNA cloning, which are not represented in this table (Reference 63).

CeN25-1) and SL2 RNAs. The recent study by MacMorris et al. indicated the molecular interaction between SL RNP proteins and some snlRNAs/SmY RNAs [48]. It is possible that the novel snlRNAs share a common pathway in several respects for their biogenesis or function with these RNAs. Several *C. elegans* RNAs of this size showed significant expressional change under stress or during development, indicating that functions of these RNAs are closely related to the condition of worms [49].

## 7.4  SL RNAs

Spliced leader RNAs (SL RNAs) are another major class of small ncRNAs of *C. elegans* (Figure 7.3E). RNomics studies have revealed the presence of at least 14 SL RNAs in the cell (Table 7.1) [43,44]. There are 22 SL RNA genes registered in

WormBase (release WS165, September 2006, http://www.wormbase.org/). SL RNAs (about 120 nt in length) function in *trans*-splicing [50–52]. The 5′ terminal 21–23-nt sequences (spliced leaders, SLs) are spliced onto the 5′ ends of mRNAs [50,51]. About 70% of *C. elegans* pre-mRNAs are subjected to *trans*-splicing according to the results of cDNA analysis of polyA RNAs [50,51,53].

Unlike most eukaryotes, *C. elegans* has a polycistronic gene organization in the genome. About 15% of the genes were estimated to form operons. More than 90% of the SL2-containing mRNAs are derived from downstream genes, located about 100–200 bp apart from the upstream genes [53,54]. This suggests that the role of SL2 RNA *trans*-splicing is in processing polycistronic pre-mRNAs into monocistronic mRNAs and in giving a cap structure to the downstream mRNAs [50,51]. In contrast, only 20 operons were found to use SL1 RNA to process polycistronic pre-mRNAs into monocistronic mRNAs. Unlike SL2 RNA *trans*-splicing, the polyadenylation site of the upstream gene is adjacent to the *trans*-splicing site of the downstream gene. Most substrates of SL1 *trans*-splicing have intronlike sequences at their very 5′ ends, which are called *outrons* [50,51,55]. SL1 also provides a cap structure to the substrate mRNA. Although the biological roles of SL RNA *trans*-splicing are not known, it may contribute to mRNA stability, polyadenylation, nuclear export, and translation in the cell, and it may have allowed operonic gene organization in the genome of *C. elegans*. A recent work has also shown that SL1 was *trans*-spliced to the primary transcript of *let-7* miRNA. This was necessary for the following step: processing of the *let-7* pri-miRNA by the *C. elegans* Drosha homologue [56].

Microarray profiling of *C. elegans* ncRNAs has also indicated that the amounts of several SL RNAs increase in embryos or worms under conditions of stress [49]. An expressional increase during later development was also shown in the case of *sls-2.18* RNA, which donates SL12 (corresponds to SL4 in Reference 57) to mRNAs. The promoter activity of this RNA gene was detected only in hypodermal cells [57]. Thus, it is expected that the functions of SL RNAs are closely related to the regulation of gene expression.

## 7.5   CONCLUSIONS

By computational searches, 2000–4000 ncRNAs were predicted to be encoded in the genome of *C. elegans*. The recent study of a whole-genome tiling microarray indicated that there were about 1200 loci that produced small nonpolyadenylated transcripts of unknown function [58]. The small ncRNAs described in this chapter are just a few of them, such as in other organisms [43,59–62]. The characteristics and function of individual ncRNAs are getting unveiled rapidly [63–68]. It is expected that these ncRNA candidates are involved in diverse physiological phenomena. Further studies on these ncRNA candidates should give rise to more paradigm shifts in understanding the complex genetic system of life.

## REFERENCES

1. Eddy, S. R., Non-coding RNA genes and the modern RNA world, *Nat. Rev. Genet.*, 30, 919, 2001.

2. Winkler, W. C., Riboswitches and the role of noncoding RNAs in bacterial metabolic control, *Curr. Op. Chem. Biol.*, 9, 594, 2005.

3. Costa, F. F., Non-coding RNAs: New players in eukaryotic biology, *Gene*, 357, 83, 2005.

4. Tycowski, K. T. et al., The ever-growing world of small nuclear ribonucleoproteins, in *The RNA World*, 3rd ed., Gesteland, R. F., Cech, T. R., and Atkins, J. F., Eds., Cold Spring Harbor Laboratory Press, New York, 2006, chap. 12.

5. Petersen, J. G., Doench, A., and Sharp, P. A., The biology of short RNAs, in *The RNA World*, 3rd ed., Gesteland, R. F., Cech, T. R., and Atkins, J. F., Eds., Cold Spring Harbor Laboratory Press, New York, 2006, chap. 19.

6. Storz, G. and Gottesman, S., Versatile roles of short RNAs, in *The RNA World*, 3rd ed., Gesteland, R. F., Cech, T. R., and Atkins, J. F., Eds., Cold Spring Harbor Laboratory Press, New York, 2006, chap. 20.

7. Spencer, R. J. and Lee, J. T., Large noncoding RNAs in mammalian gene dosage regulation, in *The RNA World*, 3rd ed., Gesteland, R. F., Cech, T. R., and Atkins, J. F., Eds., Cold Spring Harbor Laboratory Press, New York, 2006, chap. 21.

8. Stricklin, S. L. et al., *C. elegans* noncoding RNA genes (June 25, 2005), in *WormBook*, The *C. elegans* Research Community, ed., WormBook, doi/10.1895/wormbook.1.55.1, http://www.wormbook.org.

9. Griffiths-Jones, S. et al., Rfam: Annotating non-coding RNAs in complete genomes, *Nucleic Acids Res.*, 33, D121, 2005.

10. Riddle, D. L. et al., Introduction to *C. elegans*, in *C. elegans II*, Riddle, D. L., Blumenthal, T., Meyer, B. J., and Priess, J. R., Eds., Cold Spring Harbor Laboratory Press, New York, 1997, chap. 1.

11. Sulston, J., Horvitz, H. R., and Kimble, J., Cell lineage, in *The nematode Caenorhabditis elegans*, Wood, W. B. and the Community of *C. elegans* Researchers Eds., Cold Spring Harbor Laboratory Press, New York, 1988, appendix 3.

12. Schierenberg, E., Embryological variation during nematode development (January 2, 2006), in *WormBook*, The *C. elegans* Research Community, ed., WormBook, doi/10.1895/wormbook.1.55.1, http://www.wormbook.org.

13. Waterson, R. H., Sulston, J. E., and Coulson, A. R., The genome, in *C. elegans II*, Riddle, D. L., Blumenthal, T., Meyer, B. J., and Priess, J. R., Eds., Cold Spring Harbor Laboratory Press, New York, 1997, chap. 2.

14. Spieth, J., Overview of gene structure (January 18, 2006), in *WormBook*, The *C. elegans* Research Community, ed., WormBook, doi/10.1895/wormbook.1.65.1, http://www.wormbook.org.

15. The *C. elegans* Sequencing Consortium, Genome sequence of the nematode *C. elegans*: A platform for investigating biology, *Science*, 282, 2012, 1998.

16. Bieri, T. et al., WormBase: New content and better access, *Nucleic Acids Res.* D1, 2006.

17. Lee, R. C., Feinbaum, R. L., and Ambros, V., The *C. elegans* heterochronic gene *lin-4* encodes small RNAs with antisense complementarity to *lin-14*, *Cell*, 75, 843, 1993.

18. Reinhart, B. J. et al., The 21-nucleotide *let-7* RNA regulates developmental timing in *Caenorhabditis elegans*, *Nature*, 403, 901, 2000.

19. Fire, A. et al., Potent and specific genetic interference by double-stranded RNA in *Caenorhabditis elegans*, *Nature*, 391, 806, 1998.

20. Hamilton, A. J. and Baulcombe, D. C., A species of small antisense RNA in posttranscriptional gene silencing in plants, *Science*, 286, 950, 1999.

21. Zamore, P. D. et al., RNAi: Double-stranded RNA directs the ATP-dependent cleavage of mRNA at 21 to 23 nucleotide intervals, *Cell*, 101, 25, 2000.

22. Elbashir, S. M., Lendeckel, W., and Tuschl, T., RNA interference is mediated by 21- and 22-nucleotide RNAs, *Genes Dev.*, 15, 188, 2001.

23. Lagos-Quintana, M. et al., Identification of novel genes coding for small expressed RNAs, *Science*, 26, 853, 2001.

24. Lau, N. C. et al., An abundant class of tiny RNAs with probable regulatory roles in *Caenorhabditis elegans*, *Science*, 26, 858, 2001.
25. Lee, R. C. and Ambros, V., An extensive class of small RNAs in *Caenorhabditis elegans*, *Science*, 26, 862, 2001.
26. Valencia-Sanchez, M. A. et al., Control of translation and mRNA degradation by miRNAs and siRNAs, *Genes Dev.*, 20, 515, 2006.
27. Ambros, V. et al., A uniform system for microRNA annotation, *RNA*, 9, 277, 2003.
28. Grad, Y. et al., Computational and experimental identification of *C. elegans* microRNAs, *Mol. Cell*, 11, 1253, 2003.
29. Lim, L. P. et al., The microRNAs of *Caenorhabditis elegans*, *Genes Dev.*, 17, 991, 2003.
30. Ambros, V. et al., MicroRNAs and other tiny endogenous RNAs in *C. elegans*, *Curr. Biol.*, 13, 807, 2003.
31. Griffiths-Jones, S. et al., miRBase: microRNA sequences, targets and gene nomenclature, *Nucleic Acids Res.*, 34, D140, 2006.
32. Lee, R. C., Hammell, C. M., and Ambros, V., Interacting endogenous and exogenous RNAi pathways in *Caenorhabditis elegans*, *RNA*, 12, 589, 2006.
33. Ruby, J. G. et al., Large-scale sequencing reveals 21U-RNAs and additional microRNAs and endogenous siRNAs in *C. elegans*, *Cell*, 127, 1193, 2006.
34. Hokii, Y. et al., Twelve novel *C. elegans* RNA candidates isolated by two-dimensional polyacrylamide gel electrophoresis, *Gene,* 365, 83, 2006.
35. Filipowicz, W. and Pogacic, V., Biogenesis of small nucleolar ribonucleoproteins, *Curr. Op. Cell Biol.*, 14, 319, 2002.
36. Bachellerie, J.-P. and Cavaille, J., Small nucleolar RNAs guide the ribose methylations of eukaryotic rRNAs, in *Modification and Editing of RNA*, Grosjean, H. and Benne, R., Eds., ASM Press, Washington, D. C., 1998, chap. 13.
37. Ofengand, J. and Fournier, M. J., The pseudouridine residues of rRNA: Number, location, biosynthesis, and function, in *Modification and Editing of RNA*, Grosjean, H. and Benne, R., Eds., ASM Press, Washington, D.C., 1998, chap. 12.
38. Richard, P. et al., A common sequence motif determines the Cajal body-specific localization of box H/ACA scaRNAs, *EMBO J*, 22, 4283, 2003.
39. Xie, J. et al., Sno/scaRNAbase: A curated database for small nucleolar RNAs and Cajal body-specific RNAs, *Nucleic Acids Res.,* D1, 2006.
40. Torchet, C. et al., The complete set of H/ACA snoRNAs that guide rRNA pseudouridylations in *Saccharomyces cerevisiae*, *RNA*, 11, 928, 2005.
41. Higa, S. et al., Location of 2'-O-methyl nucleotides in 26S rRNA and methylation guide snoRNAs in *Caenorhabditis elegans*, *Biochem. Biophys. Res. Commun.*, 297, 1344, 2002.
42. Wachi, M. et al., Isolation of eight novel *Caenorhabditis elegans* small RNAs, *Gene*, 335, 47, 2004.
43. Deng, W. et al., Organization of the *Caenorhabditis elegans* small non-coding transcriptome: Genomic features, biogenesis, and expression, *Genome Res.,* 16, 20, 2006.
44. Zemann, A. et al., Evolution of small nucleolar RNAs in nematodes, *Nucleic Acids Res.,* 34, 2676, 2006.
44a. Sasano, Y., Hokii, Y. et al. *Biochimie,* 90, 898–907, 2008.
45. Cavaille, J. et al., Identification of brain-specific and imprinted small nucleolar RNA genes exhibiting an unusual genomic organization, *Proc. Natl. Acad. Sci. U.S.A.*, 97, 14311, 2000.
46. Kishore, S. and Stamm, S., The snoRNA HBII-52 regulates alternative splicing of the serotonin receptor 2C, *Science*, 311, 230, 2006.
47. Vitali, P. et al., ADAR2-mediated editing of RNA substrates in the nucleolus is inhibited by C/D small nucleolar RNAs, *J. Cell Biol.,* 169, 745, 2005.
48. MacMorris, M. et al., A novel family of *C. elegans* snRNPs contains proteins associated with *trans*-splicing, *RNA*, 13, 511, 2007.

49. He, H. et al., Profiling *Caenorhabditis elegans* non-coding RNA expression with a combined microarray, *Nucleic Acids Res.*, 34, 2976, 2006.
50. Hastings, K. E. M., SL *trans*-splicing: Easy come or easy go? *Trends Genet.*, 21, 240, 2005.
51. Blumenthal, T. and Steward, K., RNA processing and gene structure, in *C. elegans II*, Riddle, D. L., Blumenthal, T., Meyer, B. J., and Priess, J. R., Eds., Cold Spring Harbor Laboratory Press, New York, 1997, chap. 6.
52. Blumenthal, T., *Trans*-splicing and operons, in *WormBook*, The *C. elegans* Research Community, Ed., WormBook, doi/10.1895/wormbook.1.5.1, http://www.wormbook. org, 2005.
53. Blumenthal, T. et al., A global analysis of *Caenorhabditis elegans* operons, *Nature*, 417, 851, 2002.
54. Blumenthal, T. and Gleason, K. S., *Caenorhabditis elegans* operons: Form and function, *Nat. Rev. Genet.*, 4, 110, 2003.
55. Williams, C., Xu, L. and Blumenthal, T., SL1 *trans*-splicing and 3′-end formation in a novel class of *Caenorhabditis elegans* operon, *Mol. Cell. Biol.*, 19, 376, 1999.
56. Bracht, J. et al., *Trans*-splicing and polyadenylation of *let-7* microRNA primary transcripts, *RNA*, 10, 1586, 2004.
57. Ross, L. H., Freedman, J. H., and Rubin, C. S., Structure and expression of novel spliced leader RNA genes in *Caenorhabditis elegans*, *J. Biol. Chem.*, 270, 22066, 1995.
58. He, H. et al., Mapping the *C. elegans* noncoding transcriptome with a whole-genome tiling microarray, *Genome Res.*, 17, 1471, 2007.
59. Missal, K. et al., Prediction of structured non-coding RNAs in the genome of the nematodes *Caenorhabditis elegans* and *Caenorhabditis briggsae*, *J. Exp. Zoolog. B. Mol. Dev. Evol.*, 306, 379, 2006.
60. Washietl, S. et al., Mapping of conserved RNA secondary structures predicts thousands of functional noncoding RNAs in the human genome, *Nat. Biotechnol.*, 23, 1383, 2005.
61. McCutcheon, J. P. and Eddy, S. R., Computational identification of non-coding RNAs in *Saccharomyces cerevisiae* by comparative genomics, *Nucleic Acids Res.*, 31, 4119, 2003.
62. Hershberg, R., Altuvia, S, and Margalit, H., A survey of small RNA-encoding genes in *Escherichia coli*, *Nucleic Acids Res.*, 31,1813, 2003.
63. Huang, Z. -P. et al., A combined computational and experimental analysis of two families of snoRNA genes from *Caenorhabditis elegans*, revealing the expression and evolution pattern of snoRNAs in nematodes, *Genomics*, doi: 10.1016/j.ygeno.2006.12.002, 2006.
64. Jia, D. et al., Systematic identification of non-coding RNA 2, 2, 7-trimethylguanosine cap structures in *Canorhabditis elegans*, *BMC Mol. Biol.* 8, 86, 2007.
65. Aftab, M. N. et al., Microarray analysis of ncRNA expression patterns in *Caenorhabditis elegans* after RNAi against snoRNA associated proteins, *BMC Genomics,* 9, 278, 2008.
66. Wang, G. and Reinke, V., A *C. elegans* Piwi, PRG-1, regulates 21U-RNAs during spermatogenesis, *Curr. Biol.* 18, 861, 2008.
67. Batista, P. J. et al., PRG-1 and 21U-RNAs interact to form the piRNA complex required for fertility in *C. elegans*, *Mol. Cell* (2008), doi: 10.1016/j.molcel.2008.06.002.
68. Das, P. P. et al., Piwi and piRNAs act upstream of an endogenous siRNA pathway to suppress Tc3 transposon mobility in the *Caenorhabditis elegans* germline, *Mol. Cell* (2008), doi: 10.1016/j.molcel.2008.06.003.

# 8 MicroRNA Tales in Fly Development

*Utpal Bhadra, Sunit Kumar Singh,*
*S.N.C.V.L. Pushpavalli, Praveensingh B. Hajeri,*
*and Manika Pal-Bhadra*

## CONTENTS

## OVERVIEW

Nearly 21–25-nt-long endogenous noncoding RNAs, commonly referred as micro-RNAs (miRNAs), regulate gene expression by partial sequences similarity to the complementary mRNA. miRNAs control a wide range of biological events during animal development, including developmental timing, cell proliferation, apoptosis, cellular differentiation, fine tuning (canalization) of gene expression, stem cell differentiation, etc. Two founding members of the miRNA family were initially identified in *Caenorhabditis elegans* as genes regulating developmental timing, and since then, hundreds of microRNAs have been identified in the majority of the metazoan genomes, including worms, flies, plants, and mammals. miRNAs have diverse expression patterns and might regulate various developmental processes in fruit flies. The discovery of miRNA opens a new avenue in our understanding of complex regulatory developmental networks.

## 8.1  INTRODUCTION

The importance of the fruit fly (*Drosophila melanogaster*) for biological research has been widely emphasized in recent decades. Elaborative genetic analysis of this heterogametic insect is now offered as one of the preferred models for understanding postfertilization complex developmental processes. Immediately after fertilization, mitosis (nuclear division) starts in the single fertilized *Drosophila* egg. However, proportionate cytoplasmic division does not occur, resulting in a multinucleate cell called a *syncytium*, or *syncytial blastoderm*. The cytoplasm in the multinucleated cell allows morphogen gradients to play an important role in pattern formation. After successive and continuous 10 nuclear divisions, the nuclei migrate to the periphery of the embryo. These zygotic nuclei wait until 13 divisions before nuclei are partitioned into separate cells. This stage is the cellular blastoderm. The cells of the blastoderm are still connected to one another. However, the cellular blastoderm exhibits an extremely regular architecture, which determines the major body axes and segment boundaries. Subsequently, a cascade of genes contribute to the body plan of the *Drosophila* embryo. The synchronous function of different sets of genes harmonizes rapid cellular differentiation and organogenesis in the complex phase of developmental programs.

A set of maternal-effect genes, including *bicoid* and *nanos*, are required during oogenesis. The transcripts or proteins of these genes found in the fertilized egg are involved in initiating morphogen gradients. In the next phase, transcription factors of the maternally contributed genes control gap gene expression that subdivides the embryo along the anterior/posterior axis. The gap-genes-encoded transcription factors control pair-rule gene expression that leads to division of the embryo into pairs of segments. These pair-rule genes finally program the expression of a group of segment polarity genes for programming the anterior/posterior axis of each segment. The gap, pair-rule, and segment polarity genes are together referred as the *segmentation genes* for their active involvement in segment patterning. The different organs, including legs, wings, and antennae, develop on particular segments in flies, and this process requires the action of a set of developmental regulators, the homeotic genes. The homeotic genes control the expression of genes responsible for the development of particular organs, wings, legs, and antennae, etc. They preserve a 180-nucleotide sequence called the *homeobox* that encodes a functional domain, known as *homeodomain,* involved in DNA binding. Homeobox-containing (HOX) genes are found in a wide range of species, including worms, fish, frogs, birds, mammals, and plants. Interestingly, HOX genes are found in clusters, and the relative gene order within these clusters is conserved between organisms.

It is shown that organogenesis, polarity, segmentation, cellular differentiation, and other complex developmental processes are regulated by a vast, elaborate web of protein interactions, intercellular interactions, spatial protein concentration gradients, chromatin architecture, histone modification, and DNA methylation events. Proteins expressed by one cell may also modulate the expression of proteins in neighboring or distant cells, thereby influencing differentiation of the cells located elsewhere. Intercellular interactions have been shown to participate in cellular differentiation processes that give rise to simple pattern formations, such as the development of the eye in *D. melanogaster* and vulva in *C. elegans*. Several recent studies have demonstrated the role of DNA methylation and histone modifications (phosphorylation, acetylation, methylation, and ubiquitination) in cellular differentiation and other developmental events. These processes have been shown to turn on/off transcription of the large number of genes required/not required in a particular cell, respectively. The fruit fly *D. melanogaster* is an extraordinarily useful model organism for the analysis of developmental processes.

The genetic determinants of segmental identity in flies and its maintenance throughout development involve complex regulatory mechanisms. As reported earlier, the main genes involved in the establishment and maintenance of segmental identities are the homeotic (*Hox*) genes, whose expression patterns are established early in embryogenesis and propagated throughout development in the appropriate cell lineages [1]. *Hox* genes are organized in two clusters in flies. One cluster that was discovered earlier is the Bithorax complex (BX-C). A large genomic locus contains three homeotic genes, *Ultrabithorax (ubx), abdominal A (abd-A),* and *abdominal B (abd-B)*, which are involved in the development of the third thoracic segment and all the abdominal segments [1]. Another homeotic gene complex, called the Antennapedia Complex (ANT-C), contains five genes that are essential for the development of the head and first two thoracic segments. The initial activation or repression of these homeotic genes in the appropriate segments during early embryogenesis depends on the graded expression profiles of transcription factors and morphogens such as maternally deposited product of the *bicoid (bcd)* gene, which initiates the cascade of zygotically activated gap and pair-rule genes [1,2]. Later, *Hox* gene expression patterns persist despite the disappearance of these early regulators, suggesting that other genes may be involved in maintaining their expression. Until recently, differentiation and developmental studies have focused primarily on DNA sequences that are relatively in close proximity, upstream or downstream, to the protein-coding sequences, paying less attention to the other 98% noncoding part of the genome. Recently, an intriguing group of small, regulatory, noncoding RNAs commonly referred to as microRNAs (miRNAs) were discovered. In this chapter, we offer a comprehensive view of the diverse biological roles of miRNAs in *Drosophila* development and also highlight briefly the major contribution of miRNAs in the genetic buffering and canalization.

## 8.2 COMPREHENSIVE LINK BETWEEN MICRORNA AND DEVELOPMENT

In the current scenario, an miRNA is conventionally defined as a single-stranded RNA of 19–25 nt in length (~22 nt), which is generated by the RNase III type enzyme Dicer from an endogenous transcript pri-miRNA (primary microRNA) that contains a local hairpin structure [3,4]. These are recognized as novel agents exercising extensive posttranscriptional control over most eukaryotic genomes [5]. MicroRNAs

**TABLE 8.1**

**Biological Function of a Few miRNAs and Their Targets**

| miRNA | Target | Function/Processes Regulated | References |
|---|---|---|---|
| *let-7* | *Dappled* | Organogenesis and after terminal differentiation, specifically in neuronal tissue of the PNS and CNS | [14] |
| *mir-14* | *Ecdysone receptor* | Ecdysone signaling, fat metabolism | [15] |
| *Bantam* | *hid* | Suppression of apoptosis and promotion of cell proliferation | [16] |
| *miR-9A, miR-9B, miR-9C* | *Nerfin-1* | In CNS, nervous system development | [17] |
| *miR-34/315/305, miR-307* | *Nerfin-1* | In PNS, nervous system development | [17] |
| *mir-1* | *delta* | Differentiation of cardiac and somatic muscle progenitors, and growth | [18,19] |
| *mir-2* | *hid/ grim/reaper* | Suppression of embryonic apoptosis | [20] |
| *mir-7* | *yan* | Photoreceptor differentiation during eye development | [21] |
| *mir9a* | *senseless* | Precise specification of sensory precursor cells | [22] |
| *miR-iab-4-5p* | *Ultrabithorax* | HOX gene regulation | [23] |
| *miR-iab-8-5p* | *Ultrabithorax, abdominal A* | HOX gene regulation | [24] |

are expressed in a wide variety of organisms ranging from viruses to plants and humans. Many miRNAs are highly conserved across species [6]. In worms, they were initially found to regulate developmental timings [7], but now they are reported to be involved in a wide variety of developmental processes and have a role in fine-tuning of gene expression in the cells [5,6,8]. miRNAs and their targets appear to interact in a complex regulatory network, which leads to crucial developmental decisions. A single miRNA can bind and regulate many different mature messenger RNA (mRNA) targets. Conversely, different miRNAs can control a single mRNA target [9,10]. Recent studies have reported that intergenic sequences in the Bithorax complex of *Drosophila* are transcribed and transcription pattern of these RNAs demonstrate spatial collinearity very similar to Bithorax complex genes. It has been proposed that Antennapedia and Bithorax complex genes *Sex-comb-reduced (Scr)*, *Antennapedia (Antp), Ubx, abdominal-A (abd-A)*, and *Abdominal-B (Abd-B)* are subject to miRNA regulation [11–13]. The biological functions of a few microRNAs that regulate *Drosophila* genes are summarized in Table 8.1. However, except in a few cases, little is known about their target genes. Invertebrate model organisms such as *C. elegans* and *D. melanogaster* have been, and will be, instrumental in dissecting the versatile biological role of miRNAs.

The first evidence of a role of small RNAs in controlling gene expression originated from the discovery of two small RNAs, *lin-4* and *let-7*, in *C. elegans*. These

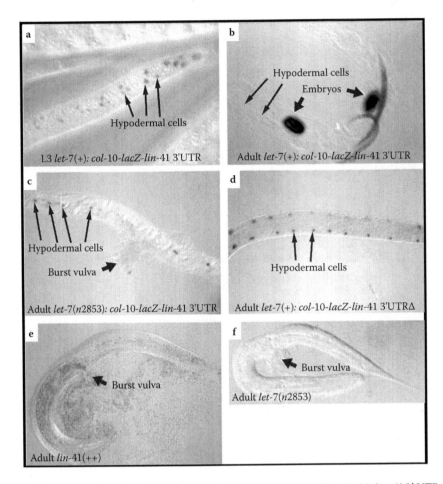

**FIGURE 8.1**  *let-7* regulation of *lin-41*. A *lac Z* reporter gene was fused with *lin-41* 3′ UTR. Lac-Z expression was visible only in the hypodermal cells of L3 larvae (1b) but not in the adult stage (1c), similar to temporal expression pattern of lin-*41*. However, *lac-Z* was expressed in adult let-7 mutants, that is, *let-7(n2853)* (1d) and even in adults, where *let-7* complementary sites in the *lin-41* 3′ UTR fused to reporter gene *lac-Z* (1e). *lin-41* gene expression levels are important in development; overexpression of *lin-41* shows phenotypes of vulval bursting similar to *let-7(n2853)* mutants (1f,g). These results imply that *let-7* regulatory RNA binds to complementary sites in *lin-41* 3′ UTR and downregulates its expression in adults. (From Reinhart, B.J. et al., *Nature*, 2000. 403(6772): 901. With permission.)

small RNAs play an important role in *C. elegans* during postembryonic developmental processes [25,26]. Cloning of the locus revealed that *lin-4* produces a 22-nucleotide noncoding RNA [25]. The negative regulation of *lin-14* by *lin-4* requires partial complementarity between *lin-4* and the 3′ untranslated regions (UTRs) of *lin-14* mRNA [27]. *let-7* functions in a similar fashion as *lin-4*, and represses the expression of the *lin-41* and *hbl-1* mRNAs by binding to their 3′ UTRs [28–31] (Figure 8.1). New miRNAs have been identified through computational analysis and cloning experiments. Many bioinformatic approaches were used to predict novel miRNAs in the genome based on the features and characteristics of known miRNAs. To date, a total of 152 miRNAs have been predicted in the *Drosophila melanogaster* genome. Fifty-five of them were confirmed by cloning or Northern blotting. On the basis of computational and experimental results, Lai et al. (2003) [32] provided a gross estimate that the *Drosophila* genome may have 110 or more miRNAs, which are nearly 1% of predicted genes in flies.

## 8.3  BIOGENESIS AND MECHANISM OF ACTION OF miRNAs

### 8.3.1  BIOGENESIS OF miRNAs

miRNA biogenesis and maturation occurs in multiple steps. The process and machinery of miRNA biogenesis are sufficiently distinct but partly resemble the synthesis of siRNA. The primary transcripts of miRNAs are synthesized from either exonic or intronic regions of protein-coding genes and noncoding regions. The majority of the miRNAs are initially expressed as precursor transcripts in a pol II dependent manner that give rise to pri-miRNAs (primary microRNA). Intronic miRNAs are produced and processed without interfering with the normal splicing of the mRNA of the gene in which they are located [33]. miRNAs are often found in clusters, which might help them to function in concert. They have been reported to express together and to be involved in regulation of related pathways, particularly in developmental programs.

Although the majority of miRNAs are transcribed by pol II, a few miRNAs are reported to be transcribed by pol III, and they mostly reside near upstream tRNA sequences, *Alu* sequences or mammalian genome-wide interspersed repeat sequences (MWIR) [34]. In viruses, however, an miRNA fused to nonfunctional viral tRNA, which is transcribed also by pol III, has been reported [35,36].

### 8.3.2  MATURATION OF miRNAs

Similar to protein coding genes, miRNAs transcribe in the nucleus as long primary transcripts (pri-miRNA). Subsequently, the cleavage of pri-miRNA by the enzyme Drosha leads to formation of stem-loop structured precursor molecules (pre-miRNAs). The pre-miRNA is ~70 nt in length. Drosha is an RNase III type endonuclease that contains two RNase III domains and a dsRNA-binding domain. Drosha cleaves close to the base of the stem and generates a 5′ phosphate and 2-nt 3′ overhangs [36]. Biochemical studies of Drosha complexes revealed that a double-stranded RNA-binding protein (dsRBP) termed Pasha/DGCR8 interacts with Drosha and guides it to the pri-miRNA transcripts [36–39]. The pre-miRNA, once formed, is exported to the cytoplasm mediated by Exportin-5. This export process requires the hydrolysis of Ran-GTP [36,40,41]. The Exportin-5 requires a 3′ overhang to interact with pre-miRNA for efficient transport [36,42]. Once exported, RNase III enzyme Dicer cleaves pre-miRNAs to synthesize 21–23 nt long mature microRNA in the cytoplasm (Figure 8.2). Flies encode two Dicers: *Dicer1 (Dcr-1)* is required for generation of mature miRNAs [36,43], whereas *Dicer-2 (Dcr-2)* is required for generation of short interfering RNAs (siRNAs) [36,44,45]. A few specific dsRBPs interact functionally with different Dicers. DCR-2 is associated with a dsRBP R2D2, for

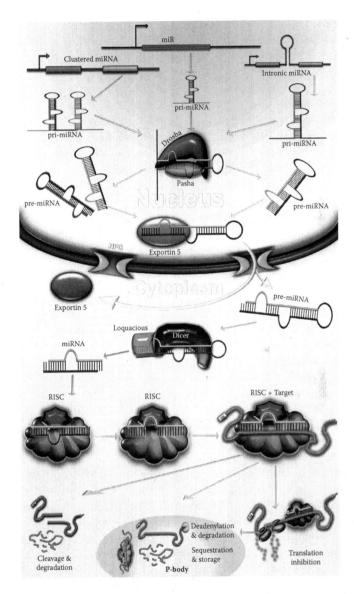

**FIGURE 8.2** *Color version of this figure follows page 238.* miRNA biogenesis and regulation. miRNAs are transcribed to produce a monocistronic or polycistronic precursor RNA called *primary*-miRNA (*pri*-miRNA). These *pri*-miRNAs are then cleaved by complex of Endonuclease III enzyme *Drosha* and *Pasha* having a double-stranded RNA-binding domain (dsRBD) to produce *pre*-miRNA of ~70 nucleotides in length. They are exported out of the nucleus to the cytoplasm by Exportin-5. Once they are in the cytoplasm, they are further processed by Dicer-1 to produce short ~21 nt mature miRNA. Mature miRNAs get incorporated into a multiple turnover protein complex called RNA Induced Silencing Complex (RISC) by a RISC loading complex and retains only one strand of miRNA called the *Guide strand*. The loaded miRISC finds its target by sequence complementarity. miRNAs regulate their targets in many different ways, such as translation inhibition, target cleavage, sequestration, and storage.

their involvement in siRNA processing [36,44], while DCR-1 pairs with *Loquacious* (*Loqs*)/R3D1-L, and is required for normal miRNA maturation [36,46–48].

In contrast to *D. melanogaster* (*Dcr-1* and *Dcr-2*) and *A. thaliana,* which carry four Dicer-like proteins (*Dcl-1-4*), *C. elegans* and mammals have only one Dicer enzyme. The distinct functional specificity of fly *dicer* molecules *Dcr-1,* which process pre-miRNAs and *Dcr-2* for generating dsRNAs, could be related to their structural properties specific to their substrates. miRNA precursors, unlike siRNA precursors, have imperfectly paired stem-loops. There is significant difference in the abundance of their precursor forms also. There are more siRNA precursors in the cell (contributed by the viral infection or loosely regulated transcription) than miRNA precursors (which are endogenously produced and regulated). The substrate specificity of Dicers is also attributed to their structural differences. *Dcr-2,* containing DExH domain, exhibits helicase activity, which is required for dsRNA processing. *Dcr-1* has a PAZ domain that can promote a link between an miRNA precursor and itself [34]. Genetic studies demonstrate that *Dcr-1* is essential, but *Dcr-2* is dispensable for miRNA silencing. Loss-of-function *Dcr-2* mutation severely impairs dsRNA processing but does not produce a significant effect on *Drosophila* development.

Following the Dicer cleavage step, miRNAs get incorporated into microribonucleic proteins (miRNPs) [36,49]. Dicer cleaves both strands of the stem of the precursor hairpin and generates miRNA duplex, but only one strand of that duplex will accumulate in miRNPs. The selection of the preferred strand of miRNA for entering in RNA-induced silencing complex (RISC) depends on the relative stability of the two ends of the miRNA duplex. It has been reported that the helicase enzyme unwinds the duplex and always incorporates the strand with the less tightly paired 5' end into RISC. Dicer itself has a DEAD-box helicase domain, but it is not clear if this DEAD-box is involved in unwinding of miRNA strands. Many other helicases, among them p68 homologue, *Gemin3* in humans and *Armitage* in *D. melanogaster,* have been implicated in miRNA function [36,49]. Members of the AGO (Argonaute) protein family are key components of miRNPs [36]. The molecular mass of this protein is nearly 100 kDa and it contains PAZ (*piwi-argonaute-zwille*) and PIWI domains [36,50]. The PIWI domain shows extensive homology to RNase H, and it facilitates the initiation of the small RNA-dependent cleavage of target RNA [36,51–53]. miRNPs have been biochemically purified from *D. melanogaster* and from HeLa cell lysates. In *D. melanogaster,* miRNAs associate with *Ago2,* the putative helicase *Dmp68,* the ribosomal proteins L5 and L11 and the fragile-X mental retardation protein dFXR, suggesting their role in mRNA translation [36,54]. Further genetic studies indicated that *Ago1* is also required for miRNA function in *D. melanogaster* [36,55].

### 8.3.3 Gene Regulation by miRNAs

MicroRNAs regulate their target genes by different mechanisms: target mRNA degradation, sequestration, and translational repression. In plants many miRNAs exhibit nearly absolute complementarity mainly at the 5' end upstream of the target genes, and they cleave their targets. On the other hand, animal miRNAs share lesser complementarity with their targets, which is restricted to the 5' end of miRNA (nucleotides ~2–8) that is, to the 3′ end of the target site. This anchoring region is referred to as *Seed region*. In the absence of high complementarity, the majority of animal miRNAs inhibit translation of their targets rather than catalyzing their cleavage [56]. Recent studies suggest that regulation by miRNAs can direct target mRNA degradation through a pathway that is distinct from small-RNA-directed endonucleolytic cleavage [31,57,58]. In human and *Drosophila* cells, the core components of the miRNA/siRNA silencing pathways along with their respective targets are found to be concentrated in the cytoplasmic foci called *P-bodies*. These are the regions where many other proteins are accumulated and actively participate in mRNA degradation reactions such as deadenylation, decapping, etc. However, the localization of miRNAs and their targets to the P-bodies is neither required for endonucleolitic cleavage nor essential for silencing. Recently, it has been reported that formation of P-bodies is a mere consequence but not the cause of silencing [59] (Figure 8.3). These P-bodies also act as temporary storage place for some mRNAs, which are later released into cytosol, enabling their translation.

It appears that miRNAs may block the translational initiation by relocalizing miRNA-programmed RISC and the target mRNA into the P-bodies and their close proximity [60]. It is also possible that the mRNAs whose translation is blocked by miRNAs become more susceptible to degradation. It is predicted from bioinformatics and experimental evidence that nearly 30% of the genes are directly regulated by miRNAs, but miRNAs have more or less influenced the expression of the entire genome [61,62]. It is reported that nearly 17% of the transcriptome is affected by *Ago-1* depletion. The discrepancies between in silico prediction and experimental evidence show the unexpected complexity and overwhelming diversity in the target selection of miRNA pools. This target selection may not be based only on the "seed" sequence similarity, but requires other parameters too. RNA editing might play a substantial role in processing and target selection. Many miRNAs and miRNA targets are reported to be edited from adenosine to inosine (A-to-I) for fine-tuning of their processing and/or picking up a target [63,64].

## 8.4  miRNAs IN *DROSOPHILA*

### 8.4.1  *DROSOPHILA* miRNAs IN CELL DEATH AND PROLIFERATION

*Drosophila* geneticists have uncovered roles for miRNAs related to cell proliferation and cell death during development, along with their important role in stress resistance and fat metabolism, etc. miRNAs are differentially expressed during *D. melanogaster* development. Cloning of small RNAs revealed that some miRNAs are exclusively expressed during defined embryonic stages. Other miRNAs are transiently functional in the larval stages, whereas others are adult specific [36,65]. Specific *Drosophila* miRNAs are extensively regulated and expressed during development, in response to hormonal signals [36,66,67]. *miR-100, miR-125,* and *let-7* have been reported to be upregulated upon induction by the insect hormone ecdysone. In contrast, *miR-34* is downregulated by this hormonal treatment. The *miR-100, miR-125,* and *let-7* are located in clusters, and this gives a hint that these miRNAs are functionally related [36,66]. Further genetic studies in flies have revealed that *bantam*

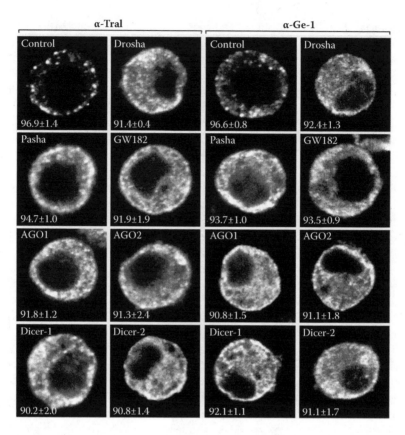

**FIGURE 8.3**   Reduced number of P-bodies in defective RNAi components. *Drosophila* S2 cells were immunostained with anti α-Tral and anti α-Ge-1 antibodies. In RNAi pathway, the number of mutant P-bodies is reduced  and P-Body proteins, which are normally accumulated in P-bodies, are delocalized. Numbers indicate the fraction of cells exhibiting a staining identical to that shown in the representative panel. Scale bar –5 μm. (From Eulalio, A., *Mol Cell Biol* doi: 10, 1128/MCB.00128-07. With permission.)

miRNA represses apoptotic events and promotes cell proliferation (Figure 8.4). On the basis of genetic screening studies for morphologically abnormal flies, Cohen and co-workers had previously identified mutations affecting the *bantam* gene [8,68]. The *bantam* was then cloned based on its mutant phenotype, and reported to produce a functional microRNA responsible for regulation of cell proliferation and cell death [8,16] (Figure 8.4). It has been reported that *bantam* is developmentally regulated and inhibits the expression of the proapoptotic gene *hid* [16,36] (Figure 8.5). Apoptosis is often related to increased and rapid cell proliferation, therefore *bantam* could play a crucial role in the process of tissue growth and homeostasis by inhibiting the apoptosis. By independent genetic screening studies, scientists identified and characterized another miRNA *miR-14* as a suppressor of cell death program in the *Drosophila* eye disc. The studies show that *miR-14* affects the expression of genes essential for cell death along with the genes responsible for fat metabolism [36,69]. Both *bantam*

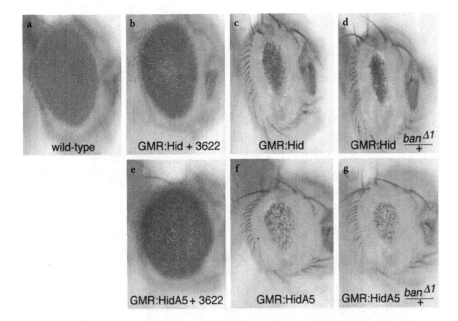

**FIGURE 8.4** *Color version of this figure follows page 238.* Endogenous expression of *bantam* in the eye imaginal disc controls the level of apoptosis induced by *hid*. The *hid* is involved in reducing cell number in the pupal eye disc. Flies expressing *hid*, *hid* A5 under GMR have small, rough-eyed phenotype (c, f), but when *bantam* was coexpressed under GMR-gal4 to suppress GMR-*hid* and GMR *hid*A5, which drives *EP(3)3622*, phenotypes slightly recover (b, e). When one copy of endogenous *bantam* was removed, phenotypes become drastic (d, g) owing to increased level of *hid*-induced apoptosis. (From Brennecke, J., *Cell* 2003. 113(1): 25. With permission.)

and *miR-14* can inhibit apoptosis induced by overexpression of cell death activators in the fly eye [8] (Figure 8.4). However, unlike bantam, *miR-14* does not appear to promote rapid cell proliferation. Loss-of-function study of *miR-14* in flies shows an increased sensitivity to different stressful conditions and also accumulation of large quantities of lipid droplets in the adipose tissues. Additional miRNAs responsible for the regulation of proapoptotic genes have also been discovered in *Drosophila* by loss-of-function studies. In this scenario, a family of miRNAs comprising *miR-2*, *miR-6*, *miR-11*, *miR-13*, and *miR-308* have been reported to be required for suppression of apoptosis during embryonic development [20]. Taken together, these findings demonstrate the potential involvement of *miR-14* and *bantam* along with other microRNAs in *Drosophila* cell proliferation and cell growth.

## 8.4.2 DROSOPHILA MIRNAS IN DEVELOPMENTAL PROCESSES

The developmental timing between distantly related worms and flies could be conserved. Two independent screening studies for isolating mutations for abnormal timings of hypodermal cell development identified loss-of-function alleles of *hbl-1*, a closest worm homologue to the *Drosophila* segmentation gene, *hunchback* [8].

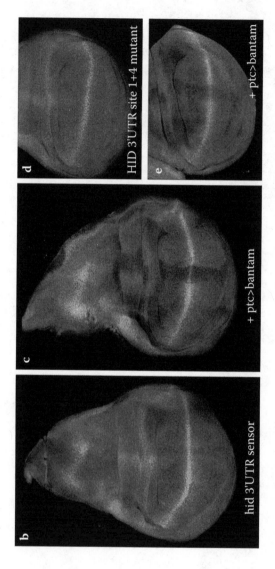

**FIGURE 8.5**  Regulation of *hid* 3' UTR by bantam miRNA. The 3' UTR of *hid* gene was tagged to Tubulin-GFP sensor gene, in which the GFP expression pattern corroborates to *hid* expression pattern (b). When two of the predicted bantam targets in *hid* 3' UTR were mutated, downregulation of GFP was less (d). The *hid* UTR sensor was downregulated when EP(3)3622 was overexpressed under ptc-Gal4 control, indicating that the *hid* UTR confers *bantam*-dependent regulation on the transgene (c,e). (From Brennecke, J., *Cell* 2003. 113(1): 25. With permission.)

Hunchback is a zinc finger transcription factor and is essential for establishing anterior-posterior differences in the *Drosophila* embryo. It also plays an important role by regulating the steps involved in the temporal sequence of cell fate choices during the central nervous system development in *Drosophila* [8,70]. *Hbl-1* is a target of *let-7* miRNA, and it is reasonable to speculate that fly *hunchback* might also be regulated by miRNAs [8]. A study based on approach of antisense-mediated depletion of miRNAs identified a wide range of miRNAs required for distinct developmental pathways during *Drosophila* development [20,36]. The *miR-9* affects cellularization, *miR-31* is required for segmentation, the *miR-310* family affects dorsal closure and *miR-2, miR-13,* and *miR-16* are involved in apoptosis and cell survival [20,36]. Genetic inactivation of *Drosophila Dicer-1* (RNase III enzyme) gene results in reduction of stem cell divisions. The transition from G1 to S phase is reported to be delayed in loss-of-function studies of *Dcr-1* in *Drosophila* during developmental divisions. This finding suggests that miRNAs are required to bypass the normal G1/S checkpoints [36,71].

Some of the recently identified miRNAs expressed in specific mesodermal and muscle tissues have thrown light on their possible role in posttranscriptional regulation during muscle development. The tissue- and stage-specific expression of many *C. elegans* miRNAs that control developmental phenotypes has been implicated in important regulation of animal development [72–74]. Among the 78 known *Drosophila* miRNAs [32,65,74,75], at least 5 are expressed in developing mesoderm and muscle tissues. Rubin, Lai, and their colleagues [74,76] identified patterns of expression of all known *Drosophila* miRNAs loci by probing for nascent transcripts with in situ hybridization during embryonic development. This study reported 5 miRNAs that are expressed predominantly in the mesoderm and muscles. Except for the intronic *miR-13b-2*, which is expressed in differentiating somatic muscles and gut endoderm, the other four miRNAs, *miR-1, miR-8, miR-184,* and *miR-316,* are expressed during specific stages of embryonic development. *miR-8* is expressed in the early mesoderm and developing muscles and also in salivary glands and ectoderm, while *miR-184* is expressed in early mesoderm and also in the early endoderm. The *miR-316* expression is specific for the visceral mesoderm and gut muscles. The expression of *miR-1* is found in the entire presumptive mesoderm. The *miR-1* is predominantly expressed in the somatic, visceral, cardiac, and pharyngeal muscles throughout embryonic development (Figure 8.6). The expression of *miR-1* occurs downstream of *twist*, which appears to be sufficient for its activation [19,74]. The early expression of *miR-1* is reported to be controlled by two enhancers. One of them is located directly upstream of the transcription start site, and the other is at about 4 kb. Both of the enhancers contain multiple *twist* binding sites that are expected to mediate transcriptional activation by the bHLH factor [19,74,77]. Additional binding sites for *Dorsal* and *Snail* proteins at the distal enhancer might contribute to the strength and temporal modulation of enhancer activity [74,77]. The late expression of *miR-1* in all myogenic cells depends on the activity of the transcription factor *Mef-2* (*Myocyte enhancer factor 2*) that is mediated by an *Mef-2* binding site, located in the proximal enhancer region. *Mef-2* binding sites are highly conserved in *Drosophila* species as in many other *twist* binding sites [19,74].

The evolutionary conservation of *miR-1* and its muscle-specific expression across species indicates that similar types of mechanisms operate during muscle

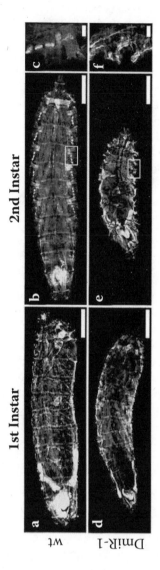

**FIGURE 8.6** miRNA-1 and muscle development. A comparison of body wall musculature of wild-type and *DmiR-1*[ko] mutant larvae: Larvae express UAS-GFP driven by muscle-specific *how*[24B] Gal4 driver. Wild-type larvae have a well-organized network of somatic musculature (1st and 2nd instar) (a, b). In *DmiR-1*[ko] mutants (1st instar), larvae are quite normal but 2nd instar exhibits severely disrupted musculature. An irregular level of GFP expression pattern is found in enlarged view (c, f) of figures b and e, respectively. (From Sokol, N.S. and Ambros, V., *Genes Dev*, 2005. 19(19): 2343. With permission.)

development. In addition, many other miRNAs, expressed mesodermally, have shown more dynamic temporal and spatial expression profiles. This functional role opens the possibility that miRNAs act as developmental regulators during specific steps of mesoderm patterning and muscle development. The same could also be true for vertebrate *miR-1* and *miR-181*, because *miR-1* and *miR-181* are not expressed in the early mesoderm but accumulate at the onset of muscle differentiation [74] (Figure 8.6). It seems that there is a functional spectrum in which miRNAs such as *Drosophila miR-1* are expressed continuously and broadly during development of muscle tissues and act through large sets of mRNAs in quality control of muscle differentiation. The dynamic or transient expression of *miR-181* miRNA acts through a more limited set of targets to promote specific steps of muscle development [74].

Genetic and bioinformatic studies suggest that the Notch signaling pathway in *Drosophila* is regulated by miRNAs. The Notch pathway is a signal transduction cascade essential for proper patterning and development of all metazoan organisms [78]. Genes are directly regulated by Notch signaling via binding sites for CSL type transcription factor [78,79]. Direct targets of Notch signaling and their biological relevance have been well studied in *Drosophila*. These genes are clustered in two genomic locations, named *Enhancer of split*-Complex [*E(spl)*-C] and the *Bearded* complex [*Brd*-C]. These complexes carry seven basic helix-loop-helix [bHLH] repressor-encoding genes and 10 *Bearded* genes, reported to regulate Notch signaling [78,80,81]. The studies based on gain-of-function alleles of bHLH repressor genes *E(spl)m8* and the Bearded family gene *Bearded (Brd)* have shown their association with loss of discrete sequence motifs from their 3′ untranslated regions (3′ UTRs). These 6–7-nt motifs, known as the GY-box (GUCUUCC), the *Brd*-box (AGCUUUA), and the K-box (cUGUGAUa), are distributed in the 3′ UTRs of E*(spl)*-C and *Brd*-C genes and mediate negative posttranscriptional regulation [78,82]. The GY-box, Brd-box, and K-box motifs are reported to be in perfect complementarity with the 5′ ends of various miRNAs in *Drosophila*. This finding suggests the direct regulatory inter-relationships between miRNAs and these boxes [78,83]. One such miRNA, miR-7, targets two *E (spl)*-C genes with GY-boxes [12,78]. Lai et al. (2007) demonstrated direct negative regulation of most members of the *E (spl)*-C and *Brd*-C genes by GY-box, Brd-box, and K-box class miRNAs [78]. *miR-7* regulates the GY-box containing 3′ UTR, *miR-4* and *miR-79* regulate Brd-box containing 3′ UTRs, and *miR-2* and *miR-11* regulate K-box containing 3′ UTRs. The presence of multiple classes of miRNA-binding sites in 3′ UTRs of most Notch target gene raises the possibility of their combinatorial regulation [78].

### 8.4.3 DROSOPHILA MIRNAS IN STEM CELL MAINTENANCE AND PROLIFERATION

Stem cells uniquely self-renew and maintain tissue homeostasis by differentiating into different cell types to replace aged or damaged cells. Self-renewal of germ line stem cells (GSCs) requires both extrinsic and intrinsic signaling mechanisms from neighboring niche cells during oogenesis of *D. melanogaster* [84,85]. Recent studies suggest that miRNA-mediated translational regulation could also be involved in control of self-renewal of *Drosophila* germinal stem cells [46,47,84]. In *Drosophila*, *Dicer-1* (*Dcr-1*) and the double-stranded RNA-binding protein Loquacious (*Loqs*) catalyze miRNA biogenesis [46–48,84]. Two *Loqs* isoforms, PA and PB, are generated by alternative splicing in fly tissues throughout development [46–48,84]. Both *Loqs*-PA and *Loqs*-PB contain three dsRNA-binding domains, but *Loqs* PA is smaller than *Loqs*-PB, by 46 amino acids. Mostly *Loqs*-PB, but to a lesser extent *Loqs*-PA, also enhance the miRNA-generating activity of *Dcr-1* by increasing affinity for pre-miRNAs [47,84]. In a recent study, Park et al. generated *Loqs* knockout (*Loqs*ko) flies by ends-out homologous recombination. They have shown that *Loqs* is essential for embryonic viability and ovarian GSC maintenance. They demonstrated that both the developmental and miRNA processing defects can be rescued by transgenic expression of *Loqs*-PB, but not *Loqs*-PA. This study provided a clear hint that *Loqs*-PB, but not *Loqs*-PA, is necessary and sufficient for *Drosophila* development and the miRNA pathway [84].

MicroRNAs have also been reported to be responsible for maintenance of stem cells in *D. melanogaster*. *Drosophila* ovaries contain three types of adult stem cells, germ-line stem cells (GSCs), escort stem cells (ESCs), and somatic stem cells (SSCs), for lifelong egg production [86–89]. By blocking of *Dcr-1* wild-type function from the stem cells, Zin et al. interpreted the role of *Dcr-1* in the maintenance of stem cells in the *Drosophila* ovary. They have reported that GSCs cannot be maintained and are rapidly lost after Dcr-1 mutation. This finding indicates that *Dcr-1* is responsible for self-renewal capacity of GSCs, but not for survival. They also reported that *bag of marbles* (*bam*) gene encodes a differentiation factor in the *Drosophila* germ line cells. They found that these genes are not upregulated in *Dcr-1*-depleted GSCs. However, the removal of *bam* does not slow down the GSC loss in the *Dcr-1* mutant background. Thus, *Dcr-1* controls GSC self-renewal by suppressing a *bam*-independent differentiation pathway [86]. This study further provided evidence that *Dcr-1* is also essential for the maintenance of SSCs in the *Drosophila* ovary and suggested that two types of stem cells of *Drosophila* ovary are maintained by *Dcr-1*-generated microRNAs (Figure 8.7). The microarray analysis of *Drosophila* cells depleted of *Drosha* and *Argonaute* proteins shows that transcripts whose levels are likely to be directly regulated by silencing pathways (upregulated transcripts) represent less than 20% of the *Drosophila* S2 cell transcriptomes [11,12,58,61,90,91], although this does not represent mRNAs whose turnover rate is affected but not their absolute quantity.

## 8.5 microRNA IN CANALIZATION AND GENETIC BUFFERING

### 8.5.1 What Is Canalization?

Genetic buffering describes phenotypic stability, and it provides the ability to withstand genetic and environmental perturbations. In literature, genetic buffering, canalization, and robustness were used as synonyms. Canalization can be defined as a "narrow canal" of genetic information through which the process of developmental program unfolds. It increases the capacity to absorb mutational variances and environmental milieus. The wall of the canal can stabilize or inhibit the variances of gene expression of different developmental processes. Canalization appears to be mediated by a buffering activity of a particular set of genes for controlling interindividual variability. It also has an important influence on evolutionary innovation.

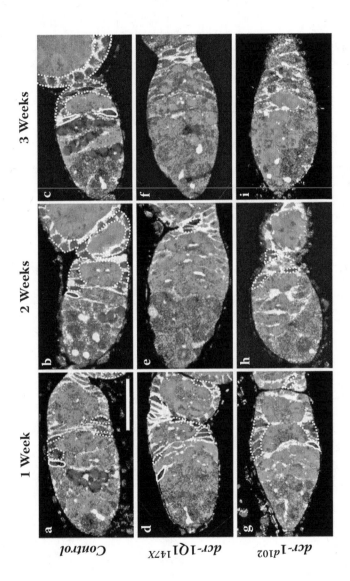

**FIGURE 8.7** *Dcr-1*-dependent miRNAs are required for maintaining SSCs in the *Drosophila* ovary. Germaria are labeled for LacZ DNA and Hu-li tai-shao (Hts), which is used as marker for GSC. SSCs are highlighted with a solid line, and progeny cells are highlighted in dashed lines (scale bar: 18.75 μm). SSCs in the control germaria were still maintained in their niche and continue to generate differentiated progeny up to 1, 2, and 3 weeks (a, b, and c). In *Dicer-1* mutants *dcr-1^{Q1147X}* and *dcr-1^{d102}*, putative SSC clones persist for 1 week and are lost subsequently in the next 2 and 3 weeks. (From Jin, Z, *Curr Biol*, 2007. 17(6): 539. Copyright Elsevier. With permission.)

The conceptual role of miRNA underlies an integrated program emerging from the accumulation of miRNAs and their targets. The network of miRNA–target pairs can be classified in two categories. In the first category, transient expression of miRNA is induced directly or indirectly by genes that inhibit target expression in the same cell where miRNA is expressed [23,91,92]. This relationship is mutually excluded either by the domain of target or its cognate miRNA. In the second type of network, coexistence and coexpression of both components suggests that two seemingly opposite processes are executed at the same time. In this case, miRNA is important for fine adjustment of the expression of their target genes. The nature of these two types of networking of miRNA might contribute to genetic buffering and canalization in two specific ways.

## 8.5.2   MiRNA in Canalization of Fly Development

The expression of *miR-1* in *Drosophila* is a novel example that can be used to understand miRNA-based canalization in the fly. miRNA-regulated canalization can be examined in wild-type *D. melanogaster,* particularly in the number of scutellar bristles in the thoracic region. Several studies on the bristle developmental process suggest that a strong and conspicuous buffering mechanism is associated with such rigid phenotype [93]. Under experimental conditions, this buffering capacity can be altered and inhibit the canalizing capacity. Once it is inherited in future generations, the number of bristles increases proportionately with different stresses and milieu.

The miRNA-regulated homeotic genes that are involved in defining the identity of different body segments and anterior/posterior axis can be evaluated for canalization. The *ultrabithorax* is one such homeotic gene that decides the fate of haltere, a balancing organ, rather than developing a second set of wings in *Drosophila*. Thus, this phenotypic rigidity is regulated by strong genetic canalization of the *Ubx* gene. The loss of *Ubx* canalization results in the loss of genetic buffering that makes haltere a newly transformed wing. However, the mechanism as to how the *Ubx* functions as a canalizing component is still not known. However, the genetic evidences suggest that the *Ubx* 3′ UTR are the target for *miR-iab-4-5p*, an important miRNA that contributes to the decanalization of *Ubx*. This pattern of miRNA and mRNA expression might provide sufficient information to understand the canalization of the *Ubx* gene. In early embryo, the expressions of miRNA *miR-iab-4-5p* and *Ubx* are mutually excluded, and it indicates that miRNA is controlling the random cleavage of leaky *Ubx* miRNA, whereas, in the later stage of development, some domain of the same sequence coexpresses *miR-iab-4-5p* and *Ubx* to reduce the noise for canalization (Figure 8.8). Therefore, the preservation of the *Ubx* phenotype unravels the riddle of miRNA function that explains the canalization. In our understanding, miRNA might generate the canalization in two alternate ways: either (1) by repressing the leaky expression of target mRNA in differentiating tissue that causes a disruptive and deleterious effect on mature miRNA, or (2) reduction of genetic noise is carried out at a posttranscriptional level by the coexpression of miRNA and their targets.

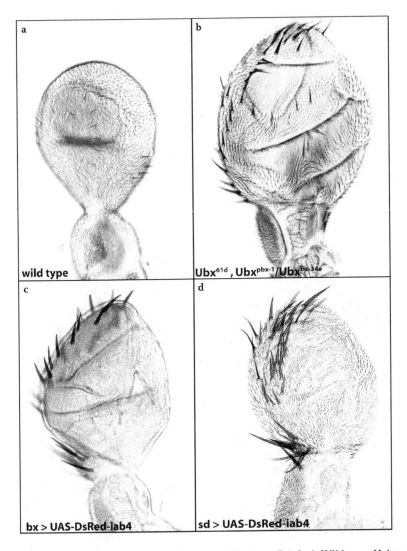

**FIGURE 8.8**  Regulation of anterior *Hox* gene *Ubx* by *miR-iab-4*. Wild-type Haltere (a) lacks a triple row of sensory bristles seen in wings. Upon misexpression of *mir-iab-4* as in *UAS-DsRed-iab-4* under *bx-Gal4* (c) or *scalloped-Gal4* (d), haltere becomes flattened and elongated in the proximal distal axis and exhibits triple rows of sensory bristles for transforming haltere to the wing. These transformations are comparable to a mild *Ubx* loss-of-function mutation (b). (From Ronshaugen, M., *Genes Dev*, 2005. 19(24): 2947. With permission.)

## 8.6   FUTURE CHALLENGES

The miRNAs are reported to be in abundance in animal cells. To understand the versatile functions of miRNAs, new experimental strategies are required. Many technologies are available for screening of genome-wide miRNA precursor sequences, and for predicting their target genes. With the help of different molecular techniques, we can determine the biological functions of genetically and biochemically characterized miRNAs that are overexpressed or underexpressed under certain conditions. Moreover, other influencing factors including environmental cues, and genetic milieus are involved in miRNA expression and their function. Such information facilitates the design of new therapeutic molecules and novel strategies to combat and protect against different inherited genetic disorders and acquired diseases. In many cases, multiple miRNAs target only one gene. Indeed, no strategy has been employed to test the combinatorial effect of multiple miRNAs; nonetheless, the influence of a single miRNA on gene expression and other phenotypic changes might open up an avenue for determination of combinatorial effects.

In vitro experiments demonstrate that the combinatorial function of different miRNAs can act synergistically to inhibit single target gene expression. However, predominant effect of one miRNA over others on the same target is rare. Conversely, many miRNAs regulate multiple target sites for contributing tissue specific expression and function. A better in-depth study at the global miRNA expression might contribute to the outcome of controlling different genetic and physiological traits. In several studies, the chemically modified or synthetic short interfering RNA have been used for effective inhibition of miRNAs and their ectopic expression. In particular, miRNA function can be inhibited by the synthetic analogs of RNA with significant amount of mismatch with the complementary sequence of miRNA. To understand the complex development of *Drosophila* and other organisms, however, a double-stranded precursor and single-stranded miRNA can be used based on their uptake into the cell and their systemic delivery in a tissue-specific manner. For such a targeted delivery system, it may be possible to conjugate with complementary modified miRNA with ligands that recognize a specific cell surface receptor and mediate a cell- or tissue-specific delivery in certain areas of an organism. In fact, a global picture of the role of miRNA during cellular differentiation and organogenesis is required to understand the role of miRNA in complex developmental programs. Basically, we need different technologies to understand the transient expression of miRNA and its different path of action and application in complex biochemical phenomena for converting a single fertilized egg to a complex organism.

## 8.7   CONCLUSIONS

We are only beginning to understand the complexities of miRNA-mediated gene regulatory networks. It is expected that the expression of more than one-third of the genes among animals are regulated by the miRNA. The understanding of the mechanism of miRNA biogenesis is required to delineate gene regulatory networks and their role in developmental processes. In-depth study is also required to identify various unknown additional factors, involved in miRNA biogenesis. The spectrum of miRNA regulation appears to be wide-ranging, from various aspects of development to adaptive responses to stress. miRNAs have been reported to control the gene activity at multiple levels, specifically chromatin organization or transcription, posttranscription, translation, and protein degradation. Many genes of signal transduction pathways and transcription factors involved in developmental processes have been reported as miRNA targets. Large-scale miRNA cloning studies have been initiated to unravel unexpected complexities and massive diversities in small regulatory RNA pathways. The study of targeting pathways of miRNAs would be helpful in understanding the new networks involved in the complex developmental processes of *Drosophila*.

## ACKNOWLEDGMENTS

We are thankful to all our laboratory members and colleagues, who have positively helped us in framing this chapter and have made many critical suggestions. We are highly thankful to Dr Rajesh Gaur for extending all kinds of help during preparation of this chapter. We are grateful enough to Dr. E. Izaurralde, Dr. S. M. Cohen, Dr. G. Ruvkun, Dr. N. S. Sokol, Dr. Ting Xie, and Dr. Eric. C. Lai for granting their kind permission to use their pictures/figures from research publications for this chapter. This work was supported by the Wellcome Trust grant of UB and MPB (GRA07006 MA, GR076395AIA), Young Investigator grant from HFSP (RGY 23/2003) and the Indo-French grant of UB (2903-2/2003-07).

## REFERENCES

1. Grimaud, C., N. Negre, and G. Cavalli, From genetics to epigenetics: The tale of Polycomb group and trithorax group genes. *Chromosome Res,* 2006. 14(4): 363–375.
2. Akam, M., The molecular basis for metameric pattern in the Drosophila embryo. *Development,* 1987. 101(1): 1–22.
3. Ambros, V. et al., A uniform system for microRNA annotation. *RNA,* 2003. 9(3): 277–279.
4. Kim, V.N., Small RNAs: Classification, biogenesis, and function. *Mol Cells,* 2005. 19(1): 1–15.
5. Pillai, R.S., MicroRNA function: Multiple mechanisms for a tiny RNA? *RNA,* 2005. 11(12): 1753–1761.
6. Bartel, D.P., MicroRNAs: Genomics, biogenesis, mechanism, and function. *Cell,* 2004. 116(2): 281–297.
7. Wightman, B., I. Ha, and G. Ruvkun, Posttranscriptional regulation of the heterochronic gene lin-14 by lin-4 mediates temporal pattern formation in *C. elegans. Cell,* 1993. 75(5): 855–862.
8. Ambros, V., MicroRNA pathways in flies and worms: Growth, death, fat, stress, and timing. *Cell,* 2003. 113(6): 673–676.
9. Lewis, B.P. et al., Prediction of mammalian microRNA targets. *Cell,* 2003. 115(7): 787–798.
10. Kim, V.N., MicroRNA biogenesis: Coordinated cropping and dicing. *Nat Rev Mol Cell Biol,* 2005. 6(5): 376–385.
11. Enright, A.J. et al., MicroRNA targets in *Drosophila. Genome Biol,* 2003. 5(1): R1.
12. Stark, A. et al., Identification of *Drosophila* MicroRNA targets. *PLoS Biol,* 2003. 1(3): E60.

13. Grun, D. et al., microRNA target predictions across seven *Drosophila* species and comparison to mammalian targets. *PLoS Comput Biol*, 2005. 1(1): e13.

14. O'Farrell, F. et al., Regulation of the *Drosophila* lin-41 homologue dappled by let-7 reveals conservation of a regulatory mechanism within the LIN-41 subclade. *Dev Dyn*, 2008. 237(1): 196–208.

15. Varghese, J. and S.M. Cohen, microRNA miR-14 acts to modulate a positive autoregulatory loop controlling steroid hormone signaling in *Drosophila*. *Genes Dev*, 2007. 21(18): 2277–2282.

16. Brennecke, J. et al., bantam encodes a developmentally regulated microRNA that controls cell proliferation and regulates the proapoptotic gene hid in *Drosophila*. *Cell*, 2003. 113(1): 25–36.

17. Kuzin, A. et al., The *Drosophila* nerfin-1 mRNA requires multiple microRNAs to regulate its spatial and temporal translation dynamics in the developing nervous system. *Dev Biol*, 2007. 310(1): 35–43.

18. Kwon, C. et al., MicroRNA1 influences cardiac differentiation in *Drosophila* and regulates Notch signaling. *Proc Natl Acad Sci USA*, 2005. 102(52): 18986–18991.

19. Sokol, N.S. and V. Ambros, Mesodermally expressed *Drosophila* microRNA-1 is regulated by Twist and is required in muscles during larval growth. *Genes Dev*, 2005. 19(19): 2343–2354.

20. Leaman, D. et al., Antisense-mediated depletion reveals essential and specific functions of microRNAs in *Drosophila* development. *Cell*, 2005. 121(7): 1097–1108.

21. Li, X. and R.W. Carthew, A microRNA mediates EGF receptor signaling and promotes photoreceptor differentiation in the *Drosophila* eye. *Cell*, 2005. 123(7): 1267–1277.

22. Li, Y. et al., MicroRNA-9a ensures the precise specification of sensory organ precursors in *Drosophila*. *Genes Dev*, 2006. 20(20): 2793–2805.

23. Ronshaugen, M. et al., The *Drosophila* microRNA iab-4 causes a dominant homeotic transformation of halteres to wings. *Genes Dev*, 2005. 19(24): 2947–2952.

24. Tyler, D.M. et al., Functionally distinct regulatory RNAs generated by bidirectional transcription and processing of microRNA loci. *Genes Dev*, 2008. 22(1): 26–36.

25. Lee, R.C., R.L. Feinbaum, and V. Ambros, The *C. elegans* heterochronic gene lin-4 encodes small RNAs with antisense complementarity to lin-14. *Cell*, 1993. 75(5): 843–854.

26. Gesellchen, V. and M. Boutros, Managing the genome: microRNAs in *Drosophila*. *Differentiation*, 2004. 72(2–3): 74–80.

27. Ha, I., B. Wightman, and G. Ruvkun, A bulged lin-4/lin-14 RNA duplex is sufficient for *Caenorhabditis elegans* lin-14 temporal gradient formation. *Genes Dev*, 1996. 10(23): 3041–3050.

28. Reinhart, B.J. et al., The 21-nucleotide let-7 RNA regulates developmental timing in *Caenorhabditis elegans*. *Nature*, 2000. 403(6772): 901–906.

29. Slack, F.J. et al., The lin-41 RBCC gene acts in the *C. elegans* heterochronic pathway between the let-7 regulatory RNA and the LIN-29 transcription factor. *Mol Cell*, 2000. 5(4): 659–669.

30. Lin, S.Y. et al., The *C. elegans* hunchback homolog, hbl-1, controls temporal patterning and is a probable microRNA target. *Dev Cell*, 2003. 4(5): 639–650.

31. Du, T. and P.D. Zamore, microPrimer: The biogenesis and function of microRNA. *Development*, 2005. 132(21): 4645–4652.

32. Lai, E.C. et al., Computational identification of *Drosophila* microRNA genes. *Genome Biol*, 2003. 4(7): R42.

33. Kim, Y.K. and V.N. Kim, Processing of intronic microRNAs. *EMBO J*, 2007. 26(3): 775–783.

34. Borchert, G.M., W. Lanier, and B.L. Davidson, RNA polymerase III transcribes human microRNAs. *Nat Struct Mol Biol*, 2006. 13(12): 1097–1101.

35. Pfeffer, S. et al., Identification of microRNAs of the herpesvirus family. *Nat Methods*, 2005. 2(4): 269–276.

36. Chen, P.Y. and G. Meister, microRNA-guided posttranscriptional gene regulation. *Biol Chem*, 2005. 386(12): 1205–1218.

37. Denli, A.M. et al., Processing of primary microRNAs by the Microprocessor complex. *Nature*, 2004. 432(7014): 231–235.

38. Gregory, R.I. et al., The Microprocessor complex mediates the genesis of microRNAs. *Nature*, 2004. 432(7014): 235–240.

39. Han, J. et al., The Drosha-DGCR8 complex in primary microRNA processing. *Genes Dev*, 2004. 18(24): 3016–3027.

40. Yi, R. et al., Exportin-5 mediates the nuclear export of pre-microRNAs and short hairpin RNAs. *Genes Dev*, 2003. 17(24): 3011–3016.

41. Lund, E. et al., Nuclear export of microRNA precursors. *Science*, 2004. 303(5654): 95–98.

42. Zeng, Y., R. Yi, and B.R. Cullen, MicroRNAs and small interfering RNAs can inhibit mRNA expression by similar mechanisms. *Proc Natl Acad Sci USA*, 2003. 100(17): 9779–9784.

43. Lee, Y.S. et al., Distinct roles for *Drosophila* Dicer-1 and Dicer-2 in the siRNA/miRNA silencing pathways. *Cell*, 2004. 117(1): 69–81.

44. Liu, Q. et al., R2D2, A bridge between the initiation and effector steps of the *Drosophila* RNAi pathway. *Science*, 2003. 301(5641): 1921–1925.

45. Pham, J.W. et al., A Dicer-2-dependent 80s complex cleaves targeted mRNAs during RNAi in *Drosophila*. *Cell*, 2004. 117(1): 83–94.

46. Forstemann, K. et al., Normal microRNA maturation and germ-line stem cell maintenance requires Loquacious, a double-stranded RNA-binding domain protein. *PLoS Biol*, 2005. 3(7): e236.

47. Jiang, F. et al., Dicer-1 and R3D1-L catalyze microRNA maturation in *Drosophila*. *Genes Dev*, 2005. 19(14): 1674–1679.

48. Saito, K. et al., Processing of pre-microRNAs by the Dicer-1-Loquacious complex in *Drosophila* cells. *PLoS Biol*, 2005. 3(7): e235.

49. Mourelatos, Z. et al., miRNPs: A novel class of ribonucleoproteins containing numerous microRNAs. *Genes Dev*, 2002. 16(6): 720–728.

50. Carmell, M.A. et al., The Argonaute family: Tentacles that reach into RNAi, developmental control, stem cell maintenance, and tumorigenesis. *Genes Dev*, 2002. 16(21): 2733–2742.

51. Parker, J.S., S.M. Roe, and D. Barford, Crystal structure of a PIWI protein suggests mechanisms for siRNA recognition and slicer activity. *EMBO J*, 2004. 23(24): 4727–4737.

52. Song, J.J. et al., Crystal structure of Argonaute and its implications for RISC slicer activity. *Science*, 2004. 305(5689): 1434–1437.

53. Ma, J.B. et al., Structural basis for 5'-end-specific recognition of guide RNA by the *A. fulgidus* Piwi protein. *Nature*, 2005. 434(7033): 666–670.

54. Ishizuka, A., M.C. Siomi, and H. Siomi, A *Drosophila* fragile X protein interacts with components of RNAi and ribosomal proteins. *Genes Dev*, 2002. 16(19): 2497–2508.

55. Okamura, K. et al., Distinct roles for Argonaute proteins in small RNA-directed RNA cleavage pathways. *Genes Dev*, 2004. 18(14): 1655–1666.

56. Olsen, P.H. and V. Ambros, The lin-4 regulatory RNA controls developmental timing in *Caenorhabditis elegans* by blocking LIN-14 protein synthesis after the initiation of translation. *Dev Biol*, 1999. 216(2): 671–680.

57. Bagga, S. et al., Regulation by let-7 and lin-4 miRNAs results in target mRNA degradation. *Cell*, 2005. 122(4): 553–563.

58. Lim, L.P. et al., Microarray analysis shows that some microRNAs downregulate large numbers of target mRNAs. *Nature*, 2005. 433(7027): 769–773.

59. Eulalio, A. et al., P-body formation is a consequence, not the cause of RNA-mediated gene silencing. *Mol Cell Biol*, 2007.

60. Pillai, R.S. et al., Inhibition of translational initiation by Let-7 MicroRNA in human cells. *Science*, 2005. 309(5740): 1573–1576.

61. Rehwinkel, J. et al., Genome-wide analysis of mRNAs regulated by Drosha and Argonaute proteins in *Drosophila melanogaster*. *Mol Cell Biol*, 2006. 26(8): 2965–2975.

62. Kawahara, Y. et al., Redirection of silencing targets by adenosine-to-inosine editing of miRNAs. *Science*, 2007. 315(5815): 1137–1140.

63. Luciano, D.J. et al., RNA editing of a miRNA precursor. *RNA*, 2004. 10(8): 1174–1177.

64. Kawahara, Y. and T. Imanishi, A genome-wide survey of changes in protein evolutionary rates across four closely related species of *Saccharomyces* sensu stricto group. *BMC Evol Biol*, 2007. 7: 9.

65. Aravin, A.A. et al., The small RNA profile during *Drosophila melanogaster* development. *Dev Cell*, 2003. 5(2): 337–350.

66. Sempere, L.F. et al., Temporal regulation of microRNA expression in *Drosophila melanogaster* mediated by hormonal signals and broad-Complex gene activity. *Dev Biol*, 2003. 259(1): 9–18.

67. Sempere, L.F. et al., The expression of the let-7 small regulatory RNA is controlled by ecdysone during metamorphosis in *Drosophila melanogaster*. *Dev Biol*, 2002. 244(1): 170–179.

68. Hipfner, D.R., K. Weigmann, and S.M. Cohen, The bantam gene regulates *Drosophila* growth. *Genetics*, 2002. 161(4): 1527–1537.

69. Xu, P. et al., The *Drosophila* microRNA Mir-14 suppresses cell death and is required for normal fat metabolism. *Curr Biol*, 2003. 13(9): 790–795.

70. Isshiki, T. et al., *Drosophila* neuroblasts sequentially express transcription factors which specify the temporal identity of their neuronal progeny. *Cell*, 2001. 106(4): 511–521.

71. Hatfield, S.D. et al., Stem cell division is regulated by the microRNA pathway. *Nature*, 2005. 435(7044): 974–978.

72. Plasterk, R.H., Micro RNAs in animal development. *Cell*, 2006. 124(5): 877–881.

73. Carthew, R.W., Gene regulation by microRNAs. *Curr Opin Genet Dev*, 2006. 16(2): 203–208.

74. Nguyen, H.T. and M. Frasch, MicroRNAs in muscle differentiation: Lessons from *Drosophila* and beyond. *Curr Opin Genet Dev*, 2006.

75. Griffiths-Jones, S. et al., miRBase: microRNA sequences, targets and gene nomenclature. *Nucleic Acids Res*, 2006. 34(Database issue): D140–144.

76. Aboobaker, A.A. et al., *Drosophila* microRNAs exhibit diverse spatial expression patterns during embryonic development. *Proc Natl Acad Sci USA*, 2005. 102(50): 18017–18022.

77. Biemar, F. et al., Spatial regulation of microRNA gene expression in the *Drosophila* embryo. *Proc Natl Acad Sci USA*, 2005. 102(44): 15907–15911.

78. Lai, E.C., B. Tam, and G.M. Rubin, Pervasive regulation of *Drosophila* Notch target genes by GY-box-, Brd-box-, and K-box-class microRNAs. *Genes Dev*, 2005. 19(9): 1067–1080.

79. Bailey, A.M. and J.W. Posakony, Suppressor of hairless directly activates transcription of enhancer of split complex genes in response to Notch receptor activity. *Genes Dev*, 1995. 9(21): 2609–2622.

80. Knust, E. et al., Seven genes of the Enhancer of split complex of *Drosophila* melanogaster encode helix-loop-helix proteins. *Genetics*, 1992. 132(2): 505–518.

81. Wurmbach, E., I. Wech, and A. Preiss, The Enhancer of split complex of *Drosophila* melanogaster harbors three classes of Notch responsive genes. *Mech Dev*, 1999. 80(2): 171–180.

82. Lai, E.C., C. Burks, and J.W. Posakony, The K box, a conserved 3' UTR sequence motif, negatively regulates accumulation of enhancer of split complex transcripts. *Development*, 1998. 125(20): 4077–4088.

83. Lai, E.C., Micro RNAs are complementary to 3' UTR sequence motifs that mediate negative post-transcriptional regulation. *Nat Genet*, 2002. 30(4): 363–364.

84. Park, J.K. et al., The miRNA pathway intrinsically controls self-renewal of *Drosophila* germline stem cells. *Curr Biol*, 2007. 17(6): 533–538.

85. Spradling, A., D. Drummond-Barbosa, and T. Kai, Stem cells find their niche. *Nature*, 2001. 414(6859): 98–104.

86. Jin, Z. and T. Xie, Dcr-1 maintains *Drosophila* ovarian stem cells. *Curr Biol*, 2007. 17(6): 539–544.

87. Decotto, E. and A.C. Spradling, The *Drosophila* ovarian and testis stem cell niches: Similar somatic stem cells and signals. *Dev Cell*, 2005. 9(4): 501–510.

88. Xie, T. et al., Intimate relationships with their neighbors: Tales of stem cells in *Drosophila* reproductive systems. *Dev Dyn*, 2005. 232(3): 775–790.

89. Lin, H., The stem-cell niche theory: lessons from flies. *Nat Rev Genet*, 2002. 3(12): 931–940.

90. Robins, H., Y. Li, and R.W. Padgett, Incorporating structure to predict microRNA targets. *Proc Natl Acad Sci USA*, 2005. 102(11): 4006–4009.

91. Stark, A. et al., Animal MicroRNAs confer robustness to gene expression and have a significant impact on 3'UTR evolution. *Cell*, 2005. 123(6): 1133–1146.

92. Hornstein, E. et al., The microRNA miR-196 acts upstream of Hoxb8 and Shh in limb development. *Nature*, 2005. 438(7068): 671–674.

93. Hornstein, E. and N. Shomron, Canalization of development by microRNAs. *Nat Genet*, 2006. 38 Suppl. 1: S20–24.

# 9 RNA Interference and microRNAs in Zebra Fish

*Alex S. Flynt, Elizabeth J. Thatcher, and James G. Patton*

## CONTENTS

## OVERVIEW

Understanding of RNA interference (RNAi) has been made possible through a variety of experimental approaches using different model organisms [1]. Some of these systems are amenable to mutational screening, facilitating forward genetic approaches and subsequent discovery of novel genes involved in the RNAi pathway. Others permit biochemical approaches producing mechanistic insight. Over the last 8 years, observations from different groups using these techniques in different systems have yielded a clear outline of the basic mechanisms behind RNAi [2]. These approaches have also uncovered a new class of genes that encode microRNAs (miRNAs), which act through an RNAi-like mechanism to regulate gene expression. Here, we will discuss advances in RNAi/miRNA research using the zebra fish model system, and discuss how this system may help answer outstanding questions in the field.

## 9.1  INTRODUCTION TO ZEBRA FISH

### 9.1.1  CHARACTERISTICS OF ZEBRA FISH AND EMBRYONIC DEVELOPMENT

Zebra fish (*Brachydanio rerio*) are a small cyprinid fish marked by a series of horizontal blue and yellow stripes, thus the zebra moniker. Sexual dimorphism is somewhat subtle—females have a white-to-silver belly, while males have an orange-colored belly (Figure 9.1a). Experimentally, zebra fish constitute an extremely accessible embryonic system [3]. Mating pairs can be placed in small tanks with a mesh for separating newly laid eggs, and breeding is triggered by a light source after a period of dark, corresponding to sunrise. Zebra fish lay freely drifting eggs that, as with most fish, are fertilized externally. Its oocytes possess a large vegetal yolk, while the embryo proper forms on the animal pole of the yolk cell.

After fertilization, six synchronous divisions occur at a regular orientation, but the cells remain continuous with the yolk via cytoplasmic bridges (Figure 9.1b). Blastula stages are marked by a loss of the regular distribution of cells and the beginning of metasynchronous division. The mid-blastula transition (MBT) signals the onset of zygotic transcription and occurs at the tenth cell cycle, whereas prior to this point, gene expression is derived exclusively from maternally contributed mRNAs [4]. At the same time the MBT occurs, an important signaling center for mesendoderm induction called the *yolk syncytial layer* (YSL) forms from marginal cells [5].

Similar to other metazoans, gastrulation begins upon the internalization of cells that give rise to mesoderm and endoderm (Figure 9.1b). This process begins at the dorsal blastoderm margin in a conspicuous region referred to as the *embryonic shield* and is comprised of several distinct cell movements: epiboly, convergence, extension, and involution [6]. After gastrulation, cells have migrated to envelop the yolk and have accumulated along the anterior–posterior axis on the dorsal side of the embryo (Figure 9.1b). Subsequently, neurulation and somitogenesis occur and are completed by 24 hours postfertilization [3]. The remainder of embryonic development occurs over a period of 4 days. Embryos hatch from their chorions during the second day after which they are free swimming but still reliant on yolk for nutrients. On the fifth day after fertilization, a protruding jaw is visible, and larva is ready to hunt prey (Figure 9.1b).

### 9.1.2  ZEBRA FISH AS AN EXPERIMENTAL SYSTEM

Due to rapid external development, it is easy to observe cellular behavior and other phenomena without disrupting zebra fish embryos. Before pigmentation occurs, after approximately the first 24 hours of development, zebra fish embryos are transparent, further facilitating observation of embryonic structures. Healthy adult fish typically lay hundreds of eggs, allowing for easy access to numerous embryos. Initial interest in zebra fish arose from its potential as a vertebrate model that was amenable to forward mutagenesis screens [7]. These screens are made possible because of the

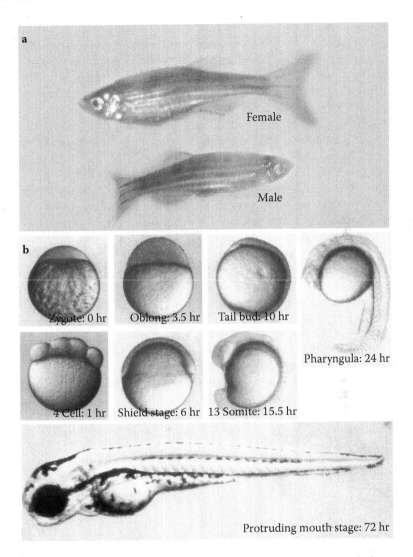

**FIGURE 9.1** Zebra fish development. (a) Adult zebra fish: female top, male bottom. (b) Developmental series showing zygote stage, 4-cell stage, oblong stage, shield stage, tail bud stage, 13-somite stage, pharyngula stage, and protruding mouth stage. The number of hours postfertilization required to reach the indicated stage is listed after the colon. (From Kimmel, C. B. et al. (1995). *Dev Dyn* 203, 253–310. With permission.)

capacity to breed and are useful for discovering genes involved in embryogenesis due to transparent and accessible embryos. Many mutants have been recovered from such screens [8].

Reverse genetic methodologies and the creation of transgenic animals are well-established technologies in zebra fish [9]. Due to the connections between the yolk cell and early dividing animal cells, it is relatively easy to deliver molecules, typically nucleic acid, through microinjection. Overexpression of specific genes can be accomplished through injection of synthetic mRNAs. The standard method for producing gene knockdown is through the use of morpholino oligonucleotides, a form of antisense technology [10]. These oligonucleotides are approximately 25 bases long and possess a modified backbone, which makes them resistant to degradation, thus prolonging the interaction between the morpholino and its target sequence. Typically, attenuation of mRNA expression is accomplished by designing morpholinos complementary to start codons and surrounding sequences. This prevents the assembly of ribosomes, inhibiting translation and ultimately gene expression. Similarly, morpholinos can be designed to hybridize to pre-mRNA splice sites, causing skipping of exons and generation of truncated proteins or destabilized mRNA [11].

In addition to the relative ease of antisense-based reverse genetics in zebra fish, methodologies for generating transgenic embryos are available. Transgenes are introduced through microinjection of DNA constructs [9]. Transgene incorporation is random and mosaic, but genetic outcrossing can result in stable transgenic lines. Successful integration into the germ line causes transmission to progeny and equal distribution throughout the animal.

Based on the preceding characteristics, zebra fish constitute an excellent system for the study of vertebrate embryonic development. However, it is not yet possible to create mutants through homologous recombination. This is due to a lack of zebra fish embryonic stem cells. Methodologies have been developed to circumvent this limitation, one of which, TILLING, is discussed in the following text [12]. Additionally, the zebra fish genome is not yet complete as of this writing. These obstacles will likely be surmounted, especially the latter, in the coming years. The strengths of the zebra fish system greatly facilitate analysis of miRNA function though simple introduction of RNAs directly into embryos via microinjection, and the ability to easily observe phenotypes in gain-of-function and loss-of-function experiments.

## 9.2  RNAi TECHNOLOGY

### 9.2.1  RNAi in Different Model Systems

Significant advances in the RNAi/miRNA field have been made using *Caenorhabditis elegans* as a model system. The initial observation of the ability of double-stranded RNA (dsRNA) to produce specific and potent silencing of homologous genes was made by Fire et al. in 1998, using *C. elegans* [13]. As part of experiments designed to produce an antisense effect with RNA complementary to an endogenous mRNA, complementary RNAs were delivered via gonadal injection. As controls, both sense RNA and a mixture of sense and antisense RNA were also injected. Unexpectedly, the mixture of sense and antisense RNA yielded the most potent knockdown of the target gene. This discovery was termed *RNA interference* (RNAi) and has been widely adopted as a standard tool for gene function analysis in *C. elegans* and many other model systems.

The ability to use long dsRNA as an effective agent of knocking down gene expression has been effective in plants and invertebrates, although the RNAi pathway in these organisms is highly divergent in terms of various components of the pathway [1]. For these organisms, triggering RNAi with long dsRNA is thought to be part of an immune response to viruses and parasitic genetic elements [14,15]. Plants seem particularly reliant on RNAi as a viral defense since siRNAs efficiently diffuse through plant tissue ahead of spreading viral infection [16]. Invertebrates also use RNAi for antiviral purposes but primarily employ it for silencing transposable elements. In *Drosophila melanogaster*, a class of repeat-associated siRNAs (rasiRNAs) have been discovered that correspond to repetitive genetic elements [17]. Invertebrate and plant cells also express miRNAs and employ the RNAi pathway to guide development.

For vertebrate cells, long dsRNA is cytotoxic and does not enter the RNAi pathway and instead triggers an interferon-mediated antiviral response through a dsRNA-dependent protein kinase (PKR). PKR binds dsRNA and phosphorylates the translation initiation factor eIF-2α [18]. This results in global inhibition of mRNA translation and ultimately cell death. Activation of RNase L/oligo(A) synthetase and Toll-like receptor 3 by long dsRNA molecules also triggers cell death [19,20]. The usefulness of RNAi as a defense against viruses in vertebrates has perhaps become reduced due to the evolution of sophisticated cell-mediated immune systems. As far as is known, vertebrates do not use RNAi as an antiviral response or to control repetitive element multiplication. Instead, the pathway is used for the production and function of miRNAs [21].

### 9.2.2 APPLICATION OF RNAi IN ZEBRA FISH

Shortly after the discovery of RNAi by Fire and Mello, long dsRNA was used successfully to knockdown genes in *C. elegans,* fruit flies, planarians, hydra, and plants [1]. Initial efforts seeking to attenuate gene expression in zebra fish through injection of long dsRNA claimed the method to be effective [22]. These experiments sought to target *no tail* (*ntl*), *floating head* (*flh*), and *pax2.1* genes that are essential for early development. The gene *ntl* is essential for mesoderm formation, and is one of the first genes expressed in the germ ring, being stimulated by signals from the YSL. Mutations in *ntl* result in a loss of notochord and severe truncation of the tail. Similarly, *flh* results in the loss of the same structures. While injections of dsRNA in zebra fish resulted in nonspecific effects at a high rate, some of the expected phenotypic effects were observed. Also, in situ staining of *ntl* mRNA suggested measurable knockdown of the gene. These results, as well as others [23,24], suggested that RNAi could be successfully used in zebra fish to specifically knock down homologous genes although the success of knockdown varied greatly.

Despite early reports, later work using long dsRNAs strongly argued that all defects in zebra fish were due to nonspecific effects [25]. Regardless of the sequence of injected dsRNA, dose-dependent toxicity was observed. Zhao et al. (2001) tested the ability of dsRNA injection to eliminate protein expression using transgenic embryos expressing GFP under the influence of the GATA-1 promoter that drives expression in blood cells. Despite severe morphological defects induced by injection of dsRNA homologous to GFP, no silencing of the GFP was observed. Additionally, injection of dsRNA, regardless of sequence, resulted in a global decrease of mRNA expression. Attempts to silence expression of *pou2* through injection of dsRNAs homologous to GFP or to *pou2* both resulted in similar decreases in *pou2* expression.

In retrospect, nonspecific effects induced by long dsRNA were predictable, considering the interferon response induced by these molecules in vertebrates. Translation arrest induced by PKR activation followed by degradation of RNAs by RNase L could easily explain these results.

### 9.2.3 SMALL HAIRPIN RNAS AND RNAI IN ZEBRA FISH

These conflicting reports have for the most part resulted in reluctance among zebra fish researchers to use RNAi to knock down gene expression. Despite this, RNAi strategies in zebra fish may be possible, provided the dsRNAs are short. Introduction of RNAs less than 30 nucleotides can apparently circumvent most PKR activation [26]. Small hairpin RNA molecules have been shown to be effective in eliciting an RNAi response in vertebrates. These molecules behave like miRNAs and are processed accordingly.

The use of small RNAs that evade PKR activation to trigger an RNAi response in zebra fish has been demonstrated, but wide adoption of the technique has not occurred. Injection of small RNAs, when introduced as small heteroduplex RNAs, produces specific RNAi in zebra fish [27]. Thus, there may be applications for RNAi technology in zebra fish, particularly through the production of transgenic lines inactivating specific genes by the synthesis of short dsRNAs encoded on the transgene.

In flies and worms, a typical RNAi strategy is to create transgenic animals that express long dsRNAs by transcription of an inverted repeat (IR) construct derived from target mRNA sequence [28]. This IR dsRNA can be driven tissue specifically, either with a specific promoter or using the GAL4 promoter where GAL4 protein is expressed in a tissue-specific manner [29,30]. Similar to invertebrate IR-expressing systems, many transgenic mice have been created that express short-hairpin RNAs (shRNAs) [31]. The dsRNA portion of the hairpin must be small enough to avoid PKR activation. This is also a viable option for cultured mammalian cells. Plasmids bearing an shRNA cassette can be either transiently or stably expressed. As of this writing, there have been no published reports using shRNA cassettes in zebra fish. Employing these constructs may be the most productive application of RNAi in zebra fish.

Currently, analysis of loss of gene function in zebra fish is accomplished using mutants, antisense morpholinos, or more infrequently, dominant negative mutants. While each of these strategies is well established, there are several limitations that could be overcome using shRNA technology. For example, mutant lines are perhaps the most desirable tools for loss-of-function analysis, but mutant alleles can be variable as to the degree of attenuation of gene function and little control can be exerted over when and where loss of function occurs. If a specific gene is required for both early and late stages of development, it is difficult to study later requirements for that gene with such a mutant. RNAi technologies may be useful for such purposes (Figure 9.2). shRNA constructs can be made that drive shRNA using heat-shock promoters to induce expression at specific times (Figure 9.2a). Similarly, expression using tissue-specific promoters such as the GAL4 system could be used to generate tissue-specific knockdown of target gene expression (Figure 9.2b).

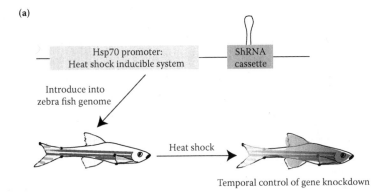

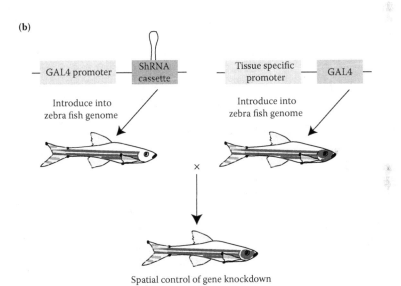

**FIGURE 9.2** Use of transgenic animals to deliver shRNAs under the control of conditional promoters. (a) Production of short hairpin RNAs (shRNAs) under the control of a heat shock promoter. (b) Production of shRNAs under the control of the GAL4 promoter. Transgenic lines are crossed with lines expressing GAL4 protein to generate tissue-specific RNA expression.

## 9.3 THE VERTEBRATE siRNA/miRNA PATHWAY

In their seminal article, Fire and co-workers (1998) proposed that the likely mechanism behind RNAi was due to RNA molecules acting as a component of nuclear feedback loops silencing transcription of homologous genes through binding to genomic DNA [13]. As an alternative, they admitted that RNAi could be achieved through a mechanism based on an RNA–RNA interaction. It is now clear that an RNA–RNA interaction is central to RNAi.

dsRNA is initially cleaved by the RNase III enzyme Dicer into small, approximately 22-nt RNA duplexes called *small-interfering RNAs* (siRNA) [32]. One of the strands is then incorporated into a multicomponent complex called the RNA-induced silencing complex (RISC) [33,34]. Argonaute proteins are core members of RISC, possessing PAZ (piwi-argonaute-zwille) and PIWI domains that physically interact with RNA molecules [35]. Argonautes bind the small RNAs and use them as specificity factors to identify mRNAs with complementary sequences. The fate of mRNAs targeted by RISC is dependent on the degree of base pairing between the small RNA and the mRNA [36–39]. Perfect pairing results in cleavage of the mRNA at the ninth nucleotide of interaction, counting from the 5′ end of the siRNA [40]. Cleavage of mRNAs is specifically carried out by Argonaute 2, also known as *slicer* [41]. If the pairing is imperfect, cleavage does not occur and instead, translation of the targeted mRNA is inhibited [37]. The precise mechanism of inhibition is unknown, but it seems that relocalization of mRNAs from polysomes to processing bodies (P-bodies) is an important aspect of this activity [42]. Immunostaining of Argonautes reveals localization to P-bodies, cytoplasmic foci that contain decapping and deadenylating enzymes and are thought to be sites of mRNA decay [43,44].

After nuclear synthesis, miRNAs follow identical cytoplasmic processing steps. Prior to nuclear export, miRNAs are cleaved from primary transcripts (pri-miRNAs) that are predominantly synthesized by RNA polymerase II [45,46]. Nuclear cleavage is carried out by another RNase III enzyme, Drosha [47]. Drosha recognizes and excises hairpin-shaped miRNA precursors (pre-miRNA) from pri-miRNAs. The pre-miRNAs are exported to the cytoplasm by Exportin 5 [48] after which final maturation by Dicer and association with RISC occurs [49].

### 9.3.1   ZEBRA FISH RNAI COMPONENTS

RNAi component composition is highly variable between phyla. For example, *D. melanogaster* has 2 Dicers and 4 Argonaute homologues, *C. elegans* has a single Dicer and 27 Argonautes, and humans encode 1 Drosha, 1 Dicer, and 4 Argonautes. Zebra fish contain two genes possessing RNAse III domains, eight genes with PAZ domains corresponding to one Drosha, one Dicer, five Argonautes, and two Piwi-like genes (Table 9.1). There are clear homologues of each of the human Argonautes, AGO1, AGO2, AGO3, and AGO4 (Table 9.1). The fifth zebra fish Argonaute is most similar to AGO3. Interestingly, the Ago1 and Ago3-2 genomic loci are juxtaposed, similar to that seen in the human genome with AGO1 and AGO3. Beyond the core Argonautes, the zebra fish genome encodes two piwi-like (piwil) proteins called Piwil1 and Piwil2. Zebra fish Piwil1 is most similar to human Piwil1, but it is also similar to human Piwil3 and Piwil4. Zebra fish piwil2 is similar to human Piwil2. Piwi proteins do not appear to be associated strictly with the RNAi pathway with apparent roles in mitotic spindle formation and also association with unique small RNAs expressed in testis called piwiRNAs (piRNAs) [50,51].

### 9.3.2   CONSERVATION OF RNAI COMPONENTS

The protein factors of the RNAi pathway possess highly conserved structural domains. One of these is the RNAse III domain that is responsible for the creation of small RNAs found in the enzymes Drosha and Dicer [52]. Bacterial RNAse III

**TABLE 9.1**

**Genomic Location and Conservation of Zebra Fish RNAi Components and Related Factors**

| ensembl ID | Gene Name | Genomic Location | Percentage Identity to Human Homologue |
|---|---|---|---|
| ENSDARG00000001129 | dicer | chr17 27.89m | 50 |
| ENSDARG00000055563 | drosha | chr12 22.48m | 71 |
| ENSDARG00000059882 | ago1 | chr16 60.61m | 92 |
| ENSDARG00000061268 | ago2 | chr19 41.70m | 83 |
| ENSDARG00000063079 | ago3-1 | chr19 9.68m | 94 |
| ENSDARG00000059888 | ago3-2 | chr16 60.53m | 81 |
| ENSDARG00000015279 | ago4 | chr19 49.36m | 96 |
| ENSDARG00000041699 | piwil1 | chr8 52.93m | 64 |
| ENSDARG00000062601 | piwil2 | chr5 11.34m | 51 |

encodes a single monomeric RNAse III domain, but catalysis requires dimer formation [53]. Eukaryotic Dicer possesses two tandem RNAse III domains allowing catalysis to be carried out by a single protein [52]. Through mutagenesis studies, three glutamic acid residues and one aspartic acid residue have been directly implicated in catalysis [54]. These amino acids coordinate two metal ions, either $Mg^{2+}$ or $Mn^{2+}$. Alignment of the zebra fish homologues of human Drosha and Dicer reveals conservation of the general domain order and critical catalytic residues. The zebra fish Dicer homologue shows the greatest sequence similarity to human Dicer in the PAZ, RNAse III, and dsRNA-binding domains. Similarly, the critical catalytic residues of the RNAse III domain are conserved. Similar to Dicer, human Drosha and zebra fish Drosha show strong conservation of domain structure with greatest similarity between the RNAse III and dsRNA-binding domains. Critical residues responsible for enacting the RNAse III activity of Drosha are also well conserved.

While RNAse III domain-containing factors are responsible for the creation of the small RNAs that participate in the RNAi pathway, Argonaute proteins, which are characterized by the presence of PAZ and Piwi domains, are the effector molecules [55]. Argonautes form the core of the RISC, directly binding mature small RNAs and, in the case of Ago2, providing enzymatic slicer activity [56]. PAZ domains are single-stranded nucleic-acid-binding motifs that associate specifically with the 2-nucleotide 3' overhangs characteristic of Dicer cleavage products [57]. Analysis of zebra fish Argonaute homologues, Ago1, Ago2, Ago3-1, Ago3-2, and Ago4, reveals considerable conservation compared to human homologues, whereas Piwil1 and Piwil2 exhibit little homology to any of the core Argonautes. Sequence analysis also shows that each zebra fish Argonaute is more similar to its human homologue than to other zebra fish paralogs. This indicates that each Argonaute evolved before the fishes split from tetrapods. The second copy of Ago3 (Ago3-1) seems to be unique to zebra fish and may encode a protein of questionable functionality, as the Piwi domain of this factor seems to be truncated. Additionally, the zebra fish homologue of Ago2 exhibits separate 20 and 10 amino acid deletions in the PAZ domain (Figure 9.3a). Zebra fish Ago2 also has a 158 amino acid insertion in the center of its Piwi domain and a 29 amino acid expansion in sequence that lies between the PAZ and Piwi domains. Despite these differences, it seems likely that the basic function of each Argonaute is well conserved. Ago1, Ago3, and Ago4 show the greatest similarity to their human counterparts, whereas Ago2 shows the greatest divergence. Critical aspartic acid, glutamine, and histidine residues are conserved in the Piwi domain from zebra fish Ago2 that would suggest slicer capability (Figure 9.3b) [56], but several critical residues are missing from the PAZ domain that are likely to abrogate small RNA binding. Overall, Ago1, Ago3-1, Ago3-2, and Ago4 are likely to participate in miRNA function similar to their human counterparts. It is unclear, however, if zebra fish Ago2 shares the slicer activity of human AGO2.

Given the previous evolutionary conservation, the basic role of RNAi is likely shared between zebra fish and humans with the possible exception of Ago2. This suggests that the main function of the RNAi pathway in zebra fish, similar to mammals, is devoted to the creation and action of miRNAs. miRNA:mRNA pairing does not typically result in slicer-mediated degradation of targeted mRNAs with the exception of *miR-196*-induced cleavage of *Hoxb8* [58]. Rare events such as *Hoxb8* cleavage may have constrained mammalian Ago2 divergence. It can only be assumed that such regulatory events may be less crucial to zebra fish development, allowing zebra fish Ago2 to drift. Nevertheless, the RNAi pathway and the role of miRNAs appear quite similar between humans and fish, suggesting that major questions best approached in zebra fish will likely translate accurately to mammalian systems and ultimately humans.

## 9.4 ZEBRA FISH DICER MUTANTS

### 9.4.1 RECOVERING A DICER1 NULL ALLELE

Analysis of the role of RNAi during zebra fish development has been facilitated by the creation of strains carrying mutations in *dicer1*. These mutations were isolated using TILLING methodology [59]. TILLING is an acronym for targeting induced local lesions in genomes, a PCR-based strategy that involves PCR amplification of genomic sequence to identify point mutations. For *dicer1*, three mutations were uncovered that produce premature stop codons in the *dicer1* open reading frame. Embryos homozygous for any of the mutations exhibit a growth arrest phenotype (Figure 9.4b). Most embryos do not survive for more than 2 weeks, show lethargic behavior, and fail to grow at normal rates beginning after the first week. In contrast to mutants identified by TILLING, injection of morpholinos complementary to *dicer1* resulted in earlier phenotypic consequences [59]. These results reflect the essential nature of Dicer1 function with survival up to 2 weeks, reflecting contributions from maternal sources. Morpholino-mediated inhibition of *dicer1* targets mRNA and is able to downregulate not only *dicer1* transcripts from zygotic sources but also maternally contributed message. Survival in morpholino-injected embryos can be attributed to the presence of maternal protein or to the incomplete elimination of *dicer1* expression.

a. PAZ domain: residues critical for small RNA binding

```
Ago2  VKFT KEI K----------          VLT R-------      QENGQTIECTVAQYF  KDKYKLVL RYPHLPCLQ
Ago3-1 VKFT KEI KGLKVEVTHCGTMRRKYRVCNVT RRPASHQTFPLQLENGQTVE   RTVAQYF  REKYNLQL KYPHLPCLQ
Ago3-2 VKFT KEI KGLKVEVTHCGTMRRKYRVCNVT RRPASHQTFPLQLENGQTVE   RTVAQYF  REKYNLQL KYPHLPCLQ
Ago4  VKFT KEI RGLKVEVTHCGQMKRKYRVCNVT RRPASHQTFPLQLENGQAMECTVAQYF  KQKYSLQL KYPHLPCLQ
Ago1  VRFT KEI KGLKVEVTHCGQMKRKYRVCNVT RRPASHQTFPLQLESGQTVECTVAQYE  KQKYNLQL KYPHLPCLQ
```

b. Piwi domain: residues critical for slicer activity

```
Ago2  VQVKNVQKTTPQTLSNLCLKINVKLGGVNNILLPQGRPLVFQQPVIFLGAD  ...  ••CATVRVQQHR••  ••IIYY RDGISE ••  ••AYYAHIVAF R
Ago3-1 VQVKNVVKTSPQTLSNLCLKINVKLGGINNILVPHQRPSVFQQPVIFLGAD  ...  ••CATVRVQRPR••  ••IIFY RDGVSE ••  ••AYYAHIVAF R
Ago3-2 VQVKNVVKTSPQTLSNLCLKINVKLGGINNILVPHQRPSVFQQPVIFLGAD  ...  ••CATVRVQRPR••  ••IIFY RDGVSE ••  ••AYYAHIVAF R
Ago4  VQVKNVVKTSPQTLSNLCLKINVKLGGINNVLVPHQRPSVFQQPVIFLGAD  ...  ••CATVRVQTSR••  ••IIYY RDGVSE ••  ••AYYA RLVAF R
Ago1  VQVKNVVKTSPQTLSNLCLKINVKLGGINNVLVPHQRSAVFQQPVIFLGAD  ...  ••CATVRVQRPR••  ••IIFY RDGVPE ••  ••AYYA RLVAF R
```

**FIGURE 9.3** Conservation of protein sequences in the zebra fish Argonautes. Alignment of zebra fish PAZ and Piwi domains. (a) Key aromatic residues in PAZ domains that interact with small RNAs are boxed. (b) Key Piwi domain residues that coordinate metal ions required for slicer activity are boxed.

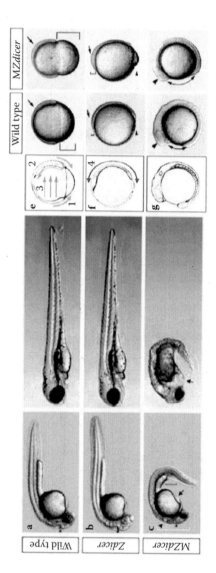

**FIGURE 9.4**  Morphogenesis defects in maternal–zygotic dicer mutants. (a) Wild type. (b) Zdicer: Zdicer mutants have no morphological defects compared to wild type at 36 and 90 hpf. (c) MZdicer mutants display morphogenesis defects. (e–g) Scheme representing cell movements (arrows) during gastrulation: (1) epiboly, (2) internalization, (3) convergence, and (4) extension—(middle) wild type; (right) MZdicer mutants. (e) 80% epiboly: arrow shows the similar extent of prechordal plate extension in wild and MZdicer mutants; bracket shows the extent of epiboly. (f) Tail bud stage. (g) 9-somite stage. (From Giraldez, A. J. et al. (2005). *Science 308,* 833–838. With permission.)

The loss of Dicer prevents miRNA maturation [59]. Consistent with this, embryos at 2 weeks postfertilization showed decreased production of mature *miR-26a* and *let-7*. Detection of *miR-26a* by Northern blotting also revealed accumulation of pre-miRNA intermediates. In addition to demonstrating an essential role for miRNAs in zebra fish development, these results indicate that zebra fish *dicer1* is not redundant and is required for maturation of miRNAs in zebra fish.

The *dicer1* null embryos exhibit a general growth arrest phenotype that is a result of attenuated miRNA expression, but the exact physiological defects that result in this phenotype remain to be determined. miRNAs are typically highly tissue specific, and if loss of specific miRNAs in multiple tissues leads to cell death, it would be expected that these tissues would exhibit greater deficiencies than others. This is not observed in *dicer1* mutant embryos implying that either different miRNAs are needed in all cells or a specific miRNA or a subset of miRNAs is required to the same degree in all cells. Requirement of a singular cellular process that is necessary to the same extent in all cell types seems unlikely. Even processes as intrinsic as metabolism would likely affect faster dividing cells more dramatically. A general requirement for miRNAs in normal development was established through analysis of dicer mutants, but understanding a requirement for specific miRNAs during zebra fish development remains.

### 9.4.2  ELIMINATING ALL DICER FUNCTION IN ZEBRA FISH EMBRYOS

To address the issue of maternal contribution of *dicer1* and to dissect the requirement for miRNA function during the earliest stages of embryogenesis, the Schier group generated a maternal–zygotic *dicer* (*MZdicer*) mutant [60] (Figure 9.4c). Mutant phenotypes that are embryonic lethal can be rescued by supplying wild-type message, typically through microinjection. Examples of mutants that can be rescued are *one-eyed pinhead* (EGF-CFC) or *swirl* (Bmp2) [61,62]. Rescued embryos that do not have an essential requirement for the specific gene during later development can be grown to adulthood, but these individuals still retain mutations and pass them on to progeny. Due to the lack of functional gene loci in the female's germ line, embryos laid by a rescued zebra fish female receive no maternal contribution of that gene.

Maternal–zygotic mutants can show drastically different phenotypes from conventional null mutants, which is well illustrated by the example of *one-eyed pinhead* (*oep*) [63]. Zygotic *oep* mutants have midline defects resulting in mild cyclopia. MZ*oep*, on the other hand, have a much more dramatic phenotype, lacking mesendoderm entirely. The same dramatic phenotypic difference was obtained when MZ*dicer1* embryos were created [60]. The approach required to generate MZ*dicer1* mutants varied significantly from the creation of MZ*oep*. Dicer function is required throughout zebra fish development and into adulthood. *oep* is only essential for mesoderm induction and dorsal–ventral patterning, both early events. Rescuing *dicer1* mutant defects would only be temporary due to the transient nature of gene expression delivered by mRNA injection. To surmount this obstacle, the Schier group employed a specialized technique called *germ-line replacement* [64]. Zebra fish germ-line cells can be visualized with a GFP marker and transplanted into wild-type embryos. The germ-line cells in the wild-type embryos are eliminated via injection of a morpholino that targets *miles apart*, a gene required for germ-line cell formation. The resulting chimeric embryo lacks the ability to contribute functional gene products to offspring, both from maternal sources or through inheritance of nonmutant loci.

*MZdicer1* embryos exhibit a much more severe phenotype compared to zygotic dicer mutants due to a complete lack of mature miRNA production [60]. Development in *MZdicer1* embryos is severely perturbed. By the end of embryogenesis, most organ systems have failed to form, resulting in embryonic lethality. *MZdicer1* embryos are able to develop a variety of structures such as eyes, heart, brain, and notochord, but, in all cases, the tissues are dysmorphic. Morphogenesis defects and developmental delay are observable from early gastrulation stages. *MZdicer1* embryos fail to coordinate gastrulation movements; specifically, internalization of mesendoderm proceeds more rapidly relative to epiboly movements resulting in precocious placement of the prechordal plate, while envelopment of yolk lags. The inability to fully extend the embryonic axis results in a failure to complete yolk plug closure and, ultimately, an accumulation of cells in the forming head. Neurulation is particularly disrupted in *MZdicer1* embryos. The neural plate forms and develops into the neural rod but fails to form the neurocoel, an evacuation that occurs along the entire anterior–posterior axis forming ventricles in the brain and the central canal in the spinal cord. Many neuronal projections were found to be misplaced or in a state of degeneration. Accompanying this, embryos exhibited little capacity to respond to physical stimuli, consistent with a dysfunctional nervous system. In addition to defects in nervous system development, *MZdicer1* embryos exhibit deficiencies in many peripheral tissues. *MZdicer1* embryos develop U-shaped somites, blood circulation is disrupted, and they have abnormal hearts. However, despite these dramatic effects on morphogenesis, many cell types are surprisingly appropriately specified. Dorsal–ventral patterning and regionalization remain intact. Marker analysis of neuronal compartments revealed that despite defects in neural development, patterning is largely unaffected. Similarly, although many organs showed abnormal development, marker analysis demonstrated that specification had occurred appropriately.

### 9.4.3 IMPLICATIONS OF THE DICER NULL PHENOTYPE

Some of the most illuminating insights concerning the role of miRNAs in zebra fish development afforded by analysis of *MZdicer1* embryos are not the finding of which tissues are perturbed, but rather those that are unaffected. In *MZdicer1* embryos, axis formation and patterning of organ system primoridia is relatively intact. Additionally, germ-line cells develop without Dicer1 function. The lack of a requirement for miRNAs during these events indicates that miRNAs are not core components of many major early signaling pathways. Dorsal–ventral patterning requires signals from Wnt, Fgf, and Bmp pathways [65]. These are master regulatory signaling pathways that participate in numerous morphogenic events in both embryos and adult animals. The importance of these pathways in animal physiology cannot be understated, mutations in these pathways typically yield lethal birth defects or diseases such as cancer. The absence of a role for miRNAs in these pathways seems to indicate that miRNA function appeared after these pathways evolved or that the function of a specific miRNA is redundant or divergent, perhaps affecting pathways in other species. While loss of miRNAs does not dramatically affect these pathways in zebra fish, it does not eliminate the possibility that miRNAs may modulate these signaling pathways, or be important agents of regulation in adult tissues.

Comparison of *MZdicer1* zebra fish mutants to other Dicer null animals reveals widely diverging functions of miRNAs. Elimination of Dicer function in mice by homologous recombination resulted in extremely early embryonic lethality [66]. Embryos fail to develop embryonic stem cells and do not progress into gastrulation stages. Similarly, *Drosophila dicer1* mutants display early defects such as ventralization [67]. It is unclear that how these discrepancies have arisen. miRNA-mediated regulation may be more susceptible to the forces of evolution than protein-coding genes. One reason behind this could be the targeting of 3' UTR sequences by miRNAs. UTR sequences show poor conservation, and a change in sequence could disrupt miRNA binding.

Another interesting aspect of the *MZdicer1* analysis is the viability of dicer-/-germ cells. Fish that possess a germ line comprised of Dicer mutant cells are fertile. This indicates that miRNAs are not required for maintenance of germ cells at least for metazoans. Both *C. elegans* and *Drosophila* Dicer mutants display defects in germ cell function [68,69]. *C. elegans* mutants are sterile, displaying both a burst vulva phenotype and improper maturation of oocytes. Fly *dicer1* mutants are embryonic lethal; however, generation of mosaic Dicer mutants revealed that Dicer is required for proper oocyte formation. While lower organisms have a requirement for Dicer function in germ cell formation, vertebrates apparently do not. Creation of the *dicer1* knockout mouse required generation of Dicer null embryonic stem cells that are viable, suggesting that miRNA function is not absolutely required for maintenance of ES cell pluripotency. Following the typical procedure of generating knockout mice, chimeric individuals were made that possessed dicer-deficient germ cells. These mice were successfully bred to generate heterozygotes. Thus, the requirement for Dicer function between different species reveals a dramatic divergence. While the Dicer null phenotypes can all be in part attributed to attenuation of miRNA function, it is unclear whether Dicer participates in other cellular process and whether some of the observed defects are a result of the loss of an unknown role for Dicer.

## 9.5   ZEBRA FISH MiRNAS

### 9.5.1   DICER MUTANTS REVEAL A ROLE FOR THE *MIR-430* FAMILY

The analysis of *dicer1* mutants revealed a general requirement for miRNAs in zebra fish development but did not describe a role for specific miRNAs. To examine specific miRNAs, direct cloning of miRNAs was performed during early development revealing a highly expressed group of miRNAs, the *miR-430* family [60]. These miRNAs begin to be expressed in stages immediately preceding gastrulation. Surprisingly, supplying individual mature members of this family through microinjection into *MZdicer1* embryos led to rescue of many of the defects observed in these embryos, including restoration of ventricle formation. The *miR-430* family is unique both in genomic organization and conservation between species. Family members are encoded in a repetitive manner from a single locus duplicated over 90 times. This locus seems to be unique to fish species being found in the genomes of *Fugu rubripes* and *Tetraodon nigroviridis*, but not other vertebrates. Other vertebrates encode *miR-430* family members or miRNAs with similar sequence, but these miRNAs are not encoded in such a highly repetitive manner, indicating a potentially unique role for *miR-430* in fish.

The *miR-430* family members target maternally contributed mRNAs [70]. The inability to fully clear these transcripts yields many of the defects associated with *MZdicer1* embryos. The composition of mRNAs expressed in embryos revealed inappropriate accumulation of maternally contributed mRNAs. Normally, these messages are targeted for degradation but are stabilized in the absence of *miR-430* due to a lack of the poly(A) tail shortening. mRNAs targeted by *miR-430* family members have decreased poly(A) tail lengths. Such destabilization of mRNAs by miRNA targeting is also seen in mammalian cells [71]. Mechanistically, a decrease of mRNA steady-state levels is thought to occur through localization of mRNAs to processing bodies.

*miR-430* is known to be expressed broadly in early zebra fish embryos. Injection of reporter mRNAs that bear sequence complementary to *miR-430* results in down-regulation of the reporter throughout the entire animal. However, not all predicted targets of *miR-430* behave in this manner. Nanos mRNA, a determinant of primordial germ cell (PGC) fate, is a target of *miR-430* but is not downregulated in PGCs [72]. The mechanism that is responsible for this behavior is not understood, and highlights the sophisticated nature of miRNA-mediated gene regulation.

### 9.5.2 MIRNA DIVERSITY IN ZEBRA FISH

As of this writing, there are 337 zebra fish miRNAs currently listed in the miRNA registry [73]. This number is derived from sequence comparisons between species and efforts to directly clone small RNAs [74,75]. Cloning of miRNAs in zebra fish revealed that very few miRNAs are expressed during early stages, but the number of miRNAs expressed increases dramatically as development progresses [76]. Initial miRNA-cloning strategies also identified 250 repeat-associated small interfering RNAs (rasiRNAs).

miRNA expression analysis is particularly amenable to microarray technology. Small RNAs can be isolated and hybridized to oligonucleotide arrays, and such experiments have confirmed the finding that the number of different miRNA expressed increases with development [77–80]. We have used an oligonucleotide array consisting of probes complementary to vertebrate miRNAs to examine miRNA expression at distinct developmental time points (Table 9.2). In contrast to previous experiments, we detected a noticeable burst in miRNA expression at the sphere stage, which coincides with the mid-blastula transition signaling the beginning of zygotic transcription [81]. This time also corresponds to high expression levels of *miR-430* and the silencing and/or degradation of maternal RNAs [70].

While it is essential toward understanding the function of miRNAs to first know their temporal expression patterns, one of the most useful breakthroughs offered by zebra fish is in situ detection of miRNA expression. The Plasterk group developed miRNA in situ technology based on LNA (locked nucleic acid) probes [80]. LNA is a

## TABLE 9.2
## Zebra Fish miRNA Expression Profiles

| microRNA | Temporal Expression | In situ Localization |
|---|---|---|
| miR-1 | Sp, 1D-5D | Body, head, and fin muscles |
| miR-133a | 256, TB-5D | Body, head, and fin muscles |
| miR-214 | Sh-5D | Ubiquitous |
| miR-124a | 1D-5D | Differentiated cells of brain, spinal cord and eyes, cranial ganglia |
| miR-145 | 1D, 2D, 4D, 5D | Gut and gall bladder, gills, swim bladder, branchial arches, fins, outflow tract of the heart, and ear |
| miR-181a | 1D-5D | Brain (tectum, telencephalon), eyes, thymic primordium, gills |
| miR-215 | Sp, 1D-5D | Gut and gall bladder |
| let-7a | 4D-5D | Brain, spinal cord |
| let-7b | Sp, TB-2D | Brain, spinal cord |
| miR-125a | 1D, 2D, 4D, 5D | Brain, spinal cord, cranial ganglia |
| miR-125b | 18S-5D | Brain, spinal cord, cranial ganglia |
| miR-200b | Sp, 1D-5D | Nose epithelium, lateral line organs, epidermis, gut (proctodeum), taste buds |
| miR-218 | 1D-4D | Brain (neurons and/or cranial nerves/ganglia in hindbrain), spinal cord |
| miR-34a | 4D-5D | Brain (cerebellum), neurons in spinal cord |
| miR-375 | 4D-5D | Pituitary gland, pancreatic islet |
| miR-99a | 256-Sp, 6S-5D | Brain (hindbrain, diencephalon), spinal cord |
| miR-100 | Sp, 1D-5D | Brain (hindbrain, diencephalon), spinal cord |
| miR-103 | 1D-5D | Brain, spinal cord |
| miR-107 | Sp, 1D-5D | Brain, spinal cord |
| miR-140* | 3D-5D | Cartilage of pharyngeal arches, head skeleton, and fins |
| miR-184 | 18S, 2D-5D | lens, hatching gland in early stages |

*Note:* A subset of zebra fish miRNAs are shown with temporal and spatial expression patterns. Temporal expression patterns were obtained through microarray experiments that spanned 12 stages of development from 256 cells through 5 dpf. In situ localizations were obtained using LNA probes. (From Wienholds et al. (2005). *Science 309*, 310–311.) Developmental stages: 256 cells, Sp: sphere, Sh: shield, TB: tail bud, 6S: 6-somite, 1D-5D: 1dpf-5dpf.

high-affinity RNA analog with a bicyclic furanose unit locked in a sugar-mimicking conformation [82]. Tissue-specific expression of many zebra fish miRNAs has been revealed using LNA probes [80]. Table 9.2 indicates the in situ expression profile of several miRNAs along with their stages of expression as determined by microarrays.

### 9.5.3 STRATEGIES FOR DETERMINING MIRNA FUNCTION IN ZEBRA FISH

Essential components of phenotypic analysis are loss- and gain-of-function experiments. While use of the MZdicer mutant provided great insight into the role of *miR-430*, it will be difficult to use this mutant to dissect the function of other specific miRNAs. Mutational approaches also present extreme challenges to understanding individual miRNA function due to the small size of this gene family. Currently, antisense technology is a standard methodology for inhibiting miRNA expression. Pairing between miRNAs and antisense oligonucleotides can effectively ablate miRNA function, especially using modified, stable antisense oligonucleotides [27,83]. The use of antisense oligonucelotides called *antagomirs* has been used in mice to inhibit miRNA expression [84]. This technique proved quite effective, resulting in complete elimination of miRNA expression. For zebra fish, antisense morpholino oligonucleotides can be used to block miRNA function (Figure 9.5).

Combining the loss of miRNA function using morpholino technology and gain of function by injection of synthetic miRNAs permits precise dissection of miRNA function. Injection of morpholinos complementary to specific miRNAs has been shown to result in potent inhibition of miRNA activity that persists for approximately 7 days. This strategy has been used to determine the role of *miR-214* in somitogenesis and the role of the *miR-375* family in pancreatic development [85,86]. Injection of morpholinos targeting *miR-214* resulted in embryos that developed "U-shaped" somites, a phenotype indicating perturbation of hedgehog (Hh) signaling [85]. To complement this analysis, injection of *miR-214* siRNAs resulted in changes in Hh-mediated cell fate in the spinal cord. Further analysis showed that *miR-214* regulates an intercellular effector of Hh signaling, *suppressor of fused su(fu)*, ensuring appropriate response to Hh stimulation in mesodermal cells. The identification of *su(fu)* as a target of *miR-214* was demonstrated with reporter constructs bearing the *su(fu)* UTR target sequence and by simultaneous inhibition of both *su(fu)* and *miR-214*. Analysis of *miR-375* showed that morpholinos can be designed to interfere with miRNA biogenesis, in addition to prevent pairing between mature miRNAs and target mRNAs [86]. The ability of morpholinos to block miRNA biogenesis can be demonstrated by Northern blotting and in situ hybridization. Morpholinos have also been shown to be an effective method for preventing specific miRNA/mRNA interactions [87]. This strategy employed morpholinos complementary to the microRNA recognition elements (MRE) of *miR-430* in the 3' UTRs of *squint* and *lefty*. This strategy permitted analysis of particular targets of *miR-430* without relying on *MZdicer* mutants, which have dramatic changes in maternal message expression that mask the contribution of specific regulatory interactions. The fact that zebra fish are amenable to antisense technology, specifically in the form of morpholinos, represents a powerful opportunity to determine miRNA function in vertebrates.

## 9.6  CONCLUSIONS

The zebra fish as a model system holds tremendous promise to elucidate the function of specific miRNAs during development. Due to imprecise pairing, algorithms designed to predict mRNA targets for specific miRNAs are limited, and experimental validation is required. The ability to analyze phenotypic consequences due to gain or loss of expression of specific miRNAs in a developmental context provides a powerful system to carry out such validation. Beyond miRNAs, introduction of

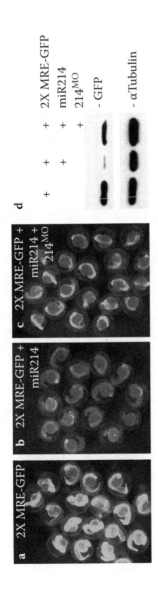

**FIGURE 9.5** Morpholino-mediated inhibition of *miR-214*. (a) Injection of zebra fish embryos with synthetic mRNA encoding GFP fused to a 3′ UTR containing tandem *miR-214* recognition elements (2XMRE). (b) Coinjection of *miR-214* with the GFP reporter mRNA as in (a). (c) Coinjection of morpholinos targeted (214^MO) to *miR-214* and *miR-214* synthetic RNAs with the GFP reporter mRNA. (d) Western blots of embryos as in a–c with antibodies against GFP or α-tubulin.

vectors expressing shRNAs, especially those controlled by tissue-specific or inducible promoters, will further increase the utility of zebra fish as a model system to study vertebrate gene function.

## REFERENCES

1. E. Bernstein, A. M. Denli, and G. J. Hannon. (2001). The rest is silence. *RNA 7*, 1509–1521.
2. D. Bartel. (2004). MicroRNAs: Genomics, biogenesis, mechanism, and function. *Cell 116*, 281–297.
3. C. B. Kimmel, W. W. Ballard, S. R. Kimmel, B. Ullmann, and T. F. Schilling. (1995). Stages of embryonic development of the zebrafish. *Dev Dyn 203*, 253–310.
4. P. Francisco. (2003). Maternal factors in zebrafish development. *Dev Dyn 228*, 535–554.
5. C. Erter, L. Solnica-Krezel, and C. V. E. Wright. (1998). Zebrafish nodal-related 2 Encodes an early mesendodermal inducer signaling from the extraembryonic yolk syncytial layer. *Dev Biol 204*, 36100–36372.
6. D. Myers, D. Sepich, and L. Solnica-Krezel. (2002). Convergence and extension in vertebrate gastrulae: Cell movements according to or in search of identity? *Trends Genet 18*, 447–455.
7. L. Solnica-Krezel, A. F. Schier, and W. Driever. (1994). Efficient Recovery of ENU-induced mutations from the zebrafish germline. *Genetics 136*, 1401–1420.
8. W. Driever, L. Solnica-Krezel, A. F. Schier, S. C. Neuhauss, J. Malicki, D. L. Stemple, D. Y. Stainier, F. Zwartkruis, S. Abdelilah, Z. Rangini, J. Belak, and C. Boggs. (1996). A genetic screen for mutations affecting embryogenesis in zebrafish. *Development 123*, 37–46.
9. P. Culp, C. Nusslein-Volhard, and N. Hopkins. (1991). High-frequency germ-line transmission of plasmid DNA sequences injected into fertilized zebrafish eggs. *PNAS 88*, 7953–7957.
10. A. Nasevicius and S. C. Ekker. (2000). Effective targeted gene "knockdown" in zebrafish. *Nat Genet 26*, 216–220.
11. G. Schmajuk, H. Sierakowska, and R. Kole. (1999). Antisense oligonucleotides with different backbones. *J Biol Chem 274*, 21783–21789.
12. E. Wienholds, F. van Eeden, M. Kosters, J. Mudde, R. H. Plasterk, and E. Cuppen. (2003). Efficient target-selected mutagenesis in zebrafish. *Genome Res 13*, 2700–7.
13. A. Fire, SiQun Xu, M. K. Montgomery, S. A. Kostas, S. E. Driver, and C. C. Mello. (1998). Potent and specific genetic interference by double-stranded RNA in *Caenorahbditis elegans*. *Nature 391*, 806–811.
14. H. Tabara, M. Sarkissian, W. G. Kelly, J. Fleenor, A. Grishok, L. Timmons, A. Fire, and C. C. Mello. (1999). The rde-1 gene, RNA interference, and transposon silencing in *C. elegans*. *Cell 99*, 123.
15. S. Hammond, A. Caudy, and G. J. Hannon. (2001). Post-transcriptional gene silencing by double-stranded RNA. *Nat Rev Genet 2*, 110–119.
16. D. S. Aaronson and C. M. Horvath. (2002). A road map for those who don't know JAK-STAT. *Science 296*, 1653–5.
17. A. A. Aravin, M. Lagos-Quintana, A. Yalcin, M. Zavolan, D. Marks, B. Snyder, T. Gaasterland, J. Meyer, and T. Tuschl. (2003). The small RNA profile during *Drosophila melanogaster* development. *Dev Cell 5*, 337.
18. S. Wu and R. J. Kaufman. (1997). A model for the double-stranded RNA (dsRNA)-dependent dimerization and activation of the dsRNA-activated protein kinase PKR. *J Biol Chem 272*, 1291–1296.

19. J. Castelli, B. Hassel, Z. Maran, J. Paranjape, J. Hewitt, X. Li, Y. Hsu, R. Silverman, and R. Youle. (1998). The role of 2′-5′ oligoadenylate-activated ribonuclease L in apoptosis. *Cell Death Differ 5*, 313–320.
20. K. Fukuda, T. Tsujita, M. Matsumoto, T. Seya, H. Sakiyama, F. Nishikawa, S. Nishikawa, and T. Hasegawa. (2006). Analysis of the interaction between human TLR3 ectodomain and nucleic acids. *Nucleic Acids Symp Ser 50*, 249–250.
21. D. P. Bartel and C. Z. Chen. (2004). Micromanagers of gene expression: the potentially widespread influence of metazoan microRNAs. *Nat Rev Genet 5*, 396–400.
22. A. Wargelius, S. Ellingsen, and A. Fjose. (1999). Double-stranded RNA induces specific developmental defects in zebrafish embryos. *Biochem Biophys Res Commun 263*, 156.
23. Y.-X. Li, M. J. Farrell, R. Liu, N. Mohanty, and M. L. Kirby. (2000). Double-stranded RNA injection produces null phenotypes in zebrafish. *Dev Biol 217*, 394.
24. A. C. Oates, A. E. E. Bruce, and R. K. Ho. (2000). Too much interference: Injection of double-stranded RNA has nonspecific effects in the zebrafish embryo. *Dev Biol 224*, 20.
25. Z. Zhao, Y. Cao, M. Li, and A. Meng. (2001). Double-stranded RNA injection produces nonspecific defects in zebrafish. *Dev Biol 229*, 215.
26. P. Stein, F. Zeng, H. Pan, and R. M. Schultz. (2005). Absence of non-specific effects of RNA interference triggered by long double-stranded RNA in mouse oocytes. *Dev Biol 286*, 464.
27. W. P. Kloosterman, E. Wienholds, R. F. Ketting, and R. H. A. Plasterk. (2004). Substrate requirements for let-7 function in the developing zebrafish embryo. *Nucleic Acids Res 32*, 6284–6291.
28. A. Piccin, A. Salameh, C. Benna, F. Sandrelli, G. Mazzotta, M. Zordan, E. Rosato, C. P. Kyriacou, and R. Costa. (2001). Efficient and heritable functional knock-out of an adult phenotype in *Drosophila* using a GAL4-driven hairpin RNA incorporating a heterologous spacer. *Nucleic Acids Res 29*, e55–5.
29. N. Perrimon. (1998). New advances in Drosophila provide opportunities to study gene functions. *PNAS 95*, 9716–9717.
30. A. Inbal, J. Topczewski, and L. Solnica-Krezel. (2006). Targeted gene expression in the zebrafish prechordal plate. *Genesis 44*, 584–588.
31. M. T. Hemann, J. S. Fridman, J. T. Zilfou, E. Hernando, P. J. Paddison, C. Cordon-Cardo, G. J. Hannon, and S. W. Lowe. (2003). An epi-allelic series of p53 hypomorphs created by stable RNAi produces distinct tumor phenotypes in vivo. *Nat Genet 33*, 396–400.
32. E. Bernstein, A. Caudy, S. Hammond, and G. Hannon. (2001). Role for a bidentate ribonuclease in the initiation step of RNA interference. *Nature 409*, 363–366.
33. P. Zamore, T. Tuschl, P. Sharp, and D. Bartel. (2000). RNAi: Double-stranded RNA directs the ATP-dependent cleavage of mRNA at 21 to 23 nucleotide intervals. *Cell 101*, 25–33.
34. S. Hammond, S. Boettcher, A. Caudy, R. Kobayashi, and G. Hannon. (2001). Argonaute2, a link between genetic and biochemical analyses of RNAi. *Science 293*, 1146–1150.
35. M. A. Carmell, Z. Xuan, M. Q. Zhang, and G. J. Hannon. (2002). The Argonaute family: Tentacles that reach into RNAi, developmental control, stem cell maintenance, and tumorigenesis. *Genes Dev 16*, 2733–2742.
36. Y. Zeng, R. Yi, and B. R. Cullen. (2003). MicroRNAs and small interfering RNAs can inhibit mRNA expression by similar mechanisms. *PNAS 100*, 9779–9784.
37. P. H. Olsen and V. Ambros. (1999). The lin-4 regulatory RNA controls developmental timing in *Caenorhabditis elegans* by blocking LIN-14 protein synthesis after the initiation of translation. *Dev Biol 216*, 671–680.
38. E. C. Lai. (2002). Micro RNAs are complementary to 3′ UTR sequence motifs that mediate negative post-transcriptional regulation. *Nat Genet 30*, 363–364.

39. J. G. Doench, C. P. Petersen, and P. A. Sharp. (2003). siRNAs can function as miRNAs. *Genes Dev 17*, 438–442.
40. B. P. Lewis, I. H. Shih, M. W. Jones-Rhoades, D. P. Bartel, and C. B. Burge. (2003). Prediction of mammalian microRNA targets. *Cell 115*, 787–798.
41. J. Liu, M. A. Carmell, F. V. Rivas, C. G. Marsden, J. M. Thomson, J. J. Song, S. M. Hammond, L. Joshua-Tor, and G. J. Hannon. (2004). Argonaute2 is the catalytic engine of mammalian RNAi. *Science 305*, 1437–1441.
42. J. Liu, M. A. Valencia-Sanchez, G. J. Hannon, and R. Parker. (2005). MicroRNA-dependent localization of targeted mRNAs to mammalian P-bodies. *Nat Cell Biol 7*, 719.
43. G. L. Sen and H. M. Blau. (2005). Argonaute 2/RISC resides in sites of mammalian mRNA decay known as cytoplasmic bodies. *Nat Cell Biol 7*, 633.
44. J. A. N. Rehwinkel, I. Behm-Ansmant, D. Gatfield, and E. Izaurralde. (2005). A crucial role for GW182 and the DCP1:DCP2 decapping complex in miRNA-mediated gene silencing. *RNA 11*, 1640–1647.
45. Y. Lee, K. Jeon, J. T. Lee, S. Kim, and V. N. Kim. (2002). MicroRNA maturation: Stepwise processing and subcellular localization. *EMBO J 21*, 4663–4670.
46. Y. Lee, M. Kim, J. Han, K. H. Yeom, S. Lee, S. H. Baek, and V. N. Kim. (2004). MicroRNA genes are transcribed by RNA polymerase II. *EMBO J 23*, 4051–4060.
47. Y. Lee, C. Ahn, J. Han, H. Choi, J. Kim, J. Yim, J. Lee, P. Provost, O. Radmark, S. Kim, and V. N. Kim. (2003). The nuclear RNase III Drosha initiates microRNA processing. *Nature 425*, 415–419.
48. E. Lund, S. Guttinger, A. Calado, J. E. Dahlberg, and U. Kutay (2004). Nuclear export of microRNA precursors. *Science 303*, 95–98.
49. K. Ikeda, M. Satoh, K. M. Pauley, M. J. Fritzler, W. H. Reeves, and E. K. L. Chan. (2006). Detection of the argonaute protein Ago2 and microRNAs in the RNA induced silencing complex (RISC) using a monoclonal antibody. *J Immunol Methods 317*, 38.
50. J. R. Kennerdell, S. Yamaguchi, and R. W. Carthew. (2002). RNAi is activated during Drosophila oocyte maturation in a manner dependent on aubergine and spindle-E. *Genes Dev 16*, 1884–1889.
51. N. C. Lau, A. G. Seto, J. Kim, S. Kuramochi-Miyagawa, T. Nakano, D. P. Bartel, and R. E. Kingston. (2006). Characterization of the piRNA complex from rat testes. *Science 313*, 363–367.
52. I. J. MacRae, K. Zhou, F. Li, A. Repic, A. N. Brooks, W. Z. Cande, P. D. Adams, and J. A. Doudna. (2006). Structural basis for double-stranded RNA processing by Dicer. *Science 311*, 195–198.
53. W. Sun, E. Jun, and A. W. Nicholson. (2001). Intrinsic double-stranded-RNA processing activity of *Escherichia coli* ribonuclease III lacking the dsRNA-binding domain. *Biochemistry 40*, 14976–14984.
54. H. Zhang, F. A. Kolb, L. Jaskiewicz, E. Westhof, and W. Filipowicz. (2004). Single processing center models for human dicer and bacterial RNase III. *Cell 118*, 57.
55. E. J. Sontheimer and R. W. Carthew. (2004). Molecular biology: Argonaute journeys into the heart of RISC. *Science 305*, 1409–1410.
56. J.-J. Song, S. K. Smith, G. J. Hannon, and L. Joshua-Tor. (2004). Crystal structure of Argonaute and its implications for RISC slicer activity. *Science 305*, 1434–1437.
57. J.-J. Song, J. Liu, N. H. Tolia, J. Schneiderman, S. K. Smith, R. A. Martienssen, G. J. Hannon, and L. Joshua-Tor. (2003). The crystal structure of the Argonaute2 PAZ domain reveals an RNA binding motif in RNAi effector complexes. *Nat Struct Mol Biol 10*, 1026.
58. S. Yekta, I. H. Shih, and D. P. Bartel. (2004). MicroRNA-directed cleavage of HOXB8 mRNA. *Science 304*, 594–596.

59. E. Weinholds, M. J. Koudijs, F. van Eeden, E. Cuppen, and R. Plasterk. (2003). The microRNA-producing enzyme Dicer1 is essential for zebrafish development. *Nat Genetics 35*, 217–218.

60. A. J. Giraldez, R. M. Cinalli, M. E. Glasner, A. J. Enright, J. M. Thomson, S. Baskerville, S. M. Hammond, D. P. Bartel, and A. F. Schier. (2005). MicroRNAs regulate brain morphogenesis in zebrafish. *Science 308*, 833–838.

61. K. Gritsman, J. Zhang, S. Cheng, E. Heckscher, W. S. Talbot, and A. F. Schier. (1999). The EGF-CFC protein one-eyed pinhead is essential for nodal signaling. *Cell 97*, 121.

62. Y. Kishimoto, K. H. Lee, L. Zon, M. Hammerschmidt, and S. Schulte-Merker. (1997). The molecular nature of zebrafish swirl: BMP2 function is essential during early dorsoventral patterning. *Development 124*, 4457–4466.

63. K. Gritsman, J. Zhang, S. Cheng, E. Heckscher, W. S. Talbot, and A. F. Schier. (1999). The EGF-CFC protein one-eyed pinhead is essential for nodal signaling. *Cell 97*, 121–132.

64. B. Ciruna, G. Weidinger, H. Knaut, B. Thisse, C. Thisse, E. Raz, and A. F. Schier. (2002). Production of maternal-zygotic mutant zebrafish by germ-line replacement. *PNAS 99*, 14919–14924.

65. A. F. Schier and W. S. Talbot. (2005). Molecular genetics of axis formation in zebrafish. *Annu Rev Genetics 39*, 561–613.

66. E. Bernstein, S. Y. Kim, M. A. Carmell, E. P. Murchison, H. Alcorn, M. Z. Li, A. M. Mills, S. J. Elledge, K. V. Anderson, and G.J. Hannon. (2003). Dicer is essential for mouse development. *Nat Genet 35*, 215–217.

67. Y. S. Lee, K. Nakahara, J. W. Pham, K. Kim, Z. He, E. J. Sontheimer, and R. W. Carthew. (2004). Distinct roles for Drosophila Dicer-1 and Dicer-2 in the siRNA/miRNA silencing pathways. *Cell 117*, 69–81.

68. S. D. Hatfield, H. R. Shcherbata, K. A. Fischer, K. Nakahara, R. W. Carthew, and H. Ruohola-Baker. (2005). Stem cell division is regulated by the microRNA pathway. *Nature 435*, 974.

69. R. F. Ketting, S. Fischer, E. Bernstein, T. Sijen, G. J. Hannon, and R. Plasterk. (2001). Dicer functions in RNA interference and in synthesis of small RNA involved in developmental timing in *C. elegans. Genes Dev 15*, 2654–2659.

70. A. J. Giraldez, Y. Mishima, J. Rihel, R. J. Grocock, S. Van Dongen, K. Inoue, A. J. Enright, and A. F. Schier. (2006). Zebrafish MiR-430 promotes deadenylation and clearance of maternal mRNAs. *Science 312*, 75–79.

71. L. Wu, J. Fan, and J. G. Belasco. (2006). MicroRNAs direct rapid deadenylation of mRNA. *PNAS 103*, 4034–4039.

72. Y. Mishima, A. J. Giraldez, Y. Takeda, T. Fujiwara, H. Sakamoto, A. F. Schier and K. Inoue. (2006). Differential regulation of germline mRNAs in soma and germ cells by zebrafish miR-430. *Curr Biol 16*, 2135.

73. S. Griffiths-Jones. (2004). The microRNA registry. *Nucleic Acids Res 32 Database issue*, D109–11.

74. M. Lagos-Quintana, R. Rauhut, W. Lendeckel, and T. Tuschl. (2001). Identification of novel genes coding for small expressed RNAs. *Science 294*, 853–858.

75. W. P. Kloosterman, F. A. Steiner, E. Berezikov, E. de Bruijn, J. van de Belt, M. Verheul, E. Cuppen, and R. H. A. Plasterk. (2006). Cloning and expression of new microRNAs from zebrafish. *Nucleic Acids Res 34*, 2558–2569.

76. P. Y. Chen, H. Manninga, K. Slanchev, M. Chien, J. J. Russo, J. Ju, R. Sheridan, B. John, D. S. Marks, D. Gaidatzis, C. Sander, M. Zavolan, and T. Tuschl. (2005). The developmental miRNA profiles of zebrafish as determined by small RNA cloning. *Genes Dev 19*, 1288–1293.

77. T. Babak, W. Zhang, Q. Morris, B. J. Blencowe, and T. R. Hughes. (2004). Probing microRNAs with microarrays: Tissue specificity and functional inference. *RNA 10*, 1813–1819.

78. L. A. Neely, S. Patel, J. Garver, M. Gallo, M. Hackett, S. McLaughlin, M. Nadel, J. Harris, S. Gullans, and J. Rooke. (2006). A single-molecule method for the quantitation of microRNA gene expression. *Nat Methods 3*, 41.
79. T. D. Schmittgen, J. Jiang, Q. Liu, and L. Yang. (2004). A high-throughput method to monitor the expression of microRNA precursors. *Nucleic Acids Res 32*, e43.
80. E. Wienholds, W. P. Kloosterman, E. Miska, E. Alvarez-Saavedra, E. Berezikov, E. de Bruijn, H. R. Horvitz, S. Kauppinen, and R. H. A. Plasterk. (2005). MicroRNA expression in zebrafish embryonic development. *Science 309*, 310–311.
81. E. J. Thatcher, A. S. Flynt, N. Li, J. R. Patton, and J. G. Patton. (2007). MiRNA expression analysis during normal zebrafish development and following inhibition of the Hedgehog and notch signaling pathways. *Dev Dyn 236*, 2172–2180.
82. J. S. Jepsen, M. D. Sorensen, and J. Wengel. (2004). Locked nucleic acid: A potent nucleic acid analog in therapeutics and biotechnology. *Oligonucleotides 14*, 130–146.
83. G. Meister, M. Landthaler, Y. Dorsett, and T. Tuschl. (2004). Sequence-specific inhibition of microRNA- and siRNA-induced RNA silencing. *RNA 10*, 544–550.
84. J. Krutzfeldt, N. Rajewsky, R. Braich, K. G. Rajeev, T. Tuschl, M. Manoharan, and M. Stoffel. (2005). Silencing of microRNAs in vivo with "antagomirs." *Nature 438*, 685.
85. A. S. Flynt, N. Li, E. J. Thatcher, L. Solnica-Krezel, and J. G. Patton. (2007). Zebrafish miR-214 modulates Hedgehog signaling to specify muscle cell fate. *Nat Genet 39*, 259.
86. W. P. Kloosterman, A. K. Lagendijk, R. F. Ketting, J. D. Moulton, and R. H. A. Plasterk. (2007). Targeted inhibition of miRNA maturation with morpholinos reveals a role for miR-375 in pancreatic islet development. *PLoS Biol 5*, e203.
87. W. Y. Choi, A. J. Giraldez, and A. F. Schier. (2007). Target protectors reveal dampening and balancing of nodal agonist and antagonist by miR-430. *Science 318,* 271–274.

# 10 Biogenesis and Function of Plant microRNAs

*Zoltan Havelda*

## CONTENTS

## OVERVIEW

The recent discovery of the miRNA-mediated gene-regulatory mechanism has had a significant impact on the understanding of developmental processes in both plants and animals. Plant-associated miRNAs were identified by a cloning method in 2002, 10 years after the discovery of animal miRNAs [1–3]. Since then, several hundred miRNAs have been identified in diverse plant species and deposited in an miRNA database (miRBase) [4–6].

This review focuses on the recent research progress on plant miRNAs, highlighting the importance of miRNA-mediated regulation in developmental processes and other biological functions.

## 10.1   INTRODUCTION

Small RNA species, the components of the RNA silencing machinery, are approximately 21–24 nucleotides in size. In plants and animals, they can be divided into two categories: small interfering (si) and micro (mi) RNAs. The difference between the two subclasses stems from the origin of small RNA molecules. siRNAs are derived from perfectly paired long double-stranded (ds) RNA molecules generated by inverted repeat sequences, RNA-dependent RNA-polymerases, or convergent transcription of genes. Virus infections and introduction of transgenes are able to mediate the induction of the siRNA-based RNA silencing pathway. In contrast, miRNAs are generated by sequential processing of genome-coded long single-stranded RNA molecules possessing the ability to form highly specific stem-loop structures.

## 10.2   BIOLOGY OF MiRNAs

### 10.2.1   BIOGENESIS

Plant miRNA genes encode capped and polyadenylated noncoding transcripts (pri-miRNAs) produced by RNA polymerase II [7,8]. Analyses of the transcription start sites of several miRNA primary transcripts revealed features consistent with an RNA polymerase II-mediated transcription [8]. Some loci showed multiple starting sites generating different transcripts, and the canonical TATA box motif was identified in most of the investigated loci [8].

The pri-miRNAs are further processed in the nucleus to shorter precursors (pre-miRNA) that contain only a stem-loop RNA structure [9]. Plant pre-miRNA stem-loops are more variable in size and usually larger than those of animal pre-miRNAs [10]. Moreover, in contrast to animal miRNA precursors, which are often found as several units clustered on a single pri-miRNA, plant miRNA precursors are mainly expressed individually to produce unique miRNAs [10]. However, physically clustered miRNAs were also detected in plants implying that these miRNAs, similar to their animal counterparts, show similar gene expression patterns and are transcribed as a polycistron [11]. While the occurrence of human miRNA genes in introns of pre-mRNAs is ubiquitous, only one intron-located miRNA gene has been identified in the model plant *Arabidopsis thaliana* so far [12].

Mature miRNAs, approximately 21–22 nucleotide (nt) in length, are processed in the nucleus from pre-miRNAs possessing specific secondary structures [1,2,13,14]. Transgenic experiments demonstrated that it is the specific secondary structure rather than the sequence of the precursors that is important for correct processing of miRNAs [15–18]. The mature miRNA can originate either from the 5′ or 3′ end of the pre-miRNA stem-loop structures [1].

Biochemical maturation of plant and animal miRNAs exhibit a number of differences. In animals, a nuclear RNase III, Drosha, performs excision of the pre-mRNA from pri-miRNA [19]. In the absence of a Drosha homologue in plants, an RNase III enzyme (DICER-LIKE 1; DCL1) homologous to animal Dicer is responsible for liberating pre-miRNAs [1,2,13,20]. The *A. thaliana* genome contains four Dicer-like genes termed *DCL1* to *4*. *DCL2* and *DCL4* produce virus-associated siRNAs [21,22], whereas *DCL3* is responsible for the production of siRNAs involved in the methylation of endogenous DNA [20,23]. *DCL4* is also involved in processing trans-acting (ta) siRNAs [24]. The role of the *DCL1* gene in miRNA maturation was demonstrated in loss-of-function *A. thaliana* mutants [25]. The *carpel factory* mutants (*caf* [26]), deficient in *DCL1*, showed severe developmental defects associated with the reduced accumulation of miRNAs [1,2]. The pivotal role of DCL1 in miRNA maturation is further supported by the fact that *dcl1* null alleles are embryo lethal [27]. An important feature of *A. thaliana*-encoded *DCL* genes is that they are able to act in a partially redundant manner [28]. A recent study identified DCL1 as a unique enzyme capable of generating miRNAs and also siRNAs, while the other DCLs produce only siRNAs [22]. An additional function of DCL1 is to execute the second cleavage resulting in the generation of miRNA/miRNA* duplex [9]. The miRNA* strand originates from the complementary region of the miRNA-containing arm of the pre-miRNA stem-loop structure. The miRNA/miRNA* duplex possesses 2-nucleotide 3' overhangs with 5' monophosphate and 3' hydroxyl ends, resembling the characteristics of the products of Dicer activity [29]. The presence of a nuclear localization signal, as well as localization assays, indicate that DCL1 exerts its activity of pri-miRNA and pre-miRNA processing in the nucleus [13].

Loss-of-function mutation of the *HYPONASTIC LEAVES1 (HYL1)* gene identified an additional factor as a crucial component of plant miRNA biogenesis [30,31]. HYL1 has two dsRNA-binding domains, with a preferential affinity to dsRNAs in vitro. This protein mainly localizes in the nucleus in rings and small bodies [30,32]. Importantly, mutation in *HYL1* induces pleiotropic developmental phenotypes similar to those of *dcl1* alleles [32]. In *hyl1* plants, several miRNAs showed reduced accumulation correlating with enhanced accumulation of uncleaved target mRNAs, and similarly to DCL1, HYL1 is not required for siRNA production either [30,31]. Biochemical characterization of HYL1 revealed that it specifically interacts with DCL1 [33] and contributes to the efficient and precise liberation of pre-miRNA from pri-miRNA [34]. SERRATE (SE) has been characterized as a general component of plant miRNA biogenesis [35,36]. Previously, it was shown that SE regulates shoot meristem and axial patterning of lateral organs by modulating the expression of some miRNAs [37]. However, a recent study revealed that SE is responsible for the accumulation of multiple miRNAs, via interacting with HYL1, to coordinate pri-miRNA processing [35].

miRNA biogenesis in plants also requires the activity of HUA ENHANCER1 (HEN1) [2,38], which contains a nuclear localization signal and two dsRNA-binding domains [2]. HEN1 was characterized as an S-adenosyl methionine (SAM)-dependent methyltransferase responsible for methylation of the 2' hydroxyl group at the 3' terminal ribose of each strand of miRNA/miRNA* duplex [39,40]. This feature of plant miRNA maturation is distinct from that of animal miRNA biogenesis where no methylation occurs. In *hen1* mutants, the size of small RNAs is increased because of a novel plant-specific 3' uridylation activity [41]. It has been proposed that methylation of miRNAs (and siRNAs) protects them from degradation or inhibits small RNAs to serve as primers for plant-encoded RNA-dependent RNA polymerase.

In animals, the pre-miRNA is exported from the nucleus to the cytoplasm by Exportin-5, which specially binds to pre-miRNAs [42,43]. In plants, the DCL1-processed miRNA/miRNA* duplexes can be exported to the cytoplasm by HASTY (HST), the plant ortholog of Exportin-5 [44]. However, alternative transport mechanisms are also likely to exist since the miRNA export is not completely eliminated in *hst* mutants [44].

## 10.2.2 ACTIVITY

One selected strand (the miRNA) of the miRNA/miRNA* duplex is exported to the cytoplasm and loaded onto the RNA-induced silencing complex (RISC), the effector component of the RNA silencing machinery, while the miRNA* strand is to be degraded. The fate of the strands depends on the thermodynamic nature of the miRNA/miRNA* duplex; the strand possessing less stably base-paired 5' end is destined to enter RISC [45,46].

The key component of RISC is a member of the Argonaute (AGO) protein family containing a central PAZ domain and a PIWI domain [47]; both domain structures have been resolved. The PAZ domain binds to the 3' end of single-stranded RNA molecules, while the PIWI domain binds to the 5' end of small RNAs [10]. It has been demonstrated that AGO1 in plants is responsible for the target cleaving activity of RISC (termed as *slicer activity*) [48,49]. Analyses of affinity-purified AGO1 revealed that it selectively associated with miRNAs, endogenous tasiRNAs, and transgene-derived siRNAs, but not with virus-specific siRNAs or siRNAs functioning in chromatin silencing [48]. Moreover, AGO1 was able to mediate the cleavage of *PHAVOLUTA* mRNA at the *mir165* target site in vitro, and this activity was dependent on the conserved amino acid residues in the PIWI domain. In *A. thaliana* [50], 10 *AGO*-like genes have been identified, and it is likely that different AGO components determine the diverse functions of the RNA silencing machinery. Consistently, AGO4 has been shown to be associated with endogenous siRNAs involved in epigenetic silencing, while AGO7 is involved in ta-siRNA biogenesis [51–53].

miRNAs exert their activity of negative regulation on target genes mainly by cleavage of the mRNA or by inhibition of its translation. In this respect, animal and plant systems show fundamental differences. In plants, the targeted mRNAs encompass perfect or near perfect complementary miRNA recognition sites [54]. The miRNA-containing RISC functions predominantly as an endonuclease bringing about the cleavage of the target mRNA [14,55]. The miRNA target sites are usually located in the coding region of plant mRNAs, although, miRNA-binding sites can also be found in the 3' unsaturated region (UTR) in some cases. The predominant occurrence of miRNA target sequences in the coding regions can be explained by the assumption that cleavage of this region is likely the most effective way to inhibit translation of the mRNA into functional protein [54]. Mutational analyses of miRNA target RNA pairing identified the central and the 5' regions but not the 3' region of the miRNAs as the crucial parts responsible for biological activity [15,56]. The mature miRNA defines the target sequence for the RISC, which catalyzes the cleavage of the target mRNA in the center of the miRNA recognition site [14]. The miRNA-mediated cleavage induces the oligo-uridylation of the 3' end of the 5' cleavage product, eventually leading to decapping and fast degradation [57]. The 3' cleavage product exhibits a slower degradation process mediated by the activity of the cytoplasmic 5' to 3' EXORIBONUCLEASE4 (XRN4) [58,59]. In contrast to what has been described here for plants, animal miRNAs are usually partially complementary to multiple binding sites residing in the 3' UTRs of the target mRNAs [10]. Complementarity is usually restricted to the 5' part of the miRNA, called *seed region*, and this region is disproportionately involved in the target RNA recognition [60–62].

Binding of the miRNA-loaded RISCs to the 3' UTR induces the inhibition of translation of the target mRNA rather than the cleavage of the mRNA [63–65]. Although in animal systems the translational inhibition is the dominant mechanism of miRNA-mediated gene regulation, several animal miRNAs have been proved to act via cleavage of target RNAs [19]. Moreover, apart from these endogenous miRNAs, virus-encoded small regulatory RNA molecules mediating cleavage have also been detected in animals [19,66]. Similar to the animal systems, exceptions exist to the general rule in plants as well. *A. thaliana APETALA2* (*AP2*) and *AP2*-like genes are regulated by miR172 primarily through translational inhibition and, to a lesser extent, by cleavage as well [67,68]. However, in contrast to the animal system, a plant miRNA, miR172, which displays almost perfect complementarity to its target site located within the coding region of the mRNA, triggers translational inhibition. This observation has been recently reinvestigated, and the findings of this study revealed an even further complicated situation demonstrating that miR172 target mRNAs are also under feedback regulation by their protein products, obscuring the effect of miR172 on its targets [69].

Mature plant miRNAs can be detected in the cytosol and also in the nucleus [44,70], suggesting that they may have nuclear function. There are two observations supporting this hypothesis. First, the target cleavage site of miR173 is located in the intron of a ta-siRNA precursor transcript [71,72] indicating its possible nuclear activity. Second, in addition to the ability of miR165/166 to cleave *PHABULOSA* (*PHB*) and *PHAVOLUTA* (*PHV*) mRNAs, they also mediate the methylation of *PHB* and *PHV* genes at positions downstream of the miRNA target site [73]. In heterozygous plants carrying a mutant *phb-d* allele that produces mRNA less efficiently cleavable by miR165/166, only the wild-type allele became methylated. Since the miRNA target sites in *PHB* and *PHV* transcripts span two exons, the miRNAs have to interact with nascent, newly processed transcripts to induce epigenetic modification of the corresponding gene. The biological role of miRNA-mediated methylation is still unclear. The schematic representation of the current model of plant miRNA biogenesis and action is shown in Figure 10.1.

## 10.2.3  MOBILITY

Transgene-induced RNA silencing, mediated by siRNAs, can act in a noncell autonomous manner as well. Local induction of transgene-associated RNA silencing can result in systemic spread of RNA silencing through the plant via plasmodesmata (cell-to-cell) or the vascular system (long distance) [74,75]. Although RNA-silencing-generated 21-nt long siRNAs were not explicitly shown to be components of systemic

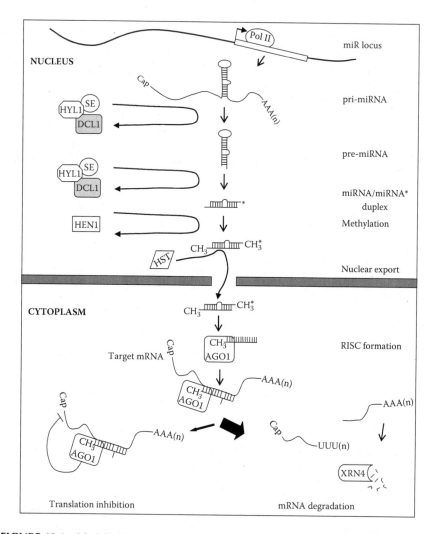

**FIGURE 10.1** Model of biogenesis and functions of plant miRNAs: The pri-miRNA transcript generated by Pol II is about 1 kb in size. The pri-miRNA is cleaved to form pre-miRNA by DCL1 and then further processed to miRNA/miRNA* duplex also by the activity of DCL1. DCL1 probably acts in close association with other factors such as HYL1, SE, and probably other proteins as well. The miRNA/miRNA* duplex is probably methylated in the nucleus by HEN1. Some of the miRNA/miRNA* duplexes may be transported to the cytoplasm by HST; however, other transport mechanisms are also likely to exist. One of the strands (the mature miRNA strand) is loaded onto the RISC complex, containing AGO1, while the other strand (the miRNA* strand) is eliminated from the system. The loaded miRNA determines the specificity of the RISC complex by guiding it to its target RNA. In plants, the active RISC predominantly executes the cleavage of its target RNA, which triggers the subsequent oligouridination of the 3′ end of the cleavage product. The cleavage products are then degraded in the cell by nucleases such as XRN4. Some miRNAs may induce translational inhibition of their target mRNAs.

signalization, they are probably able to move in a cell-to-cell manner since *dcl4* mutants, defective in the generation of 21-nt long siRNAs, showed impaired short distance signalization [76]. The ability of 21-nt siRNAs to move raises the question whether miRNAs can also exert noncell autonomous activity. In situ analyses of different miRNAs revealed a spatially and temporally highly regulated accumulation [68,70,77,78]. Experiments using transgenic plants showed near-perfect spatial overlap between the expression pattern of pre-miR171 and miR171 activity, arguing against the mobility of miR171 [15]. However, analyses of phloem sap revealed that at least some miRNAs are present in the vascular system, suggesting that they are probably able to move long distances in plants [79]. The presence of miRNAs in the vascular bundle was also confirmed by in situ hybridization [70,78]. It is possible that some miRNAs may act via transportation from a distant source to a target tissue where they exert their biological effects. Most likely, miRNAs act in association with different protein components, which may determine the mode of action and mobility of the particular miRNA. This hypothesis is supported by observations that some miRNAs were detected in vascular bundles whereas others were not, by in situ analyses, implying an active process in regulating the appearance of miRNAs in the vascular system [70].

### 10.2.4 EVOLUTION

To date, 767 plant miRNAs representing nine species have been deposited in the miRNA registry [5]. For *A. thaliana*, 118 miRNA genes have been reported, constituting 46 miRNA families based on their nucleotide sequence similarities. A particular miRNA family consists of miRNA genes encoding different pri-miRNAs, but producing identical or only slightly different mature miRNA. Rice and poplar, two other model plants with complete, publicly available genome databases, contain 182 and 213 miRNA genes, making up 46 and 33 miRNA families, respectively. Plant miRNAs can be divided into two broad categories. Conserved miRNAs are present in different species, and nonconserved miRNAs are specific to one organism [1,12,80,81]. It has been shown that evolutionarily highly conserved miRNA families are present in several species belonging to distantly related families. For example, miR159/319 is present in ten species and miR156 has been detected in 31 species [82]. Conservation reflects the role of the particular miRNA, with conserved miRNAs playing a role in basic regulatory processes via controlling genes functioning in such general mechanisms as leaf or flower development, while nonconserved miRNAs may regulate specific genes characteristic for a specific plant species. For example, in poplar, several miRNAs have been identified that are lacking in *A. thaliana* genome, possibly suggesting their tree-specific functions [83].

The evolution of miRNAs is still poorly understood; however, recent results shed light on the origin, conservation, and diversity of miRNAs. Some animal miRNAs have been shown to be evolutionarily well conserved; for example, miR7 (let-7) was detected in a wide range of animal species, including vertebrates, mollusks, arthropods, etc. [84]. In the case of animals, the precursor transcripts also show conservation between species [82]. It has been estimated that animal miRNAs already existed 420 million years ago, before the divergence of the common ancestor of metazoans [84]. Conservation in plant miRNAs has also been observed [80,85]. However, in contrast to the animal system, conservation does not typically apply to the precursor transcripts [1,82]. Several miRNAs are conserved between *A. thaliana* and *Oryza sativa* implying that they had existed before the divergence of monocot and dicot plants about 150 million years ago [86]. Floyd and Bowman [86] demonstrated that miR166-binding site was conserved in class III homeodomain-leucine zipper (HD-ZIP III) genes from bryophytes to seed plants, while regions outside the miRNA-binding site showed much less conservation. This observation indicates that plant-miRNA-mediated regulation is a very ancient trait in terms of phylogeny, dating back more than 400 million years. The date of origin of plant miRNAs is comparable to that of the origin of miRNA regulation in metazoans, but no evidence has yet been reported to prove that plant and animal miRNAs have a common ancestor.

A recent study provided data supporting a possible stepwise mechanism in the evolution of plant miRNAs. A number of *A. thaliana* miRNA genes conserved neither in monocots nor in dicots could not be assigned to any of the miRNA gene families, suggesting that they evolved recently [87]. Based on this finding, a model has been suggested in which inverted duplication of target gene sequences led to the generation of new miRNAs [87]. During the transcription of these inversely duplicated target genes, stem-loop structures are formed, which subsequently are mounted to miRNA-producing loci by the miRNA-processing machinery, eventually leading to the negative regulation of the expression of the original gene. Since the nonfunctional part of the evolving miRNA precursor is probably under a different evolutionary pressure than sequence elements encoding the miRNA/miRNA* duplex, it might be expected that conservation between the target and miRNA precursor reflects the age of duplication event. The recently evolved *A. thaliana* miRNA genes (*MIR161* and *MIR163*) possess attributes characteristic to inverted duplication. In these cases, the similarity between the miRNA gene and its target is present not only in the miRNA target site but also extends to the arms of the stem-loop structures, as would be expected for a new miRNA gene generated by target gene duplication [87]. Supporting this hypothesis, it was reported that precursors of members of some nonconserved miRNA families (miR157, miR158, miR405, and miR447) show significant sequence similarities outside the miRNA target sites as well as within the respective families [88]. It was also found that the sequence similarities between the precursors were not confined only to transcripts but extended to the promoter regions, suggesting that duplication events involved the upstream regulatory regions as well [88].

## 10.3 FUNCTIONS OF miRNAs

Most of the plant miRNAs have pivotal roles in the regulation of developmental processes by targeting transcription factors or other regulatory proteins. About two-third of the plant miRNAs are engaged in controlling plant development; however, increasing number of evidences demonstrate the role of miRNAs in other processes, such as stress responses [10]. The indispensable nature of the miRNA pathway in basic biological processes is supported by the observation that, whereas posttranscriptional gene silencing (PTGS) activity is inhibited at low temperatures, the miRNA pathway is not affected [89]. In the following text, a summary of what we have learned from recent research on the biological functions of miRNAs will be given.

## 10.3.1 Regulation of Development

The miRNA-based control of plant development involves fundamental mechanisms such as meristem boundary formation, organ separation, auxin signaling, leaf and floral development, lateral root formation, transition from juvenile-to-adult vegetative phase, etc. An interesting feature of miRNA-mediated regulation of plant development is that the activities of most miRNAs are organized in an overlapping network in which a particular miRNA may control several aspects of a plant developmental program. Moreover, a specific developmental program can be controlled by multiple miRNAs acting in coordination [90]. These regulatory networks allow the fine-tuned adjustment of mRNA levels of target genes necessary to follow a balanced developmental program. In the following sections, the current knowledge on the function of some miRNAs and their targets in plant development are outlined.

### 10.3.1.1 miR164

*CUP-SHAPED COTYLEDON (CUC)* 1, 2, and 3 belong to NAM/ATAF/CUC (NAC)-domain genes having partially overlapping functions in organ boundary formation and shoot apical meristem formation [91,92]. These genes have been shown to be expressed in the boundary of the two cotyledons in the embryo and also in the boundary of floral organ, suggesting that they play important roles in embryonic and floral development. *CUC1* and *CUC2* but not *CUC3* have been shown to be targeted by miR164 [93–95]. miR164 is encoded by three miRNA genes: *MIR164a*, *MIR164b*, and *MIR164c*. It was shown that plants overexpressing *MIR164a* and *MIR164b* phenocopied the *cuc1/cuc2* double mutant by downregulating the levels of *CUC1* and *CUC2* mRNAs [94,95]. Introduction of miR164-resistant *CUC2* or reducing the level of miR164 leads to the appearance of enlarged sepal boundary domains implying that miR164 controls the expansion of the boundaries [94]. In line with this observation, enlargement in sepal boundary domains was identified also in miRNA biogenesis mutants (*hen1*, *hyl1*, and *dcl1*) [94]. A forward genetic screen identified *MIR164c* as a gene regulating the petal number in the flower [93]. The *early extra petals1 (eep1)* mutant has been demonstrated to be a loss-of-function allele of *MIR164c*. It has been shown that *MIR164c* regulates petal number in a nonredundant manner by controlling the accumulation of transcription factors CUC1 and CUC2. Importantly, these observations indicate that closely related miRNA family members have different functions during development, although they are predicted to control the same set of target genes [93]. This phenomenon can be explained by the spatially highly regulated expression pattern of different miRNA genes.

In addition to *CUC1* and *CUC2* and other genes, *NAC1* has also been identified as a gene target for miR164 [96]. *NAC1* has a role in transduction of auxin signals required for lateral root emergence. *MIR164a* and *MIR164b* mutant plants express less miR164 and more *NAC1* mRNA and display a more intensive lateral root formation. On expression of *MIR164a* and *MIR164b* genomic sequences, the normal phenotype was restored [96]. Auxin treatment resulted in induction of miR164, which paralleled the increase in the *NAC1* mRNA cleavage; however, a cleavage-resistant form of *NAC1* mRNA was unaffected by auxin treatment.

### 10.3.1.2  miR160 and miR167

The small signaling molecule auxin elicits multiple developmental responses, such as patterning in embryogenesis, apical dominance, and cell elongation [97]. Promoters of genes regulated by auxin contain auxin-response elements (AuxRE) that bind transcription factors of the auxin response factor (ARF) family [98,99]. There are 23 members in the *ARF* gene family in *A. thaliana,* and some of them have been shown to repress or activate expression of genes carrying an AuxRE promoter element [98]. Several *ARF* genes have been shown to be controlled by miRNAs. miR160 regulates *ARF10, ARF16,* and *ARF17* via cleavage of the target mRNAs [100,101]. Transgenic plants expressing miR160 cleavage-resistant *ARF17* transcripts (where the miRNA recognition site is abolished by introducing mutations that do not influence the correct translation of the mRNA) show various developmental abnormalities including reduced petal size, abnormal stamens, and sterility [100]. These abnormalities substantiate the pivotal role of miR160-mediated *ARF17* regulation and suggest a function for *ARF17* in the regulation of early auxin-response genes. *ARF10* and *ARF16* have been identified as regulators of root cap cell formation [101]. It was shown that normal root cap development requires a finely tuned expression pattern of *ARF10* and *ARF16* achieved by independent regulation mechanism by auxin and by miR160.

miR167 is predicted to target other members of the *ARF6* and *ARF8* families [80]. Cleavage of *ARF8* was observed in cultured rice cells, and the level of miR167 was demonstrated to be controlled by auxin [102]. Moreover, it was demonstrated that a decrease in miR167 level resulted in an increase in the level of *ARF8* mRNA, and vice versa a drop in miR167 levels triggered ARF8 mRNA accumulation [102]. The cleavage of *ARF8* mRNA in *A. thaliana* was also shown; however, degradation of *ARF6* mRNA was not observed [103]. Transgenic plants overexpressing *MIR167b* showed phenotypes similar to that of *arf6/arf8* double mutants including defects in all four whorls of floral organs such as short filaments, sterile pollens etc. [103]. The authors suggested that *ARF6* is probably regulated via translational inhibition. Collectively, these data demonstrate that at least five members in the ARF gene family are regulated by miRNAs.

### 10.3.1.3  miR156

*SQUAMOSA PROMOTER BINDING PROTEIN*-box genes (*SBP*-box genes) encode plant-specific proteins that share a highly conserved DNA-binding domain (the SBP domain). In *A. thaliana*, *SBP* box-containing genes constitute a structurally heterogeneous family of 16 members known as *SPL* genes [104]. *MIR156a* is thought to target several members of the *SPL* gene family [80]. *SPL3*, *SPL4*, and *SPL5* have a target site for miR156 in their 3′ UTRs. Transgenic experiments involving constitutive expression of miR156-sensitive and miR156-insensitive forms of *SPL3*, *SPL4*, and *SPL5* demonstrated that all three genes induce vegetative phase change and flowering, and are under the control of miR156 [105]. The juvenile-to-adult transition was associated with a decrease in the level of miR156 and an increase in the level of *SPL3* mRNA. miR156 was also cloned from in vitro cultured rice embryogenic callus and exhibited an intriguing expression profile during the transition from undifferentiated to differentiated calli [106]. Moreover, miR156 is present in moss, and an *SPL* homologue in moss has also been showed to be cleaved at a position corresponding to the miRNA recognition site, suggesting that miR156 is active in lower plants [107].

### 10.3.1.4   miR165 and miR166

Genes belonging to the HD-ZIP III transcription factor families such as *PHABULOSA* (*PHB*), *PHAVULOTA* (*PHV*), and *REVOLUTA* (*REV*) control vascular bundle development and asymmetry pattern along the adaxial/abaxial (upper/lower) axis of leaves [78]. Dominant mutations in one of these three genes (*phb, phv*, and *rev*) cause polarity defects that culminate in radialization of leaves and vascular bundles in the stem [108,109]. *PHB, PHV*, and *REV* transcripts have been shown to contain sequences complementary to miR165 and miR166 differing only in one nucleotide. These are target sites for regulation by the cleavage activity of these miRNAs [56,73,78,108]. The dominant mutations (*phb, phv*, and *rev*) were mapped to the regions complementary to miR165/166. In situ analyses in maize implied that miR166 acts as a polarizing signal whose expression pattern spatially regulates the expression of *ROLLED LEAF 1*, an HD-ZIP III member that functions in defining the adaxial/abaxial asymmetry of the leaf [78].

Dominant *jabba-1D* (*jba-1D*) mutants show serious developmental defects including multiple enlarged shoot meristems, radialized leaves, and vascular abnormalities [110]. These phenotypes were caused by overexpression of *MIR166g* as a result of introduction of an enhancer element in the vicinity of this gene. Surprisingly, although miR166g overexpression in *jba-1D* plants resulted in the decrease of *PHB, PHV*, and *ATHB-15* (also known as *CORONA [CAN]*, another validated target of miR165/166), *REV* expression was increased and another validated miR165/166 target, *ATHB8*, showed no changes. A recently described gain-of-function mutation in *MIR166a* was shown to bring about the reduced accumulation of *ATHB15* transcripts associated with vascular defects [111]. It was demonstrated that miR166-mediated *ATHB15* mRNA cleavage plays a pivotal role in vascular development probably in all vascular plants. The overexpression of *MIR166a* induced different responses at the level of target genes expression; for example, transcript levels of *REV* remained unchanged [111]. These results underpin the importance of the tissue-specific expression of various members of particular miRNA families. The fact that miR165/166 are encoded by nine different loci (*MIR165a-b* and *MIR166a-g*) strongly suggests that miRNA activity is highly regulated spatially and temporally by the expression of these loci.

In addition, miR165/166 are also involved in methylation of *PHB* and *PHV* genes at sites downstream of the miRNA target site, as already discussed in Section 10.2.2 [73].

### 10.3.1.5   miR159 and miR319

*MIR-JAW* (*MIR319*) was the first miRNA gene to be identified in a mutant screen involving the introduction of enhancer elements [112]. The overexpression of miR319 results in aberrant leaf development, delayed flowering time, and greenish petals.

miR319 was shown to guide mRNA cleavage of several *TCP* transcription factor genes controlling leaf development. The overexpression of miR319–resistant *TCP* genes, carrying silent mutations within the miRNA–binding sites, partially rescued the mutant phenotype indicating the biological role of miR319–mediated regulation [112].

miR159 shares extensive sequence similarity to miR319; however, it seems that miR159 is engaged in regulation of a different set of target genes. miR159-mediated cleavage of mRNAs of *MYB33* and *MYB65* genes encoding GAMYB-related proteins was demonstrated to be important in the regulation of short-day photoperiod flowering time and anther development [113]. These *MYB* genes encode transcription factors, which are gibberellin-response regulators of other gibberellin-regulated genes [114]. It was demonstrated that miR159 level was regulated by gibberellin indicating that miR159 controls GAMYB activity in a phytohormonally regulated manner [113]. Using an miR159 cleavage-resistant MYB33:GUS fusion construct, an expanded expression pattern was detected suggesting that miR159 confines *MYB33* expression in specific tissues [115]. In addition to the role of miR159 in flower development, it has also been proposed to function in leaf development as suggested by observations when expression of an miRNA-resistant version of *MYB33*-induced curled-up leaves [112].

### 10.3.1.6   miR172

Overexpression of miR172 resulted in early flowering and defects in floral identity including absence of petals and transformation of sepals into carpels [67]. miR172 targets are members of *APETALA2* (*AP2*)-like transcription factors, including AP2, *TARGET OF EAT* (*TOE1* and *TOE2*), or *GLOSSY15* (*GL15*). It was demonstrated that AP2-like target genes function as floral repressors, indicating that miR172 regulates flowering time by downregulating AP2-like target transcripts. Interestingly, miR172 seems to be able to regulate its targets predominantly by translational inhibition rather than mRNA cleavage, as discussed in Section 10.2.2. Transition from a juvenile-to-adult leaf in maize is regulated by AP2-like gene *GL15* [116]. The accumulation level of a maize miR172 homologue has been shown to be increased during shoot development and mediated degradation of *GL15* mRNAs. These results point out that balanced regulation of *AP2*-like genes by miR172 can be a general mechanism to control vegetative phase change in plants [116].

### 10.3.2   Abiotic and Biotic Stresses

miRNAs, in addition to their pivotal role in development, are also involved in stress responses. Sequence analyses of small RNA libraries deriving from *A. thaliana* seedlings subjected to diverse abiotic stresses as well as computational approaches identified miRNAs that may play roles in plant responses to environmental stresses [12,80]. miR395 is not detectable in plants under normal circumstances; however, it is induced by stress incited by low levels of sulfate. This miRNA was shown to target ATP sulfurylases, an enzyme that catalyzes the first step of inorganic sulfate assimilation, and to be induced upon sulfate starvation, indicating its role in stress responses [80]. miR399 is also not detectable in nonstressed plants; however, phosphate starvation induces the accumulation of this miRNA [117,118]. The upregulated miRNA targets the ubiquitin-conjugating E2 enzyme by downregulating its mRNA, which is important for the plant to cope with phosphate starvation. These important results demonstrate that miRNAs have biological functions in reacting to the fluctuations in mineral nutrient availability. miR398 is another example of an miRNA responding to an environmental stimulus targeting two closely related Cu/Zn superoxide dismutases (*CSD1* and *CSD2*), which can protect the cells against superoxide radicals. In this case, the oxidative stress reduces the expression level of miR398 allowing the accumulation of *CSD1* and *CSD2* mRNAs mediating oxidative stress tolerance on this way [119]. The expression level of miR408 was induced in poplar due to mechanical stresses suggesting that it may have a function in structural and mechanical fitness of woody plants [83]. Altogether, these findings underline the indispensable role of miRNAs in sensing and moderating endogenous stress stimuli.

In plants, only one example has been described that directly implicated an miRNA in biotic stresses. It has been shown that the expression of miR393 is induced by a bacterial flagellin-derived peptide of *Pseudomonas syringae* [120]. This miRNA is able to downregulate mRNAs of F-box auxin receptors. The repression of auxin signaling results in the restriction of *P. syringae* growth indicating that miR393 may play a direct role in host defense. In addition, infection of viruses can cause disturbances in the miRNA pathway. Viruses encode suppressors to evade the RNA-silencing-based antiviral defense system of the plant [121]. Several viral-encoded suppressors interfere not only with the siRNA pathway but also with the miRNA pathway [122–125]. This can be due to the partial overlap of siRNA and miRNA pathways. Changes in host gene expression via interference with miRNA pathway, at least in the case of some viruses, can provide a molecular base for the development of disease symptoms [122,124]. Recent data indicate that the miRNA pathway may be beneficial for infection of red clover necrotic mosaic virus [126].

### 10.3.3 REGULATION OF TA-SIRNA PHASING

ta-siRNAs are a recently identified new class of plant small RNAs. They are produced by DCL4 from long, protein noncoding transcripts of *TAS* genes, which are converted into double-stranded form by the activity of RNA-dependent RNA-polymerase 6 (RDR6) [72,127,128]. In addition to RDR6, ta-siRNA biogenesis also requires enzymes involved in the miRNA pathway (AGO1, DCL1, HEN1, and HYL1) indicating a possible role for miRNAs in the generation of ta-siRNAs [72,127,128]. Indeed, ta-siRNA biogenesis is initiated by the specific miRNA-mediated cleavage of *TAS* primary transcripts [71,72]. This cleavage induces the transformation of one of the cleavage products into dsRNA, which is then processed to ta-siRNA by the activity of DCL4 [28,71,129]. Defining the initiation point of dsRNA production in pre-ta-siRNA by miRNA cleavage mediates the correct phase of ta-siRNA generation, which enables the negative regulation of target genes [71,72,127]. ta-siRNAs guide the degradation of their target transcripts, which show little similarity to the ta-siRNA-producing genes displaying a feature characteristic only for miRNAs so far. Since *rdr6* or *dcl4* mutant plants show accelerated juvenile-to-adult transition, the role of ta-siRNAs may be the regulation of this trait. Only one ta-siRNA was found to be conserved among plant species so far, which targets developmentally important *ARF* genes, *ARF3* and *ARF4* [53,130–132].

### 10.3.4 Regulation of Plant miRNA Pathway

miRNAs are engaged in many important regulatory pathways in plants including developmental processes and stress responses. The unexpectedly broad functions of miRNAs require the fine regulation of miRNA biogenesis and activity. DCL1 and AGO1 are the key components of the miRNA machinery, and recent findings revealed that they are elegantly regulated by miRNA-mediated feedback mechanisms.

In *dcl1* mutant plants, the elevated level of the *dcl1* RNA forms was observed compared to wild-type plants [133]. The increased accumulation of *DCL1* mRNA was also detected in plants, which showed defects in the miRNA pathway. miR162 has been identified as the regulator of *DCL1* mRNA mediating its cleavage. These results revealed a negative feedback mechanism regulating an enzyme involved in miRNA biogenesis by an miRNA [133].

Transgenic plants expressing a mutant but biologically active *AGO1* mRNA with resistance to miR168 cleavage displayed developmental defects characteristic to plants carrying crucial mutations in the miRNA pathway, demonstrating the importance of miRNA-mediated regulation of *AGO1* [16]. The deleterious effect of mutant *AGO1* mRNA was rescued by the introduction of an artificial miRNA targeting the mutant mRNA [16]. Moreover, new components of the miRNA-mediated feedback regulation of *AGO1* have been also described involving the AGO1-mediated post-transcriptional stabilization of miR168 and the coregulated expression of *AGO1* and *miR168* genes [134]. These results demonstrate the existence of a refined feedback regulatory loop involving several components, which allows a well-balanced control of AGO1 and miR168 accumulation. Table 10.1 summarizes the function and activity of selected miRNAs revealing the network of miRNA pathways.

## 10.4 CONCLUSIONS

The world of regulatory small RNAs has expanded with stunning velocity during recent years. About one-third of human genes are estimated to be under the control of miR-NAs [60,135]. Although in plants, the number of miRNA-regulated genes seems to be lower than that in animals, the importance of this regulation mechanism is supported by observations demonstrating its control over a wide variety of biological functions, including development, hormone and stress responses, and feedback mechanisms. Moreover, the different small RNA pathways have shown to be overlapping, uncovering a complex regulatory network [136–138]. This unexpected complexity of the world of small RNAs prognosticates a wealth of new exciting findings in future research.

## ACKNOWLEDGMENTS

The author wishes to thank Gábor Giczey and György Szittya for critical reading of the manuscript. This research was supported by a grant from the Hungarian Scientific Research Fund (OTKA; K61461). The author is a recipient of Bolyai Janos Fellowship.

**TABLE 10.1**

**List of Selected miRNAs Demonstrating Their Diverse and Overlapping Biological Functions**

| miRNA | Target Family | Leaf Dev. | Flower Dev. | Flowering Time | Auxin Response | Stress Response | miRNA ta-siRNA Biogenesis | Unknown | Reference |
|---|---|---|---|---|---|---|---|---|---|
| miR156 | SPL | | | | | | | | 105 |
| miR159 | MYB | • | • | • | | | | | 112,113 |
| miR160 | ARF | • | • | • | | | | | 100,101 |
| miR161 | PPR | | | | | | | • | 8,87 |
| miR162 | DCL1 | | | | | | • | | 133,139 |
| miR163 | SAMT | | | | | | | • | 87 |
| miR164 | NAC | • | • | | • | | | | 93–96 |
| miR165 | HD-Zip | • | | | | | | | 55,108 |
| miR167 | ARF | | • | | • | | | | 102,103 |
| miR168 | AGO1 | | | | | | • | | 16,134 |
| miR169 | HAP2 | | | | | | | • | 8 |
| miR171 | SCL | | • | | | | | | 14,15 |
| miR172 | AP2 | • | • | • | | | | | 67,116,140 |
| miR173 | TAS | | | | | | • | | 141 |
| miR390 | TAS | | | | | | • | | 141 |
| miR319 | TCP | • | • | • | | | | | 112 |
| miR399 | UBC | | | | | • | | | 117,118,142 |
| miR398 | CSD | | | | | • | | | 119 |
| miR395 | APS | | | | | • | | | 80 |

## REFERENCES

1. Reinhart, B. J., Weinstein, E. G., Rhoades, M. W., Bartel, B., and Bartel, D. P. (2002). MicroRNAs in plants. *Genes Dev 16*, 1616–1626.
2. Park, W., Li, J., Song, R., Messing, J., and Chen, X. (2002). CARPEL FACTORY, a Dicer homolog, and HEN1, a novel protein, act in microRNA metabolism in *Arabidopsis thaliana. Curr Biol 12*, 1484–1495.
3. Llave, C., Kasschau, K. D., Rector, M. A., and Carrington, J. C. (2002). Endogenous and silencing-associated small RNAs in plants. *Plant Cell 14*, 1605–1619.
4. Griffiths-Jones, S. (2004). The microRNA registry. *Nucleic Acids Res 32*, D109–111.
5. Griffiths-Jones, S., Grocock, R. J., van Dongen, S., Bateman, A., and Enright, A. J. (2006). miRBase: microRNA sequences, targets and gene nomenclature. *Nucleic Acids Res 34*, D140–144.
6. Ambros, V., Bartel, B., Bartel, D. P., Burge, C. B., Carrington, J. C., Chen, X., Dreyfuss, G., Eddy, S. R., Griffiths-Jones, S., Marshall, M., Matzke, M., Ruvkun, G., and Tuschl, T. (2003). A uniform system for microRNA annotation. *RNA 9*, 277–279.
7. Lee, Y., Kim, M., Han, J., Yeom, K. H., Lee, S., Baek, S. H., and Kim, V. N. (2004). MicroRNA genes are transcribed by RNA polymerase II. *EMBO J 23*, 4051–4060.
8. Xie, Z., Allen, E., Fahlgren, N., Calamar, A., Givan, S. A., and Carrington, J. C. (2005). Expression of *Arabidopsis* MIRNA genes. *Plant Physiol 138*, 2145–2154.
9. Kurihara, Y. and Watanabe, Y. (2004). *Arabidopsis* micro-RNA biogenesis through Dicer-like 1 protein functions. *Proc Natl Acad Sci USA 101*, 12753–12758.
10. Bartel, D. P. (2004). MicroRNAs: Genomics, biogenesis, mechanism, and function. *Cell 116*, 281–297.
11. Zhang, B., Pan, X., Cannon, C. H., Cobb, G. P., and Anderson, T. A. (2006). Conservation and divergence of plant microRNA genes. *Plant J 46*, 249–259.
12. Sunkar, R. and Zhu, J. K. (2004). Novel and stress-regulated microRNAs and other small RNAs from *Arabidopsis. Plant Cell 16*, 2001–2019.
13. Papp, I., Mette, M. F., Aufsatz, W., Daxinger, L., Schauer, S. E., Ray, A., van der Winden, J., Matzke, M., and Matzke, A. J. (2003). Evidence for nuclear processing of plant micro RNA and short interfering RNA precursors. *Plant Physiol 132*, 1382–1390.
14. Llave, C., Xie, Z., Kasschau, K. D., and Carrington, J. C. (2002). Cleavage of scarecrow-like mRNA targets directed by a class of *Arabidopsis* miRNA. *Science 297*, 2053–2056.
15. Parizotto, E. A., Dunoyer, P., Rahm, N., Himber, C., and Voinnet, O. (2004). In vivo investigation of the transcription, processing, endonucleolytic activity, and functional relevance of the spatial distribution of a plant miRNA. *Genes Dev 18*, 2237–2242.
16. Vaucheret, H., Vazquez, F., Crete, P., and Bartel, D. P. (2004). The action of ARGONAUTE1 in the miRNA pathway and its regulation by the miRNA pathway are crucial for plant development. *Genes Dev 18*, 1187–1197.
17. Schwab, R., Ossowski, S., Riester, M., Warthmann, N., and Weigel, D. (2006). Highly specific gene silencing by artificial microRNAs in *Arabidopsis. Plant Cell 18*, 1121–1133.
18. Alvarez, J. P., Pekker, I., Goldshmidt, A., Blum, E., Amsellem, Z., and Eshed, Y. (2006). Endogenous and synthetic microRNAs stimulate simultaneous, efficient, and localized regulation of multiple targets in diverse species. *Plant Cell 18*, 1134–1151.
19. Du, T. and Zamore, P. D. (2005). microPrimer: The biogenesis and function of microRNA. *Development 132*, 4645–4652.
20. Xie, Z., Johansen, L. K., Gustafson, A. M., Kasschau, K. D., Lellis, A. D., Zilberman, D., Jacobsen, S. E., and Carrington, J. C. (2004). Genetic and functional diversification of small RNA pathways in plants. *PLoS Biol 2*, E104.

21. Deleris, A., Gallego-Bartolome, J., Bao, J., Kasschau, K. D., Carrington, J. C., and Voinnet, O. (2006). Hierarchical action and inhibition of plant Dicer-like proteins in antiviral defense. *Science 313*, 68–71.

22. Bouche, N., Lauressergues, D., Gasciolli, V., and Vaucheret, H. (2006). An antagonistic function for *Arabidopsis* DCL2 in development and a new function for DCL4 in generating viral siRNAs. *EMBO J 25*, 3347–3356.

23. Pontes, O., Li, C. F., Nunes, P. C., Haag, J., Ream, T., Vitins, A., Jacobsen, S. E., and Pikaard, C. S. (2006). The *Arabidopsis* chromatin-modifying nuclear siRNA pathway involves a nucleolar RNA processing center. *Cell 126*, 79–92.

24. Willmann, M. R. and Poethig, R. S. (2005). Time to grow up: The temporal role of small RNAs in plants. *Curr Opin Plant Biol 8*, 548–552.

25. Jover-Gil, S., Candela, H., and Ponce, M. R. (2005). Plant microRNAs and development. *Int J Dev Biol 49*, 733–744.

26. Jacobsen, S. E., Running, M. P., and Meyerowitz, E. M. (1999). Disruption of an RNA helicase/RNAse III gene in *Arabidopsis* causes unregulated cell division in floral meristems. *Development 126*, 5231–5243.

27. Schauer, S. E., Jacobsen, S. E., Meinke, D. W., and Ray, A. (2002). DICER-LIKE1: Blind men and elephants in *Arabidopsis* development. *Trends Plant Sci 7*, 487–491.

28. Gasciolli, V., Mallory, A. C., Bartel, D. P., and Vaucheret, H. (2005). Partially redundant functions of *Arabidopsis* DICER-like enzymes and a role for DCL4 in producing transacting siRNAs. *Curr Biol 15*, 1494–1500.

29. Elbashir, S. M., Martinez, J., Patkaniowska, A., Lendeckel, W., and Tuschl, T. (2001). Functional anatomy of siRNAs for mediating efficient RNAi in *Drosophila melanogaster* embryo lysate. *EMBO J 20*, 6877–6888.

30. Han, M. H., Goud, S., Song, L., and Fedoroff, N. (2004). The *Arabidopsis* double-stranded RNA-binding protein HYL1 plays a role in microRNA-mediated gene regulation. *Proc Natl Acad Sci USA 101*, 1093–1098.

31. Vazquez, F., Gasciolli, V., Crete, P., and Vaucheret, H. (2004). The nuclear dsRNA binding protein HYL1 is required for microRNA accumulation and plant development, but not posttranscriptional transgene silencing. *Curr Biol 14*, 346–351.

32. Lu, C. and Fedoroff, N. (2000). A mutation in the *Arabidopsis* HYL1 gene encoding a dsRNA binding protein affects responses to abscisic acid, auxin, and cytokinin. *Plant Cell 12*, 2351–2366.

33. Hiraguri, A., Itoh, R., Kondo, N., Nomura, Y., Aizawa, D., Murai, Y., Koiwa, H., Seki, M., Shinozaki, K., and Fukuhara, T. (2005). Specific interactions between Dicer-like proteins and HYL1/DRB-family dsRNA-binding proteins in *Arabidopsis thaliana*. *Plant Mol Biol 57*, 173–188.

34. Kurihara, Y., Takashi, Y., and Watanabe, Y. (2006). The interaction between DCL1 and HYL1 is important for efficient and precise processing of pri-miRNA in plant microRNA biogenesis. *RNA 12*, 206–212.

35. Yang, L., Liu, Z., Lu, F., Dong, A., and Huang, H. (2006). SERRATE is a novel nuclear regulator in primary microRNA processing in *Arabidopsis*. *Plant J 47*(6): 841–850.

36. Lobbes, D., Rallapalli, G., Schmidt, D. D., Martin, C., and Clarke, J. (2006). SERRATE: A new player on the plant microRNA scene. *EMBO Rep 7*(10): 1052–1058.

37. Grigg, S. P., Canales, C., Hay, A., and Tsiantis, M. (2005). SERRATE coordinates shoot meristem function and leaf axial patterning in *Arabidopsis*. *Nature 437*, 1022–1026.

38. Boutet, S., Vazquez, F., Liu, J., Beclin, C., Fagard, M., Gratias, A., Morel, J. B., Crete, P., Chen, X., and Vaucheret, H. (2003). *Arabidopsis* HEN1: A genetic link between endogenous miRNA controlling development and siRNA controlling transgene silencing and virus resistance. *Curr Biol 13*, 843–848.

39. Yu, B., Yang, Z., Li, J., Minakhina, S., Yang, M., Padgett, R. W., Steward, R., and Chen, X. (2005). Methylation as a crucial step in plant microRNA biogenesis. *Science 307*, 932–935.

40. Yang, Z., Ebright, Y. W., Yu, B., and Chen, X. (2006). HEN1 recognizes 21–24 nt small RNA duplexes and deposits a methyl group onto the 2′ OH of the 3′ terminal nucleotide. *Nucleic Acids Res 34*, 667–675.

41. Li, J., Yang, Z., Yu, B., Liu, J., and Chen, X. (2005). Methylation protects miRNAs and siRNAs from a 3′-end uridylation activity in *Arabidopsis. Curr Biol 15*, 1501–1507.

42. Lund, E., Guttinger, S., Calado, A., Dahlberg, J. E., and Kutay, U. (2004). Nuclear export of microRNA precursors. *Science 303*, 95–98.

43. Yi, R., Qin, Y., Macara, I. G., and Cullen, B. R. (2003). Exportin-5 mediates the nuclear export of pre-microRNAs and short hairpin RNAs. *Genes Dev 17*, 3011–3016.

44. Park, M. Y., Wu, G., Gonzalez-Sulser, A., Vaucheret, H., and Poethig, R. S. (2005). Nuclear processing and export of microRNAs in *Arabidopsis. Proc Natl Acad Sci USA 102*, 3691–3696.

45. Khvorova, A., Reynolds, A., and Jayasena, S. D. (2003). Functional siRNAs and miRNAs exhibit strand bias. *Cell 115*, 209–216.

46. Schwarz, D. S., Hutvagner, G., Du, T., Xu, Z., Aronin, N., and Zamore, P. D. (2003). Asymmetry in the assembly of the RNAi enzyme complex. *Cell 115*, 199–208.

47. Cerutti, L., Mian, N., and Bateman, A. (2000). Domains in gene silencing and cell differentiation proteins: The novel PAZ domain and redefinition of the Piwi domain. *Trends Biochem Sci 25*, 481–482.

48. Baumberger, N. and Baulcombe, D. C. (2005). *Arabidopsis* ARGONAUTE1 is an RNA Slicer that selectively recruits microRNAs and short interfering RNAs. *Proc Natl Acad Sci USA 102*, 11928–11933.

49. Qi, Y., Denli, A. M., and Hannon, G. J. (2005). Biochemical specialization within *Arabidopsis* RNA silencing pathways. *Mol Cell 19*, 421–428.

50. Carmell, M. A., Xuan, Z., Zhang, M. Q., and Hannon, G. J. (2002). The Argonaute family: Tentacles that reach into RNAi, developmental control, stem cell maintenance, and tumorigenesis. *Genes Dev 16*, 2733–2742.

51. Zilberman, D., Cao, X., and Jacobsen, S. E. (2003). ARGONAUTE4 control of locus-specific siRNA accumulation and DNA and histone methylation. *Science 299*, 716–719.

52. Zilberman, D., Cao, X., Johansen, L. K., Xie, Z., Carrington, J. C., and Jacobsen, S. E. (2004). Role of *Arabidopsis* ARGONAUTE4 in RNA-directed DNA methylation triggered by inverted repeats. *Curr Biol 14*, 1214–1220.

53. Adenot, X., Elmayan, T., Lauressergues, D., Boutet, S., Bouche, N., Gasciolli, V., and Vaucheret, H. (2006). DRB4-dependent TAS3 trans-acting siRNAs control leaf morphology through AGO7. *Curr Biol 16*, 927–932.

54. Rhoades, M. W., Reinhart, B. J., Lim, L. P., Burge, C. B., Bartel, B., and Bartel, D. P. (2002). Prediction of plant microRNA targets. *Cell 110*, 513–520.

55. Tang, G., Reinhart, B. J., Bartel, D. P., and Zamore, P. D. (2003). A biochemical framework for RNA silencing in plants. *Genes Dev 17*, 49–63.

56. Mallory, A. C., Reinhart, B. J., Jones-Rhoades, M. W., Tang, G., Zamore, P. D., Barton, M. K., and Bartel, D. P. (2004). MicroRNA control of PHABULOSA in leaf development: Importance of pairing to the microRNA 5′ region. *EMBO J 23*, 3356–3364.

57. Shen, B. and Goodman, H. M. (2004). Uridine addition after microRNA-directed cleavage. *Science 306*, 997.

58. Souret, F. F., Kastenmayer, J. P., and Green, P. J. (2004). AtXRN4 degrades mRNA in *Arabidopsis* and its substrates include selected miRNA targets. *Mol Cell 15*, 173–183.

59. Gazzani, S., Lawrenson, T., Woodward, C., Headon, D., and Sablowski, R. (2004). A link between mRNA turnover and RNA interference in *Arabidopsis. Science 306*, 1046–1048.

60. Lewis, B. P., Burge, C. B., and Bartel, D. P. (2005). Conserved seed pairing, often flanked by adenosines, indicates that thousands of human genes are microRNA targets. *Cell 120*, 15–20.

61. Brennecke, J., Stark, A., Russell, R. B., and Cohen, S. M. (2005). Principles of microRNA-target recognition. *PLoS Biol 3*, e85.

62. Lewis, B. P., Shih, I. H., Jones-Rhoades, M. W., Bartel, D. P., and Burge, C. B. (2003). Prediction of mammalian microRNA targets. *Cell 115*, 787–798.

63. Olsen, P. H. and Ambros, V. (1999). The lin-4 regulatory RNA controls developmental timing in *Caenorhabditis elegans* by blocking LIN-14 protein synthesis after the initiation of translation. *Dev Biol 216*, 671–680.

64. Carrington, J. C. and Ambros, V. (2003). Role of microRNAs in plant and animal development. *Science 301*, 336–338.

65. Pillai, R. S., Bhattacharyya, S. N., Artus, C. G., Zoller, T., Cougot, N., Basyuk, E., Bertrand, E., and Filipowicz, W. (2005). Inhibition of translational initiation by Let-7 MicroRNA in human cells. *Science 309*, 1573–1576.

66. Sullivan, C. S., Grundhoff, A. T., Tevethia, S., Pipas, J. M., and Ganem, D. (2005). SV40-encoded microRNAs regulate viral gene expression and reduce susceptibility to cytotoxic T cells. *Nature 435*, 682–686.

67. Aukerman, M. J. and Sakai, H. (2003). Regulation of flowering time and floral organ identity by a MicroRNA and its APETALA2-like target genes. *Plant Cell 15*, 2730–2741.

68. Chen, X. (2004). A microRNA as a translational repressor of APETALA2 in *Arabidopsis* flower development. *Science 303*, 2022–2025.

69. Schwab, R., Palatnik, J. F., Riester, M., Schommer, C., Schmid, M., and Weigel, D. (2005). Specific effects of microRNAs on the plant transcriptome. *Dev Cell 8*, 517–527.

70. Valoczi, A., Varallyay, E., Kauppinen, S., Burgyan, J., and Havelda, Z. (2006). Spatiotemporal accumulation of microRNAs is highly coordinated in developing plant tissues. *Plant J 47*, 140–151.

71. Yoshikawa, M., Peragine, A., Park, M. Y., and Poethig, R. S. (2005). A pathway for the biogenesis of trans-acting siRNAs in *Arabidopsis. Genes Dev 19*, 2164–2175.

72. Allen, E., Xie, Z., Gustafson, A. M., and Carrington, J. C. (2005). microRNA-directed phasing during trans-acting siRNA biogenesis in plants. *Cell 121*, 207–221.

73. Bao, N., Lye, K. W., and Barton, M. K. (2004). MicroRNA binding sites in *Arabidopsis* class III HD-ZIP mRNAs are required for methylation of the template chromosome. *Dev Cell 7*, 653–662.

74. Voinnet, O., Vain, P., Angell, S., and Baulcombe, D. C. (1998). Systemic spread of sequence-specific transgene RNA degradation in plants is initiated by localized introduction of ectopic promoterless DNA. *Cell 95*, 177–187.

75. Palauqui, J. C., Elmayan, T., Pollien, J. M., and Vaucheret, H. (1997). Systemic acquired silencing: Transgene-specific post-transcriptional silencing is transmitted by grafting from silenced stocks to non-silenced scions. *EMBO J 16*, 4738–4745.

76. Dunoyer, P., Himber, C., and Voinnet, O. (2005). DICER-LIKE 4 is required for RNA interference and produces the 21-nucleotide small interfering RNA component of the plant cell-to-cell silencing signal. *Nat Genet 37*, 1356–1360.

77. Kidner, C. A. and Martienssen, R. A. (2004). Spatially restricted microRNA directs leaf polarity through ARGONAUTE1. *Nature 428*, 81–84.

78. Juarez, M. T., Kui, J. S., Thomas, J., Heller, B. A., and Timmermans, M. C. (2004). microRNA-mediated repression of rolled leaf1 specifies maize leaf polarity. *Nature 428*, 84–88.

79. Yoo, B. C., Kragler, F., Varkonyi-Gasic, E., Haywood, V., Archer-Evans, S., Lee, Y. M., Lough, T. J., and Lucas, W. J. (2004). A systemic small RNA signaling system in plants. *Plant Cell 16*, 1979–2000.

80. Jones-Rhoades, M. W. and Bartel, D. P. (2004). Computational identification of plant microRNAs and their targets, including a stress-induced miRNA. *Mol Cell 14*, 787–799.

81. Sunkar, R., Girke, T., Jain, P. K., and Zhu, J. K. (2005). Cloning and characterization of microRNAs from rice. *Plant Cell 17*, 1397–1411.

82. Zhang, B. H., Pan, X. P., Wang, Q. L., Cobb, G. P., and Anderson, T. A. (2005). Identification and characterization of new plant microRNAs using EST analysis. *Cell Res 15*, 336–360.

83. Lu, S., Sun, Y. H., Shi, R., Clark, C., Li, L., and Chiang, V. L. (2005). Novel and mechanical stress-responsive MicroRNAs in *Populus trichocarpa* that are absent from *Arabidopsis*. *Plant Cell 17*, 2186–2203.

84. Pasquinelli, A. E., Reinhart, B. J., Slack, F., Martindale, M. Q., Kuroda, M. I., Maller, B., Hayward, D. C., Ball, E. E., Degnan, B., Muller, P., Spring, J., Srinivasan, A., Fishman, M., Finnerty, J., Corbo, J., Levine, M., Leahy, P., Davidson, E., and Ruvkun, G. (2000). Conservation of the sequence and temporal expression of let-7 heterochronic regulatory RNA. *Nature 408*, 86–89.

85. Axtell, M. J. and Bartel, D. P. (2005). Antiquity of microRNAs and their targets in land plants. *Plant Cell 17*, 1658–1673.

86. Floyd, S. K. and Bowman, J. L. (2004). Gene regulation: Ancient microRNA target sequences in plants. *Nature 428*, 485–486.

87. Allen, E., Xie, Z., Gustafson, A. M., Sung, G. H., Spatafora, J. W., and Carrington, J. C. (2004). Evolution of microRNA genes by inverted duplication of target gene sequences in *Arabidopsis thaliana*. *Nat Genet 36*, 1282–1290.

88. Wang, Y., Hindemitt, T., and Mayer, K. F. (2006). Significant sequence similarities in promoters and precursors of *Arabidopsis thaliana* non-conserved microRNAs. *Bioinformatics 22*(21): 2585–2589.

89. Szittya, G., Silhavy, D., Molnar, A., Havelda, Z., Lovas, A., Lakatos, L., Banfalvi, Z., and Burgyan, J. (2003). Low temperature inhibits RNA silencing-mediated defence by the control of siRNA generation. *EMBO J 22*, 633–640.

90. Mallory, A. C. and Vaucheret, H. (2006). Functions of microRNAs and related small RNAs in plants. *Nat Genet 38 Suppl*, S31–36.

91. Aida, M., Ishida, T., Fukaki, H., Fujisawa, H., and Tasaka, M. (1997). Genes involved in organ separation in *Arabidopsis:* An analysis of the cup-shaped cotyledon mutant. *Plant Cell 9*, 841–857.

92. Hibara, K., Takada, S., and Tasaka, M. (2003). CUC1 gene activates the expression of SAM-related genes to induce adventitious shoot formation. *Plant J 36*, 687–696.

93. Baker, C. C., Sieber, P., Wellmer, F., and Meyerowitz, E. M. (2005). The early extra petals1 mutant uncovers a role for microRNA miR164c in regulating petal number in *Arabidopsis*. *Curr Biol 15*, 303–315.

94. Laufs, P., Peaucelle, A., Morin, H., and Traas, J. (2004). MicroRNA regulation of the CUC genes is required for boundary size control in *Arabidopsis* meristems. *Development 131*, 4311–4322.

95. Mallory, A. C., Dugas, D. V., Bartel, D. P., and Bartel, B. (2004). MicroRNA regulation of NAC-domain targets is required for proper formation and separation of adjacent embryonic, vegetative, and floral organs. *Curr Biol 14*, 1035–1046.

96. Guo, H. S., Xie, Q., Fei, J. F., and Chua, N. H. (2005). MicroRNA directs mRNA cleavage of the transcription factor NAC1 to downregulate auxin signals for *Arabidopsis* lateral root development. *Plant Cell 17*, 1376–1386.

97. Berleth, T. and Sachs, T. (2001). Plant morphogenesis: Long-distance coordination and local patterning. *Curr Opin Plant Biol 4*, 57–62.

98. Ulmasov, T., Hagen, G., and Guilfoyle, T. J. (1999). Dimerization and DNA binding of auxin response factors. *Plant J 19*, 309–319.

99. Ulmasov, T., Hagen, G., and Guilfoyle, T. J. (1997). ARF1, a transcription factor that binds to auxin response elements. *Science 276*, 1865–1868.

100. Mallory, A. C., Bartel, D. P., and Bartel, B. (2005). MicroRNA-directed regulation of *Arabidopsis* AUXIN RESPONSE FACTOR17 is essential for proper development and modulates expression of early auxin response genes. *Plant Cell 17*, 1360–1375.

101. Wang, J. W., Wang, L. J., Mao, Y. B., Cai, W. J., Xue, H. W., and Chen, X. Y. (2005). Control of root cap formation by MicroRNA-targeted auxin response factors in *Arabidopsis*. *Plant Cell 17*, 2204–2216.

102. Yang, J. H., Han, S. J., Yoon, E. K., and Lee, W. S. (2006). Evidence of an auxin signal pathway, microRNA167-ARF8-GH3, and its response to exogenous auxin in cultured rice cells. *Nucleic Acids Res 34*, 1892–1899.

103. Ru, P., Xu, L., Ma, H., and Huang, H. (2006). Plant fertility defects induced by the enhanced expression of microRNA167. *Cell Res 16*, 457–465.

104. Cardon, G., Hohmann, S., Klein, J., Nettesheim, K., Saedler, H., and Huijser, P. (1999). Molecular characterisation of the *Arabidopsis* SBP-box genes. *Gene 237*, 91–104.

105. Wu, G. and Poethig, R. S. (2006). Temporal regulation of shoot development in *Arabidopsis thaliana* by miR156 and its target SPL3. *Development 133*, 3539–3547.

106. Xie, K., Wu, C., and Xiong, L. (2006). Genomic organization, differential expression and interaction of SPL transcription factors and microRNA156 in rice. *Plant Physiol 142*(1): 280–293.

107. Arazi, T., Talmor-Neiman, M., Stav, R., Riese, M., Huijser, P., and Baulcombe, D. C. (2005). Cloning and characterization of micro-RNAs from moss. *Plant J 43*, 837–848.

108. Emery, J. F., Floyd, S. K., Alvarez, J., Eshed, Y., Hawker, N. P., Izhaki, A., Baum, S. F., and Bowman, J. L. (2003). Radial patterning of *Arabidopsis* shoots by class III HD-ZIP and KANADI genes. *Curr Biol 13*, 1768–1774.

109. McConnell, J. R., Emery, J., Eshed, Y., Bao, N., Bowman, J., and Barton, M. K. (2001). Role of PHABULOSA and PHAVOLUTA in determining radial patterning in shoots. *Nature 411*, 709–713.

110. Williams, L., Grigg, S. P., Xie, M., Christensen, S., and Fletcher, J. C. (2005). Regulation of *Arabidopsis* shoot apical meristem and lateral organ formation by microRNA miR166g and its AtHD-ZIP target genes. *Development 132*, 3657–3668.

111. Kim, J., Jung, J. H., Reyes, J. L., Kim, Y. S., Kim, S. Y., Chung, K. S., Kim, J. A., Lee, M., Lee, Y., Narry Kim, V., Chua, N. H., and Park, C. M. (2005). microRNA-directed cleavage of ATHB15 mRNA regulates vascular development in *Arabidopsis* inflorescence stems. *Plant J 42*, 84–94.

112. Palatnik, J. F., Allen, E., Wu, X., Schommer, C., Schwab, R., Carrington, J. C., and Weigel, D. (2003). Control of leaf morphogenesis by microRNAs. *Nature 425*, 257–263.

113. Achard, P., Herr, A., Baulcombe, D. C., and Harberd, N. P. (2004). Modulation of floral development by a gibberellin-regulated microRNA. *Development 131*, 3357–3365.

114. Gubler, F., Raventos, D., Keys, M., Watts, R., Mundy, J., and Jacobsen, J. V. (1999). Target genes and regulatory domains of the GAMYB transcriptional activator in cereal aleurone. *Plant J 17*, 1–9.

115. Millar, A. A. and Gubler, F. (2005). The *Arabidopsis* GAMYB-like genes, MYB33 and MYB65, are microRNA-regulated genes that redundantly facilitate anther development. *Plant Cell 17*, 705–721.

116. Lauter, N., Kampani, A., Carlson, S., Goebel, M., and Moose, S. P. (2005). microRNA172 down-regulates glossy15 to promote vegetative phase change in maize. *Proc Natl Acad Sci USA 102*, 9412–9417.

117. Chiou, T. J., Aung, K., Lin, S. I., Wu, C. C., Chiang, S. F., and Su, C. L. (2006). Regulation of phosphate homeostasis by MicroRNA in *Arabidopsis*. *Plant Cell 18*, 412–421.

118. Fujii, H., Chiou, T. J., Lin, S. I., Aung, K., and Zhu, J. K. (2005). A miRNA involved in phosphate-starvation response in *Arabidopsis*. *Curr Biol 15*, 2038–2043.

119. Sunkar, R., Kapoor, A., and Zhu, J. K. (2006). Posttranscriptional induction of two Cu/Zn superoxide dismutase genes in *Arabidopsis* is mediated by downregulation of miR398 and important for oxidative stress tolerance. *Plant Cell 18*, 2051–2065.

120. Navarro, L., Dunoyer, P., Jay, F., Arnold, B., Dharmasiri, N., Estelle, M., Voinnet, O., and Jones, J. D. (2006). A plant miRNA contributes to antibacterial resistance by repressing auxin signaling. *Science 312*, 436–439.

121. Silhavy, D. and Burgyan, J. (2004). Effects and side-effects of viral RNA silencing suppressors on short RNAs. *Trends Plant Sci 9*, 76–83.

122. Kasschau, K. D., Xie, Z., Allen, E., Llave, C., Chapman, E. J., Krizan, K. A., and Carrington, J. C. (2003). P1/HC-Pro, a viral suppressor of RNA silencing, interferes with *Arabidopsis* development and miRNA unction. *Dev Cell 4*, 205–217.

123. Mallory, A. C., Reinhart, B. J., Bartel, D., Vance, V. B., and Bowman, L. H. (2002). A viral suppressor of RNA silencing differentially regulates the accumulation of short interfering RNAs and micro-RNAs in tobacco. *Proc Natl Acad Sci USA 99*, 15228–15233.

124. Chapman, E. J., Prokhnevsky, A. I., Gopinath, K., Dolja, V. V., and Carrington, J. C. (2004). Viral RNA silencing suppressors inhibit the microRNA pathway at an intermediate step. *Genes Dev 18*, 1179–1186.

125. Dunoyer, P., Lecellier, C. H., Parizotto, E. A., Himber, C., and Voinnet, O. (2004). Probing the microRNA and small interfering RNA pathways with virus-encoded suppressors of RNA silencing. *Plant Cell 16*, 1235–1250.

126. Takeda, A., Tsukuda, M., Mizumoto, H., Okamoto, K., Kaido, M., Mise, K., and Okuno, T. (2005). A plant RNA virus suppresses RNA silencing through viral RNA replication. *EMBO J 24*, 3147–3157.

127. Vazquez, F., Vaucheret, H., Rajagopalan, R., Lepers, C., Gasciolli, V., Mallory, A. C., Hilbert, J. L., Bartel, D. P., and Crete, P. (2004). Endogenous trans-acting siRNAs regulate the accumulation of *Arabidopsis* mRNAs. *Mol Cell 16*, 69–79.

128. Peragine, A., Yoshikawa, M., Wu, G., Albrecht, H. L., and Poethig, R. S. (2004). SGS3 and SGS2/SDE1/RDR6 are required for juvenile development and the production of trans-acting siRNAs in *Arabidopsis*. *Genes Dev 18*, 2368–2379.

129. Xie, Z., Allen, E., Wilken, A., and Carrington, J. C. (2005). DICER-LIKE 4 functions in trans-acting small interfering RNA biogenesis and vegetative phase change in *Arabidopsis thaliana*. *Proc Natl Acad Sci USA 102*, 12984–12989.

130. Williams, L., Carles, C. C., Osmont, K. S., and Fletcher, J. C. (2005). A database analysis method identifies an endogenous trans-acting short-interfering RNA that targets the *Arabidopsis* ARF2, ARF3, and ARF4 genes. *Proc Natl Acad Sci USA 102*, 9703–9708.

131. Garcia, D., Collier, S. A., Byrne, M. E., and Martienssen, R. A. (2006). Specification of leaf polarity in *Arabidopsis* via the trans-acting siRNA pathway. *Curr Biol 16*, 933–938.

132. Fahlgren, N., Montgomery, T. A., Howell, M. D., Allen, E., Dvorak, S. K., Alexander, A. L., and Carrington, J. C. (2006). Regulation of AUXIN RESPONSE FACTOR3 by TAS3 ta-siRNA affects developmental timing and patterning in *Arabidopsis*. *Curr Biol 16*, 939–944.

133. Xie, Z., Kasschau, K. D., and Carrington, J. C. (2003). Negative feedback regulation of Dicer-Like1 in *Arabidopsis* by microRNA-guided mRNA degradation. *Curr Biol 13*, 784–789.

134. Vaucheret, H., Mallory, A. C., and Bartel, D. P. (2006). AGO1 homeostasis entails coexpression of MIR168 and AGO1 and preferential stabilization of miR168 by AGO1. *Mol Cell 22*, 129–136.

135. Krek, A., Grun, D., Poy, M. N., Wolf, R., Rosenberg, L., Epstein, E. J., MacMenamin, P., da Piedade, I., Gunsalus, K. C., Stoffel, M., and Rajewsky, N. (2005). Combinatorial microRNA target predictions. *Nat Genet 37*, 495–500.

136. Vazquez, F. (2006). *Arabidopsis* endogenous small RNAs: Highways and byways. *Trends Plant Sci 11*(9): 460–468.
137. Vaucheret, H. (2006). Post-transcriptional small RNA pathways in plants: Mechanisms and regulations. *Genes Dev 20*, 759–771.
138. Bonnet, E., Van de Peer, Y., and Rouze, P. (2006). The small RNA world of plants. *New Phytol 171*, 451–468.
139. Hirsch, J., Lefort, V., Vankersschaver, M., Boualem, A., Lucas, A., Thermes, C., d'Aubenton-Carafa, Y., and Crespi, M. (2006). Characterization of 43 non-protein-coding mRNA genes in *Arabidopsis*, including the MIR162a-derived transcripts. *Plant Physiol 140*, 1192–1204.
140. Mlotshwa, S., Yang, Z., Kim, Y., and Chen, X. (2006). Floral patterning defects induced by *Arabidopsis* APETALA2 and microRNA172 expression in *Nicotiana benthamiana*. *Plant Mol Biol 61*, 781–793.
141. Vaucheret, H. (2005). MicroRNA-dependent trans-acting siRNA production. *Sci STKE 2005*, pe43.
142. Bari, R., Datt Pant, B., Stitt, M., and Scheible, W. R. (2006). PHO2, microRNA399, and PHR1 define a phosphate-signaling pathway in plants. *Plant Physiol 141*, 988–999.

# 11 Endogenous Small RNA Pathways in *Arabidopsis*

*Manu Agarwal, Julien Curaba, and Xuemei Chen*

## CONTENTS

## OVERVIEW

Small RNAs of 21–24 nucleotides (nt) have been found to play regulatory roles at the transcriptional and posttranscriptional levels in diverse eukaryotic organisms. Small RNAs are classified into microRNAs (miRNAs) and small interfering RNAs (siRNAs) based on their modes of biogenesis. miRNAs are derived from single-stranded precursor RNAs that form imperfect hairpin structures. siRNAs are produced from long double-stranded RNAs formed between two overlapping antisense RNAs or synthesized from single-stranded RNAs by the activities of RNA-dependent RNA polymerases. Plants appear to make extensive use of endogenous small RNAs to regulate gene expression in developmental processes and in response to abiotic and biotic stresses. Plants also use endogenous small RNAs to silence transposons to ensure genome stability.

## 11.1 INTRODUCTION

A few years ago, proteins were thought to be the major "effector molecules" involved in gene regulation. With the discovery of RNAi and small RNAs, RNA has ceased to be just a messenger from genome to proteome. It has become clear that endogenous small RNAs play an imperative role in transcriptional as well as posttranscriptional gene regulation in plants and animals. In the last decade or so, momentous steps have been made toward understanding the genesis and function of endogenous small RNAs. Plants are particularly rich in endogenous small RNAs, and plant researchers have made groundbreaking discoveries in the field of small RNAs. In this chapter we will discuss the types of endogenous small RNAs and the pathways for the biogenesis of these molecules in the model plant *Arabidopsis thaliana*.

## 11.2  MICRORNAs (MIRNAs)

miRNAs are 20–24-nt RNAs that are the products of non-protein-coding genes (Figure 11.1). miRNAs recognize target mRNAs through sequence complementarity and regulate target mRNAs by guiding their cleavage or translation inhibition. Nuclear single-stranded hairpin RNAs are the precursor molecules that give rise to both animal and plant miRNAs. There are minor differences in the processing of the long primary transcript (pri-miRNA) between animals and plants (reviewed in Ambros, 2004; Bartel, 2004; Du and Zamore, 2005; Kim, 2005; Vaucheret, 2006). In animals, different miRNAs tend to be clustered on the same precursor RNA, but in plants, each miRNA is derived from an individual precursor. Another difference in pri-miRNAs between plants and animals is the length; plant pri-miRNAs generally are longer than their animal counterparts. The overall process of generating the mature miRNA from pri-miRNA is similar in animals and plants; that is, pri-miRNA is first processed into pre-miRNA, which is then processed into the mature miRNA. In animals, two RNAse III enzymes are involved in the process. Drosha cleaves on either arm of the stem loop of a pri-miRNA leading to the formation of the intermediate pre-miRNA. The pre-miRNA is then exported to the cytoplasm by Exportin 5 followed by its cleavage by another RNAse III enzyme Dicer to purge the loop and give rise to the miRNA/miRNA* duplex. In plants, only one RNAse III enzyme, known as DICER-LIKE1 (DCL1), is involved in processing both pri- and pre-miRNAs, and plant miRNAs are exclusively nuclearly processed (Park et al., 2002; Reihart et al., 2002; Papp et al., 2003). A small number of *Arabidopsis* miRNAs depend on DICER-LIKE4 (DCL4) instead of DCL1 for their biogenesis (Rajagopalan et al., 2006).

Both animal and plant RNAse III enzymes require RNA-binding proteins for the biogenesis of miRNAs. In *Drosophila*, Drosha interacts with Pasha and Dicer interacts with Loquacious for cleaving the pri- and pre-miRNAs, respectively (Denli et al., 2004; Forstemann et al., 2005; Gregory et al., 2004; Yeom et al., 2006). In plants, HYL1, a member of the double-stranded RNA-binding (DRB) family of proteins, interacts with DCL1 and is important for efficient and precise processing of pri-miRNAs (Kurihara et al., 2006; Dong et al., 2008). HYL1 also participates in the processing of pre-miRNAs to miRNA/miRNA* duplexes (Wu et al., 2007).

Apart from DCL1 and HYL1, another protein named SERRATE (SE) is involved in miRNA biogenesis in *Arabidopsis*. Strong mutant alleles in *SE* (*se-2* and *se-3*) were initially observed to show severe defects in leaf development with leaf curling, loss of asymmetric differentiation of abaxial and adaxial cell types, and development of trumpet shaped or radial leaves (Grigg et al., 2005). These developmental defects were correlated with reduced levels of miR165 and miR166 and increased expression

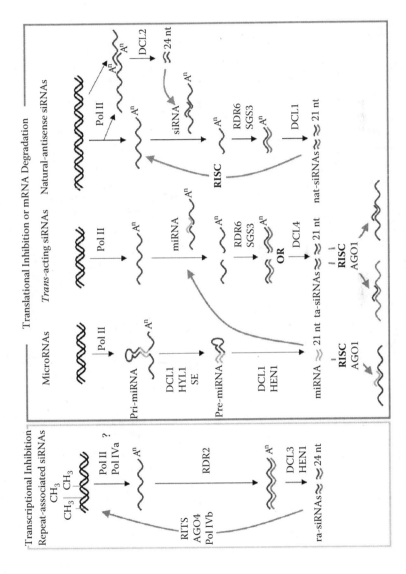

**FIGURE 11.1** A diagram of the four small RNA pathways in *Arabidopsis*.

of an adaxial determinant gene *PHABULOSA*, a known target of miR165/166 (Grigg et al., 2005; Mallory et al., 2004). The weaker *se-1* allele also has pleiotropic defects that overlap with those of mutants defective in the miRNA biogenesis pathway. The similarity in the phenotypes of *se-1* and mutants involved in the miRNA pathway led to investigations of *SE* being involved in general miRNA processing (Lobbes et al., 2006; Yang, L., et al., 2006). *SE* was found to be critical for accumulation of multiple miRNAs. It was shown to physically interact with HYL1 and to act in concert with DCL1 in the processing of pri-miRNAs. A whole-genome tiling array-based transcriptome analysis revealed that SE also plays a role in pre-mRNA splicing (Laubinger et al., 2008).

The *Arabidopsis* nuclear mRNA cap-binding complex (CBC), which is a heterodimeric complex composed of the CBP20 and CBP80 subunits, promotes the processing of some but not all pri-miRNAs (Gregory et al., 2008; Laubinger et al., 2008). Mutations in the two genes result in increased pri-miRNA levels and reduced mature miRNA levels for some miRNAs. The CBC is conserved in fungi, plants, and animals; therefore, a role of the CBC in pri-miRNA processing in animals is also likely.

Once the miRNAs are synthesized, they are methylated at their 3′ ends by HEN1. The *hen1* mutants, initially identified as enhancers of *hua1 hua2* (Chen et al., 2002), had developmental defects similar to *dcl1* mutants, suggesting that HEN1 might be involved in miRNA biogenesis. Indeed, HEN1 was found to be required in vivo for the accumulation of small RNAs investigated to date. Plants carrying strong *hen1* alleles have reduced abundance of miRNAs (Park et al., 2002), trans-acting siRNAs (ta-siRNAs, Vazquez et al., 2004), repeat associated siRNAs (ra-siRNAs; Xie et al., 2004), and sense transgene siRNAs (Boutet et al., 2003). Although the levels of naturally occurring antisense transcript siRNAs (nat-siRNAs) and inverted repeat transgene siRNAs are not affected, these small RNAs become larger in size in *hen1* mutants (Borsani et al., 2005; Li et al., 2005). HEN1 has a C-terminal methyltransferase domain and an N-terminal, putative dsRNA-binding motif. HEN1 protein produced in *E. coli* can methylate miRNA/miRNA* and siRNA/siRNA* duplexes in a sequence-independent manner in vitro (Yang, Z., et al., 2006; Yu et al., 2005). HEN1 is required for the methylation of miRNAs and siRNAs in vivo (Yu et al., 2005). Though the exact role of methylation of miRNAs and siRNAs at their ends is not known, it is conceivable that methylation on the 3′ terminal ribose is a deterrent for enzymes that utilize the 3′ hydroxyl groups in the miRNA/miRNA* duplex or in the mature small RNAs. In the absence of methylation, enzymes such as ligases, terminal transferases, exonucleases, or polymerases might act on the duplex or the mature small RNAs and form undesired new molecules. One of the protective effects of methylation appears to be the inhibition of uridylation of 3′ ends of the miRNAs and siRNAs. Besides being reduced in abundance, miRNAs and siRNAs are heterogeneous in size in *hen1* mutant plants. The reason for their heterogeneity was found to be additional nucleotides (primarily U's) at their 3′ ends (Li et al., 2005). Uridylation of miRNAs may guide degradation of miRNAs and therefore lead to the reduced abundance of miRNAs observed in the *hen1* mutants. A similar activity causes the addition of U's to the 3′ ends of the 5′ cleavage products after miRNA-mediated cleavage of target mRNAs in plants (Shen and Goodman, 2004).

The miRNA/miRNA* duplex is exported to the cytoplasm where the miRNA is loaded into a ribonucleoprotein complex that is essentially similar to the RNA-induced silencing complex (RISC, Hammond et al., 2001). Alternatively, RISC loading occurs in the nucleus followed by export of the RISC complex into the cytoplasm. In animals, Exportin 5 exports pre-miRNAs into the cytoplasm. HASTY, the *Arabidopsis* ortholog of Exportin 5, is probably involved in the export of miRNA/miRNA* duplexes or miRISCs (Park et al., 2005).

The core component of the RISC is believed to be an Argonaute protein as purified RISC from flies or mammals always had a member of the Argonaute family (Hammond et al., 2001; Ishizuka et al., 2002; Mourelatos et al., 2002). Argonaute proteins contain conserved PAZ and PIWI domains. The structure of a PIWI domain was found to adopt a similar fold as RNAseH, leading to the conclusion that Argonaute proteins possess siRNA-directed RNA cleavage (so-called slicer) activity. There are ten *ARGONAUTE* (*AGO*) genes in *Arabidopsis* and *AGO1* is the founding member of the family (Bohmert et al., 1998; Morel et al., 2002). Strong *ago1* alleles were first isolated in a screen for mutants displaying developmental defects (Bohmert et al., 1998). A screen for mutants deficient in posttranscriptional gene silencing (PTGS) also identified additional *ago1* alleles (Fagard et al., 2000; Morel et al., 2002). The earliest indication of *AGO1* being involved in miRNA biogenesis was the overlapping development defects of *ago1*, *dcl1*, *hen1*, and *hyl1* mutants. AGO1 is likely the slicer in the miRISC complex; immunoprecipitated AGO1 complex has miRNA-mediated mRNA cleavage activity (Baumberger and Baulcombe, 2005; Qi et al., 2005). Apart from having the slicer activity, AGO1 appears to play a role in miRNA stabilization and/or production as miRNAs are generally reduced in abundance or undetectable in the strong *ago1-3* mutant (Vaucheret et al., 2004). A recent study showed that AGO1 also mediates the translation inhibition activity of miRNAs (Brodersen et al., 2008).

Functional studies on plant miRNAs have revealed that the majority of the evolutionarily conserved ones target transcription factors regulating important processes in development, such as phase change, leaf morphogenesis and polarity, root initiation, flowering and floral patterning, etc. (for details refer to reviews by Chen 2005 and Jones-Rhoades et al., 2006). The role of miRNAs in adaptation to environmental perturbations recently came into the picture with the discovery of miR399-regulating phosphate homeostasis (Fujii et al., 2005) and miR398-regulating oxidative stress responses (Sunkar et al., 2006).

## 11.3 TRANS-ACTING siRNAs

A class of endogenous siRNAs that is very similar to miRNAs was discovered in plants. This class of siRNAs is known as trans-acting siRNAs (ta-siRNAs) by virtue of its ability to repress the expression of genes that bear little resemblance to the genes from which they derive. There are no reports of ta-siRNAs from animals; therefore, this pathway seems to be plant specific. miRNAs and ta-siRNAs share some similarities: (1) both come from non-protein-coding genes and appear to obey the same asymmetry rule that determines which strand of the duplex is incorporated in RISC, (2) both are endogenous small RNAs that regulate their targets in trans, and

(3) both target multiple genes (usually within a single gene family; Peragine et al., 2004; Vazquez et al., 2004).

The biogenesis of ta-siRNAs is shown in Figure 11.1. The initial step in the production of the ta-siRNAs is the cleavage of a transcript from a *TAS* locus by an miRNA. The cleavage triggers the production of ta-siRNAs that are in phase with the cleavage site in the *TAS* transcript. The *TAS* genes are non-protein-coding genes such that the transcripts of these genes only serve as precursors for the ta-siRNAs. Six *TAS* loci have been identified in *Arabidopsis*: *TAS1a, 1b, 1c, TAS2* and *TAS3* (Allen et al., 2005), and *TAS4* (Rajagopalan et al., 2006). Three transcripts from *TAS1a, 1b,* and *1c* loci are sources for ta-siRNAs that target a group of related genes (At4g29760, At4g29770, and At5g18040) of unknown function. *TAS2* transcript is the source of siRNAs that target pentatricopeptide repeat (*PPR*) genes At1g63130, At1g63230, and At1g62930 (Allen et al., 2005; Yoshikawa et al., 2005) and *TAS3* transcript is the source of ta-siRNAs that target auxin response factors (Adenot et al., 2006; Fahlgren et al., 2006; Garcia et al., 2006; Hunter et al., 2006). A ta-siRNA from the *TAS4* locus targets three MYB transcription factors (Rajagopalan et al., 2006).

With cloning followed by deep sequencing of small RNAs from *Arabidopsis*, 11 loci were found to produce clusters of phased small RNAs, a feature of ta-siRNA generating loci (Lu et al., 2006). Intriguingly, At1g63130, a target of a ta-siRNA from the *TAS2* locus, is itself a source of phased siRNAs. The production of these siRNAs is probably triggered by the cleavage of At1g63130 RNA by the ta-siRNA from *TAS2* (Chen et al., 2007).

The biogenesis of ta-siRNAs involves a number of proteins (such as SGS3, RDR6, and DCL4) known to mediate posttranscriptional gene silencing (Figure 11.1). After the transcription of the *TAS* locus, the *TAS* transcripts are targeted for cleavage by miR173, miR390, or miR828 in a manner similar to other miRNA targeted transcripts (Allen et al., 2005; Yoshikawa et al., 2005; Rajagopalan et al., 2006). The cleaved products are potentially protected by SGS3 from further degradation and converted into double-stranded molecules by RDR6, an RNA-dependent RNA polymerase. This double-stranded RNA is a substrate for DCL4, which produces ta-siRNAs in 21-nt increments from the end that is defined by miRNA-guided cleavage such that the ta-siRNAs are in phase with one another. The 3′ cleavage products of *TAS1a–c* and *TAS2,* and the 5′ cleavage product of *TAS3* are processed more efficiently than their respective counterparts (Allen et al., 2005; Yoshikawa et al., 2005; Axtell et al., 2006). Dependence of ta-siRNAs on the miRNA biogenesis genes can be attributed to the fact that cleavage of the *TAS* transcript by miRNAs sets the stage for synthesis of ta-siRNAs. Mutations in *AGO7* or *DRB4* result in loss of ta-siRNAs from *TAS3*, indicating that AGO7 and DRB4 act in ta-siRNA biogenesis in addition to SGS3, RDR6, and DCL4 (Adenot et al., 2006).

*TAS3* siRNAs alone account for the leaf morphology defect observed in *rdr6* and *sgs3* mutants since hypomorphic mutants in *RGR6* and *SGS3* that do not exhibit developmental defects are defective in the production of *TAS1* and *TAS2* siRNAs but accumulate wild-type levels of *TAS3* siRNAs (Adenot et al., 2006).

It is intriguing why only the *TAS* transcripts are further processed into phased ta-siRNAs while many transcripts cleaved by miRNAs do not. One model to explain this was put forward by Herve Vaucheret. He proposed that the entire pathway for ta-siRNA biogenesis occurs inside the nucleus, while miRNA-mediated cleavage of mRNAs likely occurs in the cytoplasm (Vaucheret, 2006). According to this model, the entire ta-siRNA biogenesis machinery (including the identified components like SGS3, DCL4, RDR6, miR173, miR390, and miR828, and *TAS* loci transcripts) should be present inside the nucleus. Alternatively, an intermediate in the ta-siRNA biogenesis pathway is imported into the nucleus. DCL4 has been shown to be localized in the nucleus (Hiraguri et al., 2005). There are other indications of the potential presence of the components inside the nucleus. For example, accumulation of miR173 and miR390 is still detectable in *hasty* mutants (Xie et al., 2005a), which are potentially defective in nuclear export of miRNAs. However, it cannot be ruled out that export of these molecules is carried out by another mechanism.

Recently, a molecular feature that allows a cell to discriminate between loci that are merely miRNA targets and loci that are subsequently used for the production of secondary siRNAs was found in both moss and diverse plant species (Axtell et al., 2006). David Bartel's group observed that the *TAS3* locus is conserved in many plants including the moss *Physcomitrella patens* (*PpTAS*). They observed that all four *TAS* loci in moss had two regions for miR390 binding. Both regions are sites of miR390-mediated cleavage of the *TAS* transcript and the intervening cleavage product is the substrate for ta-siRNA biogenesis. *Arabidopsis TAS3* (*AtTAS3*) locus was also found to have two miR390-binding sites although *AtTAS3* is cleaved by miR390 at the 3′ site only (Axtell et al., 2006). Although the 5′ site is not cleaved, it appears to bind to the miR390 silencing complex, and both miR390 complementary sites are required for full *AtTAS3* function (Axtell et al., 2006). Other genes that had dual small RNA-binding sites included *ARF* (which are targets of ta-siRNAs derived from *AtTAS3*) and *PPR* (target of ta-siRNAs derived from *AtTAS1a-c* and *AtTAS2*) and phased secondary RNAs were found to originate from the regions flanked by the two sites. Therefore, dual small RNA complementary sites appear to predispose the intervening region towards phased siRNA production. However, this may not be the only mechanism to mark ta-siRNA precursors since other *Arabidospis TAS* loci (*TAS1a-c, TAS2*) lack the dual-binding sites for miR173.

The 3′ miR390-binding site can be replaced by another miRNA-binding site without affecting the biogenesis of ta-siRNAs from *TAS3*, but the 5′ site cannot be replaced by another miRNA-binding site (Montgomery et al., 2008). AGO7 displays a specific affinity for miR390, and plays an unknown but indispensable role at the 5′ miR390-binding site in the *TAS3* RNA (Montgomery et al., 2008). While a few Argonaute proteins have been shown to select their resident miRNAs according to the nature of the 5′ nucleotide (Mi et al., 2008), AGO7's affinity for miR390 cannot be solely explained by the 5′ nucleotide of miR390 (Montgomery et al., 2008).

## 11.4   siRNAs DERIVED FROM NATURAL ANTISENSE TRANSCRIPTS (NAT-siRNAs)

There are a number of genes in diverse organisms that give rise to transcripts that can partially pair with other transcripts because of sequence complementarity. These transcript pairs can either be transcribed from partially overlapping loci (hence known as *cis*-antisense pairs) or from distinct loci (known as trans-antisense pairs). *Arabidopsis* has 1340 potential *cis*-antisense pairs (Wang et al., 2005) and rice has at least 687 bidirectional transcript pairs (Osato et al., 2003). Similar spatial or temporal presence of these transcripts will enable base pairing between the two RNAs and provide a fertile ground for the production of small RNAs using the cellular RNAi machinery. Hence, presence of these pairs may be a major mechanism of gene regulation in diverse organisms. More than 100 nat-siRNAs that map to the overlap regions of pairs of *cis*-antisense genes have been identified in *Arabidopsis* (Katiyar-Agarwal et al., 2006). Though a large number of siRNAs have been sequenced (ASRP: Gustafson et al., 2005; MPSS: Nakano et al., 2006) from plants grown under normal conditions, the repertoire of these is far from complete as the following studies show that small RNAs from *cis*-antisense pairs are induced when plants are challenged with abiotic and biotic stresses.

The first small RNA that derives from the overlapping region of a *cis*-antisense transcript pair was discovered by Jian-Kang Zhu's group. The siRNA was named nat-siRNA as it is derived from a naturally occurring antisense transcript pair (Borsani et al., 2005; Figure 11.1). The two transcripts responsible for genesis of this nat-siRNA, SRO5 and P5CDH, have an overlap of 760 nt at their 3′ ends. Salt stress induces the SRO5 transcript and potentially results in the formation of a double-stranded RNA between the induced SRO5 mRNA and the constitutively present P5CDH transcript. This paired region gives rise to at least one 24-nt long nat-siRNA whose formation is correlated with salt stress induction of SRO5. Biogenesis of P5CDH-SRO5 nat-siRNA depends on DCL2, RDR6, SGS3, and NRPD1a, a subunit of a plant specific, putative DNA-dependent RNA polymerase. This siRNA, then, guides the cleavage of P5CDH mRNA, which sets the phase for subsequent generation of 21-nt siRNAs from the cleaved end. The generation of phased 21-nt siRNAs is dependent on DCL1 as well as the components required for 24-nt siRNA production—that is, DCL2, RDR6, SGS3, and NRPD1a. Whether these proteins participate directly in the production of 21-nt siRNAs is not clear. The 21-nt siRNAs then further guide the cleavage of more P5CDH transcripts.

Another nat-siRNA, nat-siRNAATGB2, was recently discovered and found to play a role in plant innate immunity (Katiyar-Agarwal et al., 2006). *ATGB2* nat-siRNA was specifically induced by bacterial pathogen *Pseudomonas syringae* carrying the effector avrRpt2. This siRNA is derived from the overlap region of a Rab2-like small GTP-binding protein gene (*ATGB2*) and a PPR protein-like gene (*PPRL*), and its biogenesis is dependent on DCL1, RDR6, NRPD1a, SGS3, HEN1, and HYL1. ATGB2 nat-siRNA contributes to RPS2-mediated race-specific resistance by repressing PPRL, a putative negative regulator of the RPS2 resistance pathway (Katiyar-Agarwal et al., 2006).

## 11.5 REPEAT-ASSOCIATED siRNAs (ra-siRNAs)

Cytosine methylation is a gene regulation mechanism prevalent in eukaryotic organisms, and it protects the genome by inactivating selfish DNA elements such as transposable elements (Chan et al., 2005). In plants, DNA methylation occurs in three contexts: CG, CNG, and CNN (where N is A, C, or T). One of the primary reasons that account for the establishment of cytosine methylation is RNA-directed DNA methylation (RdDM). RdDM is a process in which the "aberrant" RNA generated by methylated DNA template is a substrate for formation of ra-siRNAs, which further guide the methylation machinery to the DNA template from which they are formed and promote more cytosine methylation. Therefore, ra-siRNAs are part of a positive feedback loop wherein DNA methylation at a locus results in formation of ra-siRNAs, which then guide more cytosine methylation at this locus. A surveillance mechanism probably comprising of DNA glycosylases (e.g., ROS1 and DME: Gong et al., 2002; Gehring et al., 2006; Zhu et al., 2007) and perhaps as yet unknown factors cause active demethylation to maintain an equilibrium of DNA methylation.

Genetic and molecular analysis has revealed important factors involved in the production of ra-siRNAs and established a framework of understanding of ra-siRNA biogenesis and function (Figure 11.1). Precursor transcripts are converted into long dsRNAs by RDR2, an RNA-dependent RNA polymerase. The long dsRNAs are then processed by DCL3 into siRNAs, which are methylated by HEN1. One of the strands is then incorporated into the RNA-induced transcriptional silencing (RITS) complex, the main player of which is ARGONAUTE4 (AGO4). RITS complex is thought to guide homology-dependent recruitment of de novo methyltransferases such as DRM2 (for domains-rearranged methyltransferase) (Cao et al., 2000), which then methylates cytosines. Genetic lesions in RDR2 and DCL3 affect ra-siRNAs accumulation from the four canonical endogenous loci that are associated with detectable levels of siRNAs: AtSN1, cluster 2, 5S rDNA, and siRNA02 (Xie et al., 2004). Mutations in *AGO4* result in loss of siRNAs from only two of the four loci (Zilberman et al., 2004), thereby indicating that other Argonaute proteins may also bind 24-nt ra-siRNAs. One such candidate protein is AGO6, which plays a partially redundant role with AGO4 in transcriptional gene silencing (Zheng et al., 2007).

The plant specific RNA polymerase IV (pol IV) is also required for self-sustained ra-siRNA production (Herr et al., 2005; Kanno et al., 2005; Onodera et al., 2005; Pontier et al., 2005). The largest subunit of Pol IV is encoded either by *NRPD1a* or *NRPD1b*, and the second largest subunit is encoded by *NRPD2a*. Biochemical studies identified two distinct pol IV complexes (Pontier et al., 2005). A complex comprising of NRPD1a and NRPD2a, known as pol IVa, is required for production and/or amplification of ra-siRNAs and DNA methylation (Herr et al., 2005; Onodera et al., 2005; Pontier et al., 2005). Another complex comprising of NRPD1b and NRPD2a, known as pol IVb, is required for DNA methylation and ra-siRNA accumulation from a subset of the loci requiring pol IVa (Kanno et al., 2005; Pontier et al., 2005). Recent studies by two independent groups have revealed the subcellular location for ra-siRNA production (Li et al., 2006; Pontes et al., 2006). Pol IVa is located in the bulk of the nucleoplasm and presumably generates precursor RNAs, which by some unknown mechanism are moved to nucleolar-associated processing centers known as Cajal bodies. In these bodies, the single-stranded RNA template is converted into dsRNA by RDR2, and the dsRNA is processed by DCL3 to generate ra-siRNAs. The siRNAs are then loaded into RITS in Cajal bodies (Li et al.,

2006; Pontes et al., 2006). However, the Cajal bodies are not essential for ra-siRNA production (Li et al., 2008).

## 11.6   REDUNDANCY IN SMALL RNA PATHWAYS

There are multiple members of *DCL* (4), *DRB* (6), *RDR* (6), and *AGO* (10) gene families in *Arabidopsis*. Genetic, molecular, and biochemical studies have provided considerable evidence that although individual family members have specialized functions, redundancy exists within most families. In contrast to *DCL*, *DRB*, and *AGO* family members that tend to have functional overlaps, *RDR* family members appear to have distinct functionality. Among the six *RDR* genes, *RDR3, 4,* and *5* are clustered at a single locus, and there are no reports yet on their characterization (Herr, 2005). *RDR1* is known to play a role in antiviral defense, but it is not yet clear whether *RDR1* acts in small RNA biogenesis (Yu et al., 2003). *RDR2* and *RDR6* are known to have distinct functions in small RNA biogenesis and cannot substitute for each other. *RDR2* is involved in the production of ra-siRNAs (Xie et al., 2004), and *RDR6* is involved in the production of ta-siRNAs and siRNAs from sense transgenes (Boutet et al., 2003; Vazquez et al., 2004; Yoshinawa et al., 2005).

### 11.6.1   DCL Family

The initial studies on the redundant nature of DCL proteins arose from the fact that none of the dcl mutants were isolated in a forward genetic screen for reactivation of posttranscriptional gene silencing. The initial reports on redundancy in the DCL family came from studies on *Neurospora* when deletions in the two dicers resulted in more proficient PTGS than singly mutated genes (Catalanotto et al., 2004). Similar combinatorial mutations of DCLs in *Arabidopsis* have provided direct evidence on their redundant nature in different small RNA pathways.

To date there is no evidence of redundancy among DCLs in the miRNA pathway. It appears that *DCL1* is the predominant member that processes miRNA precursors. Weak *dcl1* alleles fail to accumulate most miRNAs whereas dcl2 dcl3 dcl4 triple mutants still accumulate most miRNAs (Park et al., 2002; Reinhart et al., 2002; Henderson et al., 2006; Lu et al., 2006). dcl1 null mutants are embryo lethal indicating that other DCLs cannot substitute for the essential function of DCL1 in miRNA production (McElver et al., 2001; Schwartz et al., 1994). A few miRNAs that are derived from precursors with long hairpin structures are generated by DCL4 (Rajagopalan et al., 2006).

DCL4 is the main DCL for ta-siRNAs production (Gasciolli et al., 2005; Xie et al., 2005b). *DCL2* or *DCL3* does not play an essential role in ta-siRNA production as *dcl2* or *dcl3* single or double mutants do not affect ta-siRNA accumulation (Vazquez et al., 2004; Allen et al., 2005; Gasciolli et al., 2005). Although *DCL1* is required for ta-siRNA accumulation, its role is likely indirect as mutations in *DCL1* affect the accumulation of miRNAs that guide the cleavage of *TAS* precursors. Genetic lesions in *DCL4* lead to reduced accumulation (in the case of *dcl4-1*) or the absence of 21-nt ta-siRNAs (in the case of *dcl4-2*) (Gasciolli et al., 2005; Xie et al., 2005b). In addition to 21-nt siRNAs, 22-nt and 24-nt siRNAs are also found for the *TAS3* locus and their production requires *DCL2* and *DCL3*, respectively (Xie et al., 2005b; Henderson et al., 2006). The redundancy between *DCL2*, *DCL3*, and *DCL4* in the production of endogenous siRNAs (not necessarily ta-siRNAs) is also evident by the stronger developmental phenotypes of their double or triple mutants as compared to the single mutants (Gasciolli et al., 2005; Henderson et al., 2006).

Redundancy also exists in the ra-siRNA pathway. DCL3 is the enzyme responsible for the production of 24-nt siRNA02, a RDR2 dependent ra-siRNA (Xie et al., 2004). However, in the absence of *DCL3*, alternative forms of siRNA02 of 21–23 nt in length are produced, indicating that other dicers can function in the absence of *DCL3* (Gasciolli et al., 2005). In the *dcl2 dcl3* double mutant, only the 21-nt form of siRNA02 is detected, whereas in the *dcl3 dcl4* double mutant, only 22–23-nt forms are detected. This indicates that in the absence of *DCL3*, *DCL4* and *DCL2* are able to produce 21 and 22–23-nt long ra-siRNAs, respectively. Deep sequencing of small RNAs from a *dcl3* mutant confirmed the replacement of 24-nt siRNAs by 21–22-nt siRNAs at numerous loci (Kasschau et al., 2007). Development of stochastic developmental defects in the *dcl2 dcl3* or *dcl3 dcl4* double and *dcl2 dcl3 dcl4* triple mutants after three to five generations is also indicative of redundancy in these enzymes. These defects are not visible in the *dcl3* mutant plants (Henderson et al., 2006) and are reminiscent of the defects visible after progressive demethylation over successive generations in *ddm1* mutants (Kakutani et al., 1996). This was recently confirmed at the molecular level; non-CpG methylation at some loci is strongly altered in *dcl2 dcl3 dcl4* triple mutants, whereas it is only weakly affected in *dcl3* single mutants (Henderson et al., 2006).

## 11.6.2 DRB Family

The production of certain types of small RNAs by DCLs requires DRB proteins. As discussed earlier, interaction between DCL1 and HYL1 (also known as DRB1) is important for the precise processing of the pri-miRNA into pre-miRNA in miRNA maturation (Kurihara et al., 2006; Dong et al., 2008). DCL1 can cleave pre-miRNAs even in the absence of HYL1, which may be due to other DRB protein(s) that can compensate for HYL1 deficiency (Vazquez, 2006). In vitro pull down assays show that apart from HYL1, DCL1 can also interact weakly with DRB2 and DRB5. Similarly, DCL3 can weakly interact with HYL1, DRB2, and DRB5 (Hiraguri et al., 2005). However, analysis of drb2 drb3 drb4 and drb2 drb3 drb5 triple mutants does not support a role of DRB2, DRB3, DRB4, or DRB5 in miRNA biogenesis (Curtin et al., 2008).

*TAS3* siRNA production in leaves is dependent on *DRB4* (a DRB protein that interacts with DCL4). However, the ta-siRNAs from *TAS3* locus in flowers or *TAS1* and *TAS2* siRNAs are independent of *DRB4* (Adenot et al., 2006). The other *DRBs* may account for the incomplete requirement for *DRB4* in ta-siRNA biogenesis. However, introduction of multiple mutations in DRB2, 3, and 5 genes into the drb4 background did not lead to any reduction in ta-siRNA abundance, suggesting that DRB2, 3, and 5 do not contribute to ta-siRNA biogenesis (Curtin et al., 2008).

### 11.6.3 ARGONAUTE FAMILY

There are 10 members of the Argonaute family in *Arabidopsis* including the founding member *AGO1*. Structurally, Argonaute proteins have an N-terminal PAZ domain (for Piwi/Argonaute/Zwille) for RNA-binding and an RNase H-like PIWI domain at the C-terminus (Herr, 2005). Studies on human Ago2 and *Arabidopsis* AGO1 and AGO4 have identified key residues in these proteins that are indispensable for their activities. Of particular interest is a DDH triad that forms the basis of the slicer activity (Liu et al., 2004; Baumberger and Baulcombe, 2005; Rivas et al., 2005; Qi et al., 2006). Amino acid analysis of the *Arabidopsis* PIWI domain has indicated that many of the Argonaute family members contain the catalytic triad (Herr, 2005). AGO1, AGO4, and AGO7 have been established as bona fide slicers (Baumberger and Baulcombe, 2005; Qi et al., 2005, 2006; Montgomery et al., 2008).

Both functional specialization and redundancy exist in the Argonaute family. So far only *AGO1*, *AGO4*, *AGO6*, *AGO7*, and *AGO10* in *Arabidopsis* have been studied at functional levels. It is generally believed that *AGO1* mediates the function of most miRNAs. miRNAs are associated with AGO1 in vivo (Baumberger and Baulcombe, 2005; Qi et al., 2005, 2006) and AGO1 immunoprecipitates cause miRNA-guided cleavage of mRNA targets (Baumberger and Baulcombe, 2005; Qi et al., 2005). *AGO1* and the closely related *AGO10* may share redundant functions. A double mutant of *ago1* and *pnh* (*ago10*) has synergistic developmental phenotypes (Lynn et al., 1999). *AGO10*, in the absence of *AGO1*, can down-regulate the expression of the leaf polarity determinant gene *PHABULOSA* through miR165-guided cleavage (Kidner and Martienessen, 2004). Both AGO1 and AGO10 mediate translational inhibition of target mRNAs by miRNAs (Brodersen et al., 2008). *Arabidopsis* AGO4 binds 23–24-nt small RNAs and many of these small RNAs map to repeat regions in the genome (Qi et al., 2006). *AGO6* has partially overlapping functions in transcriptional gene silencing with *AGO4;* mutations in *AGO4* and *AGO6* together lead to more extensive loss of DNA methylation than either single mutant (Zheng et al., 2007). Mutations in *AGO7* cause vegetative phase change defects and absence of ta-siRNAs from the *TAS3* locus. The over-accumulation of *TAS3* siRNA target RNAs in *ago7* likely underlies the developmental defects of *ago7* mutants (Adenot et al., 2006). AGO7 specializes in miR390 binding and triggers the production of ta-siRNAs from *TAS3* (Montgomery et al., 2008). AGO1 is responsible for the production of ta-siRNAs from other loci. AGO1 also binds ta-siRNAs in vivo and therefore presumably mediates the functions of these ta-siRNAs (Baumberger and Baulcombe, 2005).

## 11.7  CONCLUSIONS

Plants employ diverse classes of small RNAs to accomplish diverse biological functions. miRNAs and ta-siRNAs repress target mRNAs through transcript cleavage or translational inhibition and are instrumental in developmental patterning. Some miRNAs also regulate metabolic processes in response to abiotic stresses. Repeat-associated siRNAs target homologous DNA loci for DNA methylation or histone H3 lysine 9 methylation to lead to heterochromatin formation and transcriptional gene silencing. Natural *cis*-antisense siRNAs cause target transcript cleavage in response to abiotic and biotic stresses. The biogenesis of the different classes of small RNAs involves a core group of proteins such as RDR (except for miRNAs), DCL, HEN1, and AGO. These proteins (except for HEN1) belong to gene families and different members within a family tend to have specialized functions in one or more small RNA biogenesis pathways.

## ACKNOWLEDGMENTS

Research in the Chen lab is supported by grants from the National Institutes of Health (2 R01 GM61146) and National Science Foundation (MCB-0718029).

## REFERENCES

Adenot, X., Elmayan, T., Lauressergues, D., Boutet, S., Bouche, N., Gasciolli, V., and Vaucheret, H., *DRB4*-dependent *TAS3* trans-acting siRNAs control leaf morphology through *AGO7*, *Curr. Biol.*,16, 927–932, 2006.

Allen, E., Xie, Z., Gustafson, A.M., and Carrington, J.C., microRNA-directed phasing during transacting siRNA biogenesis in plants, *Cell*, 121, 207–221, 2005.

Ambros, V., The functions of animal microRNAs, *Nature*, 431, 350–355, 2004.

Axtell, M.J., Jan, C., Rajagopalan, R., and Bartel, D.P., A two-hit trigger for siRNA biogenesis in plants, *Cell*, 127, 565–577, 2006.

Bartel, D.P., MicroRNAs: genomics, biogenesis, mechanism, and function, *Cell*, 116, 281–297, 2004.

Baumberger, N. and Baulcombe, D.C., *Arabidopsis* ARGONAUTE1 is an RNA slicer that selectively recruits microRNAs and short interfering RNAs, *Proc. Natl. Acad. Sci.*, 102, 11928–11933, 2005.

Bohmert, K., Camus, I., Bellini, C., Bouchez, D., Caboche, M., and Benning, C., *AGO1* defines a novel locus of *Arabidopsis* controlling leaf development, *EMBO J.*, 17, 170–180, 1998.

Borsani, O., Zhu, J., Verslues, P.E., Sunkar, R., and Zhu, J.K., Endogenous siRNAs derived from a pair of natural cis-antisense transcripts regulate salt tolerance in *Arabidopsis*, *Cell*, 123, 1279–1291, 2005.

Boutet, S., Vazquez, F., Liu, J., Beclin, C., Fagard, M., Gratias, A., Morel, J.B., Crete, P., Chen, X., and Vaucheret, H., *Arabidopsis HEN1*: a genetic link between endogenous miRNA controlling development and siRNA controlling transgene silencing and virus resistance. *Curr. Biol.*, 13, 843–848, 2003.

Brodersen, P., Sakvarelidze-Achard, L., Bruun-Rasmussen, M., Dunoyer, P., Yamamoto, Y.Y., Sieburth, L., and Voinnet, O., Widespread translational inhibition by plant miRNAs and siRNAs, *Science*, 320, 1185–1190, 2008.

Cao, X., Springer, N.M., Muszynski, M.G., Phillips, R.L., Kaeppler, S., and Jacobsen, S.E., Conserved plant genes with similarity to mammalian de novo DNA methyltransferases, *Proc. Natl. Acad. Sci.*, 97, 4979–4984, 2000.

Catalanotto, C., Pallotta, M., ReFalo P., Sachs M.S., Vayssie L., Macino G., and Cogoni, C., Redundancy of the two dicer genes in transgene-induced posttranscriptional gene silencing in *Neurospora crassa*, *Mol. Cell. Biol.* 24, 2536–2545, 2004.

Chan S.W., Henderson I.R., and Jacobsen S.E., Gardening the genome: DNA methylation in *Arabidopsis thaliana*, *Nat. Rev. Genet.*, 6, 351–360, 2005.

Chen X., MicroRNA biogenesis and function in plants, *FEBS Lett.*, 579, 5923–5931, 2005.

Chen, H.-M., Li, Y.-H., and Wu, S.-H., Bioinformatic prediction and experimental validation of a microRNA-directed tandem trans-acting siRNA cascade in *Arabidopsis*, *Proc. Nat. Acad. Sci.*, 104, 3318–3323, 2007.

Chen, X., Liu, J., Cheng, Y., and Jia, D., *HEN1* functions pleiotropically in *Arabidopsis* development and acts in C function in the flower, *Development*, 129, 1085–1094, 2002.

Curtin, S.J., Watson, J.M., Smith, N.A., Eamens, A.L., Blanchard, C.L., and Waterhouse, P.M., *FEBS Letters* 582, 2753–2760, 2008.

Denli, A.M., Tops, B.B., Plasterk, R.H., Ketting, R.F., and Hannon, G.J., Processing of primary microRNAs by the Microprocessor complex, *Nature*, 432, 231–235, 2004.

Dong, Z., Han, M.H., and Fedoroff, N., The RNA-binding proteins HYL1 and SE promote accurate in vitro processing of pri-miRNA by DCL1, *Proc. Natl. Acad. Sci.* 105, 9970–9975, 2008.

Du, T. and Zamore, P.D., microPrimer: the biogenesis and function of microRNA, *Development*, 132, 4645–4652, 2005.

Fagard, M., Boutet, S., Morel, J.B., Bellini, C., and Vaucheret, H., AGO1, QDE-2, and RDE-1 are related proteins required for post-transcriptional gene silencing in plants, quelling in fungi, and RNA interference in animals, *Proc. Natl. Acad. Sci.*, 97, 11650–11654, 2000.

Fahlgren, N., Montgomery, T.A., Howell, M.D., Allen, E., Dvorak, S.K., Alexander, A.L., and Carrington, J.C., Regulation of *AUXIN RESPONSE FACTOR3* by *TAS3* ta-siRNA affects developmental timing and patterning in *Arabidopsis*, *Curr. Biol.*, 16, 939–944, 2006.

Forstemann, K., Tomari, Y., Du, T., Vagin, V.V., Denli, A.M., Bratu, D.P., Klattenhoff, C., Theurkauf, W.E., and Zamore, P.D., Normal microRNA maturation and germ-line stem cell maintenance requires Loquacious, a double-stranded RNA-binding domain protein, *PLoS Biol.*, 3, e236, 2005.

Fujii, H., Chiou, T.J., Lin, S.I., Aung, K., and Zhu, J.K., A miRNA involved in phosphate-starvation response in *Arabidopsis*, *Curr. Biol.*, 15, 2038–2043, 2005.

Garcia, D., Collier, S.A., Byrne, M.E., and Martienssen, R.A., Specification of leaf polarity in *Arabidopsis* via the trans-acting siRNA pathway, *Curr. Biol.*, 16, 933–938, 2006.

Gasciolli, V., Mallory, A.C., Bartel, D.P., and Vaucheret, H., Partially redundant functions of *Arabidopsis* DICER-like enzymes and a role for DCL4 in producing trans-acting siRNAs, *Curr. Biol.*, 15, 1494–1500, 2005.

Gehring, M., Huh, J.H., Hsieh, T.F., Penterman, J., Choi, Y., Harada, J.J., Goldberg, R.B., and Fischer, R.L., DEMETER DNA glycosylase establishes MEDEA polycomb gene self-imprinting by allele-specific demethylation, *Cell*, 124, 495–506, 2006.

Gong, Z., Morales-Ruiz, T., Ariza, R.R., Roldan-Arjona, T., David, L., and Zhu, J.K., ROS1, a repressor of transcriptional gene silencing in *Arabidopsis*, encodes a DNA glycosylase/lyase, *Cell*, 111, 803–814, 2002.

Gregory, R.I., Yan, K.P., Amuthan, G., Chendrimada, T., Doratotaj, B., Cooch, N., and Shiekhattar, R., The Microprocessor complex mediates the genesis of microRNAs, *Nature*, 432, 235–240, 2004.

Gregory, B.D., O'Malley, R.C., Lister, R., Urich, M.A., Tonti-Filippini, J., Chen, H., Millar, A.H., and Ecker, J.R., A link between RNA metabolism and silencing affecting *Arabidopsis* development, *Dev. Cell*, 14, 854–866, 2008.

Grigg, S.P., Canales, C., Hay, A., and Tsiantis, M., *SERRATE* coordinates shoot meristem function and leaf axial patterning in *Arabidopsis*, *Nature*, 437, 1022–1026, 2005.

Gustafson, A.M., Allen, E., Givan, S., Smith, D., Carrington, J.C., and Kasschau, K.D. ASRP: the *Arabidopsis* small RNA project database, *Nucleic Acids Res.*, 33, D637–640, 2005.

Hammond, S.M., Boettcher, S., Caudy, A.A., Kobayashi, R., and Hannon, G.J., Argonaute2, a link between genetic and biochemical analyses of RNAi, *Science*, 293, 1146–1150, 2001.

Henderson, I.R., Zhang, X., Lu, C., Johnson, L., Meyers, B.C., Green, P.J., and Jacobsen, S.E., Dissecting *Arabidopsis thaliana* DICER function in small RNA processing, gene silencing and DNA methylation patterning, *Nat. Genet.*, 38, 721–725, 2006.

Herr, A.J., Jensen, M.B., Dalmay, T., and Baulcombe, D.C., RNA polymerase IV directs silencing of endogenous DNA, *Science*, 308, 118–120, 2005.

Herr, A.J., Pathways through the small RNA world of plants. *FEBS Lett.*, 579, 5879–5888, 2005.

Hiraguri, A., Itoh, R., Kondo, N., Nomura, Y., Aizawa, D., Murai, Y., Koiwa, H., Seki, M., Shinozaki, K., and Fukuhara, T., Specific interactions between Dicer-like proteins and HYL1/DRB-family dsRNA-binding proteins in *Arabidopsis thaliana*, *Plant Mol. Biol.*, 57, 173–188, 2005.

Hunter, C., Willmann, M.R., Wu, G., Yoshikawa, M., de la Luz Gutierrez-Nava, M., and Poethig, S.R., Trans-acting siRNA-mediated repression of *ETTIN* and *ARF4* regulates heteroblasty in *Arabidopsis*, *Development*, 133, 2973–2981, 2006.

Ishizuka, A., Siomi, M.C., and Siomi, H.A., Drosophila fragile X protein interacts with components of RNAi and ribosomal proteins, *Genes Dev.*, 16, 2497–2508, 2002.

Jones-Rhoades M.W., Bartel D.P., and Bartel B., microRNAs and their regulatory roles in plants, *Annu. Rev. Plant Biol.*, 57, 19–53, 2006.

Kakutani, T., Jeddeloh, J.A., Flowers, S.K., Munakata, K., and Richards, E.J., Developmental abnormalities and epimutations associated with DNA hypomethylation mutations, *Proc. Natl. Acad. Sci.*, 93, 12406–12411, 1996.

Kanno, T., Huettel, B., Mette, M.F., Aufsatz, W., Jaligot, E., Daxinger, L., Kreil, D.P., Matzke, M., and Matzke, A.J., Atypical RNA polymerase subunits required for RNA-directed DNA methylation, *Nat. Genet.* 37, 761–765, 2005.

Kasschau, K.D., Fahlgren, N., Chapman, E.J., Sullivan, C.M., Cumbie, J.S., Givan, S.A., and Carrington, J.C., Genome-wide profiling and analysis of *Arabidopsis* siRNAs, *PLoS Biol.*, 5(3), e57, 2007.

Katiyar-Agarwal, S., Morgan, R., Dahlbeck, D., Borsani, O., Villegas, A. Jr., Zhu, J.K., Staskawicz, B.J., and Jin, H., A pathogen-inducible endogenous siRNA in plant immunity, *Proc. Natl. Acad. Sci.*, 103, 18002–18007, 2006.

Kidner, C.A. and Martienssen, R.A., Spatially restricted microRNA directs leaf polarity through *ARGONAUTE1*, *Nature*, 428, 81–84, 2004.

Kim, V.N., MicroRNA biogenesis: coordinated cropping and dicing, *Nat. Rev. Mol. Cell Biol.*, 6, 376–385, 2005.

Kurihara, Y., Takashi, Y., and Watanabe, Y., The interaction between DCL1 and HYL1 is important for efficient and precise processing of pri-miRNA in plant microRNA biogenesis, *RNA*, 12, 206–212, 2006.

Laubinger, S., Sachsenberg, T., Zeller, G., Busch, W., Lohmann, J.U., Rätsch, G., and Weigel, D. Dual roles of the nuclear cap-binding complex and SERRATE in pre-mRNA splicing and microRNA processing in *Arabidopsis thaliana*, *Proc. Natl. Acad. Sci.*, 105, 8795–8800, 2008.

Li, C.F., Pontes, O., El-Shami, M., Henderson, I.R., Bernatavichute, Y.V., Chan, S.W., Lagrange, T., Pikaard, C.S., and Jacobsen, S.E., An ARGONAUTE4-containing nuclear processing center colocalized with Cajal bodies in *Arabidopsis thaliana*, *Cell*, 126, 93–106, 2006.

Li, C.F., Henderson, I.R., Song, L., Fedoroff, N., Lagrange, T., and Jacobsen, S.E., Dynamic regulation of ARGONAUTE4 within multiple nuclear bodies in *Arabidopsis thaliana*, *PLoS Genet.*, 4(2), e27, 2008.

Li, J., Yang, Z., Yu, B., Liu, J., and Chen, X., Methylation protects miRNAs and siRNAs from a 3′-end uridylation activity in *Arabidopsis*, *Curr. Biol.*, 15, 1501–1507, 2005.

Liu, J., Carmell, M.A., Rivas, F.V., Marsden, C.G., Thomson, J.M., Song, J.J., Hammond, S.M., Joshua-Tor, L., and Hannon, G.J., Argonaute2 is the catalytic engine of mammalian RNAi, *Science* 305, 1437–1441, 2004.

Lobbes, D., Rallapalli, G., Schmidt, D.D., Martin, C., and Clarke, J., *Serrate*: A new player on the plant microRNA scene, *EMBO Rep.*, 7, 1052–1058, 2006.

Lu, C., Kulkarni, K., Souret, F.F., MuthuValliappan, R., Tej, S.S., Poethig, R.S., Henderson, I.R., Jacobsen, S.E., Wang, W., Green, P.J., and Meyers, B.C., MicroRNAs and other small RNAs enriched in the *Arabidopsis* RNA-dependent RNA polymerase-2 mutant, *Genome Res.*, 16, 1276–1288, 2006.

Lynn, K., Fernandez, A., Aida, M., Sedbrook, J., Tasaka, M., Masson, P., and Barton, M.K., The *PINHEAD/ZWILLE* gene acts pleiotropically in *Arabidopsis* development and has overlapping functions with the *ARGONAUTE1* gene, *Development*, 126, 469–481, 1999.

Mallory, A.C., Reinhart, B.J., Jones-Rhoades, M.W., Tang, G., Zamore, P.D., Barton, M.K., and Bartel, D.P., MicroRNA control of *PHABULOSA* in leaf development: importance of pairing to the microRNA 5' region, *EMBO J.*, 23, 3356–3364, 2004.

McElver, J., Tzafrir, I., Aux, G., Rogers, R., Ashby, C., Smith, K., Thomas, C., Schetter, A., Zhou, Q., Cushman, M.A., Tossberg, J., Nickle, T., Levin, J.Z., Law, M., Meinke, D., and Patton, D., Insertional mutagenesis of genes required for seed development in *Arabidopsis thaliana*, *Genetics*, 159, 1751–1763, 2001.

Mi, S., Cai, T., Hu, Y., Chen, Y., Hodges, E., Ni, F., Wu, L., Li, S., Zhou, H., Long, C., Chen, S., Hannon, G.J., and Qi, Y., Sorting of small RNAs into Arabidopsis argonuate complexes is directed by the 5' terminal nucleotide, *Cell*, 133, 116–127, 2008.

Montgomery, T.A., Howell, M.D., Cuperus, J.T., Li, D., Hansen, J.E., Alexander, A.L., Chapman, E.J., Fahlgren, N., Allen, E., and Carrington, J.C., *Cell*, 133, 128–141, 2008.

Morel, J.B., Godon, C., Mourrain, P., Beclin, C., Boutet, S., Feuerbach, F., Proux, F., and Vaucheret, H. Fertile hypomorphic *ARGONAUTE* (*ago1*) mutants impaired in post-transcriptional gene silencing and virus resistance, *Plant Cell*, 14, 629–639, 2002.

Mourelatos, Z., Dostie, J., Paushkin, S., Sharma, A., Charroux, B., Abel, L., Rappsilber, J., Mann, M., and Dreyfuss, G., miRNPs: a novel class of ribonucleoproteins containing numerous microRNAs, *Genes Dev.*, 16, 720–728, 2002.

Nakano, M., Nobuta, K., Vemaraju, K., Tej, S.S., Skogen, J.W., and Meyers, B.C., Plant MPSS databases: signature-based transcriptional resources for analyses of mRNA and small RNA, *Nucleic Acids Res.*, 34, D731–735, 2006.

Onodera, Y., Haag, J.R., Ream, T., Nunes, P.C., Pontes, O., and Pikaard, C.S., Plant nuclear RNA polymerase IV mediates siRNA and DNA methylation-dependent heterochromatin formation, *Cell*, 120, 613–622, 2005.

Osato, N., Yamada, H., Satoh, K., Ooka, H., Yamamoto, M., Suzuki, K., Kawai, J., Carninci, P., Ohtomo, Y., Murakami, K., Matsubara, K., Kikuchi, S., and Hayashizaki, Y., Antisense transcripts with rice fulllength cDNAs, *Genome Biol.*, 5, R5, 2003.

Papp, I., Mette, M.F., Aufsatz, W., Daxinger, L., Schauer, S.E., Ray, A., van der Winden, J., Matzke, M., and Matzke, A.J., Evidence for nuclear processing of plant microRNA and short interfering RNA precursors, *Plant Physiol.*, 132, 1382–1390, 2003.

Park, M.Y., Wu, G., Gonzalez-Sulser, A., Vaucheret, H., and Poethig, R.S., Nuclear processing and export of microRNAs in *Arabidopsis*, *Proc. Natl. Acad. Sci.*, 102, 3691–3696, 2005.

Park, W., Li, J., Song, R., Messing, J., and Chen, X., CARPEL FACTORY, a Dicer homolog, and HEN1, a novel protein, act in microRNA metabolism in *Arabidopsis thaliana*, *Curr. Biol.*, 12, 1484–1495, 2002.

Peragine, A., Yoshikawa, M., Wu, G., Albrecht, H.L., and Poethig, R.S., *SGS3* and *SGS2/SDE1/RDR6* are required for juvenile development and the production of trans-acting siRNAs in *Arabidopsis*, *Genes Dev.*, 18, 2368–2379, 2004.

Pontes, O., Li, C.F., Nunes, P.C., Haag, J., Ream, T., Vitins, A., Jacobsen, S.E., and Pikaard, C.S., The *Arabidopsis* chromatin-modifying nuclear siRNA pathway involves a nucleolar RNA processing center, *Cell*, 126, 79–92, 2006.

Pontier, D., Yahubyan, G., Vega, D., Bulski, A., Saez-Vasquez, J., Hakimi, M.A., Lerbs-Mache, S., Colot, V., and Lagrange, T., Reinforcement of silencing at transposons and highly repeated sequences requires the concerted action of two distinct RNA polymerases IV in *Arabidopsis*, *Genes Dev.*, 19, 2030–2040, 2005.

Qi, Y., Denli, A.M., and Hannon, G.J., Biochemical specialization within *Arabidopsis* RNA silencing pathways, *Mol. Cell*, 19, 421–428, 2005.

Qi, Y., He, X., Wang, X.J., Kohany, O., Jurka, J., and Hannon, G.J., Distinct catalytic and non catalytic roles of ARGONAUTE4 in RNA-directed DNA methylation, *Nature*, 443, 1008–1012, 2006.

Rajagopalan, R., Vaucheret, H., Trejo, J., and Bartel, D.P., A diverse and evolutionarily fluid set of microRNAs in *Arabidopsis thaliana*, *Genes Dev.*, 20, 3407–3425, 2006.

Reinhart, B.J., Weinstein, E.G., Rhoades, M.W., Bartel, B., and Bartel, D.P., MicroRNAs in plants, *Genes Dev.*, 16, 1616–1626, 2002.

Rivas, F.V., Tolia, N.H., Song, J.J., Aragon, J.P., Liu, J., Hannon, G.J., and Joshua-Tor, L., Purified Argonaute2 and an siRNA form recombinant human RISC, *Nat. Struct. Mol. Biol.*, 12, 340–349, 2005.

Schwartz, B.W., Yeung, E.C., and Meinke, D.W., Disruption of morphogenesis and transformation of the suspensor in abnormal suspensor mutants of *Arabidopsis*, *Development*, 120, 3235–3245, 1994.

Shen, B. and Goodman, H.M., Uridine addition after microRNA-directed cleavage, *Science*, 306, 997, 2004.

Sunkar, R., Kapoor, A., and Zhu, J.K., Posttranscriptional induction of two Cu/Zn superoxide dismutase genes in *Arabidopsis* is mediated by downregulation of miR398 and is important for oxidative stress tolerance, *Plant Cell*, 18, 2051–2065, 2006.

Vaucheret, H., Vazquez, F., Crete, P., and Bartel, D.P., The action of ARGONAUTE1 in the miRNA pathway and its regulation by the miRNA pathway are crucial for plant development, *Genes Dev.*, 18, 1187–1197, 2004.

Vaucheret, H., Post-transcriptional small RNA pathways in plants: Mechanisms and regulations, *Genes Dev.*, 20, 759–7771, 2006.

Vazquez, F., Vaucheret, H., Rajagopalan, R., Lepers, C., Gasciolli, V., Mallory, A.C., Hilbert, J.L., Bartel, D.P., and Crete, P., Endogenous trans-acting siRNAs regulate the accumulation of *Arabidopsis* mRNAs, *Mol. Cell*, 16, 69–79, 2004.

Vazquez, F., *Arabidopsis* endogenous small RNAs: Highways and byways, *Trends Plant Sci.*, 11, 460–468, 2006.

Wang, X.J., Gaasterland, T., and Chua, N.H., Genome-wide prediction and identification of cis natural antisense transcripts in *Arabidopsis thaliana*, *Genome Biol.*, 6, R30, 2005.

Wu, F., Yu, L., Cao, W., Mao, Y., Liu, Z., and He, Y., The N-terminal double-stranded RNA binding domains of *Arabidopsis* HYPONASTIC LEAVES1 are sufficient for pre-microRNA processing, *Plant Cell*, 19, 914–925, 2007.

Xie, Z., Allen, E., Fahlgren, N., Calamar, A., Givan, S.A., and Carrington, J.C., Expression of *Arabidopsis* miRNA genes, *Plant Physiol.*, 138, 2145–2154, 2005a.

Xie, Z., Allen, E., Wilken, A., and Carrington, J.C., *DICER-LIKE 4* functions in trans-acting small interfering RNA biogenesis and vegetative phase change in *Arabidopsis thaliana*, *Proc. Natl. Acad. Sci.*, 102, 12984–12989, 2005b.

Xie, Z., Johansen, L.K., Gustafson, A.M., Kasschau, K.D., Lellis, A.D., Zilberman, D., Jacobsen S.E., and Carrington, J.C., Genetic and functional diversification of small RNA pathways in plants. *PLoS Biol.*, 2, E104, 2004.

Yang, L., Liu, Z., Lu, F., Dong, A., and Huang, H., *SERRATE* is a novel nuclear regulator in primary microRNA processing in *Arabidopsis*. *Plant J.*, 47, 841–850, 2006a.

Yang, Z., Ebright, Y.W., Yu, B., and Chen, X., HEN1 recognizes 21–24 nt small RNA duplexes and deposits a methyl group onto the 2' OH of the 3' terminal nucleotide, *Nucleic Acids Res.*, 34, 667–675, 2006b.

Yeom, K.H., Lee, Y., Han, J., Suh, M.R., and Kim, V.N., Characterization of DGCR8/Pasha, the essential cofactor for Drosha in primary miRNA processing, *Nucleic Acids Res.*, 34, 4622–4629, 2006.

Yoshikawa, M., Peragine, A., Park, M.Y., and Poethig, R.S., A pathway for the biogenesis of transacting siRNAs in *Arabidopsis*, *Genes Dev.*, 19, 2164–2175, 2005.

Yu, B., Yang, Z., Li, J., Minakhina, S., Yang, M., Padgett, R.W., Steward, R., and Chen, X., Methylation as a crucial step in plant microRNA biogenesis, *Science*, 307, 932–935, 2005.

Yu, D., Fan, B., MacFarlane, S.A., and Chen, Z., Analysis of the involvement of an inducible *Arabidopsis* RNA-dependent RNA polymerase in antiviral defense, *Mol Plant Microbe Interact.*, 16, 206–216, 2003.

Zheng, X., Zhu, J., Kapoor, A., and Zhu, J., Role of *Arabidopsis AGO6* in siRNA accumulation, DNA methylation and transcriptional gene silencing, *EMBO J.*, 26, 1691–1701, 2007.

Zhu, J., Kapoor, A., Sridhar, V.V., Agius, F., and Zhu, J.K., The DNA glycosylase/lyase ROS1 functions in pruning DNA methylation patterns in *Arabidopsis*, *Curr. Biol.*, 17, 54–59, 2007.

Zilberman, D., Cao, X., Johansen, L.K., Xie, Z., Carrington, J.C., and Jacobsen, S.E., Role of *Arabidopsis ARGONAUTE4* in RNA-directed DNA methylation triggered by inverted repeats, *Curr. Biol.*, 14, 1214–1220, 2004.

# 12 How to Assay microRNA Expression
## *A Technology Guide*

*Mirco Castoldi, Vladimir Benes, and*
*Martina U. Muckenthaler*

## CONTENTS

## OVERVIEW

MicroRNAs (miRNAs) represent a class of short, noncoding RNAs that control gene expression posttranscriptionally. miRNAs regulate numerous cellular processes as diverse as differentiation, proliferation, apoptosis, self-renewal, stress response, and metabolism. Moreover, cellular miRNAs are important for the replication of pathogenic viruses, and small RNAs are also encoded by the genomes of several viruses to facilitate viral replication by suppressing cellular genes. Detection of differential expression of miRNAs in many cases has established the basis for miRNA functional analysis and specific miRNA expression patterns can provide valuable diagnostic and prognostic indications, for example, in the context of human malignancies. Different methodologies have been applied to profile miRNA expression, including miRNA cloning approaches, Northern blotting with radiolabeled probes, quantitative PCR-based (qPCR) amplification of precursor or mature miRNAs, as well as microarray-based profiling technologies. Here we provide an overview of some of the technologies available for miRNA expression analysis.

## 12.1   INTRODUCTION

Sequences of 541 human miRNAs are currently deposited in the miRNA database (miRBase release 10.1, December 2007[1]), but according to bioinformatic predictions, the actual number of miRNAs encoded in the human genome could be in the thousands.[2] This chapter focuses on technologies to assay miRNA expression.

What can be learned from miRNA expression profiles? Alteration of miRNA expression in a disease compared to a healthy state and/or correlation of miRNA expression with clinical parameters such as disease progression or therapy response may indicate that miRNA profiles can serve as clinically relevant biomarkers.[3–5] In addition, the information of differential miRNA expression in cellular processes such as differentiation,[6,7] proliferation, or apoptosis[8] is an important first step for further functional characterization that may allow determination of which disease-causing genes are specifically regulated by miRNAs or vice versa, which genes regulate miRNA expression. Whatever the question you would like to address, the precise information on miRNA expression in a specific cell type or tissue is often considered an important first step. Currently, a range of methods is available for the isolation and profiling of miRNAs. This chapter aims to provide an overview of these technologies.

## 12.2   HISTORICAL PERSPECTIVE ON NONCODING RNA

The central dogma of molecular biology defines how stored genetic information (DNA) is translated into effectors (proteins). It states that DNA makes RNA and RNA makes protein, and that information cannot move backward (e.g., from protein to RNA or from RNA to DNA, James Watson[9,10]). This dogma was challenged by several discoveries such as alternative splicing,[11] retroviruses,[12,13] and prions.[14] During the last decade, the "dogma" has come under scrutiny once more by the discovery of microRNA[15] and other noncoding RNA.[16,17] More than 100,000 transcriptional units are detected across the genome,[18] and transcription not only occurs from those regions that encode proteins (the human genome contains between 20,000 and 25,000 protein coding genes[19]) but also within regions of the genome (e.g., intronic or intergenic regions) that were commonly termed "junk" DNA.[20,21] A proportion of these transcriptional units codes for RNA transcripts that regulate conventional protein-coding mRNAs.[16] miRNAs are included within this category.

The first miRNA (at the time referred to as small temporal RNA), lin-4, was identified by Victor Ambros's group at Harvard Medical School in 1993 while studying the developmental consequences of a mutation in the worm *Caenorhabditis elegans*.[22] It turned out that the correct development of the worm depended on a small noncoding RNA. However, it was not until 2001 that researchers realized that hundreds of miRNAs are found in worms, insects, and mammals. At this point, researchers started to call these regulatory molecules "microRNAs."[23]

## 12.3   WHAT ARE miRNAs?

miRNAs constitute a class of short regulatory RNAs that control gene expression posttranscriptionally. miRNAs are initially transcribed as long precursor RNA molecules (pri-miRNA) that are successively processed by large RNA–protein complexes with RNase III-like activities termed Drosha and Dicer into their mature forms of ~22 nucleotides[24] (exception: miRNAs localized within introns are processed by the spliceosome and therefore have been termed miRtrons[25,26]). Cloning efforts and bioinformatic predictions suggest that miRNAs may regulate up to 30% of mammalian genes.[2] miRNAs control mRNA expression at the level of turnover and/or translation via base pairing, usually to the 3′ untranslated regions.[15] Hundreds of distinct miRNA genes are now known to exist. While for a large number of miRNAs expression is detectable across various tissues, for others expression is spatially or temporally contained to defined developmental stages or tissue types.[27] Overall, little is known about the biological function of animal miRNAs, but recent studies suggest important regulatory roles in a broad range of biological processes including developmental timing,[28] cellular differentiation,[6] proliferation,[29] apoptosis,[30] oncogenesis,[31] insulin secretion,[32] and cholesterol biosynthesis.[33,34] In most cases, however, the specific miRNA target genes are not known.

Both the qualitative and quantitative expression of miRNA are expected to regulate the transcriptome of a given cell or tissue. Therefore, the accurate profiling of miRNA expression represents an important tool to investigate physiological and pathological states of organisms.

## 12.4 WHICH METHODS ARE AVAILABLE FOR miRNA EXPRESSION PROFILING?

miRNA expression can be assessed by various technologies, including Northern blotting with radiolabeled probes,[35] qPCR-based amplification of precursor and mature miRNAs,[36,37] single molecule detection in liquid phase,[38,39] oligonucleotide microarrays spotted onto glass surface,[40–44] oligonucleotide macroarrays spotted on membranes,[45] and detection of miRNA by in situ hybridization.[46,47]

The specificity and accuracy of miRNA-specific expression profiling techniques are challenged (1) by the short length of the mature miRNA (~22 nt), (2) by the presence of not yet fully processed precursor forms that also contain the sequence of the mature miRNA, and (3) by the presence of closely related miRNA family members within the mammalian genome that often differ by as little as a single nucleotide, (4) by the absence of any structure that could serve for capturing these molecules similar to poly A tail or cap, and (5) by the hugely heterogeneous base composition of miRNA sequences. Therefore, protocols for miRNA isolation, enrichment, labeling, detection, and quantification have to be optimized to take these miRNA-specific features into consideration (for a summary of the methods, see Table 12.1).

## 12.5 ISOLATION AND FRACTIONATION OF SMALL RNA

The reliability of miRNA expression profiling largely depends on the quality of the total RNA used as input material. Thus, a robust method for RNA isolation is essential. Historically, miRNA isolation for miRNA expression profiling studies relied on protocols developed for miRNA cloning.[48] These protocols are complex and involve alternating steps of miRNA purification (mainly from polyacrylamide gels [PAG]) and enzymatic reactions.[49] Later on, less complex methods, such as the differential

**TABLE 12.1**

**Summary of the miRNA Analysis Methods Described in the Chapter**

| | Method | Described in | Substrate | Amount of RNA Required | References |
|---|---|---|---|---|---|
| | | Direct Labeling of miRNAs Using | | | |
| 1 | T4 RNA ligase (miChip) | 12.6.1.1.1 and 12.7.1.1 | Total RNA | 2.5 µg Total RNA | Castoldi et al.[40] |
| 2 | PAP (miRvana - Ambion) | 12.7.1.2 | Enriched small RNA | 10 µg Total RNA | Shingara et al.[67] |
| 3 | PAP (Ncode - Invitrogen) | 12.7.1.2 | Enriched small RNA | 10 µg Total RNA | Ncode, http://www.Invitrogen.com |
| 4 | Quantum dots | 12.7.1.3 | Enriched small RNA | 0.2 µg Enriched, small RNA | Liang et al.[43] |
| 5 | chemical modification of miRNAs (Ulysis, Invitrogen) | 12.7.1.3 | Enriched small RNA | 7 µg Total RNA | Babak et al.[55] |
| 6 | the ILLUMINATE labeling kit | 12.7.1.4 | Total RNA | 1 µg Total RNA | http://www.bioventures.com |
| | | Indirect Labeling of miRNA by | | | |
| 7 | PCR using RNA linkers | 12.7.2.1 | Enriched small RNA | 10 µg Total RNA | Miska et al.[49] |
| 8 | PCR using RNA linker with T7 adapter | 12.7.2.1 | Enriched small RNA | 3 µg Enriched, small RNA | Barad et al.[56] |
| 9 | PCR using random priming | 12.7.2.2 | Total RNA | 20 µg Total RNA | Sun et al.[68] |
| 10 | PCR using RNA linkers and Tm normalization | 12.6.1.1 and 12.7.2.1 | Enriched, small RNA | 10 µg Total RNA | Baskerville et al.[53] |

*Continued*

**TABLE 12.1 (*Continued*)**
**Summary of the miRNA Analysis Methods Described in the Chapter**

| | Method | Described in | Substrate | Amount of RNA Required | References |
|---|---|---|---|---|---|
| | **miRNA Expression Profiling in Liquid Phase** | | | | |
| 11 | Detection of individual miRNAs using invasive oligonucleotides | 12.8 | Total RNA | 50 ng Total RNA/miRNA | Allawi et al.[69] |
| 12 | Detection of miRNAs using xMAP technology | 12.8 | Enriched small RNA | 1 μg Total RNA | Lu et al.,[39] Barad et al.[56] |
| 13 | Direct detection of miRNAs in liquid phase | 12.8 | Total RNA | 500 ng Total RNA | Neely et al.[38] |
| | **miRNA Expression Profiling by qPCR** | | | | |
| 14 | Detection of miRNA precursors by qPCR | 12.9.1 | Total RNA | 1 μg Total RNA/genome wide pre-miRNA-RT | Jiang et al.[58] |
| 15 | Ambion platform | 12.9.2.1 | Total RNA | 100 ng Total RNA/miRNA-RT | Wang et al.[86] |
| 16 | Invitrogen platform | 12.9.2.2 | Total RNA | 1 μg Total RNA/genome wide miRNA-RT | Shi et al.,[75] Invitrogen.com |
| 17 | ABI platform | 12.9.2.3 | Total RNA | 10 ng Total RNA/miRNA-RT | Chen et al.[36] |
| 18 | ABI platform with multiplexed RT-step | 12.9.3 | Total RNA | As low as 0.01 pg total RNA/genome wide miRNA-RT | Tang et al.[78] |
| 19 | Using DNA-LNA modified primers | 12.9.4 | Total RNA | 100 ng Total RNA/miRNA-RT | Raymond et al.[62] |
| | **Detection of miRNA Using Other Techniques** | | | | |
| 20 | Northern blot | 12.10 | Total RNA | 10 to 20 μg Total RNA | Valoczi et al.[35] |
| 21 | Oligonucleotide macroarray | 12.6.2.2 | Enriched small RNA | 5 to 10 μg of Total RNA | Krichevsky et al.[45] |
| 22 | Detection of miRNA using RAKE | 12.6.2.1 | Enriched small RNA | 2 μg Total RNA | Nelson et al.[82,44] |

enrichment of small nucleic acids by polyethylene glycol (PEG) precipitation[42] or size fractionation using silica-based columns with a specific cut-off[45] were introduced. However, some of these methods, although faster than PAGE enrichment, can be prone to a significant loss of miRNA (unpublished observation).

Because miRNAs play such a prominent role in the posttranscriptional regulation of gene expression, miRNA profiling is now widely utilized by molecular biologists and clinicians. Thus, robust and efficient protocols for the purification of miRNAs needed to be developed.

### 12.5.1 TOTAL RNA EXTRACTION USING PHENOL-CONTAINING REAGENTS

RNA extraction using acid phenol[50,51] is a straightforward, robust methodology to isolate total RNA. In addition to the small RNA fraction it also recovers mRNA and rRNA. Thus, the total RNA recovered is not only suitable for miRNA profiling but can be used in parallel to generate mRNA expression profiles as well. Several commercial reagents such as Trizol (Invitrogen), Quiazol (Qiagen), or Tri reagent (a product with this name is available from different companies such as Sigma, Ambion, and others) are based on this principle. The actual reagent is a monophasic solution of acid phenol and guanidine isothiocyanate. During sample homogenization, the integrity of the RNA is maintained, while cells are disrupted and cellular components are dissolved. Addition of chloroform followed by centrifugation separates the solution into an aqueous phase and an organic phase. RNA remains exclusively in the aqueous phase, while DNA and proteins stay in the organic phase. Finally, RNA is precipitated by the addition of an alcohol (such as isopropanol or ethanol) followed by centrifugation. Please note that phenol-extracted RNA may display an additional absorbance peak at 230 nm. This peak indicates phenol-derived contaminants within the RNA preparation that may impair miRNA expression profiling. To clean the RNA preparation from these contaminants, we suggest adding a vol/vol of chloroform, mixing by hand shaking, and spinning for 10 minutes at maximum speed in the cold room. Repeat chloroform extraction twice, and then perform standard ethanol precipitation to remove all the traces of phenol from the sample.

### 12.5.2 ELUTION OF SMALL RNAS FROM GEL SLABS

A current benchmark method for miRNA enrichment utilizes total RNA separation by polyacrylamide gel electrophoresis (PAGE). Low-molecular-weight RNAs are excised from the gel, eluted, and subsequently concentrated.[49] Although this method is still widely applied, the experimental procedure is time consuming and usually requires large amounts of starting material.

The same principle is now applied in a commercial product (FlashPAGE®; Ambion). FlashPAGE is a miniaturized cartridge-based PAGE that isolates small single-stranded nucleic acids by an electrophoresis step and thus separates mature miRNAs from precursor miRNAs and larger mRNAs. Enriched miRNAs are then directly collected from the cartridge.

### 12.5.3 METHODS FOR miRNA ENRICHMENT

Some methods used for miRNA expression analysis require the separation of mature miRNAs from miRNA precursor and high-molecular-weight RNA. Therefore, protocols were developed to enrich small RNA either directly from cell and tissue extracts (e.g., miRvana®, Ambion, or miRacle®, Stratagene) or from previously purified total RNA. These methodologies include miRNA enrichment by solid-phase-based (e.g., column-based) methodologies whereby small RNAs are separated from large RNAs through their capacity to bind with a column (e.g., available from Qiagen, Ambion, and Stratagene). An additional technique used for small RNA enrichment is based on microfiltration. Here, high-molecular-weight RNA will be retained in the resin of the column, while low-molecular-weight RNA can be collected within the flow through fraction (see Reference 45, microcon centrifugal units Ym 100, Millipore). Finally, small RNA can also be enriched and separated from high-molecular-weight RNA using reverse-phase chromatography by HPLC.[52]

Some of these methods (e.g., combination of Qiagen RNeasy® and Qiagen MiniElute®, microfiltration by Ym 100 centrifugal units and reverse-phase chromatography) additionally allow for the isolation of the high molecular weight RNA fraction, which can be used for mRNA analysis from the identical starting material. Such methodologies are particularly interesting if the starting material is limiting.

As an alternative to commercial kits and to PAGE fractionation described earlier, small RNAs can also be enriched through a selective-precipitations step. For example, an aqueous solution containing a blended mixture of PEG and salt will effectively precipitate the high-molecular-weight RNA, while the low-molecular-weight RNA will remain in the aqueous phase.[42]

## 12.6 miRNA EXPRESSION PROFILING USING MICROARRAYS

Microarray-based miRNA profiling assays constitute an efficient methodology to screen for the expression of a large number of miRNAs in a parallel fashion. The general principles of miRNA and mRNA expression profiling are similar: (1) Capture probes complementary to the mRNA/miRNA target sequences are immobilized on glass surfaces; (2) mRNAs/miRNAs contained within total RNA of interest are labeled with fluorescent dyes and are subsequently hybridized to the microarray; and (3) signals obtained after hybridization and specific for each capture probe are measured and related to the amount of mRNA/miRNA present within the total RNA.

In contrast to mRNA expression profiling, which assays for target sequences of up to several thousand nucleotides of length, the flexibility in capture-probe design to detect miRNA expression is hampered by the small size of miRNAs and the existence of closely related family members that differ by as little as one nucleotide. Various miRNA-specific microarray platforms have been developed in recent years. Some of the available alternatives will be discussed in this section.

### 12.6.1 EXPERIMENTAL DESIGNS FOR miRNA-SPECIFIC MICROARRAY PLATFORMS

Most of the microarray platforms currently available[44,49,53–56] rely on DNA oligonucleotides that are either complementary to the sequence of the mature miRNA (see References 49,54–56; DNA oligonucleotides of approximately 20–25 nucleotides) or to the sequence of the mature miRNA plus additional nucleotides flanking the mature miRNA sequence that are derived from the stem-loop sequences of the miRNA precursor form (see Reference 57; DNA oligonucleotides of approximately 40 nucleotides). In contrast to 20-mer oligonucleotide capture probes, 40-mer DNA oligonucleotides bind more efficiently to most surfaces (for microarray surface selection, see Section 12.6.3), but their utility is hampered by the fact that they recognize both the mature miRNA form and its precursor. Because miRNA biogenesis is a regulated process and the amount of mature miRNA is not always proportional to the respective precursor form, it may be difficult to interpret the signals obtained.[58]

### 12.6.1.1 miRNA-Specific Microarray Platforms That Rely on $T_m$ Normalization of the Capture Probes

An additional level of complexity derives from the fact that the predicted melting temperatures ($T_m$'s) of mature miRNA duplexes vary between 45°C and 74°C.[1] It is therefore impossible to establish hybridization conditions so that all mature miRNAs are recognized by DNA capture probes with equal sensitivity and specificity. A solution to this problem is the use of $T_m$-adjusted capture probes. Three independent principles are currently applied to achieve $T_m$ adjustment of the capture probes:

1. To reduce the large range of $T_m$ values across the miRNome, DNA capture probes with predicted $T_m$ values above 55°C are shortened such that their $T_m$ is reduced, while the capture probe with a predicted Tm below 55°C remain unaltered (MIRmax microarray platform; http://cord.rutgers.edu/mirmax/). To increase specificity, the capture probe sequences are organized in trimeric repeats to generate 60-nucleotide-long DNA oligomers that are spotted onto microarrays.[59]
2. In the second approach, DNA capture probes with a predicted $T_m$ above 55°C (versus their target miRNAs) are shortened until a calculated $T_m$ of 55°C is reached. By contrast, capture probes with a predicted $T_m$ below 55°C are artificially elongated (up to 5 nucleotides) to increase their predicted Tm. The sequence of the elongated nucleotides corresponds to an adapter that is ligated to miRNAs prior to a reverse transcription and labeling step[53] (see Section 12.7.2.1 for a description of the principle).
3. In the third approach, DNA capture probes are modified by the incorporation of locked nucleic acid (LNA) monomers as well as sequence length variation (see Section 12.6.1.1.1 for a description of the principle; see Reference 40; the LNA-modified capture probes are commercially available as miRCURY® probe sets, Exiqon).

#### 12.6.1.1.1 LNA Modified Capture Probes

LNA is a synthetic RNA analog characterized by increased thermostability of nucleic acid duplexes when LNA monomers are introduced into oligonucleotides.[60] It has been shown that each incorporated LNA monomer increases the $T_m$ of a DNA/RNA hybrid by 2–10°C.[35] As results of these properties, LNA-modified capture probes can be designed in such a way that a uniform $T_m$ can be achieved for the genome-wide set of miRNAs by adjusting the LNA content and the length of the individual capture probes. Tm normalization of capture probes permits establishing hybridization conditions suitable for all miRNAs. Hybridization conditions optimized for high-stringency binding not only accurately detect the expression of miRNAs with high sensitivity, but also increase the hybridization specificity for related miRNA family members that often differ by as little as a single nucleotide.[1] This principle has been applied by Castoldi et al.[40,61] to generate the miChip microarray platform. In addition to its use on microarrays, LNA-modified oligonucleotides have also been successfully used for miRNA detection by Northern blotting,[35] flow cytometry,[38] qPCR,[62] and ISH (in situ hybridization[46]).

## 12.6.2 ALTERNATIVE PRINCIPLES FOR miRNA PROFILING USING MICROARRAYS

### 12.6.2.1 RNA-Primed Array-Based Klenow Enzyme (RAKE) Microarray Platform

An alternative approach to detect miRNA expression is by the RNA-primed array-based Klenow enzyme (RAKE) microarray platform. This approach was developed by Nelson and co-workers[44] and relies on the property of the *E. coli* DNA polymerase I Klenow fragment to elongate DNA-RNA hybrids. In brief, miRNAs are hybridized to surface-bound complementary DNA capture probes. After hybridization, the array is washed and processed for miRNA detection, whereby as a first step, nonhybridized ssDNA is removed by Exonuclease I digestion; then, the DNA capture probes designed to have a common nucleotide sequence at their 5′ end are elongated using the Klenow enzyme and biotin-conjugated nucleotides. Subsequent incubation of the arrays with fluorophore-conjugated streptavidin allows for specific miRNA detection.

## 12.6.3 MICROARRAY SURFACES COMMONLY USED FOR miRNA PROFILING

Microarrays can be fabricated using various technologies, including photolithography with prefabricated masks or with dynamic micromirror devices, ink-jet printing, or electrochemistry on microelectrode arrays. However, most of the currently available miRNA-specific microarray platforms apply contact-printing technology. For the purpose of contact-printing, microarray surfaces can be roughly divided into two different types, three-dimensional surfaces, such as *N*-hydroxysuccinimide coated slides (CodeLink®, GE HealthCare, formerly Amersham) and two-dimensional (2D) surfaces such as Aminosilane (3-Aminopropyl trimethoxysilane) coated slides (e.g., GAPS2®, Corning) and Epoxide (3-Glycidoxypropyl trimethoxysilane) coated slides (e.g., UltraGaps®, Corning). The *N*-hydroxysuccinimide coated slide surface offers the following advantages over other surfaces: capture probes are bound to the slide surface (1) in a length-independent manner, (2) with high density, (3) and only by the 3′ or 5′ end—therefore, they are free to float, facilitating the hybridization with the labeled probe, which must contain an amino group at either the 3′ or 5′ end.

Alternatively, capture probes can be immobilized on nylon membranes rather than glass slides to generate oligonucleotide macroarrays (macroarrays are defined by a spot size of mm rather than μm and are printed on porous membranes, e.g., nylon). Macroarrays are cheaper alternatives to glass-spotted microarrays, and the membrane can be stripped and rehybridized several times. Similar to Northern blotting, macroarrays are usually hybridized to radioactively labeled miRNA probes[45,63] but have the advantage over Northern analysis that a single RNA sample can be analyzed for a large panel of miRNAs.

### 12.6.4 FLUORESCENT MOLECULES COMMONLY USED FOR MIRNA PROFILING

Cyanine dyes (Cy3 [absorption wavelength 554 nm, emission wavelength 570 nm] and Cy5 [absorption wavelength of 649 nm, emission wavelength of 666 nm]) are commonly used for miRNA labeling in microarray experiments. However, a significant disadvantage of Cy5 dyes is the decreased stability and fluorescence intensity when ozone levels are high. Because Cy5 dyes are affected more strongly by ozone compared to Cy3 dyes, this may introduce artifacts within dual-color microarray experiments. These artifacts are recognized if the Cy3 and Cy5 dyes are switched between the experimental and control sample and the resulting data are compared. Alexa-fluorophores (Alexa-555 [absorption wavelength 556 nm, emission wavelength 572 nm] and Alexa-647 [absorption wavelength 647 nm, emission wavelength 666 nm]) can be used as alternatives. Alexa-dyes are more resistant to photo-bleaching compared to Cy-dyes[64] and Alexa-647 (Cy5 replacement) remains unaffected by high ozone levels.[65] However, Alexa-dyes so far have failed to replace the Cy-dyes because Alexa-dyes compared to Cy-dyes exert lower hydrophobicity; it is therefore difficult to separate coupled from uncoupled dyes by RP-HPLC. Moreover, Cy-dyes can directly be incorporated during oligonucleotide synthesis, which is one great advantage. Because of a lower chemical resistance of Alexa-dyes, this modification can only be introduced postsynthetically into oligonucleotides (Dr. Hüseyin Aygün; Biospring GmbH, personal communications).

## 12.7 MIRNA LABELING PROTOCOLS FOR SUBSEQUENT HYBRIDIZATION TO MICROARRAY

Contrary to the detection of miRNAs by RAKE (see Section 12.6.2.1), the majority of miRNA profiling protocols require miRNA labeling preceding their hybridization to the array. Often, the fluorescent dye(s) are introduced by enzymatic reactions or chemical modification of specific ribonucleotides. The following paragraphs focus on labeling methodologies that utilize either direct (Sections 12.7.1.1 to 12.7.1.4) or indirect methods (Sections 12.7.2.1 and 12.7.2.2) to append detection molecules to miRNAs.

### 12.7.1 DIRECT LABELING OF MIRNAS

Here, in this chapter we will focus on four principles to directly label miRNAs: (1) ligation of a fluorescently labeled linker to the miRNA, (2) addition of fluorescently labeled nucleotides to the miRNA 3′ end using poly(A) polymerase, (3) attachment of fluorophores or reactive groups to the miRNA through chemical modification of specific nucleotides, and (4) Illuminate. (Schematic representation of direct labeling methods is sketched in Figure 12.1.)

### 12.7.1.1 Direct Labeling of miRNAs Using Fluorescent RNA-Linkers

This protocol exploits the properties of the single-strand specific T4 RNA ligase to efficiently ligate the 5′ end (phosphate) of a donor ribonucleotide to the 3′-end (-OH) of an acceptor RNA molecule.[66] Specifically, a short Cy-dye labeled RNA linker (donor) is ligated to the single-stranded miRNA (acceptor) (schematic representation in Figure 12.1a). It is noteworthy that T4 RNA ligase ligates Cy-dye-labeled RNA linkers to mature miRNAs with higher efficiency than to the miRNA precursor forms (unpublished observations). Because usually only a single fluorescent dye is conjugated to the RNA linker molecule, this protocol requires sufficient starting material (2–3 ug total RNA) to also detect miRNAs that are expressed at low levels.[40]

### 12.7.1.2 Direct miRNA Labeling Utilizing miRNA Tailing by Poly(A) Polymerase

The enzymatic addition of homopolymeric ribonucleotide tails to miRNA 3′ ends is catalyzed by Poly(A) polymerase (PAP), whereby PAP preferentially adds adenines to ribonucleotide tails. Because PAP is able to add several fluorescently labeled adenines to a single RNA molecule, this in principle will enable sensitive miRNA detection. However, PAP preferentially uses high-molecular-weight RNA templates as substrate and, thus, the reaction may be biased against miRNAs. It is thus recommended that only the low-molecular-weight RNA fraction is used as input material. In addition, the determination of the tailing efficiency is problematic, which may result in unaccounted for chip-to-chip variability. Two protocols based on miRNA labeling by PAP are available. Shingara and co-workers[67] use miRvana (Ambion)-enriched miRNAs as starting material for a tailing reaction that adds unmodified and amino-modified nucleotides to the miRNA 3′ ends. In a second step, the amino-modified nucleotides are coupled to monoreactive NHS-ester dyes (such as Cy3 and Cy5 from GE HealthCare) prior to microarray hybridization (schematic representation in Figure 12.1b). A reagent kit utilizing this protocol is commercially available (miRvana miRNA labeling kit, Ambion). A further commercial product that utilizes PAP enzymatic activity to label miRNAs is available from Invitrogen (NCode). Here, detection is based on Alexa-dyes (Invitrogen). To preserve the sensitivity of the system Alexa-dyes are used for miRNA detection in a posthybridization step (schematic representation in Figure 12.1c).

### 12.7.1.3 Direct miRNA Labeling by Chemical Modification

Enzymatic reactions may not be equally efficient, depending on the tissue from which miRNAs are extracted. Direct chemical modification of miRNAs is thought to overcome this problem. To efficiently label miRNAs contained in a total RNA pool, Babak and co-workers[55] rely on a platinum-based chemical labeling reagent to attach Alexa-546 or Alexa-647 fluorescent dyes directly to guanidine (G) residues

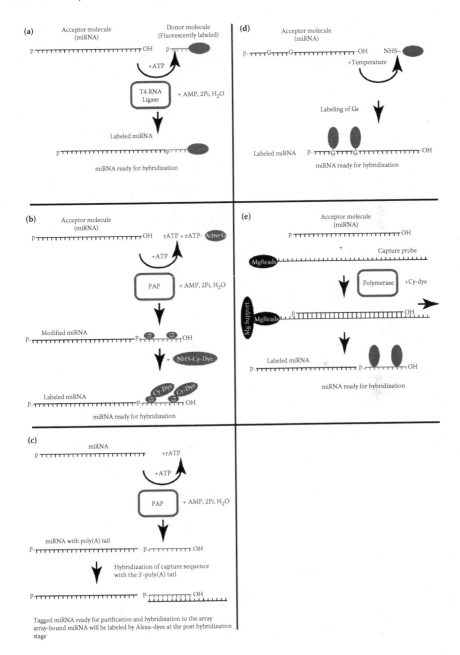

**FIGURE 12.1** Schematic representations of direct miRNA labeling techniques (for a detailed description, see text). (a) Direct labeling of miRNAs using fluorescent RNA-linkers (12.7.1.1). (b) Direct miRNA labeling utilizing miRNA tailing by Poly(A) polymerase (12.7.1.2). (c) Addition of homopolymeric ribonucleotide tails to miRNAs using PAP for posthybridization labeling (12.7.1.2). (d) Direct labeling of miRNAs by chemical modification of specific nucleotides (12.7.1.3). (e) Direct labeling of miRNAs using magnetic beads (Illuminate, 12.7.1.4).

within the miRNA sequence (Ulysis, Alexa fluor, Invitrogen; see Reference 55, schematic representation in Figure 12.1d). All miRNAs identified so far indeed contain at least one G residue within the mature miRNA sequence. Approximately 7μg of total RNA are required per reaction. Whether or not guanidine chemical modification affects the hybridization step is unclear.

An alternative principle to chemically modify miRNAs is based on the Quantum Dots technology (QD, Invitrogen). QD is a novel type of fluorescence dye with several advantages over classical organic dyes.[43] QDs are semiconductor nanostructures, and one of the optical features of QD is coloration. While the material that makes up a QD defines its intrinsic energy signature, significant in terms of coloration is the QD size. Large QD are red, while small QD are blue. Through QD size modification it is possible, at least in theory, to generate an unlimited number of colors. To ensure miRNA labeling, the 3' end of the previously enriched low-molecular-weight RNAs are first periodated-oxidated and subsequently conjugated to biotin through a condensation reaction.[43] The biotin-labeled miRNAs are hybridized to a miRNA-specific microarray platform. For detection, the arrays are incubated with a solution containing streptavidin conjugated QDs (Invitrogen) that will bind to the biotin-tagged miRNA.

Protocols applying direct chemical modification of miRNAs are still hampered by (1) low selectivity for miRNAs (e.g., high-molecular-weight RNAs are modified equally well), (2) limited space for the addition of fluorophores, and (3) apparent low sensitivity, because large amounts (10 to 20 μg) of starting material is required for most protocols.

### 12.7.1.4    miRNA Labeling Using Illuminate Magnetic Beads

In this alternative approach a mixture of capture probes complementary to known miRNAs are conjugated to magnetic beads (Illuminate μRNA labeling kit, http://www.bioventures.com). The capture probes are designed such that in addition to a region that is 100% complementary to the mature miRNAs, they contain additional sequences at their 3' and 5' ends that are common to all the capture probes. In solution, the capture probes are hybridized to total RNA, and then the 3' ends of the capture-probe-bound miRNAs are extended by an enzymatic reaction with a nucleotide mixture containing Cy-dyes. Finally, the magnetic bead-bound capture probes are collected using a magnetic base. The labeled miRNAs are eluted from the corresponding capture probes and hybridized to a miRNA specific microarray platform (schematic representation in Figure 12.1e).

### 12.7.2    INDIRECT LABELING OF MIRNAS

In the following paragraphs, we will discuss two methodologies for indirect labeling of miRNAs: (1) PCR-based amplification of miRNAs and (2) reverse transcription of miRNAs using random primers. (Schematic representations of indirect labeling methods are sketched in Figure 12.2.)

### 12.7.2.1    Indirect miRNA Labeling by PCR-Based Amplification

Most miRNA labeling protocols require several micrograms of starting material for sensitive detection on a microarray. However, for clinical applications using patient

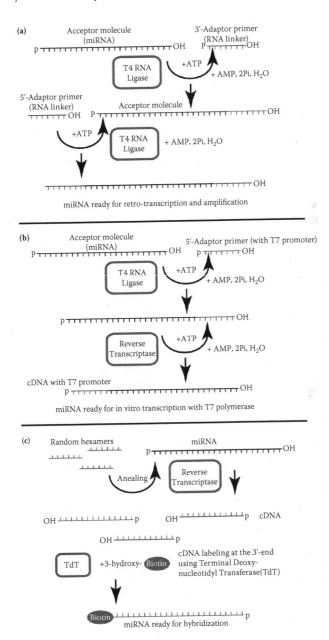

**FIGURE 12.2** Schematic representations of indirect miRNA labeling techniques (for detailed description, see text). (a) Indirect labeling of miRNAs using 3'- and 5'-end adaptor primers for cDNA synthesis and amplification (12.7.2.1). (b) Indirect labeling of miRNAs using a 3' end composite primer containing a T7 promoter for cDNA and cRNA synthesis (12.7.2.1). (c) Indirect labeling of miRNAs using random hexamer primed cDNA synthesis and terminal deoxynucleotide transferase with Biotin-ddUTP (12.7.2.2).

biopsies the starting material can be limiting. Miska and co-workers[49] adapted methodology that was originally developed for miRNA cloning approaches to fluorescently label miRNAs for microarray-based expression profiling (schematic representation in Figure 12.2a). In a first step, adaptor molecules are sequentially ligated to the 3' and 5' ends of miRNAs using T4 RNA ligase, and the ligation products are purified from PAGEs. miRNAs with adaptors at both ends are then reverse-transcribed and amplified by PCR, whereas for the synthesis of the amplified sense strand a Cy3-conjugated amplification primer is used. The labeled sense strand is then hybridized to a microarray with immobilized antisense capture probes. While this method allows for the sensitive detection of miRNAs on a microarray, accuracy of miRNA detection may be impaired by suboptimal efficiency of enzymatic reverse transcription and/or PCR amplification.

Based on this labeling strategy Baskerville et al.[53] established an miRNA-specific microarray platform that enables $T_m$ normalization of capture probes (see also Section 12.6.1.1). Capture probes with a predicted Tm above 55°C are shortened to match the Tm. Capture probes with a predicted $T_m$ below 55°C are designed with additional nucleotides (up to five) to match the adapter sequence that was ligated to the miRNAs prior to the reverse transcription step.[53]

Related to these principles Barad et al.[56] developed an miRNA specific microarray platform (MIRchip) that relies on adapter ligation to previously enriched, small RNA. In contrast to the methods described previously, however, one of the adaptor primers contains a binding sequence for T7 polymerase. Thus, following reverse transcription, the resulting cDNA can be amplified by T7 polymerase, and Cy-dye-labeled ribonucleotides are incorporated into the cRNAs during this process (schematic representation in Figure 12.2b[56]).

### 12.7.2.2   Indirect miRNA Labeling by Reverse Transcription Using Random Primers

To overcome work-intensive protocols that rely on linker ligation to miRNAs, Yingqing Sun and co-workers[68] apply random hexamer-primed, reverse transcription of mature miRNAs. After RNA digestion, the cDNAs are tailed using terminal deoxynucleotide transferase and Biotin-dUTP and hybridized to a microarray platform. After hybridization, slides are washed, followed by two consecutive incubations with R-Phycoerythrin-streptavidin (Invitrogen) to amplify the signal obtained (schematic representation in Figure 12.2c).

## 12.8   MiRNA EXPRESSION PROFILING IN LIQUID PHASE

Single molecule detection in liquid phase[38,39,56,69] is a recently described alternative to miRNA expression profiling using microarrays. Four independent platforms are currently available:

1. Allawi et al.[69] described a fluorimetric, plate-based detection method for individual miRNAs, called the Invader™ assay. Two miRNA specific oligonucleotides, termed probe and invasive oligonucleotides, are annealed to

the 5′ end (probe oligonucleotide) and 3′ end (invasive oligonucleotide) of the mature miRNA. The oligos are designed in such a way that the 3′ end of the "invasive" oligonucleotide overlaps with the central region of the probe oligonucleotide, and so that the 5′ end of the probe oligonucleotide forms a protruding structure known as the 5′ flap. The probe oligonucleotide contains a fluorescent (F) molecule within the 5′ flap region and a quencher (Q) molecule adjacent to the area that overlaps between the probe and the invasive oligonucleotide; Q and F form a FRET (Fluorescence resonance energy transfer) pair. If the F and Q molecules are in close proximity, no fluorescence will be emitted. The overlapping sequence between the probe and the invader oligonucleotide can be targeted by nuclease digestion, releasing the flap containing the F molecule from the Q. The emitted fluorescence can be quantified using a fluorescence plate reader and related to the number of miRNA molecules present. Advantages of the invader assay include direct miRNA detection in cell lysates and the low amount of starting material (50–100 ng of total RNA) required.

2. Similar to the methods described in (1) Neely et al.[38] developed a simplified protocol for miRNA expression profiling in liquid phase. The general principle underlying this technique is the single molecule detection-laser-induced fluorescence (LIF; see References 70, 71). Briefly, two miRNA-specific LNA-modified DNA capture probes conjugated to different fluorophores are hybridized to the 5′ and 3′ end of each miRNA. The fluorophores can generate hundreds of different colors. Upon hybridization in solution, a combination of two-fluorophore-labeled miRNAs and a variable amount of "free" capture probes that did not find a target miRNA are obtained. Those capture probes that did not find an miRNA target are blocked by the activity of a quencher present in the hybridization solution. The mixture is then passed through a multifluidic, multicolor confocal laser system capable of counting individual molecules as they flow through the system. The read-out allows the determination of the amount and specificity of the miRNAs contained in the sample.

3. With the purpose to validate miRNA microarray analysis Barad et al.[56] implemented the mirMASA system that is based on Luminex xMAP technology (http://gene.genaco.com/technology.html[72,59]), which uses an "in-suspension array" for multiplex detection of miRNAs in solution. The actual method is a combination of the principles explained in (1) and (2) (see preceding text). Each miRNA is recognized by two specific LNA-modified capture oligonucleotides. One capture oligonucleotide is covalently coupled to color-coded microspheres, thus identifying the respective miRNA by a color code. The second-capture oligonucleotide is coupled to phycoerythrin, which enables quantifying of a specific miRNA target. After the two LNA-modified capture oligonucleotides are hybridized to the respective miRNA in solution, the mixture is passed through a microfluidics system equipped with a dual-laser detection device to decipher the different fluorescent emissions. The study by Lu et al.[39] implemented xMAP technology.

## 12.9  miRNA EXPRESSION PROFILING USING QUANTITATIVE REAL-TIME PCR

miRNA-specific real-time quantitative PCR (qPCR) is frequently used to validate data obtained from microarray experiments. At first glance, major advantages of this technology over microarrays are (1) the speed of the assay, (2) the increased sensitivity with which miRNAs expressed at low levels can be measured, (3) the extended dynamic range compared to microarray analysis, and (4) the requirement for low amounts of starting material (10 ng/reaction). If, however, the aim is to perform "genome-wide" miRNA profiling of all 541 human miRNAs included in miRBase (release 10.1; December 2007[1]) by qPCR, together with technical replicates and nontemplate controls (NTC), according to the number of technical replicas and controls, 4 to 6 384 well plates will be required for each analyzed sample. This will bring the amount of total RNA required to more than 10 µg and the overall cost and time required for miRNA profiling by qPCR well above the ones required for microarray analysis.

A major challenge of miRNA analysis by qPCR resides in the small size of the mature miRNA. The absence of a common anchor sequence, such as a poly(A) tail, makes it necessary that miRNA specific primers have to be designed for each miRNA analyzed. Currently, six different principles to quantify miRNA expression by qPCR analyses are available (schematic representation of qPCR methods are sketched in Figure 12.3).

### 12.9.1  DETECTION OF miRNA PRECURSORS BY qPCR

Schmittgen et al.[73] have developed a technique for miRNA precursor detection by qPCR. Originally, it was thought that this technique could ultimately substitute for the detection of mature miRNAs. However, it is now recognized that miRNA processing consists of several regulated steps, that the ratio between the precursor and the mature miRNA form can be different from one, and that this ratio may be affected in a pathological state.[58] The underlying experimental principle is similar to the one of standard qPCR analysis of mRNAs with the exception that miRNA precursor from specific primers are used to prime the reverse transcription step. Extensive removal of the genomic DNA is a prerequisite for this method because the amplification primers will also recognize genomic DNA contaminants. In addition, the enrichment for the low molecular weight RNA fraction is recommended.[73]

### 12.9.2  DETECTION OF MATURE miRNAs BY qPCR

#### 12.9.2.1  miRNA Detection Using miRNA Specific Primers for Reverse Transcription

Despite the short length of mature miRNAs, specific complementary primers can be annealed to prime a reverse transcription step. The resulting cDNA is then used as a substrate for a qPCR reaction with one miRNA specific primer and a

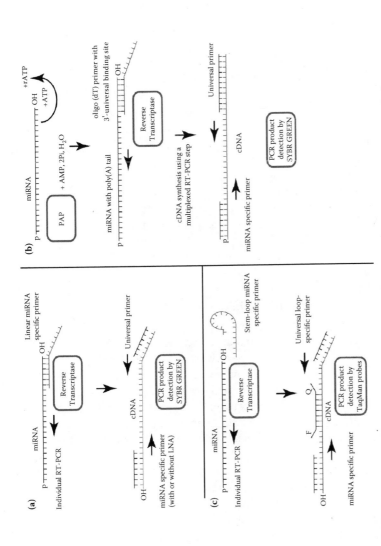

**FIGURE 12.3** Schematic representations of qPCR techniques for the amplification of mature miRNAs: (a) Amplification of mature miRNAs using linear miRNA specific primers for reverse transcription and amplification product detection by SYBR Green (Sections 12.9.2.1 and 12.9.4). (b) Amplification of mature miRNAs using miRNA specific stem-loop primers for reverse transcription and PCR amplification product detection by TaqMan probes (Section 12.9.2.2). (c) Amplification of mature miRNAs with a 3′ end synthetic poly(A) tail added for oligo d(T) primed reverse transcription and qPCR detection by SYBR Green (12.9.3).

second universal primer, whose binding site is included within the initial primer that catalyzed the reverse transcription step. SYBR green incorporated into the amplification products enables detection. Approximately 50 ng starting material is required for each qPCR reaction. The disadvantage of this method consists in (1) a lack of discrimination between mature, precursor, and primary miRNA, and (2) the absence of a multiplexed reverse transcription step.[74] A commercial product available from Ambion is based on a similar approach (schematic representation in Figure 12.3a).

## 12.9.2.2  miRNA Detection Using a Synthetic Poly(A) Tail for Reverse Transcription

In this alternative approach, PAP is employed to add a homopolymeric adenosine tail to miRNAs.[75] Then, reverse transcription is primed by a composite primer that contains poly(dT) residues at its 3′ end and a specific universal primer binding site at its 5′ end. The resulting cDNA is then used as a template for qPCR analysis using an miRNA-specific primer and the universal primer matching the sequence of the reverse transcription primer. One of the advantages offered by this method is that the introduction of a sequence common to all miRNAs (the poly[A] tail) allows the introduction of a multiplexed reverse transcription step facilitating the analysis (schematic representation in Figure 12.3b). This product is commercially available from Invitrogen (NCode SYBR Green miRNA qRT-PCR Kit® Invitrogen; note that products based on a similar approach have also been developed by Qiagen and Stratagene).

## 12.9.2.3  miRNA Detection Using Stem-Loop Primers for Reverse Transcription

A third approach is based on special stem-loop structured primers that in part are complementary to the miRNA and that contain an additional sequence with a universal primer-binding site. Due to their structure and sequence these primers can prime a reverse transcription reaction only from mature miRNAs. The resulting cDNAs are amplified by qPCR with a miRNA-specific and a universal primer. In this case, the detection of the amplification product is not based on the intercalation of SYBRGreen, but on the TaqMan principle[76]; the issue of the miRNA length constraint is dealt with by utilizing so-called 3′-minor groove binders (MGB).[77] TaqMan probes hybridize to the central region of the amplified PCR product and carry both a fluorescent reporter and a quencher in adjacent positions. Close proximity of the fluorescent reporter to its quencher molecule prevents the emission of fluorescence. If the TaqMan probe is degraded by the 5′–3′ exonuclease activity of the polymerase used for PCR amplification, the fluorescent reporter will be separated from its quencher, and fluorescence will be released. The emission of fluorescence is proportional to the amount of PCR products generated, and this will allow accurate quantitation of the initial and final quantities of the cDNA. The method is sensitive, only approximately 10 ng of total RNA are required as starting material, and the data obtained highly correlate with data obtained from miChip analysis (see 12.6.1.1.1; unpublished observation). The qPCR system from Applied Biosystems (AB, see Reference 36) is based upon this principle (schematic representation in Figure 12.3c).

### 12.9.3 MiRNA Detection Using Multiplex Primers for Reverse Transcription

Genome-wide analysis of miRNA expression using qPCR is both time and reagent consuming. Tang and co-workers[78] described an approach to reduce both the handling time of the samples and the amount of material required for genome-wide analysis of miRNAs through the introduction of a multiplexed reverse transcription step (e.g., miRNAs are reverse transcribed in the same tube). For this purpose they modified the approach described by Chen et al. (see Section 12.9.2.3 and Reference 36) to apply the multiplex RT-PCR principle to the amplification of miRNAs. In a first step, a mixture of stem-loop RT-primers hybridizes to the mature miRNAs to enable a multiplexed reverse transcription step. Then, the resulting cDNA is amplified in the presence of low amounts of qPCR primers (pre-PCR; this step is also carried out in a single tube). Subsequently, the pre-PCR amplification product is diluted, and a fraction of the dilution is used for qPCR in 96 or 384 well plates using qPCR primers and TaqMan probes (AB).

### 12.9.4 MiRNA Detection Using DNA-LNA Modified Linear Primers

Raymond et al.[62] used LNA-modified DNA primers for miRNA specific qPCR analysis. Similar to the qPCR system introduced by Ambion (see Section 12.9.2.1), the reverse transcription of mature miRNAs is primed using composite DNA primers that contain complementarities to each miRNA at their 5′ ends and a universal primer binding site at the 3′ end. The subsequent qPCR analysis is performed using both LNA-modified (see 6.1.1.1) universal and miRNA specific primers, which results in increased accuracy and sensitivity of the amplification step. According to the authors,[62] however, a large amount of starting material (~500 ng/RT) is required for the reverse transcription reaction (schematic representation in Figure 12.3a).

## 12.10 MiRNA DETECTION BY NORTHERN BLOTTING

miRNA-specific Northern blot analysis will yield quantitative and qualitative information. However, it is both time and reagent consuming, and not the method of choice if the starting material is limiting. For Northern analysis usually 10 to 30 μg of total RNA are size separated on a denaturing polyacrylamide gel. Subsequently, the nucleic acids are transferred to a nylon or nitrocellulose membrane by semi-dry or wet blotting. UV exposure or baking of the membrane, respectively, covalently binds nucleic acids to the membrane. Then, the membranes are hybridized to radiolabeled DNA or LNA-modified probes that are complementary to the miRNA of interest. LNA modifications within the radiolabeled probe will increase sensitivity approximately 10-fold compared to unmodified DNA oligonucleotide probes.[35] Depending on the abundance of the miRNA of interest the increased sensitivity allows a reduction of the total RNA separated by PAGE.[35] In addition, if LNA-modified DNA oligonucleotides are used as probes for membrane hybridization they also can be $T_m$ normalized, enabling the use of identical hybridization and washing conditions (see Reference 40) for different miRNAs.

## 12.11  MiRNA LOCALIZATION BY IN SITU HYBRIDIZATION

While miRNA-specific expression profiles yield information about miRNA-specific expression in a given tissue or cell type at best, in situ hybridization techniques are required to map the cellular or intracellular localization of a miRNA. In particular during development miRNAs are often differentially expressed[6,15] and localize to specific cellular subtypes or even subcellular regions.[79,80] During the in situ hybridization (ISH) process, a radioactively or digoxigenin-labeled oligonucleotide probe is hybridized to tissue sections or whole mounts (e.g., embryos). Recent reports show[46,81] that the use of LNA-modified DNA probes considerably increases the sensitivity and specificity of an ISH reaction. This methodology has been applied to map the temporal localization of miRNAs during zebra fish development[46] and for the differential localization of miRNA in the human brain.[82]

## 12.12  MiRNA CLONING

Originally, miRNAs were identified by cloning and sequencing approaches.[48] For this procedure, total RNA is prepared, and the sample is enriched for small RNAs. Then a 3' end RNA-linker (with a phosphate group at the 5'-end) is ligated to the 3' end hydroxyl group of the miRNA and the ligation product is purified using PAGE. Subsequently, a 5' end RNA-linker (lacking a phosphate group at the 5' end) is ligated to the 5' end phosphate group of the miRNA followed by a second PAGE purification. The linker-modified miRNAs are then amplified by PCR utilizing universal primer binding sites located within the attached linkers. After amplification, the PCR products can either be cloned directly into TOPO vectors® (Invitrogen), or can be digested using restriction enzyme binding sites present within the attached linker sequences and then cloned into the vector of choice. The vectors are amplified in bacteria, and plasmid DNA is then extracted from individual bacterial colonies. Sequencing of vector insertions will identify the cloned miRNAs.

Recently, a modified approach for large scale cloning of miRNAs was described.[83] The method is named miRNA analysis of gene expression (miRage) and follows a similar principle as described for miRNA cloning (see earlier text). Briefly, small RNAs are isolated from PAAGs and ligated to 3'- and 5'-end linkers. cDNA synthesis, PCR amplification of the complex cDNA mixture, tag purification, concatenation, cloning and sequencing are analogous to conventional SAGE analysis (serial analysis of gene expression; see References 84, 85). Using this approach as many as 35 miRNA tags can be identified in a single sequencing reaction.

## 12.13  CONCLUSIONS

Information on miRNA expression in cells and tissues is an essential first step to understanding the role of miRNAs during development, differentiation, or disease. However, it is considered only a starting point for most projects. Bioinformatic target gene prediction and functional validation of putative miRNA target genes as well as the definition of the role of the miRNA of interest in physiology and disease will have to follow to draw conclusions about the relevance of miRNA expression profiles.

Novel techniques for miRNA profiling become available on an (almost) daily basis. While finalizing this chapter the following reports have been published and are cited here for completeness:

Direct and sensitive miRNA profiling from low-input total RNA[86]
A microRNA detection system based on padlock probes and rolling circle amplification[87]
Real-time PCR profiling of 330 human micro-RNAs[88]
A novel microarray approach reveals new tissue-specific signatures of known and predicted mammalian microRNAs[89]

## ACKNOWLEDGMENTS

This work was supported by a Cancer Research Net grant (BMBF (NGFN) 201GS0450) to Martina U. Muckenthaler

## REFERENCES

1. Griffiths-Jones, S., miRBase: The microRNA sequence database, *Methods Mol Biol* 342, 129–138, 2006.
2. Lewis, B. P., Burge, C. B., and Bartel, D. P., Conserved seed pairing, often flanked by adenosines, indicates that thousands of human genes are microRNA targets, *Cell* 120 (1), 15–20, 2005.
3. Zhang, B. et al., microRNAs as oncogenes and tumor suppressors, *Dev Biol* 302 (1), 1–12, 2007.
4. Mattie, M. D. et al., Optimized high-throughput microRNA expression profiling provides novel biomarker assessment of clinical prostate and breast cancer biopsies, *Mol Cancer* 5, 24, 2006.
5. Lee, E. J. et al., Expression profiling identifies microRNA signature in pancreatic cancer, *Int J Cancer*, 2006.
6. Song, L. and Tuan, R. S., MicroRNAs and cell differentiation in mammalian development, *Birth Defects Res C Embryo Today* 78 (2), 140–149, 2006.
7. Chen, C. Z. et al., MicroRNAs modulate hematopoietic lineage differentiation, *Science* 303 (5654), 83–86, 2004.
8. Hwang, H. W. and Mendell, J. T., MicroRNAs in cell proliferation, cell death, and tumorigenesis, *Br J Cancer* 94 (6), 776–780, 2006.
9. Crick, F., Central dogma of molecular biology, *Nature* 227 (5258), 561–563, 1970.
10. Thieffry, D. and Sarkar, S., Forty years under the central dogma, *Trends Biochem Sci* 23 (8), 312–316, 1998.
11. Berk, A. J. and Sharp, P. A., Spliced early mRNAs of simian virus 40, *Proc Natl Acad Sci USA* 75 (3), 1274–1278, 1978.
12. Baltimore, D., RNA-dependent DNA polymerase in virions of RNA tumour viruses, *Nature* 226 (5252), 1209–1211, 1970.
13. Temin, H. M. and Mizutani, S., RNA-dependent DNA polymerase in virions of Rous sarcoma virus, *Nature* 226 (5252), 1211–1213, 1970.
14. Prusiner, S. B. et al., Further purification and characterization of scrapie prions, *Biochemistry* 21 (26), 6942–6950, 1982.

15. de Moor, C. H., Meijer, H., and Lissenden, S., Mechanisms of translational control by the 3′ UTR in development and differentiation, *Semin Cell Dev Biol* 16 (1), 49–58, 2005.

16. Mattick, J. S., Challenging the dogma: the hidden layer of non-protein-coding RNAs in complex organisms, *Bioessays* 25 (10), 930–939, 2003.

17. Watanabe, T. et al., Analysis of small RNA profiles during development, *Methods Enzymol* 427C, 155–169, 2007.

18. Beaton, M. J. and Cavalier-Smith, T., Eukaryotic non-coding DNA is functional: Evidence from the differential scaling of cryptomonad genomes, *Proc Biol Sci* 266 (1433), 2053–2059, 1999.

19. Stein, L. D., Human genome: End of the beginning, *Nature* 431 (7011), 915–916, 2004.

20. Ohno, S., So much "junk" DNA in our genome, *Brookhaven Symp Biol* 23, 366–370, 1972.

21. Gibbs, W. W., The unseen genome: gems among the junk, *Sci Am* 289 (5), 26–33, 2003.

22. Lee, R. C., Feinbaum, R. L., and Ambros, V., The *C. elegans* heterochronic gene lin-4 encodes small RNAs with antisense complementarity to lin-14, *Cell* 75 (5), 843–854, 1993.

23. Ambros, V., microRNAs: tiny regulators with great potential, *Cell* 107 (7), 823–826, 2001.

24. Bartel, D. P., MicroRNAs: genomics, biogenesis, mechanism, and function, *Cell* 116 (2), 281–297, 2004.

25. Ruby, J. G., Jan, C. H., and Bartel, D. P., Intronic microRNA precursors that bypass Drosha processing, *Nature* 448 (7149), 83–86, 2007.

26. Okamura, K. et al., The mirtron pathway generates microRNA-class regulatory RNAs in Drosophila, *Cell* 130 (1), 89–100, 2007.

27. Gaur, A. et al., Characterization of microRNA expression levels and their biological correlates in human cancer cell lines, *Cancer Res* 67 (6), 2456–2468, 2007.

28. Pasquinelli, A. E. and Ruvkun, G., Control of developmental timing by microRNAs and their targets, *Annu Rev Cell Dev Biol* 18, 495–513, 2002.

29. Hayashita, Y. et al., A polycistronic microRNA cluster, miR-17–92, is overexpressed in human lung cancers and enhances cell proliferation, *Cancer Res* 65 (21), 9628–9632, 2005.

30. Jovanovic, M. and Hengartner, M. O., miRNAs and apoptosis: RNAs to die for, *Oncogene* 25 (46), 6176–6187, 2006.

31. Esquela-Kerscher, A. and Slack, F. J., Oncomirs—microRNAs with a role in cancer, *Nat Rev Cancer* 6 (4), 259–269, 2006.

32. Plaisance, V. et al., MicroRNA-9 controls the expression of Granuphilin/Slp4 and the secretory response of insulin–producing cells, *J Biol Chem* 281 (37), 26932–26942, 2006.

33. Esau, C. et al., miR-122 regulation of lipid metabolism revealed by in vivo antisense targeting, *Cell Metab* 3 (2), 87–98, 2006.

34. Krutzfeldt, J. et al., Silencing of microRNAs in vivo with "antagomirs," *Nature* 438 (7068), 685–689, 2005.

35. Valoczi, A. et al., Sensitive and specific detection of microRNAs by northern blot analysis using LNA-modified oligonucleotide probes, *Nucleic Acids Res* 32 (22), e175, 2004.

36. Chen, C. et al., Real-time quantification of microRNAs by stem-loop RT-PCR, *Nucleic Acids Res* 33 (20), e179, 2005.

37. Duncan, D. D. et al., Absolute quantitation of microRNAs with a PCR-based assay, *Anal Biochem* 359 (2), 268–270, 2006.

38. Neely, L. A. et al., A single-molecule method for the quantitation of microRNA gene expression, *Nat Methods* 3 (1), 41–46, 2006.

39. Lu, J. et al., MicroRNA expression profiles classify human cancers, *Nature* 435 (7043), 834–838, 2005.

40. Castoldi, M. et al., A sensitive array for microRNA expression profiling (miChip) based on locked nucleic acids (LNA), *RNA* 12 (5), 913–920, 2006.

41. Davison, T. S., Johnson, C. D., and Andruss, B. F., Analyzing micro-RNA expression using microarrays, *Methods Enzymol* 411, 14–34, 2006.

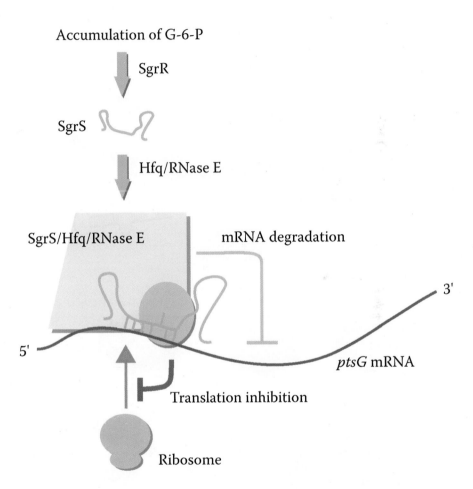

**FIGURE 4.3** Model for gene silencing by SgrS. SgrS induced in response to the accumulation of glucose-6-phosphate forms a ribonucleoprotein complex with Hfq (blue) and RNase E (yellow) and acts on the ribosome-binding site of *ptsG* mRNA through imperfect base pairing. The base pairing and recruitment of Hfq/RNase E on the target mRNA lead to translation inhibition and RNase E-dependent rapid degradation of the target mRNA.

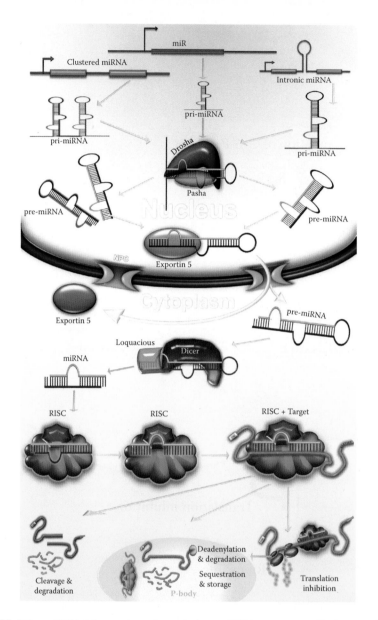

**FIGURE 8.2** miRNA biogenesis and regulation. miRNAs are transcribed to produce a monocistronic or polycistronic precursor RNA called *primary*-miRNA (*pri*-miRNA). These *pri*-miRNAs are then cleaved by complex of Endonuclease III enzyme *Drosha* and *Pasha* having a double-stranded RNA-binding domain (dsRBD) to produce *pre*-miRNA of ~70 nucleotides in length. They are exported out of the nucleus to the cytoplasm by Exportin-5. Once they are in the cytoplasm, they are further processed by Dicer-1 to produce short ~21 nt mature miRNA. Mature miRNAs get incorporated into a multiple turnover protein complex called RNA Induced Silencing Complex (RISC) by a RISC loading complex and retains only one strand of miRNA called the *Guide strand*. The loaded miRISC finds its target by sequence complementarity. miRNAs regulate their targets in many different ways, such as translation inhibition, target cleavage, sequestration, and storage.

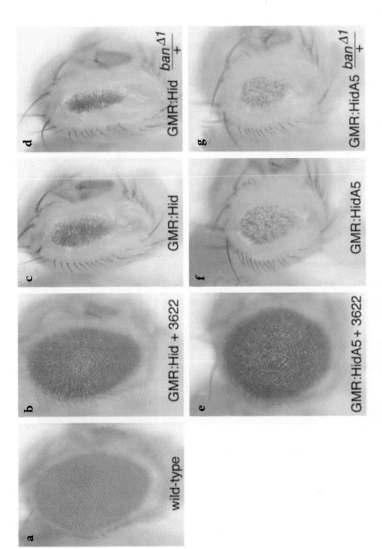

**FIGURE 8.4** Endogenous expression of *bantam* in the eye imaginal disc controls the level of apoptosis induced by *hid*. The *hid* is involved in reducing cell number in the pupal eye disc. Flies expressing *hid*, *hid* A5 under GMR have small, rough-eyed phenotype (c, f), but when *bantam* was coexpressed under GMR-gal4 to suppress GMR-*hid* and GMR *hidA5*, which drives *EP(3)3622*, phenotypes slightly recover (b, e). When one copy of endogenous *bantam* was removed, phenotypes become drastic (d, g) owing to increased level of *hid*-induced apoptosis. (From Brennecke, J., *Cell*, 2003. 113(1): 25. With permission.)

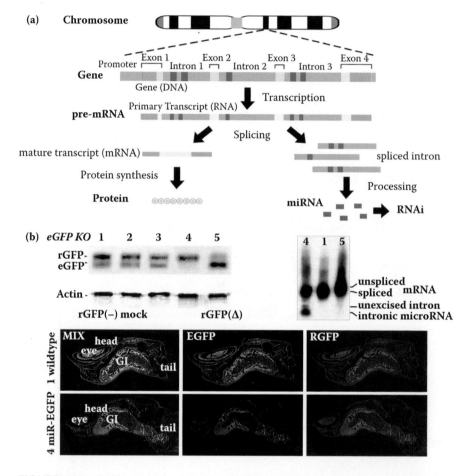

**FIGURE 15.2** Mechanisms for intronic miRNA biogenesis and relative gene silencing. (a) Intronic miRNA is generated as a part of precursor messenger RNA (pre-mRNA), containing protein-coding exons and noncoding introns. The introns are spliced out of pre-mRNA and further excised into small miRNA-like molecules capable of inducing RNA interference, while the exons are ligated to form a mature mRNA for protein translation. (b) The transfection of intronic miRNA into the green fluorescent protein (eGFP)-expressing Tg(UAS:gfp) strain zebra fish elicited a strong gene-specific silencing effects (>80% suppression, left lane 4). The hairpin-like pre-miRNA directed against *eGFP* was placed in the intron region of a coral reef red fluorescent protein (*rGFP*) pre-mRNA. Northern blot analysis (right) showed that spliced miRNAs were only generated by correct intron insertion, but not rGFP(–) and rGFPΔ.

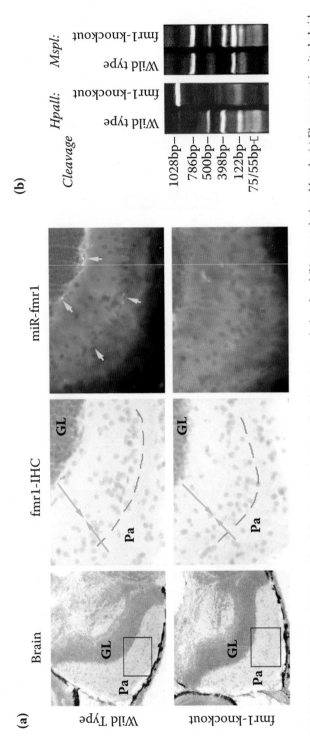

**FIGURE 15.7** r(CGG) miRNA-mediated *fmr1* gene silencing at the (a) posttranscriptional and (b) transcriptional levels. (a) Fluorescent in situ hybridization (FISH) analysis of *FMR1* mRNA and anti-*FMR1* miRNA (miR-*FMR1*) expressions in day-7 transgenic zebra fish brains. Fluorescent labels are shown in neurons (green EGFP), anti-*fmr1* miRNA (red RGFP), and *fmr1* mRNA (blue DAPI). Gray arrows indicate the dendro-dendritic contacts between lateral pallium and neocortical neurons. (b) Methylation site cleavage assay with *HpaII* and *MspI* restriction enzymes shows the changes of CpG methylation patterns in the *fmr1* 5′ UTR r(CGG) expansion region of the wild-type versus loss-of-*fmr1*-function zebra fish brains.

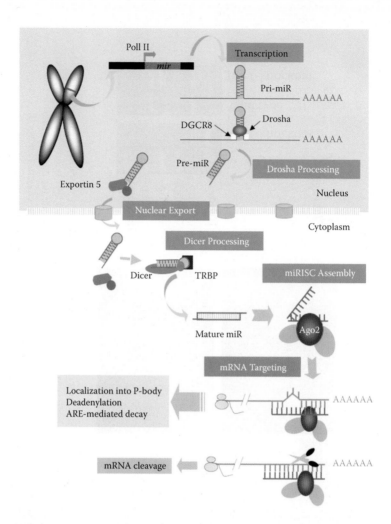

**FIGURE 19.1** microRNA biogenesis. The major steps in the biogenesis pathway are high-lighted in blue boxes. A microRNA gene is transcribed by RNA polymerase II to generate a long primary miRNA transcript (pri-miR) that contains an miRNA sequence embedded in an ~70-nucleotide stem-loop structure. The RNAse III enzyme Drosha, with the help of its RNA-binding partner, DGCR8, cleaves at the base of the stem-loop structure to liberate the precursor miRNA (pre-miR), which is then exported into the cytoplasm by Exportin 5 nuclear transporter. In the cytoplasm, the second RNAse III enzyme, Dicer, coordinates with TRBP (HIV-1 TAR RNA Binding Protein) to process pre-miR to generate mature miRNA upon cleaving off the loop. The Drosha and the Dicer cleavages specify the two ends of the mature miRNA. The miRNA strand in the mature miRNA duplex gets incorporated into a multiprotein complex referred to as miRISC. The miRNA strand guides the miRISC to identify its targets through complementary base pairing between the miRNA and the 3′ UTR of target mRNAs. The 5′ end (nucleotides 2–7) defines the "seed" region of an miRNA that is critical to target recognition. The fate of the mRNA targeted by an miRNA is dictated by the degree of base pairing between the two. Near-perfect base pairing between miRNA and mRNA leads to message cleavage by the nuclease "slicer" associated with hAgo-2 protein within the miRISC and subsequently degraded rapidly. Imperfect base pairing within these two entities guides the target message for translation inhibition through one of several mechanisms, as indicated.

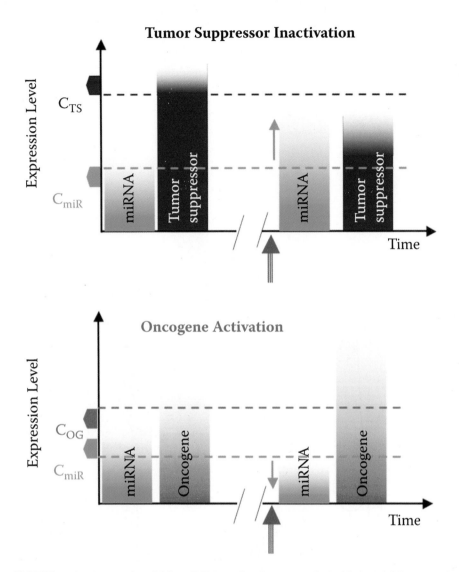

**FIGURE 19.5** A general model for miRNA-mediated oncogenesis. In this model, it is assumed that human tumors could originate in the absence of activated or inactivated mutations within genes that could function as oncogenes or tumor suppressors, respectively. It is proposed that the tumor formation is initiated or maintained by changing the dosage of gene expression of certain genes through changes in microRNA expression. Those genes that changed their expression levels in response to deregulated miRNA levels may exert oncogenic and tumor suppressor activities. However, other genetic changes could work in tandem with the miRNA-mediated gene dosage changes to contribute to oncogenesis. The upper panel illustrates how the overexpression of certain miRNAs beyond a certain threshold value decreases the expression of genes with tumor suppressor activity. Tumor formation occurs when their expression level decreases beyond a threshold value. The lower panel shows the opposite effect, in which the expression of certain miRNAs falls below a threshold level. This change causes an upregulation above a threshold level of a number of genes with oncogenic activity. The red arrow denotes an event that contributes to the change in miRNA expression.

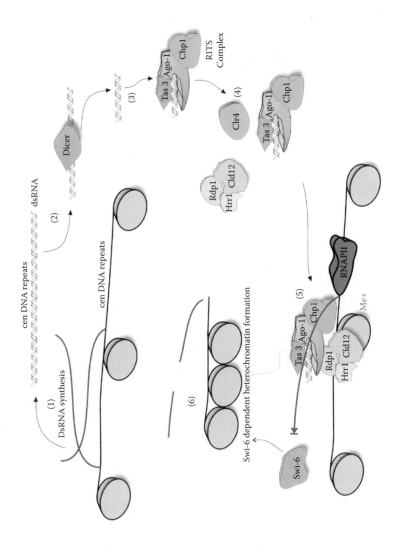

**FIGURE 21.1** TGS in *S. pombe*. DsRNAs are generated from the transcription of centromeric DNA repeats (cen DNA repeats, (1)), which are then processed by Dicer (2) to 21-22 bp siRNAs (3). Next, the dicer processed siRNAs are loaded into the RITS complex (4). The RITS complex then interacts with the RNA dependent RNA polymerase complex (RDRC, (5)) and the histone methyltransferase Clr4, which can then lead to H3K9 methylation (5) and silencing of the siRNA targeted cen DNA repeat regions and/or swi-6 dependent heterochromatin formation (6). The siRNA targeted cen DNA repeat RNAs are also targeted by the action of Ago-1 and sliced (6).

42. Thomson, J. M. et al., A custom microarray platform for analysis of microRNA gene expression, *Nat Methods* 1 (1), 47–53, 2004.
43. Liang, R. Q. et al., An oligonucleotide microarray for microRNA expression analysis based on labeling RNA with quantum dot and nanogold probe, *Nucleic Acids Res* 33 (2), e17, 2005.
44. Nelson, P. T. et al., Microarray-based, high-throughput gene expression profiling of microRNAs, *Nat Methods* 1 (2), 155–161, 2004.
45. Krichevsky, A. M. et al., A microRNA array reveals extensive regulation of microRNAs during brain development, *RNA* 9 (10), 1274–1281, 2003.
46. Kloosterman, W. P. et al., In situ detection of miRNAs in animal embryos using LNA-modified oligonucleotide probes, *Nat Methods* 3 (1), 27–29, 2006.
47. Deo, M. et al., Detection of mammalian microRNA expression by in situ hybridization with RNA oligonucleotides, *Dev Dyn* 236 (3), 912, 2007.
48. Lau, N. C. et al., An abundant class of tiny RNAs with probable regulatory roles in *Caenorhabditis elegans*, *Science* 294 (5543), 858–862, 2001.
49. Miska, E. A. et al., Microarray analysis of microRNA expression in the developing mammalian brain, *Genome Biol* 5 (9), R68, 2004.
50. Chomczynski, P. and Sacchi, N., Single-step method of RNA isolation by acid guanidinium thiocyanate-phenol-chloroform extraction, *Anal Biochem* 162 (1), 156–159, 1987.
51. Chomczynski, P., A reagent for the single-step simultaneous isolation of RNA, DNA and proteins from cell and tissue samples, *Biotechniques* 15 (3), 532–534, 536–537, 1993.
52. Dickman, M. J. and Hornby, D. P., Enrichment and analysis of RNA centered on ion pair reverse phase methodology, *RNA* 12 (4), 691–696, 2006.
53. Baskerville, S. and Bartel, D. P., Microarray profiling of microRNAs reveals frequent coexpression with neighboring miRNAs and host genes, *RNA* 11 (3), 241–247, 2005.
54. Lim, L. P. et al., Microarray analysis shows that some microRNAs downregulate large numbers of target mRNAs, *Nature* 433 (7027), 769–773, 2005.
55. Babak, T. et al., Probing microRNAs with microarrays: Tissue specificity and functional inference, *RNA* 10 (11), 1813–1819, 2004.
56. Barad, O. et al., MicroRNA expression detected by oligonucleotide microarrays: System establishment and expression profiling in human tissues, *Genome Res* 14 (12), 2486–2494, 2004.
57. Liu, C. G. et al., An oligonucleotide microchip for genome-wide microRNA profiling in human and mouse tissues, *Proc Natl Acad Sci USA* 101 (26), 9740–9744, 2004.
58. Jiang, J. et al., Real-time expression profiling of microRNA precursors in human cancer cell lines, *Nucleic Acids Res* 33 (17), 5394–5403, 2005.
59. Goff, L. A. et al., Rational probe optimization and enhanced detection strategy for microRNAs using microarrays, *RNA Biol* 2 (3), 2005.
60. Braasch, D. A. and Corey, D. R., Locked nucleic acid (LNA): fine-tuning the recognition of DNA and RNA, *Chem Biol* 8 (1), 1–7, 2001.
61. Castoldi, M. et al., miChip: A microarray platform for expression profiling of micro-RNAs based on locked nucleic acid (LNA) oligonucleotide capture probes, *Methods* 43 (2), 146–152, 2007.
62. Raymond, C. K. et al., Simple, quantitative primer-extension PCR assay for direct monitoring of microRNAs and short-interfering RNAs, *RNA* 11 (11), 1737–1744, 2005.
63. Monticelli, S. et al., MicroRNA profiling of the murine hematopoietic system, *Genome Biol* 6 (8), R71, 2005.
64. Branham, W. S. et al., Elimination of laboratory ozone leads to a dramatic improvement in the reproducibility of microarray gene expression measurements, *BMC Biotechnol* 7, 8, 2007.
65. Ballard, J. L. et al., Comparison of Alexa Fluor and CyDye for practical DNA microarray use, *Mol Biotechnol* 36 (3), 175–183, 2007.

66. Romaniuk, P. J. and Uhlenbeck, O. C., Joining of RNA molecules with RNA ligase, *Methods Enzymol* 100, 52–59, 1983.

67. Shingara, J. et al., An optimized isolation and labeling platform for accurate microRNA expression profiling, *RNA* 11 (9), 1461–1470, 2005.

68. Sun, Y. et al., Development of a micro-array to detect human and mouse microRNAs and characterization of expression in human organs, *Nucleic Acids Res* 32 (22), e188, 2004.

69. Allawi, H. T. et al., Quantitation of microRNAs using a modified Invader assay, *RNA* 10 (7), 1153–1161, 2004.

70. Yanagida, T. et al., Single molecule analysis of the actomyosin motor, *Curr Opin Cell Biol* 12 (1), 20–25, 2000.

71. Li, H. et al., Ultrasensitive coincidence fluorescence detection of single DNA molecules, *Anal Chem* 75 (7), 1664–1670, 2003.

72. Yang, L., Tran, D. K., and Wang, X., BADGE, Beads Array for the Detection of Gene Expression, a high-throughput diagnostic bioassay, *Genome Res* 11 (11), 1888–1898, 2001.

73. Schmittgen, T. D. et al., A high-throughput method to monitor the expression of microRNA precursors, *Nucleic Acids Res* 32 (4), e43, 2004.

74. Wang, X., Systematic identification of microRNA functions by combining target prediction and expression profiling, *Nucleic Acids Res* 34 (5), 1646–1652, 2006.

75. Shi, R. and Chiang, V. L., Facile means for quantifying microRNA expression by real-time PCR, *Biotechniques* 39 (4), 519–525, 2005.

76. Kuimelis, R. G. et al., Structural analogues of TaqMan probes for real-time quantitative PCR, *Nucleic Acids Symp Ser* (37), 255–256, 1997.

77. Kutyavin, I. V. et al., 3′-minor groove binder-DNA probes increase sequence specificity at PCR extension temperatures, *Nucleic Acids Res* 28 (2), 655–661, 2000.

78. Tang, F. et al., MicroRNA expression profiling of single whole embryonic stem cells, *Nucleic Acids Res* 34 (2), e9, 2006.

79. Chan, S. P. and Slack, F. J., microRNA-Mediated Silencing Inside P-Bodies, *RNA Biol* 3 (3), 97–100, 2006.

80. Liu, J. et al., MicroRNA-dependent localization of targeted mRNAs to mammalian P-bodies, *Nat Cell Biol* 7 (7), 719–723, 2005.

81. Obernosterer, G. et al., Post-transcriptional regulation of microRNA expression, *RNA* 12 (7), 1161–1167, 2006.

82. Nelson, P. T. et al., RAKE and LNA-ISH reveal microRNA expression and localization in archival human brain, *RNA* 12 (2), 187–191, 2006.

83. Cummins, J. M. et al., The colorectal microRNAome, *Proc Natl Acad Sci USA* 103 (10), 3687–3692, 2006.

84. Adams, M. D., Serial analysis of gene expression: ESTs get smaller, *Bioessays* 18 (4), 261–262, 1996.

85. Powell, J., Enhanced concatemer cloning-a modification to the SAGE (Serial Analysis of Gene Expression) technique, *Nucleic Acids Res* 26 (14), 3445–3446, 1998.

86. Wang, H., Ach, R. A., and Curry, B., Direct and sensitive miRNA profiling from low-input total RNA, *RNA* 13 (1), 151–159, 2007.

87. Jonstrup, S. P., Koch, J., and Kjems, J., A microRNA detection system based on padlock probes and rolling circle amplification, *RNA* 12 (9), 1747–1752, 2006.

88. Lao, K. et al., Real time PCR profiling of 330 human micro-RNAs, *Biotechnol J* 2 (1), 33–35, 2007.

89. Beuvink, I. et al., A novel microarray approach reveals new tissue-specific signatures of known and predicted mammalian microRNAs, *Nucleic Acids Res*, Methods online, March 13, 2007.

# 13 Methods to Quantify microRNA Gene Expression

*Lori A. Neely*

## CONTENTS

## OVERVIEW

Mounting evidence suggests small RNAs regulate the fundamental cellular processes of development, differentiation, metabolism, and apoptosis. Aberrant small RNA expression has been identified in numerous disease states; thus, accurate small RNA quantification is critically important for defining and characterizing disease and normal conditions. The biological question as well as the sample amount and quality will ultimately limit the researcher to one or two profiling technologies; here, we review methods for small RNA quantification highlighting the strengths and weaknesses of each approach.

## 13.1 INTRODUCTION

The expression of short non-protein-coding RNAs has been detected within cellular nuclei, cytoplasm, and even organelles.[1] Once deemed cellular detritus and washed away from RNA preparations, the functions of many of these molecules remain a mystery. Mounting evidence suggests that two classes of these short RNAs (the miRNAs and siRNAs) play commanding roles in the regulation of gene expression, and while the biogenesis of each of these small RNAs is distinct, their functions are similar. MicroRNAs (miRNAs) are endogenously expressed, 21–24-nucleotide long, noncoding RNA transcripts that posttranscriptionally modulate gene expression. These tiny RNAs regulate a number of fundamental cellular processes, including cellular differentiation, proliferation, apoptosis, and metabolism. In silico analyses suggest 30% of human genes are regulated by miRNAs.[2] Because the expressed miRNAs serve as both determinants of cellular identity and dictators of cell cycle progression, it is no surprise that aberrant miRNA expression has been observed in number of human malignancies.[3–5]

Sequences that encode miRNAs are arranged within the human genome as clusters or as single entities within introns or intergenic sequences. miRNAs are transcribed by RNA polymerase II into a long primary RNA transcript (pri-miRNA).[6] A nuclear RNAse III endonuclease called Drosha along with the DCGR8 double-stranded RNA-binding protein processes the primary miRNA into a hairpin precursor.[7] This hairpin is exported from the nucleus into the cytoplasm through a Ran-GTP dependent exportin-5 pathway. Once in the cytosol, the hairpin becomes bound by a dynamic protein complex called the microRNP (miRNP) whose exact composition at each phase in miRNA maturation/function remains elusive; however, data suggests these complexes contain Paz/Piwi domain (PPD) proteins of the Argonaute family. The human genome encodes four Argonaute proteins (Ago1-4). Maturation of the miRNA transcript occurs through cleavage catalyzed by an RNAse III endonuclease (DICER). The Paz domain of the Argonaute proteins forms an oligonucleotide-binding cleft to bind the 3′ end of the mature miRNA.[8] The PIWI domain binds the 5′ end of the miRNA and contains endonucleolytic activity and is responsible for cleavage of messenger RNA targets; however, this "slicer" activity has only been observed for human Ago2.[9]

Through partial sequence complementarity, the miRNAs direct the docking of the miRNP to their messenger RNA targets. miRNA-binding sites have been identified in the 3′ untranslated regions of messenger RNA[10]; however, since only 6–8 nucleotide base pairs dictate the specificity of this interaction, informatically predicting mRNA targets of miRNA regulation is challenging. The fate of the messenger RNA is partially dictated by the extent of sequence complementary between the miRNA and mRNA. If this sequence complementarity is large, then the messenger RNA is cleaved by "slicer" and subsequently degraded. Lower sequence complementarity leads to repression of messenger RNA translation, although recent evidence suggests that degradation of the mRNA may also ensue.[11–13] This degradation may occur following miRNA-induced deadenylation of the mRNA.[14] miRNA-bound human Ago2 has been shown to localize to processing bodies (P-bodies: sites of mRNA degradation) in human cells.[15] Whether the miRNP bound message is merely sequestered within the P body from ribosome access, immediately degraded, or sequestered and subsequently degraded is unknown. However, a recent study provides evidence that these messages may not be degraded within the P bodies. In this work, relief from mir-122-mediated translational repression of cationic amino acid transporter (CAT1) expression was shown to occur through the binding of a stress activated protein (HuR) to the 3′

UTR of the CAT1 transcript.[16] Upon HuR binding, the transcripts relocated from the P-bodies to polysomes. This is a particularly attractive model for stress-activated derepression as it would provide a rapid "on switch" enabling the cell to rapidly synthesize proteins needed to tolerate the stress without requiring de novo transcription. Together, the data suggest there are multiple pathways through which miRNAs can regulate gene expression, indicating that our present understanding of miRNA function is naïve at best.

A critical step in the functional characterization of miRNAs is identifying where miRNAs are transcribed and quantifying these expressed transcripts. The short nature of the transcripts makes the application of standard molecular biology techniques challenging and modifications to these techniques necessary. Here, we review methods used for the quantification of miRNAs, beginning with the more conventional method of Northern blotting and ending with a new single molecule method for miRNA quantification.

## 13.2 QUANTIFICATION OF microRNA GENE EXPRESSION

### 13.2.1 Northern Blotting and RNase Protection Assays (RPA)

Northern blotting and RNase protection assays (RPA assays) have been utilized for decades to quantify messenger RNAs. It is thus no surprise that Northern blotting is still considered the "gold standard" for miRNA quantification. In both methods, a radiolabeled miRNA complementary probe is hybridized to total RNA. In the case of Northern blotting, this hybridization occurs after the total RNA is electrophoresed through a denaturing polyacrylamide gel and then transferred to a nylon membrane. Electrophoretic size separation allows precise discrimination of the hairpin precursor molecules from the mature miRNA transcripts. While Northern blotting is a standard method for quantification of miRNAs, it is relatively insensitive and requires microgram quantities of total RNA per well (5–30 ug). In addition, Northern blotting cannot discriminate between miRNA family members that differ by a single nucleotide. Quantitative Northern blot data indicate the detection limit is approximately 1 femtomole and the dynamic range is 2 to 3 logs.[17] Since Northern blotting requires several long incubation steps, 2 days is typically required to obtain quantifiable data.

RNase protection assays are approximately 100-fold more sensitive than Northern blotting. This increased sensitivity is due to the solution-based hybridization of the radiolabeled probe to miRNA target, as well as the avoidance of sample loss due to membrane transfer. Once the probe is hybridized to the miRNA target, RNase is added to the sample to exonucleolytically degrade all single stranded RNAs. The sample is then electrophoresed on a polyacrylamide gel. Multiplexing is possible with the RPA assay by employing miRNA probes of different lengths. RNase protection assays typically require 50–500 nanograms of total RNA with the capability to detect 10 attomoles of miRNA. Similar to Northern blotting the dynamic range of RPA assays is 2–3 logs. The specificity of the RPA is high since RNase cleavage/degradation will occur at single nucleotide mismatches. RPA assays are relatively complex and require precise titration of RNases. Too little RNase will result in incomplete digestion of single stranded RNAs; too much could result in complete digestion of all RNA molecules. In addition, several days are required to obtain quantifiable expression data.

## 13.2.2  CLONING

The early days of miRNA discovery relied heavily on directional cloning of short RNAs. One of the obstacles to cloning miRNAs is that they possess a 5' phosphate and 3' hydroxyl, which must be modified to prevent self-ligation. While each of these methods differed slightly from laboratory to laboratory, most followed the same stepwise process outlined in Figure 13.1: polyacrylamide gel purification of the small RNA fraction from total RNA, ligation of adaptor oligonucleotides to the 5' end and the 3' end of the small RNAs, reverse transcription followed by PCR amplification of the resulting cDNA, and subsequent cloning of the amplicons as single units or concatemers into plasmid vectors. The sequences of the small RNA clones are then determined. Hundreds of miRNAs in a variety of organisms (worm, mouse, human) were discovered through cloning with tissue-specific cloning providing early information on functionality. Relative miRNA transcript levels were reflected by the number of clones obtained within a tissue library. In addition, by comparing cloned sequences to the genomes from a diverse number of organisms, many miRNAs were putatively identified through evolutionary conservation. Confirmation of miRNA expression is often verified using Northern blotting. The sensitivity limits imposed by Northern blotting was seldom problematic as the cloning process facilitated discovery of only the more abundantly expressed miRNAs. For example, 41% of the clones derived from HeLa cell RNA were variants of let-7 and 45% of the clones derived from human heart RNA encoded mir-1.[18,19] This data suggests novel discovery through direct cloning is saturable; however, a more recent study that used massively parallel signature sequencing (MPSS) to characterize the small RNAs of the *Arabidopsis* transcriptome uncovered 7,549 new small RNA sequences in a second round of sequencing, where the first identified 75,000 nonredundant RNA clones.[20] This indicates the cloning in this case was far from saturated and demonstrates the power of highly multiplexed sequencing platforms for small RNA discovery.

**FIGURE 13.1   (See facing page.)** A small RNA cloning protocol. (1) Isolate total RNA. (2) Enrichment of small RNAs through denaturing polyacrylamide gel electrophoresis. (3) T4 RNA ligase catalyzes ligation of a "donor" adaptor oligonucleotide to the 3' end of the small RNAs. This phosphorylated donor oligonucleotide bears a 3'-end-blocking group to prevent self-ligation and ligation to the 5' end of the small RNAs. In this example, the donor oligonucleotide is designed to contain a cloning vector compatible restriction enzyme recognition sequence. (4) Gel purification removes excess donor oligonucleotides. (5) T4 RNA ligase catalyzes ligation of the acceptor oligonucleotide on the 5' end of the purified small RNAs. In this example, the acceptor oligonucleotide is designed to contain a cloning vector compatible enzyme restriction recognition sequence. (6) Gel purification removes excess acceptor oligonucleotides. (7) RT-PCR amplify using primers complementary to donor and acceptor oligonucleotides. (8) Digest the amplicons with a restriction enzyme that cleaves at the recognition sequences engineered into the donor and acceptor oligonucleotides. (9) T4 DNA ligase catalyzes ligation of the PCR fragments into concatemers or ligates the digested amplicons directly into the cloning vector of choice.

**Cloning Protocol**   1. Isolate total RNA

1. Enrich for small RNAs

3. Ligation of 3' donor oligonucleotide with 3' blocking group

4. Gel purify ligation product

5. Ligation of 5' acceptor oligonucleotide

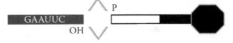

6. Gel purify ligation product

7. RT-PCR amplify small RNA

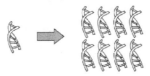

8. Digest amplicons with restriction enzyme

9. Concatemerize or directly clone

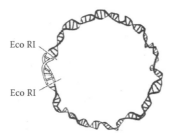

Eco RI

Eco RI

**FIGURE 13.1**

### 13.2.3 MICROARRAYS

To date, no miRNA quantification technique can match the throughput and multi-plexing capabilities of the microarray. The microarray process can be divided into four basic steps: sample preparation, hybridization, detection, and data analysis.

The short length of the miRNA made modifications to the sample preparation and hybridization steps unavoidable. The diminutive size of the miRNA makes amplification and/or labeling with fluorophores technically challenging. In most microarray protocols, enrichment or purification of the small RNA fraction from total RNA is required. This is to reduce the background "noise" attributable to nonspecific or specific binding of higher-molecular-weight RNAs on the chip. Several groups then used modifications of the cloning strategy typically retaining the enrichment, adaptor ligation, and reverse transcription steps to generate cDNA and then amplify this cDNA either linearly[21] or exponentially.[22–26] Wary of skewing miRNA expression data through amplification biases, several groups have devised microarray protocols that do not require amplification[27–29]; however, many still require up-front enrichment for small RNAs. Exogenous spike-in controls can be utilized to account for variability in small RNA retention across different samples.

Direct labeling of the miRNA may be conducted by chemical modification of the chemically reactive guanosines (ULISYS kit) or RNA ligation of a fluorescent "tail."[30] It should be noted in the direct labeling of miRNAs that those lacking guanosines will not be fluorescently labeled. A modification of the direct labeling strategy entails the enzymatic incorporation of modified nucleotides bearing chemically reactive groups or biological ligands (i.e., amine-reactive nucleotides, biotinylated nucleotides) followed by indirect labeling with organic fluorophores or fluorophore conjugated avidin. In protocols utilizing amplification fluorescent nucleotides can be incorporating into the amplicon or the PCR primers can be synthesized to contain fluorophores or fluorophore modifiable groups.

The short nature of miRNAs make them difficult to profile using conventional DNA capture probes since the melting temperatures of the resulting duplexes will have a broad distribution across the array. Several groups have responded by incorporating thermal/thermodynamic stabilizing modified nucleotides into their microarray capture probes; these include 2′ O-methyl DNA capture probes as well as locked nucleic acid (LNA)/DNA chimeric capture probes. Other groups have utilized longer capture probes consisting of pre-miRNA complementary sequences or tandem arrays of mature miRNA complementary sequences.[23,26]

No miRNA-specific method for microarray data analysis has been developed; therefore, researchers have employed normalization methods commonly used for mRNA expression profiling. For a more detailed review on miRNA microarrays, please see Reference 31. Normalization must be conducted to minimize variation stemming from nonbiological sources. Normalization also enables the use of parametric statistics such as the paired Student's $t$-test. The most commonly used normalization method involves background subtraction, averaging of replicate spots, normalization to the median of expressed miRNAs, and binary log transformation.

Because our knowledge on the number of human miRNAs and their expression levels is limited, there are no miRNA-specific controls (i.e., features hybridizing to miRNAs whose expression profile do not change) present on the array. These controls would help assess the retention of small RNAs through the sample prep process and provide a method to stabilize normalization. Consequently, the U6 small nucleolar RNA and 5S ribosomal RNA are often used as controls on miRNA microarrays. Specificity controls, including capture probes bearing one or two base changes, are helpful in assessing the array's ability to discriminate among miRNA family members. Intra-assay variability is measured through the use of multiple capture probe spots per miRNA. Sensitivities vary among the different protocols with the greatest reported at ~100 attomoles. The amount of input total RNA required per chip ranges from 100 ng to 10 μg, depending on the nature of the capture oligonucleotides and the sample preparation protocol, and the hybridization volume. The dynamic range also varies but is typically 2–3 orders.

Statistical analyses of expression data begins by imposing order with a clustering algorithm. Unsupervised hierarchical clustering is often employed to uncover patterns buried in a sea of data. miRNAs whose expression patterns are similar will group closest on the dendrogram or tree graph. One of the earliest discoveries obtained through microarray profiling of miRNA expression across scores of tissues (in contrast to Northern blots conducted on only a few tissues) was that the expression of most miRNAs is not confined to a single tissue and that the range of expression levels for a single miRNA varies broadly across numerous different tissues.[22] Clearly, the data implicate miRNAs in the establishment and maintenance of tissue-specific gene expression as related tissues possess similar miRNA expression profiles and therefore cluster closely together on the dendrogram.[22] Most published miRNA expression data have been generated with RNA isolated from cellularly heterogeneous tissues. Ideally, miRNA expression profiles should be obtained for each cell type. Laser capture microdissection to obtain pure cell populations makes this possible; however, amplification is required to obtain sufficient quantities of RNA for profiling. In a recent study, as little as 500 pg of enriched miRNA was subjected to linear amplification to generate 750 ng of RNA. This RNA was then fluorescently labeled and hybridized to a microarray. The expression results were then compared with those obtained with 150 nanograms of unamplified (but enriched) miRNAs and good correlation was observed ($R^2 = 0.875$).[21]

Microarrays are unmatched in their ability to generate thousands of data points. Gleaning statistically significant and biologically relevant expression data from the morass of variables and variability is the challenge. A hindrance to the comparison of array data has been the lack of standardization in the profiling process. Methods for sample preparation and data analyses vary from laboratory to laboratory with the resulting miRNA expression data showing little correlation.

### 13.2.4    Bead-Based Suspension Arrays

Polystyrene beads optically encoded with different fluorophores or ratios of two distinct fluorophores (Luminex, Inc., Austin, Texas) can be derivatized with miRNA complementary capture probes. This generates a "bead array" in which the color of each bead is distinct for each miRNA. In the simplest application of this method the miRNA are hybridized to the fluorescent beads with a second biotinylated detection oligo, then hybridized to the bound miRNAs. The subsequent addition of phycoerythrin streptavidin then allows detection and relative quantification of the amount of bead-bound miRNA.[23,32] The advantages of this approach over microarrays are that solution-phase hybridizations occur more rapidly, reproducibly, and discriminatively than solid-phase hybridizations. In addition, modest multiplexing has been demonstrated (5–6 miRNAs), thereby conserving sample material[23]; however, it remains unclear how this multiplexing impacts the sensitivity and quantitation accuracy of each individual assay. The disadvantages are that the sensitivity of the current assay is limited to low picomolar concentrations of miRNAs, the linear dynamic range is two logs,[32] and multiplexing to the extent possible with microarrays is not achievable.

## 13.2.5   REAL-TIME REVERSE TRANSCRIPTION POLYMERASE CHAIN REACTION (RT-PCR)

The most sensitive technique to quantify miRNAs is reverse transcription PCR (real time RT-PCR). The development of reliable and robust methods for the amplification of such short templates was challenging. Early on, RT-PCR quantification was confined to the longer hairpin precursors,[33] but soon techniques were devised to lengthen the miRNA during the reverse transcription step to enable subsequent PCR amplification. The first method employs reverse transcription of the miRNA with a primer designed to create a 5′ overhang or "tail" upon hybridization to the 3′ end of the miRNA.[34] PCR amplification of the resulting cDNA is conducted with the forward primer consisting of universal DNA oligo complementary to the tail and the reverse primer consisting of an LNA/DNA chimeric oligonucleotide. Real-time detection of the amplicons is conducted by monitoring SYBR green fluorescent emission.

An adaptation of Tm Biosciences (Toronto, Ontario) hairpin capture probes that were designed to contain a stem-loop structure, in which base stacking effects increase both the affinity and thermal stability of the hybridized duplexes,[35] was used to generate Applied Biosystem's (Foster City, California) stem-loop RT primers.[36] Within this assay a stem-loop primer is engineered to contain miRNA complementary sequences within the last 6–8 nucleotides. Following hybridization, reverse transcription generates a longer cDNA amplicon. This cDNA is then exponentially amplified using a forward primer complementary to the 5′ end of the miRNA and a reverse primer complementary to the stem-loop RT primer. A TaqMan probe hybridizes to the cDNA sequences present at the junction of the RT-primer and 3′ end of the miRNA. Upon amplification, the 5′ nuclease activity of Taq polymerase releases the fluorophore from the TaqMan probe, resulting in fluorescence emission.

A clear drawback to this method is that three primers and one probe need to be designed to amplify and detect a single miRNA. However, the addition of the TaqMan probe does confer an additional layer of specificity to the reactions, which is absent in SYBR green detection. However, it is important to note that the specificity of a SYBR green-based assay may be monitored through melt curve analysis. High sensitivity can be achieved with RT-PCR with data obtained from as little as 25 picograms of total RNA.[36] The dynamic range of the assay is quite broad, spanning 6 to 7 logs.[34,36]

Accurate quantification requires efficient reverse transcription of all miRNA sequences with the assumption being that the short miRNAs do not form stable secondary structures within the conditions used in the RT step, and that the efficiency of reverse transcription is similar for rare and abundant targets. Given the length of the longer RT primers, controls must be designed to ensure the primer does not interact with other RNA sequences or with other primers. The variability in the results obtained from RT-PCR in different laboratories or even within the same laboratory by different operators has been documented.[14,37] Tissue RNA samples must be devoid of enzyme inhibitors to enable efficient reverse transcription and amplification steps.[38] At the early stages in the reactions subtle variations in the thermal cycling conditions, reaction composition, as well as nonspecific priming can have a dramatic effect on the amount of amplified target. In instances where relative quantitation is desired, an endogenous control is needed, and care must be taken to ensure the amplification efficiencies of both the endogenous control and the target are the same. Despite these challenges, amplification is required when source material (number of cells) is a limitation, and the availability of commercially available RT-PCR kits should reduce the variability associated with primer design. Recent studies have shown that RT-PCR quantification of miRNAs can be conducted with RNA isolated from individual cells.[39] This should refine tissue-specific miRNA gene expression profiles and eliminate the errors introduced by the averaging of miRNA expression across different cell types.

## 13.2.6   INVADER ASSAY

In the standard mRNA Invader™ assay (Third Wave Technologies, Inc., Madison, Wisconsin), two primers (the invasive primer and the probe) are designed with complementarity to the target of interest. Upon hybridization, the two primers partially overlap, with one of the primers (the probe) bearing a 5′ overhang called the 5′ FLAP (see Figure 13.2). This generates a structure that is recognized by the 5′ nuclease Cleavase, which then catalyzes the cleavage of the probe releasing the FLAP. The released FLAP then acts in a secondary reaction as an invader primer on a synthetic target, which is also hybridized to another probe oligo. This probe oligo has been synthesized with a fluorophore and a close-proximity quencher such that upon cleave of the overlap the fluorophore is released and fluorescence emission occurs. The fluorescence signal increases in proportion to the number of target molecules. Thus, linear signal amplification is obtainable with this method, which makes it quite sensitive, and given the requirement for not only dual hybridization of the primers but also generation of the correct overlapping structure, the method is also very specific. To apply the Invader assay to miRNA quantification required modifications that would stabilize the duplex formed by the invasive primer and probe. Only ~11 nucleotide bases are available for hybridization to each oligo per miRNA, thus the inventors engineered the oligos to bear a 2′ O-methyl stem-loop structure,[40] thereby taking advantage of the stability conferred through base stacking, much like the TaqMan assay. Using this method quantification of miRNAs is possible in as little as 50 ng of total RNA. The linear dynamic range of the assay spans 3 logs. Multiplexing is possible; however, there are limitations as each miRNA would require a spectrally distinguishable fluorophore.

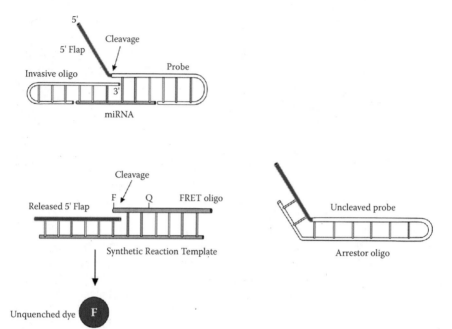

**FIGURE 13.2** The miRNA Invader assay uses a 2′ O-methylated stem-loop stabilized inva-sive primer and probe along with a 5′ cleavase enzyme. (a) The primary reaction consists of the hybridization of an invasive primer and a probe to their complementary miRNA. To stabilize the short duplexes generated upon hybridization of the invasive primer and probe to their com-plementary miRNA, a 2′ O-methylated stem loop is incorporated into the ends of the oligos. The probe is designed such that upon hybridization next to the invasive primer, a 5′ overlap called the 5′ flap is generated. This structure is recognized by the 5′ nuclease cleavase, which catalyzes cleavage of the overhang releasing the FLAP. (b) The flap becomes an invasive oligo in the signal amplification step. The flap hybridizes to a complementary synthetic template (the secondary reaction template) along with another probe oligonucleotide. In this case the probe oligonucleotide contains 5′ fluorophore and a quencher in close proximity. Again, the 5′ flap overlap is generated but this time cleavage via cleavase will release the fluorophore from its neighboring quencher resulting in fluorescent signal. A probe-complementary 2′ O-methylated oligonucleotide is also included in the signal amplification reaction to sequester uncleaved probes and prevent them from binding to the secondary reaction template.

### 13.2.7 A SINGLE MOLECULE METHOD FOR MIRNA QUANTIFICATION

A single molecule approach has been applied to quantify miRNAs.[41] In the Direct™ miRNA assay (U.S. Genomics, Inc., Woburn, Massachusetts) two spectrally dis-tinguishable fluorescent LNA/DNA 10-mer probes hybridize to the miRNA (see Figure 13.3a). The inclusion of LNA bases within the probe increases the thermal stability of the LNA/RNA hybrids enabling hybridization at higher temperatures (55°C), thereby minimizing nonspecific hybridization. The use of these short probes also maximizes specificity, as single base mismatches have profound affects on the stability of the duplex. Following hybridization, the temperature is decreased to 40°C, and synthetic DNA oligonucleotides labeled with quencher molecules are

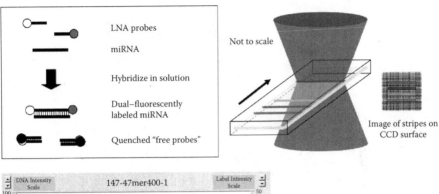

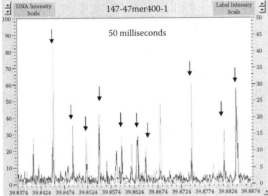

**FIGURE 13.3** A single molecule two-color coincident detection strategy is employed to quantify miRNAs. miRNAs are detected and counted as coincident bursts in photon emission emanating from two different fluorescent dyes. The steps of the assay are shown in (a) where miRNAs are hybridized in solution to two spectrally distinguishable fluorescent probes (in this case Oyster 556 and Oyster 656) and flowed by vacuum pressure through a capillary containing a series of femtoliter laser focal volumes. Complementary DNA probes bearing fluorescence quencher molecules are hybridized to the remaining unbound fluorescent probes to minimize coincident events that could be created by free probes simultaneously entering the laser interrogation spots. (b) Laser illumination volumes are focused into "stripes" and arrayed as shown within a microfluidic channel. For clarity we show only the illumination of the second red stripe. The 532 nm laser interrogation volume is focused 2 microns downstream from the first 633 nm interrogation volume (Red 1) while the second 633 nm focal volume (Red 2) is focused 8 microns downstream from the first red focal volume. The data analysis software uses a standard cross-correlation algorithm to measure the flow velocity between the two red laser interrogation volumes. This algorithm also accounts for the two micron off-set between the Red 1 and Green laser interrogation volumes and synchronizes the data traces such that a single dual color fluorescently tagged molecule will appear as a coincident peak in fluorescence emission shown in a Trilogy™ screen shot in (c) with arrows highlighting the coincident peaks in photon emission.

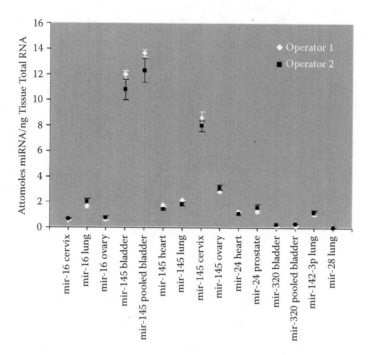

**FIGURE 13.4** Precision of the direct miRNA assay. Quantification of a selection of miRNAs from several human tissues was conducted by two different operators on two different Trilogy 2020 instruments. The average miRNA expression from three independent assays in attomoles per nanograms was plotted +/- the standard deviations. The coefficients of variation between the expression averages obtained by the two different operators were all less than 20%.

hybridized to the remaining "free" LNA probes. This quenching reaction reduces the background fluorescence, thereby increasing assay sensitivity. Quenched reactions are then diluted twofold and interrogated on a microfluidic, multicolor laser platform capable of counting individual molecules as they flow at high velocity through the system (Trilogy 2020™, U.S. Genomics).

miRNAs are detected and quantified by the bursts in fluorescence emission emanating from the two hybridized probes. The laser focal volumes are arrayed within the microfluidic channel as shown in Figure 13.4b. Using a method previously described,[42] cross-correlation between the two red channels is applied to measure the flow velocity of the fluorescently labeled molecules. We focus the 633 nm laser two micrometers upstream of the 532 nm laser. The off-set eliminates spectral cross-talk. The data analysis software adjusts for the offset and synchronizes the data traces, allowing a dual fluorescently labeled molecule to be counted as a coincident event (Figure 13.4c). The data analysis method counts the number of coincident events above an established threshold. A novel feature of our data analysis method is an estimation of the number of random coincidences expected on the basis of the raw data. This estimate is subtracted from the raw coincidence count to yield of the number of coincidences caused by dual-tagged molecules, and by inference,

**TABLE 13.1**

**Accurate Quantifying of miRNAs by Direct miRNA Assay in as Little as 10 ng of Tissue Total RNA**

| miRNA | Tissue | Average attomoles/nanogram ± Standard Deviation | | CV |
| --- | --- | --- | --- | --- |
| | | 10 ng | 25 ng | |
| mir-17-5p | Lung | 0.517 ± 0.108 | 0.570 ± 0.030 | 7% |
| mir-21 | Bladder | 3.963 ± 0.146 | 4.164 ± 0.132 | 4% |
| | Lung | 4.839 ± 0.237 | 4.319 ± 0.444 | 11% |
| | Kidney | 3.214 ± 0.224 | 2.830 ± 0.192 | 13% |
| mir-22 | Brain | 0.760 ± 0.019 | 0.626 ± 0.121 | 14% |
| | Kidney | 1.140 ± 0.131 | 0.899 ± 0.101 | 17% |
| mir-24 | Brain | 0.703 ± 0.117 | 0.566 ± 0.018 | 15% |
| | Sk. Muscle | 1.194 ± 0.269 | 1.036 ± 0.023 | 10% |
| mir-126 | Kidney | 1.10 ± 0.05 | 0.97 ± 0.17 | 9% |
| | Heart | 1.885 ± 0.353 | 1.455 ± 0.302 | 18% |
| mir-143 | Colon | 7.795 ± .0154 | 5.716 ± 0.178 | 22% |
| mir-145 | Heart | 2.457 ± 0.217 | 1.748 ± 0.079 | 24% |
| mir-205 | Cervix | 0.662 ± 0.029 | 0.608 ± 0.039 | 6% |
| mir-221 | Heart | 0.28 ± 0.042 | 0.315 ± 0.025 | 8% |

*Note:* Mir-17-5p, mir-21, mir-22, mir-24, mir-126, mir-143, mir-145, mir-191, mir-205, and mir-221 levels were quantified in 10 and 25 nanograms of tissue total RNA. All data was normalized to attomoles of target miRNA per nanogram of tissue total RNA and presented as the mean of three independent experiments ± the standard deviation.

the concentration of the analyte. A detailed description of this correction method is presented in Reference 43.

Table 13.1 shows the levels of miRNA expression obtained for a representative selection of miRNAs. These miRNAs were profiled in tissues where previous results indicated their expression was present at low, moderate, and high levels. Data in the table is the average in attomoles of miRNA per nanogram total RNA from three independent experiments. The data in the table shows precise quantification of low, moderate, and abundant miRNAs is obtainable in as little as 10 nanograms of tissue total RNA. The coefficient of variation (standard deviation/mean) is indicated for each experiment. The assay is capable of quantification of ~2 attomoles of miRNA per well with no amplification/enrichment required. The linear dynamic range is 3 logs. Multiplexing is currently limited to two miRNA assays per well. The precision of the assay when conducted by two different operators is shown in Figure 13.4. The expression levels of mir-16, mir-145, mir-320, mir-142-3p, mir-24, and mir-28 were quantified in a number of human tissues (bladder, cervix, ovary, prostate, heart, and lung). The assays were conducted using the same lot of RNA, but the time that elapsed between the two operators' experiments varied from a few weeks to several months, and in most cases different instruments were used by the two operators.

The quality of the data is unique, with signals emanating from single fluorophores detected and counted. This enables the measurement of subtle changes in gene expression that cannot be detected using assays that measure bulk fluorescence. No preparative enrichment, ligation, or target/signal amplification steps are required, and there is no sample clean-up step. This lack of sample manipulation limits the points at which variability could be introduced into the assay, thereby adversely affecting quantification accuracy.

## 13.3  CONCLUSIONS

Critical to our understanding of miRNA function is the need to accurately quantify the expressed transcripts. Here, we present a number of methods for miRNA quantification. These techniques all differ in their complexity, sensitivity, dynamic response, variability, multiplex capabilities, and cost. The quantitative nature of the data generated differs as well with most inferring concentrations based on bulk fluorescent or radioactive emissions.

Many of these methods are new and lack a large body of data to characterize their performance; however, over time the merits of each will be revealed. The choices available allow the investigator to prioritize which assay features are most important (i.e., accuracy of quantitation, multiplex capabilities, sample consumption) in the generation of data to address their scientific question.

## REFERENCES

1. Lung, B., Zemann, A., Madej, M. J., Scheulke, M, Tehritz, S., Ruf, S., Bock, R., and Huttenhofer, A. Identification of small non-coding RNAs from mitochondria and chloroplasts. *Nucleic Acids Res* **34**, 3842–3852 (2006).
2. Lewis, B. P., Burge, C. B., and Bartel, D. P. Conserved seed pairing, often flanked by adenosines, indicates that thousands of human genes are miRNA targets. *Cell* **120**, 15–20 (2005).
3. Michael, M. Z., O' Connor, S. M., van Holst Pellekaan, N. G., Young, G. P., and James, R. J. Reduced accumulation of specific microRNAs in colorectal neoplasia. *Mol Cancer Res* **12**, 882–891 (2003).
4. Calin, G. A., Dumitru, C. D., Shimizu, M., Bichi, R., Zupo, S., Noch, E., Aldler, H., Rattan, S., Keating, M., Rai, K., Rassenti, L., Kipps, T., Negrini, M., Bullrich, F., and Croce, C. M. Frequent deletions and down-regulation of micro- RNA genes miR15 and miR16 at 13q14 in chronic lymphocytic leukemia. *Proc Natl Acad Sci* **99**, 15524–15529 (2002).
5. Calin, G. A., Liu, C-G., Sevignani, C., Ferracin, M., Felli, N., Dumitru, C. D., Shimizu, M., Cimmino, A., Zupo, S., Dono, M., Dell'Aquila, M., Alder, H., Rassenti, L., Kipps, T. J., Bullrich, F., Negrini, M., and Croce, C. MicroRNA profiling reveals distinct signatures in B cell chronic lymphocytic leukemias. *Proc Natl Acad Sci* **101**, 11755–11760 (2004).
6. Lee, Y., Kipyoung, J., Lee, J. T., Kim, S., and Kim V. N. MicroRNA maturation: stepwise processing and subcellular localization. *EMBO J* **21**, 4663–4670 (2002).
7. Lee, Y., Ahn, C., Han, J., Choi, H., Kim, J., Yim, J., Lee, J., Provost, P., Radmark, O., Kim, S., and Kim, V. N. The nuclear RNase III Drosha initiates microRNA processing. *Nature* **425**, 415–419 (2003).

8. Song, J. J., Liu, J., Tolia, N. H., Schneiderman, J., Smith, S. K., Martienssen, R. A., Hannon, G. J., and Joshua-Tor, L. The crystal structure of the Argonaute2 PAZ domain reveals and RNA binding motif in RNAi effector complexes. *Nat Struct Biol* **10**, 1026–1032 (2003).

9. Liu, J., Carmell, M. A., Rivas F. V., Marsden, C. G., Thomson, J. M., Song, J. J., Hammond, S. M., Joshua-Tor, L., and Hannon, G. J. Argonaute2 Is the Catalytic Engine of Mammalian RNAi. *Science* **305**, 1437–1441 (2004).

10. Lai, E. Micro RNAs are complementary to 3′ UTR sequence motifs that mediate negative post-transcriptional regulation. *Nature* **30**, 363,364 (2002).

11. Lim, L. P., Lau, N. C., Garrett-Engle, P., Grimson, A., Schelter, J. M., Castle, J., Bartel, D. P., Linsley, P. S., and Johnson, J. M. Microarray analysis shows some miRNAs downregulate large numbers of target mRNAs. *Nature* **433**, 769–773 (2005).

12. Jing, Q., Huang, S., Guth, S., Zarubin, T., Motoyama, A., Chen, J., Di adova, F., Lin, S. C., Gram, H., and Han, J. Involvement of miRNA in AU rich element-mediated mRNA stability. *Cell* **120**, 623–634 (2005).

13. Bagga, S., Bracht, J., Hunter, S., Massirer, K., Holtz, J., Eachus, R., and Pasquinelli, A. E. Regulation by let-7 and lin-4 miRNAs results in target mRNA degradation. *Cell* **122**, 553–563 (2005).

14. Boulfer, P., Lo, C. F., Grimawde, D., Barragan, E., Diverio, D. Cassinat, B., Chomienne, C., Gonzalez, M., Colomer, D., Gomez, M. T., Marugan, I., Roamn, J., Delgado, M. D., Garcia-Marco, J. A., Bornstein, R., Vizmanos, J. L., Martinez, B., Jansen, J. Villegaras, A., de Blas, J. M., Cabello, P., and Sanz, M. A. Variability in the levels of PML-RARa fusion transcripts detected by the laboratories participating in an external quality control program using several reverse transcription polymerase chain reaction protocols. *Haematologica* **86**, 570–576 (2001).

15. Behm-Ansmant, R. J., Doerks, T., Stark, A., Bork, P., and Izaurralde, E. mRNA degradation by miRNAs and GW182 requires both CCR4:NOT deadenylase and DCP1:DCP2 decapping complexes. *Genes Dev* **20**, 1885–1898 (2006).

16. Bhattahcaryya, S. N., Habermacher, R., Martine, U., Closs, E. I., and Fillopowiscz, W. Relief of microRNA-mediated translational repression in human cells subjected to stress. *Cell* **126**, 1111–1124 (2006).

17. Lee, R. C., and Ambros, V. An Extensive Class of Small RNAs in *Caenorhabditis elegans*. *Science* **294**, 862–865 (2001).

18. Lagos-Quintana, M., Rauhut, R., Lendeckel, W., and Tuschl, T. Identification of novel genes coding for small expressed RNAs. *Science* **294**, 853–858 (2001).

19. Lagos-Quintana, M., Rauhut, R., Meyer, J., Lendeckel, W., and Tuschl, T. Identification of tissue-specific microRNAs from mouse. *Curr Biol* **12**, 735–739 (2002).

20. Lu, C., Tej, S. S., Luo, S., Haudenschild, C. D., Meyers, B. C., and Green, P. J. Elucidation of the small RNA component of the transcriptome. *Science* **309**, 1567–1569 (2005).

21. Mattie, M. D., Benz, C. C., Bowers, J., Sensinger, K., Wong, L., Scott, G. K., Fedele, V., Ginsinger, D. G., Getts, R. C., and Haqq, C. M. Optimized high-throughput microRNA expression profiling provides novel biomarker assessment of clinical prostate and breast cancer biopsies. *Mol Cancer* **5** (2006).

22. Baskerville, S. and Bartel, D. P. Microarray profiling of microRNAs reveals frequent coexpression with neighboring miRNAs and host genes. *RNA*, 241–247 (2005).

23. Barad, O., Meiri, E., Avniel, A., Aharonov, R., Barilai, A., Bentwich, I., Einav, U., Gilad, S., Hurban, P., Karov, Y., Lobenhofer, E., Sharon, E., Shiboleth, Y., Shtutman, M., Bentwich, Z., and Einat, P. MicroRNA expression detection by oligonucleotide microarrays: System establishment and expression profiling in human tissues. *Genome Res* **14**, 2486–2494 (2004).

24. Ambros, V., Lee, R. C., Lavanway, A., Williams, P. T., and Jewell, D. MicroRNAs and other tiny endogenous RNAs in *C. elegans*. *Cell* **13**, 1–20 (2003).

25. Miska, E. A., Alvarez-Saavedra, E., Townsend, M., Yoshii, A., Sestan, N., Rakic, P., Constantine-Patton, M., and Horvitz, R. Microarray analysis of microRNA expression in the developing mammalian brain. *Genome Biol* **5**, 68–68.13 (2004).

26. Liu, C. G., Calin, G. A., Meloon, B., Galiel, N., Sevignani, C., Ferracin, M., Dumitru, C. D., Shimuzi, M., Zupo, S., Dono, M., Alder, H., Bullrich, F., Negrini, M., and Croce, C. M. An oligonucleotide microchip for genome-wide microRNA profiling in human and mouse tissues. *Proc Natl Acad Sci* **101**, 9740–9744 (2004).

27. Nelson, P. T., Baldwin, D. A., Scearce, L. M., Oberholtzer, J. C., Tobias, J. W., and Mourelatos, Z. Microarray-based, high-throughput gene expression profiling of micro-RNAs. *Nat Methods* **1**, 1–7 (2004).

28. Babak, T., Zhang, W., Morris, Q., Blencowe, B. J., and Hughes, T. R. Probing micro-RNAs with microarrays: Tissue specificity and functional inference. *RNA* **10**, 1813–1819 (2004).

29. Shingara, J., Keiger, K., Shelton, J., Laosinchai-Wolf, W., Powers, P., Conrad, R., Brown, D., and Labourier, E. An optimized isolation and labeling platform for accurate microRNA expression profiling. *RNA* **11**, 1461–1470 (2005).

30. Thomson, J. M., Parker, J., Perou, C. M., and Hammond, S. A custom microarray platform for analysis of microRNA gene expression. *Nat Methods* **1**, 1–7 (2004).

31. Davison, T. S., Johnson, C. D., and Andruss, B. F. Analyzing micro-RNA expression using microarrays. *Methods Enzymol* **411**, 14–34 (2006).

32. Lu, J., Getz, G., Miska, E. A., Alvarez-Saavedra, E., Lamb, J., Peck, D., Sweet-Cordero, A., Ebert, B. L., Mak, R. H., Ferrando, A. A., Downing, J. R., Jacks, T., Horvitz, H. R., and Golub, T. R. MicroRNA expression profiles classify human cancers. *Nature* **435**, 834–838 (2005).

33. Schmittgen, T. D., Jiang, J., Liu, Q, and Yang, L. A high-throughput method to monitor the expression of microRNA precursors. *Nucleic Acids Res* **32**, e43 (2004).

34. Raymond, C. K., Roberts, B. S., Garret-Engele, P., Lim, L. P., and Johnson, J. M. Simple, quantitative primer-extension PCR assay for direct monitoring of microRNAs and short-interfering RNAs. *RNA* **11**, 1737–1744 (2005).

35. Riccelli, P. V., Merante, F., Leung, K. T., Bortolin, S., Zastawny, R. L., Jeneczko, R., and Benight, A. S. Hybridization of single-stranded DNA targets to complementary DNA probes: comparison of hairpin versus linear capture probes. *Nucleic Acids Res* **29**, 996–1004 (2001).

36. Chen, C., Ridzon, D. A., Broomer, A. J., Zhou, Z., Lee, D. H., Nguyen, J. T., Barbisin, M., Xu, N. L., Mahuvakar, V. R., Andersen, M. R., Lao, K. Q., Livak, K. J., and Guegler, K. J. Real-time quantification of microRNAs by stem-loop RT-PCR. *Nucleic Acids Res* **33**, e179 (2005).

37. Bustin, S. A. Quantification of mRNA using real-time reverse transcription PCR (RT-PCR): trends and problems. *J Mol Endocrinol* **29**, 23–39 (2002).

38. Tichopad, A., Didier, A., and Pfaffl, M. W. Inhibition of real-time RT-PCR quantification due to tissue-specific contaminants. *Mol Cell Probes* **18**, 45–50 (2004).

39. Tang, F., Hajkova, P., Barton, S. C., Lao, K., and Surani, M. A. MicroRNA expression profiling of single whole embryonic stem cells. *Nucleic Acids Res* **34**, e9 (2006).

40. Allawi, H. T., Dahlberg, J. E., Olson, S., Lund, E., Olson, M., Ma, W., Takova, T., Neri, B. P., and Lyamichev, V. I. Quantitation of microRNAs using a modified Invader assay. *RNA* **10**, 1153–1161 (2004).

41. Neely, L. A., Patel, S., Garver, J., Gallo, M., Hackett, M., McLaughlin, S., Nadel, M., Harris, J., Gullans, S., and Rooke, J. A single-molecule method for the quantitation of microRNA gene expression. *Nat Methods* **3**, 41–46 (2006).

42. Brinkmeier, M., Dorre, K., Stephan, J., and Eigen, M. Two beam cross correlation: A method to characterize transport phenomena in micrometer-sized structures. *Anal Chem* **71**, 609–616 (1999).

43. D'Antoni, C. M., Fuchs, M., Harris, J. L., Ko, H. P., Meyer, R. E., Nadel, M. E., Randall, J. D., Rooke, J. E., and Nalefski, E. A. Rapid quantitative analysis using a single molecule counting approach. *Anal Biochem* **1**, 97–109 (2006).

# 14 Regulation of Alternative Splicing by microRNAs

*Rajesh K. Gaur*

## CONTENTS

## OVERVIEW

MicroRNAs (miRNAs) are single-stranded noncoding RNA molecules of about ~22 nucleotides in length that regulate gene expression in diverse organisms. Although the underlying mechanisms of miRNA-mediated regulation of gene expression are not fully defined, these molecules normally function by either suppressing translation or destabilization of mRNAs possessing complementary target sequences. Recent studies have demonstrated the role of miRNAs in virtually every aspect of cell function. For example, they alter the expression of genes involved in cell growth and differentiation, transcription, and alternative splicing. This chapter provides an overview of the role of miRNAs in regulation of gene expression via modulation of alternative splicing.

## 14.1 INTRODUCTION

In order to perform diverse and complex cellular processes ranging from cell proliferation to differentiation, multicellular organisms express a variety of proteins in a spatial and temporal manner. Since the human genome contains a far lower number of genes than its proteome, other mechanisms must exist to produce protein diversity. Indeed, a variety of mechanisms such as multiple transcription start sites, alternative splicing, polyadenylation, RNA editing, and posttranslation modification can increase protein complexity from a single gene.[1] Alternative splicing has the potential to generate multiple splice variants, and the expression of a particular splice isoform can vary significantly in cell or tissue specific manner.[2–4] Thus, alternative splicing is an important mechanism that plays a vital role in regulation of gene expression. It is now widely accepted that alternative pre-mRNA splicing is the most important source of protein diversity in vertebrates.[1,5–7] It has been estimated that 35–60% of human genes generate transcripts that are alternatively spliced,[6,8] and 70–90% of alternative splicing decisions result into the generation of proteins with diverse functions.[9,10] This may explain that although organisms such as *Drosophila* and *Caenorhabditis elegans* are significantly less complex than humans, the number of protein-coding genes between these organisms is not very different.[11]

While the role of alternative pre-mRNA splicing in the expansion of organismal complexity is well established, the newly discovered functional interactions between miRNAs and splicing regulators has added an additional layer of complexity to an already elaborate process of gene expression in metazoans.[12–14] Interestingly, on the one hand, miRNAs have been shown to regulate gene expression by controlling alternative splicing; on the other hand, a splicing factor has been demonstrated to be involved in the processing of miRNA.[15,16] This chapter provides an overview of the role of miRNAs in establishing alternative splicing program that governs cellular differentiation in mammalian cells. A brief introduction to pre-mRNA splicing, and the role of splicing factors in the regulation of alternative splicing is also provided.

## 14.2   SPLICING AND ITS ROLE IN THE REGULATION OF GENE EXPRESSION

More than half of the protein-coding genes in metazoans are transcribed as precursor mRNAs in which the coding portion (exons) of the messenger RNA (mRNA) is interrupted by the intervening sequences (introns). Pre-mRNA splicing, which performs the excision of introns and ligation of exons, is an essential step in the pathway of gene expression. The chemistry of pre-mRNA splicing is well defined and consists of two sequential transesterification reactions catalyzed by the spliceosome, a dynamic ribonucleoprotein (RNP) assembly.[17] In the first step of the splicing, the spliceosome performs the cleavage at the 5′ splice site, generating two splicing intermediates: a linear first exon RNA, and an intron second exon RNA in a lariat configuration. The second step of the splicing involves the nucleophillic attack of the 3′-hydroxyl group of the last nucleotide in the first exon at the phosphodiester bond separating the intron and the second exon (3′ splice site) enabling the joining of two exons and the release of the intron as a lariat.[18–21] Within the spliceosome, a number of RNA–RNA and RNA–protein interactions involving five small nuclear RNAs (U1, U2, U4, U5, and U6) and many snRNP and non-snRNP proteins mediate the removal of introns and joining of exons.[20–25]

In higher eukaryotes, three distinct sequences direct the splicing reaction: the 5′ splice site (/GURAGY), the branchpoint sequence (BPS; YNYURAC), and the 3′ splice site (YAG/), where a slash (/) denotes a splice site; N denotes any nucleotide, R denotes purine, and Y denotes pyrimidine; and underlining indicates the conserved nucleotide. During the early stages of the spliceosome assembly the 5′ and the 3′ end of the intron are recognized by intermolecular base pairing between U1 snRNA and the 5′ splice site[26–28] and by the binding of U2AF to the poly(Y) tract/3′ ss AG,[29–31] respectively. Later in the spliceosome assembly, the U1 snRNA–5′ splice site base pairing is disrupted, and the 5′ splice site is bound by U6 snRNA.[32–37] The branchpoint adenosine is selected in part by intermolecular base pairing between the BPS and U2 snRNA, and the RS domain of U2AF[65] stabilizes this interaction.[38,39] In addition, a one-step assembly of the spliceosome has also been reported.[40,41]

Pre-mRNAs can also undergo alternative splicing to generate variant mRNAs with diverse, and often antagonistic functions.[2,42,43] Regulation of gene expression by alternative splicing is especially common for genes that are involved in nervous system development.[44] Recent studies have indicated that subtle changes in the expression of splicing regulatory proteins can modulate expression of genes that are important for neuronal development.[45–47] Importantly, a defect in the process of alternative splicing is the cause of several neurological diseases and genetic disorders.[48–53] Moreover, certain forms of cancer have been linked to unbalanced isoform expression from genes involved in cell cycle regulation or angiogenesis.[54–60] In the following sections, I describe a splicing regulatory protein that controls the alternative splicing of a number of developmentally regulated genes, and whose own expression is under the control of miRNAs.

## 14.2.1 MULTIFUNCTIONAL PROTEIN PTB: A MASTER SPLICING REGULATOR

More than two decades of research has led to the identification and characterization of numerous splicing factors that regulate alternative splicing by controlling splice site choice. The tissue- or cell-specific expression of alternative splicing factors determines cell fate by modulating the expression of essential genes. However, the mechanisms that regulate the expression of splicing regulators are not known. PTB (polypyrimidinetract-binding protein), also known as hnRNP I (heterogeneous nuclear ribonucleoprotein I) or PTBP1, is undoubtedly one of the most extensively studied and well-characterized splicing factors. PTB is a multifunctional RNA-binding protein known to control an array of cellular processes such as alternative RNA splicing, polyadenylation, mRNA stability, and translation.[61] PTB was originally identified as a splicing factor that binds to the poly(Y)-tract sequence, which often precedes the 3′ splice site.[62] Since the binding of essential splicing factor U2AF to the poly-(Y)-tract is required for the early recognition of 3′ splice site, PTB has been proposed to function as a splicing repressor by interfering with U2AF-poly(Y)-tract interaction.[63–65] Subsequent studies have revealed additional mechanisms of PTB splicing regulation. For example, PTB binding to pre-mRNA can inhibit spliceosome assembly at a step following recognition of 3′ splice site.[66–69] In addition, PTB affords regulation of alternative splicing by interacting with splicing coregulators,[70,71] and can function as an antirepressor of splicing.[72]

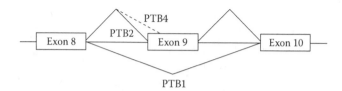

**FIGURE 14.1** Alternative splicing of PTB pre-mRNA. PTB pre-mRNA undergoes alternative splicing producing three splice variants. Exclusion of exon 9 generates PTB1 mRNA, whereas its inclusion via the use of two alternative 3' splice donor sites produces PTB2 and PTB4 splice isoforms. In addition, PTB autoregulates its own expression by controlling the splicing of alternative exon 11 (not shown) whose skipping causes a frame shift and introduction of premature translation termination codon. The PTB mRNA lacking exon 11 is subjected to nonsense-mediated decay.

## 14.2.2 PTB SPLICE ISOFORMS AND HOMOLOGUES

PTB transcripts undergo alternative splicing to generate three main isoforms[62,73] (Figure 14.1). While PTB is the shortest splice variant generated owing to the skipping of exon 9, inclusion of exon 9 via activation of one of two competing 3' splice sites produces PTB2 and PTB4 isoforms.[74,75] The exclusion of exon 11 generates a frame-shifted mRNA that undergoes degradation by nonsense-mediated decay (NMD). Interestingly, skipping of exon 11 in PTB mRNA is controlled by PTB protein thus establishing an autoregulatory feedback loop.[76] Recently, a fourth alternatively spliced isoform of PTB, known as PTB-T has been identified. This isoform is generated due to the alternative splicing of exons 2–10 and expresses a PTB protein lacking RRM1 and RRM2.[77] Cell-specific expression of these isoforms has been suggested to play an important role in pre-mRNA splicing, internal ribosome entry site (IRES)-dependent translation, and mRNA stability.[75,77,78]

At least two PTB homologues are expressed in mammals, nPTB (also known as brPTB) or PTB2, and ROD1 (regulator of differentiation). Whereas expression of ROD1 is restricted to haemopoietic cells,[79] nPTB is expressed in brain, testis, and muscle, but its expression has also been observed in other tissues.[80–82] Although PTB and nPTB are encoded by different genes, they are highly homologous and share many similarities at structural and functional levels.[83] For example, both are RNA-binding proteins containing four RNA recognition motifs (RRMs) and bind to similar RNA motifs, albeit with different affinities.[81,84] Despite remarkable structural similarities, these proteins display distinct splicing regulatory properties, apparently because of the differences in their affinity for various cellular factors.[80,82,85,86] Several studies have indicated that nPTB, just like PTB, uses alternative splicing and NMD to generate different mRNA isoforms.[76,87–91]

## 14.3 MIRNA TARGETING OF REGULATORS OF ALTERNATIVE SPLICING TRIGGERS CELL DIFFERENTIATION

Previous studies have indicated that PTB functions as a repressor of nervous system-specific splicing in nonneuronal cells where it is expressed at higher levels.

Mechanistically, it binds to the pyrimidine-rich sequences in pre-mRNAs and promotes the exclusion of neuron-specific alternative exons.[67,69,92] The downregulation of PTB in neuronal cells prevent the skipping of neuron-specific alternative exons in mature mRNA.[82,93] Thus, a splice switch from nonneuronal to neuronal-specific alternative splicing program ensures development of nervous system. However, the molecular basis of PTB downregulation in neuronal cells was not known until after the landmark discoveries from the laboratories of Tom Maniatis and Dough Black (see below). These authors demonstrated for the first time that miRNAs could influence nervous system and muscle development by triggering large-scale changes in gene expression.[13,14] Although the detailed understanding of this process remains to be elucidated, miRNAs promote cell differentiation by reprogramming alternative splicing by modulating the expression of master splicing repressor PTB.

## 14.3.1   ROLE OF miRNAs IN MUSCLE DEVELOPMENT

It is well established that alternative splicing regulator PTB plays an important role in the alternative splicing of muscle- or neuron-specific gene transcripts. The PTB regulation of c-src N1, $\alpha$-actinin SM, $\alpha$-tropomyosin exon 2, NMDAR1 exon 5, and cardiac troponin T exon 5 are some of the most extensively studied examples.[86,94–97] However, what regulates the regulator of alternative splicing was a mystery, especially during development of embryonic muscle. Black and co-workers demonstrated that a muscle-specific microRNA (miR-133) reprograms the alternative splicing of pre-mRNAs required for muscle maturation.[14] Because of the overlapping function of PTB and nPTB, it was suspected that nPTB could be the target of miR-133-dependent muscle-cell differentiation. To test this hypothesis, Boutz et al.[14] used C2C12 muscle cells as an in vitro model of muscle development. The C2C12 cells are derived from adult mouse satellite cells and can be induced to differentiate to myotubes under suitable culture conditions. Western blot and RT-PCR experiments performed on proliferating and differentiated C2C12 cells indicated that while the levels of nPTB and PTB proteins were down in differentiated cells, only PTB mRNA but not the nPTB mRNA was downregulated. This raised the possibility that a microRNA-dependent translation repression of nPTB mRNA may be operative.

Experiments performed with muscle- and nonmuscle-specific cell lines of human and mouse origin reveled that expression of miR-133 and miR-1/206 (expressed from a common precursor) is restricted to muscle-derived cell lines, and an increase in their expression coincides with differentiating C2C12 myoblasts (Figure 14.2). Together, these results suggest that downregulation of nPTB could be the result of miR-133 targeting of nPTB mRNA.

### 14.3.1.1   miR-133: Direct Repressor of nPTB mRNA Translation

If miR-133 is responsible for downregulation of nPTB mRNA translation, then the introduction of artificial miR-133 into proliferating myoblasts should affect the expression of nPTB protein. Two sets of experiments confirmed the direct relationship between miR-133 expression and nPTB downregulation. First, cotransfection of HEK 293 cells with a luciferase reporter (in which human nPTB 3′ UTR was inserted next to the *Renilla* luciferase coding sequence) along with miR-133 RNA

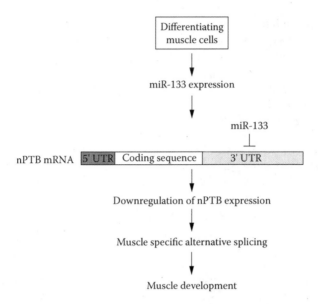

**FIGURE 14.2**  Muscle development controlled by miR-133. In differentiating muscle cells, expression of miR-133 downregulates nPTB by binding to two sites within its 3′ UTR. A decrease in nPTB levels promotes muscle-specific alternative splicing and muscle development.

exhibited >50% reduction in luciferase expression. In contrast, the expression from reporter carrying mutations in the microRNA response element (MREs) showed weak response to miR-133. Importantly, luciferase expression from the reporter that lacks nPTB 3′ UTR remained unaltered. Blocking of miR-133 function by the use of complementary locked nucleic acid (LNA) in differentiating C2C12 cells led to an increase in the level of nPTB expression. Significantly, the combination of LNA oligonucleotides targeting miR-133, miR-1, and miR-206 showed increased nPTB levels as compared to the miR-133 LNA alone. Surprisingly, although the expression of PTB protein is not lost during differentiation, LNA treatment displayed a similar increase in the levels of PTB, suggesting that miR-1 and its sequence homologue miR-206 may also contribute to downregulation of PTB and nPTB during muscle development. In contrast, LNA-mediated block of miR-124 had no effect on nPTB expression. Clearly, additional experiments will be required to understand why decrease in the expression of both nPTB and PTB coincides with muscle development. In summary, these data suggest that the miR-133-mediated repression of nPTB reprograms the splicing of muscle specific alternative exons during muscle development (Figure 14.2).

### 14.3.2   ROLE OF MIR-124 IN NEURONAL DIFFERENTIATION

Initiation and maintenance of differentiation program requires global changes in gene expression at various levels including the reprogramming of alternative splicing.[2,98] Given that the vast majority of alternative splicing events occur in the nervous system, differential expression of splicing regulatory proteins is a common theme during neuronal differentiation.[1,45,46,49,99] Recent studies have identified PTB as a key splicing regulatory protein, which reprograms the splicing of a large group of exons during neuronal development. It has been demonstrated that a decrease in the level of PTB expression in neuronal cells correlates with the induction of neuronal-specific alternative pre-mRNA splicing.[100] In contrast, nonneuronal cells express higher levels of PTB, which regulates the alternative splicing of nPTB pre-mRNA by promoting the skipping of exon 10. The skipping of exon 10 in nPTB mRNA introduces a premature stop codon targeting nPTB mRNA to the nonsense-mediated decay pathway (Figure 14.3). Importantly, the RNAi-mediated knockdown of PTB protein triggers neuronal-specific alternative splicing.[13,14,100]

Although a number of miRNAs are expressed in neuronal cells,[101–104] their exact role and targets are not known.[105–107] It has recently been demonstrated that miR-124 plays an important role in neuronal differentiation (see following text). MiR-124 is one of the most conserved and highly expressed miRNAs in neurons.[108] Studies in the human and mouse indicate that miR-124 is encoded by three distinct genes.[109,110] In addition, the expression of miR-124 is restricted to neurons and in developing neurons, its level of expression increases with time.[101,103,111] Importantly, forced expression of miR-124 in nonneuronal cells has been demonstrated to downregulate a large number of nonneural mRNAs.[112] In contrast, antisense RNA-mediated inhibition of miR-124 increases the levels of nonneuronal gene transcripts in primary cortical neurons.[110] Finally, PTB is the predicted target of miR-124.[14] Together, these observations suggest a link between miR-124, PTB, and nervous system development. To investigate the relationship between miR-124 and development of the nervous system, Makeyev et al.[13] performed a series of experiments to demonstrate that expression of mir-124 in neuronal cells indeed decrease the levels of PTB protein by directly binding to the 3′ UTR of PTB mRNA. These authors showed that a decrease in PTB expression is necessary and sufficient to trigger a switch in the pattern of alternative pre-mRNA splicing from nonneuronal to neuronal. Mechanistically, PTB promotes the exclusion of a cassette exon in the pre-mRNA of nPTB such that the exon skipped nPTB mRNA is subjected to nonsense-mediated decay (Figure 14.3). Although miR-124 overexpression in P19 cell was not sufficient to induce neuronal differentiation, a significant increase in retinoic-acid-induced differentiation was observed. Interestingly, mir-124 also downregulates the expression of nPTB albeit to a lesser degree as compared to PTB. In summary, miRNAs promote cell differentiation by controlling the expression of splicing regulators, which in turn triggers tissue-specific alternative splicing.

## 14.4   REGULATION OF miRNA EXPRESSION

It is well established that miRNAs regulate gene expression by binding to target mRNAs and controlling protein production or inducing mRNA cleavage. The central question in the field is, what regulates the expression of miRNAs? Although miRNA processing is a complex multistep phenomenon, Guil and Caceres[15] have found the answer for the production of at least one miRNA, miR-18a. Using an in vivo cross-linking and immunoprecipitation (CLIP) assay, these authors demonstrated that hnRNP A1, a multifunctional RNA-binding protein, is involved in the processing

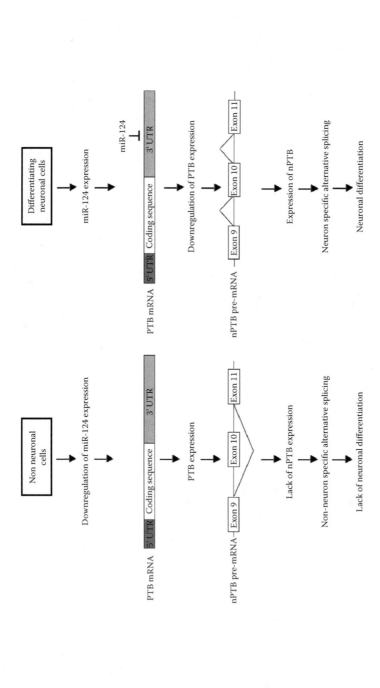

**FIGURE 14.3** MiR-124 regulates neuronal differentiation. Both neuronal and nonneuronal cells produce nPTB protein. The expression of miR-124 in neuronal cells downregulates PTB protein by targeting its mRNA, which allows correct splicing of nPTB mRNA and protein. The expression of nPTB triggers neuron-specific alternative splicing and neuronal differentiation. In contrast, lack of miR-124 expression in nonneuronal cells allows expression of PTB, which suppresses splicing of alternative exon 10 in nPTB. The skipping of exon 10 in nPTB mRNA introduces a frame shift and a premature termination stop codon leading to mRNA degradation by NMD.

of miR-18a, a member of the polycistronic *mir-17-92* cluster. In a series of elegant experiments, the authors showed that specific binding of hnRNP A1 is required for the processing of miR-18a in HeLa extracts, and this interaction takes place prior to the cropping by Drosha. Thus, it appears that hnRNP A1 facilitate Drosha-mediated processing of pre-miR-18a. This was confirmed by RNA interference-mediated knockdown of hnRNP A1, which suppressed the production of pre-miR-18a. By contrast, depletion of hnRNP A1 had no effect on the processing of pri23-24.2 substrate. Multiple lines of evidence indicate that hnRNP A1-mediated processing of miR-18a is highly specific. First, hnRNP A1 influences only the processing of pri-miR-18a without displaying any effect on the neighboring miRNA hairpins in the cluster, suggesting the sequence-specific nature of this processing. This is consistent with the observation that swapping miR-18a and miR-27 hairpins rendered miR-18a processing hnRNP A1 independent. Second, hnRNP A1 had no effect on the processing of a highly similar miR-18b encoded in a homologous cluster. Third, while RNAi-dependent knockdown of hnRNP A1 resulted in lower levels of both pre-miR-18a and processed miR-18a, its overexpression led to an increased level of miR-18a. Finally, cotransfection of miR-18a expression plasmid with a luciferase reporter in which the target site for miR-18a was inserted downstream of firefly luciferase ORF showed a threefold decrease in luciferase activity. By contrast, RNAi-mediated depletion of hnRNP A1 exhibited derepression of the luciferase mRNA with a simultaneous increase in luciferase activity. Taken together, these observations indicate that the RNA binding protein hnRNP A1 plays a direct role in the processing of miR-18a.

## 14.5 CONCLUSIONS

A delicate balance between cell growth, differentiation, and cell death controls various aspects of development. To accomplish this highly complex task, a large number of genes need to be switched on/off or produce proteins with altogether different properties. Undoubtedly, microRNA-mediated reprogramming of alternative splicing has the ability to expand the cell's proteome, which is required to carry out specialized cellular function. Given that miRNAs play an important role in almost every aspect of cell function, the PTB/nPTB switch could be one of the many targets of miRNA-mediated control of alternative splicing.

## ACKNOWLEDGMENTS

I would like to thank Marieta Gencheva for comments and Faith Osep for administrative assistance. This work was supported in part by Department of Defense (DOD; CDMRP) grant BC023235 to RKG.

## REFERENCES

1. Maniatis, T. and Tasic, B. Alternative pre-mRNA splicing and proteome expansion in metazoans. *Nature* 2002; 418(6894): 236–243.
2. Black, D. L. Mechanisms of alternative pre-messenger RNA splicing. *Annu Rev Biochem* 2003; 72: 291–336.

3. House, A. E. and Lynch, K. W. Regulation of alternative splicing: more than just the ABCs. *J Biol Chem* 2008; 283(3): 1217–1221.

4. Wang, Z. and Burge, C. B. Splicing regulation: from a parts list of regulatory elements to an integrated splicing code. *RNA* 2008; 14(5): 802–813.

5. Brett, D., Pospisil, H., Valcarcel, J., Reich, J., and Bork, P. Alternative splicing and genome complexity. *Nat Genet* 2002; 30(1): 29–30.

6. Mironov, A. A., Fickett, J. W., and Gelfand, M. S. Frequent alternative splicing of human genes. *Genome Res* 1999; 9(12): 1288–1293.

7. Modrek, B. and Lee, C. A genomic view of alternative splicing. *Nat Genet* 2002; 30(1): 13–19.

8. Johnson, J. M., Castle, J., Garrett-Engele, P., Kan, Z., Loerch, P. M., Armour, C. D., Santos R., Schadt, E. E., Stoughton, R., and Shoemaker, D. D. Genome-wide survey of human alternative pre-mRNA splicing with exon junction microarrays. *Science* 2003; 302(5653): 2141–2144.

9. Kan, Z., Rouchka, E. C., Gish, W. R., and States, D. J. Gene structure prediction and alternative splicing analysis using genomically aligned ESTs. *Genome Res* 2001; 11(5): 889–900.

10. Modrek, B., Resch, A., Grasso, C., and Lee, C. Genome-wide detection of alternative splicing in expressed sequences of human genes. *Nucleic Acids Res* 2001; 29(13): 2850–2859.

11. Ureta-Vidal, A., Ettwiller, L., and Birney, E. Comparative genomics: genome-wide analysis in metazoan eukaryotes. *Nat Rev Genet* 2003; 4(4): 251–262.

12. Makeyev, E. V. and Maniatis, T. Multilevel regulation of gene expression by microRNAs. *Science* 2008; 319(5871): 1789–1790.

13. Makeyev, E. V., Zhang, J., Carrasco, M. A., and Maniatis, T. The MicroRNA miR-124 promotes neuronal differentiation by triggering brain-specific alternative pre-mRNA splicing. *Mol Cell* 2007; 27(3): 435–448.

14. Boutz, P. L., Chawla, G., Stoilov, P., and Black, D. L., MicroRNAs regulate the expression of the alternative splicing factor nPTB during muscle development. *Genes Dev* 2007; 21(1): 71–84.

15. Guil, S. and Caceres, J. F. The multifunctional RNA-binding protein hnRNP A1 is required for processing of miR-18a. *Nat Struct Mol Biol* 2007; 14(7): 591–596.

16. Nielsen, A. F., Leuschner, P. J., and Martinez, J. Not miR-ly a splicing factor: hnRNP A1 succumbs to microRNA temptation. *Nat Struct Mol Biol* 2007; 14(7): 572–573.

17. Nilsen, T. W. The spliceosome: the most complex macromolecular machine in the cell? *Bioessays* 2003; 25(12): 1147–1149.

18. Nilsen, T. W. RNA-RNA interactions in the spliceosome: Unraveling the ties that bind. *Cell* 1994; 78: 1–4.

19. Green, M. R. Biochemical mechanisms of constitutive and regulated pre-mRNA splicing. *Annu Rev Cell Biol* 1991; 7: 559–599.

20. Kramer, A. The structure and function of proteins involved in mammalian pre-mRNA splicing. *Annu Rev Biochem* 1996; 65: 367–409.

21. Moore, J. J., Query, C. C, and Sharp, P. A. Splicing of precursors to mRNA by the spliceosome. In Gesteland, R. F. and Atkins, J. R. (Ed.), *The RNA World*. New York: Cold Spring Harbor; 1993.

22. Will, C. L. and Luhrmann, R. Protein functions in pre-mRNA splicing. *Curr Opin Cell Biol* 1997; 9(3): 320–328.

23. Lamond, A. I. Splicing in stereo. *Nature* 1993; 365: 294–295.

24. Jurica, M. S. and Moore, M. J. Pre-mRNA splicing: awash in a sea of proteins. *Mol Cell* 2003; 12(1): 5–14.

25. Newman, A. Activity in the spliceosome. *Curr Biol* 1994; 4: 462–464.

26. Zhuang, Y. and Weiner, A. M. A compensatory base change in U1 snRNA suppresses a 5′ splice site mutation. *Cell* 1986; 46: 827–835.
27. Seraphin, B., Kretzner, L., and Rosbash, M. A U1 snRNA: pre-mRNA base pairing interaction is required early in yeast spliceosome assembly but does not uniquely define the 5′ cleavage site. *EMBO J* 1988; 7: 2533–2538.
28. Siliciano, P. G. and Guthrie, C. 5′ splice site selection in yeast: genetic alterations in base-pairing with U1 reveal additional requirements. *Genes Dev* 1988; 2: 1258–1267.
29. Ruskin, B., Zamore, P. D., and Green, M. R. A factor, U2AF, is required for U2 snRNP binding and splicing complex assembly. *Cell* 1988; 52: 207–219.
30. Zamore, P. D., Patton, J. G., and Green, M. R. Cloning and domain structure of the mammalian splicing factor U2AF. *Nature* 1992; 355(6361): 609–614.
31. Wu, S., Romfo, C. M., Nilsen, T. W., and Green, M. R. Functional recognition of the 3′ splice site AG by the splicing factor U2AF35. *Nature* 1999; 402: 832–835.
32. Sawa, H. and Abelson, J. Evidence for a base-pairing interaction between U6 small nuclear RNA and the 5′ splice site during the splicing reaction in yeast. *Proc Natl Acad Sci USA* 1992; 89: 11269–11273.
33. Wassarman, D. A. and Steitz, J. A. Interactions of small nuclear RNA's with precursor messenger RNA during in vitro splicing. *Science* 1992; 257: 1918–1925.
34. Kandels-Lewis, S. and Seraphin, B. Involvement of U6 snRNA in 5′ splice site selection. *Science* 1993; 262: 2035–2039.
35. Konforti, B. B., Koziolkiewicz, M. J., and Konarska, M. M. Disruption of base pairing between the 5′ splice site and the 5′ end of U1 snRNA is required for spliceosome assembly. *Cell* 1993; 75: 863–873.
36. Lesser, C. and Guthrie, C. Mutations in U6 snRNA that alter splice site specificity: implications for the active site. *Science* 1993; 262: 1982–1988.
37. Sontheimer, E. and Steitz, J. A. The U5 and U6 small nuclear RNAs as active site components of the spliceosome. *Science* 1993; 262: 1989–1996.
38. Gaur, R. K., Valcarcel, J., and Green, M. R. Sequential recognition of the pre-mRNA branch point by U2AF[65] and a novel spliceosome-associated 28-kDa protein. *RNA* 1995; 1: 407–417.
39. Valcarcel, J., Gaur, R. K., Singh, R., and Green, M. R. Interaction of U2AF65 RS region with pre-mRNA branch point and promotion of base pairing with U2 snRNA. *Science* 1996; 273(5282): 1706–1709.
40. Malca, H., Shomron, N., and Ast, G. The U1 snRNP base pairs with the 5′ splice site within a penta-snRNP complex. *Mol Cell Biol* 2003; 23(10): 3442–3455.
41. Stevens, S. W., Ryan, D. E., Ge. H. Y., Moore, R. E., Young, M. K., Lee, T. D, and Abelson, J. Composition and functional characterization of the yeast spliceosomal penta-snRNP. *Mol Cell* 2002; 9(1): 31–44.
42. Graveley, B. R. Alternative splicing: increasing diversity in the proteomic world. *Trends Genet* 2001; 17(2): 100–107.
43. Claverie, J. M. Gene number. What if there are only 30,000 human genes? *Science* 2001; 291(5507): 1255–1257.
44. Lee, C. J. and Irizarry, K. Alternative splicing in the nervous system: an emerging source of diversity and regulation. *Biol Psychiatry* 2003; 54(8): 771–776.
45. Black, D. L. and Grabowski, P. J. Alternative pre-mRNA splicing and neuronal function. *Prog Mol Subcell Biol* 2003; 31: 187–216.
46. Lipscombe, D. Neuronal proteins custom designed by alternative splicing. *Curr Opin Neurobiol* 2005; 15(3): 358–63.
47. Li, Q., Lee, J. A., and Black, D. L. Neuronal regulation of alternative pre-mRNA splicing. *Nat Rev Neurosci* 2007; 8(11): 819–31.
48. Ranum, L. P. and Cooper, T. A. RNA-mediated neuromuscular disorders. *Annu Rev Neurosci* 2006; 29: 259–77.

49. Licatalosi, D. D. and Darnell, R. B. Splicing regulation in neurologic disease. *Neuron* 2006; 52(1): 93–101.

50. Faustino, N. A. and Cooper, T. A. Pre-mRNA splicing and human disease. *Genes Dev* 2003; 17(4): 419–37.

51. Garcia-Blanco, M. A., Baraniak, A. P., and Lasda, E. L. Alternative splicing in disease and therapy. *Nat Biotechnol* 2004; 22: 535–546.

52. Nissim-Rafinia, M. and Kerem, B. Splicing regulation as a potential genetic modifier. *Trends Genet* 2002; 18(3): 123–7.

53. Pagani, F. and Baralle, F. E. Genomic variants in exons and introns: identifying the splicing spoilers. *Nat Rev Genet* 2004; 5: 389–396.

54. Gaur, R. K. RNA interference: a potential therapeutic tool for silencing splice isoforms linked to human diseases. *Biotechniques* 2006; Suppl: 15–22.

55. Krajewska, M., Fenoglio-Preiser, C. M., Krajewski, S., Song, K., Macdonald, J. S., Stemmerman, G., and Reed, J. C. Immunohistochemical analysis of Bcl-2 family proteins in adenocarcinomas of the stomach. *Am J Pathol* 1996; 149(5): 1449–1457.

56. Krajewska, M., Krajewski, S., Epstein, J. I., Shabaik, A., Sauvageot, J., Song, K., Kitada, S., and Reed, J. C. Immunohistochemical analysis of bcl-2, bax, bcl-X, and mcl-1 expression in prostate cancers. *Am J Pathol* 1996; 148(5): 1567–1576.

57. Novak, U., Grob, T. J., Baskaynak, G., Peters, U. R., Aebi, S., Zwahlen, D., Tschan, M. P., Kreuzer, K. A., Leibundgut, E. O., Cajot, J. F., and others. Overexpression of the p73 gene is a novel finding in high-risk B-cell chronic lymphocytic leukemia. *Ann Oncol* 2001; 12(7): 981–986.

58. Steinman, H. A., Burstein, E., Lengner, C., Gosselin, J., Pihan, G., Duckett, C. S., and Jones, S. N. An alternative splice form of Mdm2 induces p53-independent cell growth and tumorigenesis. *J Biol Chem* 2004; 279(6): 4877–4886.

59. Xerri, L., Parc, P., Brousset, P., Schlaifer, D., Hassoun, J., Reed, J. C., Krajewski, S., and Birnbaum, D. Predominant expression of the long isoform of Bcl-x (Bcl-xL) in human lymphomas. *Br J Haematol* 1996; 92(4): 900–906.

60. Venables, J. P. Aberrant and alternative splicing in cancer. *Cancer Res* 2004; 64(21): 7647–7654.

61. Sawicka, K., Bushell, M., Spriggs, K. A., and Willis, A. E. Polypyrimidine-tract-binding protein: a multifunctional RNA-binding protein. *Biochem Soc Trans* 2008; 36(Pt 4): 641–647.

62. Garcia-Blanco, M. A., Jamison, S. F., and Sharp, P. A. Identification and purification of a 62,000-dalton protein that binds specifically to the polypyrimidine tract of introns. *Genes Dev* 1989; 3(12A): 1874–1886.

63. Singh, R., Valcarcel, J., and Green, M. R. Distinct binding specificities and functions of higher eukaryotic polypyrimidine tract-binding proteins. *Science* 1995; 268(5214): 1173–1176.

64. Amir-Ahmady, B., Boutz, P. L., Markovtsov, V., Phillips, M. L., and Black, D. L. Exon repression by polypyrimidine tract binding protein. *RNA* 2005; 11(5): 699–716.

65. Lin, C. H. and Patton, J. G. Regulation of alternative 3′ splice site selection by constitutive splicing factors. *RNA* 1995; 1(3): 234–245.

66. Singh, R. and Valcarcel, J. Building specificity with nonspecific RNA-binding proteins. *Nat Struct Mol Biol* 2005; 12(8): 645–653.

67. Sharma, S., Falick, A. M., and Black, D. L. Polypyrimidine tract binding protein blocks the 5′ splice site-dependent assembly of U2AF and the prespliceosomal E complex. *Mol Cell* 2005; 19(4): 485–496.

68. Izquierdo, J. M., Majos, N., Bonnal, S., Martinez, C., Castelo, R., Guigo, R., Bilbao, D., and Valcarcel, J. Regulation of Fas alternative splicing by antagonistic effects of TIA-1 and PTB on exon definition. *Mol Cell* 2005; 19(4): 475–484.

69. Spellman, R. and Smith, C. W. Novel modes of splicing repression by PTB. *Trends Biochem Sci* 2006; 31(2): 73–76.

70. Gromak, N., Rideau, A., Southby, J., Scadden, A. D., Gooding, C., Huttelmaier, S., Singer, R. H., and Smith, C. W. The PTB interacting protein raver1 regulates alpha-tropomyosin alternative splicing. *EMBO J* 2003; 22(23): 6356–6364.

71. Rideau, A. P., Gooding, C., Simpson, P. J., Monie, T. P., Lorenz, M., Huttelmaier, S., Singer, R. H., Matthews, S., Curry, S., and Smith, C. W., A peptide motif in Raver1 mediates splicing repression by interaction with the PTB RRM2 domain. *Nat Struct Mol Biol* 2006; 13(9): 839–848.

72. Paradis, C., Cloutier, P., Shkreta, L., Toutant, J., Klarskov, K., and Chabot, B., hnRNP I/PTB can antagonize the splicing repressor activity of SRp30c. *RNA* 2007; 13(8): 1287–1300.

73. Ghetti, A., Pinol-Roma, S., Michael, W. M., Morandi, C., and Dreyfuss, G., hnRNP I, the polypyrimidine tract-binding protein: distinct nuclear localization and association with hnRNAs. *Nucleic Acids Res* 1992; 20(14): 3671–3678.

74. Robinson, F. and Smith, C. W., A splicing repressor domain in polypyrimidine tract-binding protein. *J Biol Chem* 2006; 281(2): 800–806.

75. Wollerton, M. C., Gooding, C., Robinson, F., Brown, E. C., Jackson, R. J., and Smith, C. W., Differential alternative splicing activity of isoforms of polypyrimidine tract binding protein (PTB). *RNA* 2001; 7(6): 819–832.

76. Wollerton, M. C., Gooding, C., Wagner, E. J., Garcia-Blanco, M. A., and Smith, C. W., Autoregulation of polypyrimidine tract binding protein by alternative splicing leading to nonsense-mediated decay. *Mol Cell* 2004; 13(1): 91–100.

77. Hamilton, B. J., Genin, A., Cron, R. Q., and Rigby, W. F., Delineation of a novel pathway that regulates CD154 (CD40 ligand) expression. *Mol Cell Biol* 2003; 23(2): 510–525.

78. Wagner, E. J. and Carstens, R. P., Garcia-Blanco M. A. A novel isoform ratio switch of the polypyrimidine tract binding protein. *Electrophoresis* 1999; 20(4–5): 1082–1086.

79. Yamamoto, H., Tsukahara, K., Kanaoka, Y., Jinno, S., and Okayama, H., Isolation of a mammalian homologue of a fission yeast differentiation regulator. *Mol Cell Biol* 1999; 19(5): 3829–3841.

80. Polydorides, A. D., Okano, H. J., Yang, Y. Y., Stefani, G., and Darnell, R. B., A brain-enriched polypyrimidine tract-binding protein antagonizes the ability of Nova to regulate neuron-specific alternative splicing. *Proc Natl Acad Sci USA* 2000; 97(12): 6350–6355.

81. Markovtsov, V., Nikolic, J. M., Goldman, J. A., Turck, C. W., Chou, M. Y., and Black, D. L., Cooperative assembly of an hnRNP complex induced by a tissue-specific homolog of polypyrimidine tract binding protein. *Mol Cell Biol* 2000; 20(20): 7463–7479.

82. Ashiya, M. and Grabowski, P. J., A neuron-specific splicing switch mediated by an array of pre-mRNA repressor sites: evidence of a regulatory role for the polypyrimidine tract binding protein and a brain-specific PTB counterpart. *RNA* 1997; 3(9): 996–1015.

83. Kikuchi, T., Ichikawa, M., Arai, J., Tateiwa, H., Fu, L., Higuchi, K., and Yoshimura, N., Molecular cloning and characterization of a new neuron-specific homologue of rat polypyrimidine tract binding protein. *J Biochem* 2000; 128(5): 811–821.

84. Oberstrass, F. C., Auweter, S. D., Erat, M., Hargous, Y., Henning, A., Wenter, P., Reymond, L., Amir-Ahmady, B., Pitsch, S., and Black, D. L. and others. Structure of PTB bound to RNA: specific binding and implications for splicing regulation. *Science* 2005; 309(5743): 2054–2057.

85. Underwood, J. G., Boutz, P. L., Dougherty, J. D., Stoilov, P., and Black, D. L., Homologues of the *Caenorhabditis elegans* Fox-1 protein are neuronal splicing regulators in mammals. *Mol Cell Biol* 2005; 25(22): 10005–10016.

86. Chan, R. C. and Black, D. L., The polypyrimidine tract binding protein binds upstream of neural cell-specific c-src exon N1 to repress the splicing of the intron downstream. *Mol Cell Biol* 1997; 17(8): 4667–4676.

87. Rahman, L., Bliskovski, V., Reinhold, W., and Zajac-Kaye, M., Alternative splicing of brain-specific PTB defines a tissue-specific isoform pattern that predicts distinct functional roles. *Genomics* 2002; 80(3): 245–249.

88. Gooding, C., Clark, F., Wollerton, M. C., Grellscheid, S. N., Groom, H., and Smith, C. W., A class of human exons with predicted distant branch points revealed by analysis of AG dinucleotide exclusion zones. *Genome Biol* 2006; 7(1): R1.

89. Tateiwa, H., Gotoh, N., Ichikawa, M., Kikuchi, T., and Yoshimura, N., Molecular cloning and characterization of human PTB-like protein: a possible retinal autoantigen of cancer-associated retinopathy. *J Neuroimmunol* 2001; 120(1–2): 161–169.

90. Lareau, L. F., Green, R. E., Bhatnagar, R. S., and Brenner, S. E., The evolving roles of alternative splicing. *Curr Opin Struct Biol* 2004; 14(3): 273–282.

91. Ni, J. Z., Grate L., Donohue, J. P., Preston, C., Nobida, N., O'Brien, G., Shiue, L., Clark, T. A., Blume, J. E., and Ares, M., Jr., Ultraconserved elements are associated with homeostatic control of splicing regulators by alternative splicing and nonsense-mediated decay. *Genes Dev* 2007; 21(6): 708–718.

92. Wagner, E. J. and Garcia-Blanco, M. A., Polypyrimidine tract binding protein antagonizes exon definition. *Mol Cell Biol* 2001; 21(10): 3281–3288.

93. Lillevali, K., Kulla, A., and Ord, T., Comparative expression analysis of the genes encoding polypyrimidine tract binding protein (PTB) and its neural homologue (brPTB) in prenatal and postnatal mouse brain. *Mech Dev* 2001; 101(1–2): 217–220.

94. Charlet, B. N., Logan, P., Singh, G., and Cooper, T. A., Dynamic antagonism between ETR-3 and PTB regulates cell type-specific alternative splicing. *Mol Cell* 2002; 9(3): 649–658.

95. Zhang, L., Liu, W., and Grabowski, P. J., Coordinate repression of a trio of neuron-specific splicing events by the splicing regulator PTB. *RNA* 1999; 5(1): 117–130.

96. Southby, J., Gooding, C., and Smith, C. W., Polypyrimidine tract binding protein functions as a repressor to regulate alternative splicing of alpha-actinin mutally exclusive exons. *Mol Cell Biol* 1999; 19(4): 2699–2711.

97. Gooding, C., Roberts, G. C., and Smith, C. W., Role of an inhibitory pyrimidine element and polypyrimidine tract binding protein in repression of a regulated alpha-tropomyosin exon. *RNA* 1998; 4(1): 85–100.

98. Pascual, M., Vicente, M., Monferrer, L., and Artero, R., The Muscle blind family of proteins: an emerging class of regulators of developmentally programmed alternative splicing. *Differentiation* 2006; 74(2–3): 65–80.

99. Ule, J. and Darnell, R. B., RNA binding proteins and the regulation of neuronal synaptic plasticity. *Curr Opin Neurobiol* 2006; 16(1): 102–110.

100. Boutz, P. L., Stoilov, P., Li, Q., Lin, C. H., Chawla, G., Ostrow, K., Shiue, L., Ares, M., Jr., and Black, D. L., A post-transcriptional regulatory switch in polypyrimidine tract-binding proteins reprograms alternative splicing in developing neurons. *Genes Dev* 2007; 21(13): 1636–1652.

101. Krichevsky, A. M., King, K. S., Donahue, C. P., Khrapko, K., and Kosik, K. S., A microRNA array reveals extensive regulation of microRNAs during brain development. *RNA* 2003; 9(10): 1274–1281.

102. Kim, J., Krichevsky, A., Grad, Y., Hayes, G. D., Kosik, K. S., Church, G. M., and Ruvkun, G., Identification of many microRNAs that copurify with polyribosomes in mammalian neurons. *Proc Natl Acad Sci USA* 2004; 101(1): 360–365.

103. Miska, E. A., Alvarez-Saavedra, E., Townsend, M., Yoshii, A., Sestan, N., Rakic, P., Constantine-Paton, M., and Horvitz, H. R., Microarray analysis of microRNA expression in the developing mammalian brain. *Genome Biol* 2004; 5(9): R68.

104. Sempere, L. F., Freemantle, S., Pitha-Rowe, I., Moss, E., Dmitrovsky, E., and Ambros, V., Expression profiling of mammalian microRNAs uncovers a subset of brain-expressed microRNAs with possible roles in murine and human neuronal differentiation. *Genome Biol* 2004; 5(3): R13.

105. Ashraf, S. I., McLoon, A. L., Sclarsic, S. M., and Kunes, S., Synaptic protein synthesis associated with memory is regulated by the RISC pathway in *Drosophila*. *Cell* 2006; 124(1): 191–205.

106. Kosik, K. S., The neuronal microRNA system. *Nat Rev Neurosci* 2006; 7(12): 911–920.

107. Schratt, G. M., Tuebing, F., Nigh, E. A., Kane, C. G., Sabatini, M. E., Kiebler, M., and Greenberg, M. E., A brain-specific microRNA regulates dendritic spine development. *Nature* 2006; 439(7074): 283–289.

108. Lagos-Quintana, M., Rauhut, R., Yalcin, A., Meyer, J., Lendeckel, W., and Tuschl, T., Identification of tissue-specific microRNAs from mouse. *Curr Biol* 2002; 12(9): 735–739.

109. Deo, M., Yu, J. Y., Chung, K. H., Tippens, M., and Turner, D. L., Detection of mammalian microRNA expression by in situ hybridization with RNA oligonucleotides. *Dev Dyn* 2006; 235(9): 2538–2548.

110. Conaco, C., Otto, S., Han, J. J., and Mandel, G., Reciprocal actions of REST and a microRNA promote neuronal identity. *Proc Natl Acad Sci USA* 2006; 103(7): 2422–2427.

111. Smirnova, L., Grafe, A., Seiler, A., Schumacher, S., Nitsch, R., and Wulczyn, F. G., Regulation of miRNA expression during neural cell specification. *Eur J Neurosci* 2005; 21(6): 1469–1477.

112. Lim, L. P., Laum, N. C., Garrett-Engele, P., Grimson, A., Schelter, J. M., Castle, J., Bartel, D. P., Linsley, P. S., and Johnson, J. M. Microarray analysis shows that some microRNAs downregulate large numbers of target mRNAs. *Nature* 2005; 433(7027): 769–773.

# 15 Recent Progress in Polymerase II-Mediated Intronic microRNA Expression Systems

*Shi-Lung Lin and Shao-Yao Ying*

## CONTENTS

## OVERVIEW

Nearly 97% of the human genome is noncoding DNA, which varies from one species to another, and changes in these sequences frequently manifest clinical and circumstantial malfunction. Numerous noncoding DNA regions have been recently found to encode microRNA (miRNA) genes, which are responsible for RNA-mediated gene silencing through RNA interference (RNAi)-like pathways. miRNAs, 17 to 25 nucleotide single-stranded RNAs that are capable of regulating targeted gene transcripts with high complementarity, are useful for the development of new therapies against cancer polymorphism and viral mutation. This flexible characteristic is different from double-stranded siRNAs (small interfering RNAs), which require completely matched complementarity for targeted gene silencing. miRNAs were first discovered in *Caenorhabditis elegans* as native small RNAs that modulate a wide range of genetic regulatory pathways during embryonic development. Varieties of miRNAs have been widely reported in plants, animals, and even microbes. Intronic miRNA is a new class of miRNAs derived from the processing of gene introns transcribed by RNA polymerases type II (Pol-II). Because its biogenesis requires RNA splicing and is regulated by a nonsense-mediated decay (NMD) mechanism, the intronic miRNA is safe, effective, and powerful for development of new tools regulating specific gene functions of interest in vitro and in vivo. We have presented the first Pol-II-mediated intronic miRNA expression system in 2002 and subsequently demonstrated its special RNAi effects in many vertebrate models. Transgene-like chicken embryos, adult mice, and transgenic zebra fish models have been developed for studying various human diseases and neuropathological disorders. These advances suggest that the use of this novel miRNA expression system in animal models may provide significant insight into the miRNA functions and the mechanisms in related diseases.

## 15.1 INTRODUCTION

The first microRNA (miRNA) molecules, *lin-4* and *let-7*, were identified in 1993 [1]; since then, the advance of small RNA research has gradually progressed in identifying more miRNA species and understanding their biogenesis, functionality, and target gene regulation. Most of these early miRNAs were located in the noncoding regions between genes and transcribed by yet-to-be-determined promoters; these are intergenic miRNA. All miRNAs studied at this stage were recognized as intergenic miRNA until 2003, when Ambros et al. [2] discovered some tiny noncoding RNAs derived from the intron regions of gene transcripts in *Caenorhabditis elegans*. At the same time, Lin et al. [3] proved the biogenesis and gene silencing mechanism of these intron-derived noncoding RNAs, providing the first functional evidence for a new miRNA category, intronic miRNA. Table 15.1 lists several intronic miRNAs currently identified in *C. elegans*, mouse, and human genomes [2,4] and some of their functions related to the miRNA-mediated RNA interference (RNAi) pathway.

Intron occupies the largest proportion of noncoding sequences in a gene. Gene transcription generates precursor messenger RNA (pre-mRNA), which contains four major parts including 5′ untranslational region (UTR), protein-coding exon, noncoding intron, and 3′ UTR. In broad definition, both 5′ and 3′ UTR are a kind of intron extension; however, their processing during mRNA maturation is slightly different from the intron located between two protein-coding exons, also termed the *in-frame intron*. The in-frame intron can be as big as several 10-kb nucleotides and was thought to be a huge genetic waste in gene transcripts; however, this misconception was changed after the finding of the intronic miRNA. miRNA, sized about 17–25 nucleotides in length, is capable of either directly degrading its targeted messenger RNA (mRNA) or suppressing the protein translation of its targeted mRNA, depending on the complementarity between the miRNA and its target. In this way, intronic miRNA functions similar to the previously described intergenic miRNAs, but differs from them in its unique requirement of RNA splicing for biogenesis [3,5]. Because introns naturally contain multiple translational stop codons for recognition by the intracellular nonsense-mediated decay (NMD) system [6,7], most of the intron parts can be quickly degraded after RNA splicing to prevent excessive accumulation,

## TABLE 15.1
## Intronic miRNAs in *C. elegans*, Mouse, and Human Genomes

| miRNA | Species | Host Gene (Intron) [#] | Target Gene(s) |
|---|---|---|---|
| miR-2a, -b2 | Worm | Spi | |
| miR-7b | Mammal | Pituitary gland specific factor 1A (2) [NM174947] | Paired mesoderm homeobox protein 2b; HLHm5 |
| miR-10b | Mammal | Homeobox protein HOX-4 (4) | |
| miR-11 | *Drosophila* | E2f | |
| miR-13b2 | *Drosophila* | Cg7033 | |
| miR-15b, -16-2 | Mammal | Chromosome-associated polypeptide C | |
| miR-25, -93, -106b | Mammal | CDC47 homologue (13) | |
| miR-26a1, -26a2, -26b | Vertebrate | Nuclear LIM interactor-interacting factor 1, 2, 3 | |
| miR-28 | Human | LIM domain-containing preferred translocation partner in lipoma [NM005578] | |
| miR-30c1, -30e | Mammal | Nuclear transcription factor Y subunit γ (5) | Transcription factor HES-1; PAI-1 mRNA-binding protein |
| miR-33 | Vertebrate | Sterol regulatory element binding protein-2 (15) | RNA-dependent helicase p68; NAG14 protein |
| miR-101b | Human | RNA 3′ terminal phosphate cyclase-like protein (8) | |
| miR-103, -107 | Human | Pantothenate kinase 1, 2, 3 | |
| miR-105-1, -105-2, -224 | Mammal | γ-Aminobutyric-acid receptor α-3 subunit precursor, epsilon subunit precursor | |
| miR-126, -126* | Mammal | EGF-like, Notch4-like, NEU1 protein (6) [NM178444] | |
| miR-128b | Mammal | Camp-regulated phospho-protein 21 (11) | |
| miR-139 | Mammal | Cgmp-dependent 3′,5′-cyclic phosphodiesterase (2) | |
| miR-140 | Human | NEDD4-like ubiquitin-protein ligase WWP2 (15) | |
| miR-148b | Mammal | Coatomer ζ-1 subunit | |
| miR-151 | Mammal | | |

*Continued*

**TABLE 15.1 (*Continued*)**
**Intronic miRNAs in *C. elegans*, Mouse, and Human Genomes**

| miRNA | Species | Host Gene (Intron) [#] | Target Gene(s) |
|-------|---------|------------------------|----------------|
| miR-152 | Human | Coatomer ζ-2 subunit | N-myc proto-oncogene protein; noggin precursor |
| miR-153-1, -153-2 | Human | Protein-tyrosine phosphatase N precursors | |
| miR-208 | Mammal | Myosin heavy chain, cardiac muscle α isoform (28) | |
| miR-218-1, -218-2 | Human | Slit homologue proteins [NM003062] | |

which is toxic to the cells. It has been measured that approximately 10% of a spliced intron is spared from NMD digestion and further exported to cytoplasm with a moderate half-life, suggesting the potential source for intronic miRNA generation [8].

Our previous studies have demonstrated that effective mature miRNAs can be derived from the intron regions of mammalian genes [3,5]. As shown in Figure 15.1, these intronic miRNAs are transcribed by type-II RNA polymerases (Pol-II) and excised by spliceosomal components and other *Dicer*-like RNaseIII endonucleases to form mature miRNAs. Because such an intronic miRNA biogenesis pathway is coordinately regulated by intracellular Pol-II transcription, RNA splicing and NMD mechanisms, the gene silencing effects of these intronic miRNA are safe, effective, and powerful for the development of new tools regulating specific gene functions of interest [9,10]. Utilizing this Pol-II-mediated intronic miRNA expression system, we have first demonstrated the targeted RNAi effects of synthetic miRNAs in many mouse and human cell lines in vitro [3,9–12] and mouse skin, chicken embryo and zebra fish in vivo [13–15]. Furthermore, Zhou et al. [16] and Chung et al. [17] have recently observed that both intergenic and intronic miRNAs possess the same RNAi effectiveness while the intronic expression of miRNA allows coexpression of a protein marker with the miRNA. Because there are currently over 1000 native miRNA species found in vertebrates and many more new miRNA homologues continue to be identified, conceivably it is possible to utilize the intronic miRNA as a transgenic tool for generating target-specific loss-of-gene-function animal strains and cell lines for evaluating the targeted gene functions of interest. Indeed, we have developed the first miRNA-mediated loss-of-*FMR1*-function zebra fish strain as a viable model for studying the human fragile X mental retardation syndrome.

## 15.2 DEFINITION AND BIOGENESIS OF INTRONIC miRNA

The definition of intronic miRNAs is based on two factors: first, they must share the same promoter with their encoded genes; second, their miRNA precursors (i.e., pri-miRNA and pre-miRNA) are spliced out of the transcripts of their encoded genes and further processed into mature miRNAs. Although some of the currently identified miRNAs are encoded in the genomic intron region of a gene, they are in the opposite

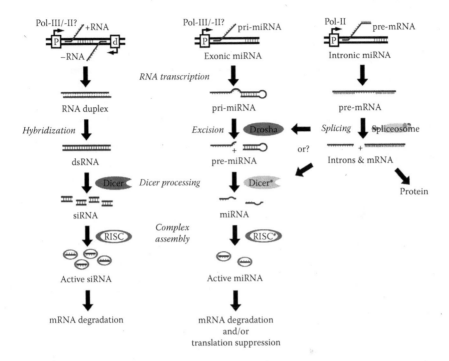

**FIGURE 15.1**   Comparison of biogenesis and RNAi mechanisms among siRNA, intergenic (exonic) miRNA, and intronic miRNA. SiRNA is likely formed by two perfectly complementary RNAs transcribed from two different promoters (remains to be determined) and further processing into 19-22 bp duplexes by RNase III-familial endonucleases, *Dicer*. The biogenesis of intergenic miRNAs (e.g., *lin*-4 and *let*-7) involves a long transcript precursor (pri-miRNA), which is probably generated by Pol-II or Pol-III RNA promoters, while intronic miRNAs are transcribed by the Pol-II promoters from its encoded genes and coexpressed in the intron regions of the gene transcripts (pre-mRNA). After pre-mRNA splicing, the spliced intron may function as a pri-miRNA for intronic miRNA generation. In the nucleus, the pri-miRNA is excised by Drosha-like RNases to form a hairpin-like pre-miRNA template, and then exported to cytoplasm for further processing by Dicer* to form mature miRNAs. The *Dicers* for siRNA and miRNA pathways may be different. All three small regulatory RNAs are finally incorporated into an RNA-induced silencing complex (RISC), which contains either strand of siRNA or the single-strand of miRNA. The effect of miRNA is considered to be more specific and less adverse than that of siRNA because only one strand is involved. On the other hand, siRNAs primarily trigger mRNA degradation, whereas miRNAs can induce either mRNA degradation or suppression of protein synthesis, depending on the sequence complementarity to the target gene transcripts.

orientation to the gene transcript; therefore, these miRNAs are not intronic miRNAs because they neither share the same promoter with the gene nor need to be released from the gene transcript by RNA splicing. For the transcription of these miRNAs, their promoters are located in the antisense direction to the gene, likely using the gene transcript as a potential target for the antisense miRNAs. A good example is *let-7c*, which was found to be an intergenic miRNA located in the antisense region of a gene intron. Current computer programs for miRNA prediction cannot distinguish the intronic miRNAs from the intergenic miRNAs. Because intronic miRNAs are encoded in the gene transcript precursors (pre-mRNAs) and share the same promoter with the encoded gene transcripts, the miRNA prediction programs tend to classify the intronic miRNAs as intergenic miRNAs located in the exonic regions of genes. Given that the mechanisms underlying the biogenesis of intronic and exonic miRNAs are different, these two types of miRNAs may function as different gene regulators in the adjustment of cellular physiology. Thus, an miRNA-prediction program including database of noncoding sequences located in the entire genomic pre-mRNA regions is urgently needed for thoroughly screening and understanding the distribution and variety of hairpin-like intronic miRNAs in the genomes.

The process of intronic miRNA biogenesis in vertebrates involves five steps (Figure 15.1). First, a long primary precursor miRNA (pri-miRNA) is transcribed by type II RNA polymerases (Pol-II) encoded in the intron region of a gene transcript such as pre-mRNA [3,18]. Second, the long pri-miRNA is spliced out of the encoded pre-mRNA by spliceosomal components. Third, the spliced intron is further processed by *Drosha*-like RNase III endonucleases to form a mature precursor miRNA (pre-miRNA) [3,19–21]. Fourth, the pre-miRNA is exported out of the nucleus by Ran-GTP and a receptor Exportin-5 [22,23]. In the cytoplasm, *Dicer*-like nucleases cleave the pre-miRNA to form a mature miRNA. Lastly, the mature miRNA is incorporated into a ribonuclear particle (RNP), which becomes the RNA-induced silencing complex (RISC), capable of executing RNAi-related gene silencing [24,25]. Although some similarities between the siRNA and miRNA pathways have been observed, the RNAi agent maturation and RISC assembly of siRNA and miRNA are different and remain to be determined. The characteristics of *Dicer* and RISC in the miRNA mechanism have been reported to be distinct from that of siRNA [26]. We have recently observed that the stem-loop structure of pre-miRNA was involved in the strand selection for mature miRNA during RISC assembly in zebra fish [13]. These findings suggest that the prefect duplex structure of siRNA may not be essential for the assembly of miRNA-associated RISC in vivo. Most miRNA biogenesis previously proposed were based on the siRNA pathway; one must distinguish their individual properties and differences in order to understand the evolutional and functional relationship between these two RNAi pathways. In addition, these differences may provide a clue to the prevalence of native siRNAs in invertebrates, but rarely in mammals.

## 15.2.1 INTRONIC MIRNA AND MICROSATELLITE-ASSOCIATED DISEASES

The majority of human gene transcripts contain introns, which vary from species to species, and changes in these non-protein-coding sequences are often observed in clinical and circumstantial malfunction. Numerous introns are recently found to encode miRNAs, which are involved in RNAi-related chromatin-silencing mechanisms. Over 90 intronic miRNAs have been identified using the bioinformatic approaches to date, but the function of the vast majority of these molecules remains to be determined [4]. According to the strictly expressive correlation of intronic miRNAs to their encoded genes, one may speculate that the levels of condition-specific, time-specific, and individual-specific gene expression are determined by interactions of distinctive miRNAs on single or multiple gene modulation. This interpretation accounts for more accurate genetic expression of various traits and any dysregulation of the interactions, resulting in genetic diseases. For instance, monozygotic twins frequently demonstrate slightly, but definitely distinguishing, disease susceptibility and physiological behavior. For instance, a microsatellite-like CCTG expansion in the intron 1 of a zinc finger protein ZNF9 gene has been correlated to type 2 myotonic dystrophy in one of the twins with higher susceptibility [27]. Since this expansion motif obtained high affinity to certain RNA-binding proteins, the RNA-interfering role of intron-derived expansion fragments remains to be elucidated.

Another well-established example involving intronic expansion fragments in its pathogenesis is fragile X syndrome (FXS). FXS is the most common form of inherited mental retardation, with an estimated prevalence of 30% of total human mental retardation disorders, and is also among the most frequent single-gene disorders [28]. Recent studies have found that the *FMR1* gene is transcriptionally inactivated in 99% of severe FXS patients by the expansion and probably methylation of (CGG) trinucleotide repeats (r[CGG]) located in the *FMR1* 5′ UTR [29–31]. The lengths and levels of expression of the r(CGG) expansion have actually affected the *FMR1* gene expression. FXS mental retardation usually results from loss of the FMR1 protein due to complete silencing of *FMR1* expression by expansions of >200 CGG repeats, whereas expression of intermediate expansions of 70–120 repeats causes fragile X tremor ataxia syndrome (FXTAS) [32]. Two theories have been proposed to explain the pathological mechanism of *FMR1* gene silencing in FXS. First, Handa et al. [33] suggested that noncoding RNA transcripts transcribed from the *FMR1* 5′ UTR r(CGG) expansion can fold into RNA hairpins and are further processed by RNase *Dicer* to form miRNA-like molecules directed against the FMR1 expression. Second, Jin et al. [34] proposed that RNAi-mediated gene methylation occurs in the CpG regions of the *FMR1* r(CGG) expansion, which are targeted by hairpin RNAs derived from the 3′ UTR of the *FMR1* expanded allele transcript. The *Dicer*-processed hairpin RNA may trigger the formation of RNA-induced initiator of transcriptional gene silencing (RITS) on the homologous r(CGG) sequences and leads to heterochromatin repression of the *FMR1* locus. *FMR1* encodes an RNA-binding protein, FMRP or FMR1, which is associated with polyribosome assembly in an RNP-dependent manner and capable of suppressing translation through an RNAi-like pathway that is important for neuronal development and plasticity. FMRP also contains a nuclear localization signal (NLS) and a nuclear export signal (NES) for shuttling certain mRNAs or even miRNAs between nucleus and cytoplasm [35,36]. Therefore, the excessive expression of r(CGG)-derived miRNAs during embryonic brain development may cause early *FMR1* gene inactivation, leading to the pathogenesis of FXS.

These examples suggest that natural evolution gives raise to more complexity and more of a variety of introns in higher animals and plants for coordinating their vast gene expression volumes and interactions; therefore, a microsatellite-like intron expansion or deletion may cause dysregulation of intronic miRNA or miRNA–target interaction in that specific region and lead to genetic diseases, such as myotonic dystrophy and fragile X mental retardation.

## 15.2.2 SYNTHETIC INTRONIC miRNA

To understand the disease caused by dysregulation of intronic miRNAs, an artificial expression system is needed to recreate the function and mechanism of the miRNA in vitro and in vivo. The same approach may be used to design and develop therapies for various diseases. Previously, several vector-based RNAi systems were developed, using direct shRNA expression by type-III RNA polymerases (Pol-III) [37–40]. Several studies have succeeded in maintaining constant gene silencing effects using the Pol-III-directed shRNA expression systems in vivo [41,42]; nevertheless, these systems do not work in nonmammalian animal models [13,15]. These approaches did not provide tissue-specific RNAi efficacy in the targeted cell population, because of the ubiquitous existence of Pol-III activities. Moreover, the leaky read-through effect of Pol-III-directed transcription often occurs on a short DNA template in the absence of proper termination and large dsRNA products longer than the desired 17–25 base-pairs (bp) can be synthesized and cause unexpected interferon cytotoxicity in vertebrates [43,44]. These problems can also result from the competition between a Pol-III promoter and another vector promoter (i.e., LTR and CMV promoters). Sledz et al. [45] and Lin et al. [46] have both reported that the utilization of high siRNA doses (e.g., >250 nM in human T cells) caused strong cytotoxicity similar to that of using long dsRNA. High siRNA/shRNA concentrations generated by the Pol-III-directed RNAi systems could oversaturate the cellular miRNA pathway and thus produce global miRNA inhibition and cell death [47]. In contrast, the Pol-II-directed intronic miRNA expression systems did not show these problems because of their precise regulation under cellular RNA splicing and nonsense-mediated decay mechanisms [15–17,48], which degrade excessive RNA accumulation to prevent cytotoxicity.

Based on the intronic miRNA biogenesis mechanism shown in Figure 15.2a, we have developed the first Pol II-mediated miRNA-based small hairpin RNA (shRNA) expression system for introducing RNAi activities in vitro and in vivo [3]. The design is relied on a recombinant gene construct containing one or more artificial splicing-competent introns (SpRNAi). Structurally, the SpRNAi consists of several consensus nucleotide elements such as a 5′-splice site, a branch-point domain, a poly-pyrimidine tract and a 3′-splice site (Figure 15.3). In addition, a pre-miRNA or pre-miRNA cluster insert is placed within the SpRNAi sequence between the 5′-splice site and the branch-point domain. This portion of the intron would normally form a lariat structure during RNA splicing and processing. The spliceosomal U2 and U6 snRNPs, both helicases, may be involved in the unwinding and excision of the lariat RNA fragment into pre-miRNA; however, the detailed processing remains to be elucidated. Moreover, the SpRNAi contains a multiple translation stop codon motif (T codons) in its 3′ proximal region to facilitate the accuracy of RNA splicing, which if present in a pre-matured mRNA, would signal the diversion of the splicing-defective mRNA to the nonsense-mediated decay (NMD) pathway and thus eliminate the excess pre-matured mRNA in the cell.

The SpRNAi construct can be further incorporated into any gene or vector for coexpression with the gene or vector in a cell or an organism, preferably in a reporter gene. After such coexpression, the SpRNAi intron is released by intracellular splice-osomal excision, and then its encoded pre-miRNA insert is processed to form

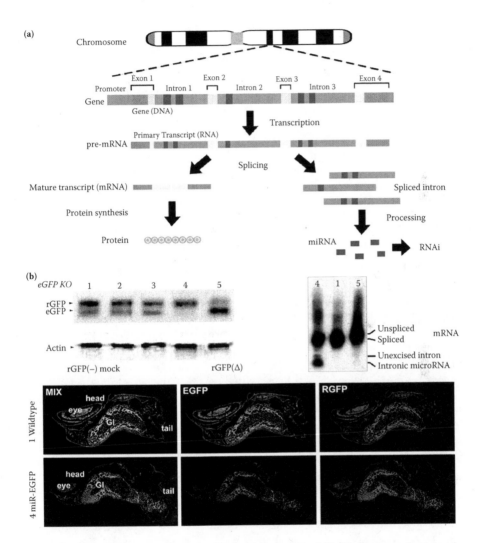

**FIGURE 15.2**  *Color version of this figure follows page 238.* Mechanisms for intronic miRNA biogenesis and relative gene silencing. (a) Intronic miRNA is generated as a part of precursor messenger RNA (pre-mRNA), containing protein-coding exons and noncoding introns. The introns are spliced out of pre-mRNA and further excised into small miRNA-like molecules capable of inducing RNA interference, while the exons are ligated to form a mature mRNA for protein translation. (b) The transfection of intronic miRNA into the green fluorescent protein (eGFP)-expressing Tg(UAS:gfp) strain zebra fish elicited a strong gene-specific silencing effects (>80% suppression, left lane 4). The hairpin-like pre-miRNA directed against *eGFP* was placed in the intron region of a coral reef red fluorescent protein (*rGFP*) pre-mRNA. Northern blot analysis (right) showed that spliced miRNAs were only generated by correct intron insertion, but not rGFP(–) and rGFPΔ.

**Pre-mRNA construct with SpRNAi?**

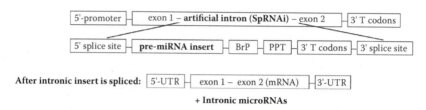

FIGURE 15.3   Schematic construct of the artificial *SpRNAi* intron in a recombinant gene *SpRNAi-RGFP* for intracellular expression and processing. The components of the Pol-II-mediated *SpRNAi* expression system include several consensus nucleotide elements consisting of a 5′-splice site, a branch-point domain (BrP), a poly-pyrimidine tract (PPT), a 3′-splice site, and a pre-miRNA insert located between the 5′-splice site and the BrP domain. The expression of this recombinant gene is under the regulation of either a mammalian Pol-II RNA promoter or a compatible viral promoter for cell-type-specific effectiveness. Mature miRNAs are released from the *SpRNAi* intron by RNA splicing and further *Dicer* processing.

mature miRNA, which is capable of triggering the desired gene silencing effect on targeted genes, whereas the exon transcripts of the reporter gene are linked together to form mature mRNA for translating the reporter protein for indicating the location of the mature miRNA. For example, we have inserted the *SpRNAi* intron in the *Dra*II cleavage site of a red fluorescent membrane protein (*rGFP*) gene, derived from mutated chromoproteins of coral reef *Heteractis crispa*, so as to form a recombinant *SpRNAi-rGFP* gene. The *Dra*II cleavage at the *rGFP* 208th nucleotide site generates an AG–GN nucleotide break with three recessing nucleotides in each end, which then serves as 5′- and 3′-splice sites, respectively, after the *SpRNAi* insertion. Because this intronic insertion disrupts the fluorescent function of rGFP protein, it becomes possible to determine the occurrence of intron splicing and *rGFP* mRNA maturation through the appearance of red fluorescent emission around the membrane surface of the transfected cells. The *rGFP* sequence also provides multiple exonic splicing enhancers (ESEs) to increase RNA splicing efficiency.

Using the *SpRNAi* introns carrying various hairpin-like miRNA precursors (pre-miRNA), we have successfully generated mature miRNA molecules with full capacity of triggering RNAi-like gene silencing in many mouse and human cell lines in vitro [3,9–12] and in mouse, chicken embryo and zebra fish in vivo [11,13–15]. After testing various intronic secondary structures as determined by Western blot analysis of the targeted gene protein expression, we observed that the production of intron-derived hairpin RNA fragments triggers the most strong suppression of targeted genes with high complementarity (Figure 15.2b, lane 4), whereas off-target hairpin RNA inserts—that is, empty intron rGFP(–) (lane 1), *HIV-p24* intron (lane 2), *integrin* β1 intron (lane 3) and splicing-defective intron rGFP(Δ) (lane 5)—present no effects on the targeted gene, for example, eGFP. No silencing effect was detected on off-target genes, such as rGFP, GAPDH, and β-actin, suggesting the high gene-specificity of intronic miRNA-directed RNAi. We also detected that the sizes of these splicing-processed intron fragments are ranged about 17–25 nucleotides (nt) approximate to the newly identified intronic small RNAs in *C. elegans*. Moreover, the intronic miRNAs isolated by guanidinium-chloride ultracentrifugation can elicit a strong, but short-term, gene silencing effect on the target genes, indicating their special RNAi inducibility and target specificity.

## 15.3    TRANSGENE-LIKE GENE SILENCING IN MICE

To evaluate the efficacy and safety of intronic miRNA in animals, we have tested the vector-based intronic miRNA transfection in mice as previously described [15]. As shown in Figure 15.4, patched albino (white) skins of melanin-knockout mice (W-9 black) were created by a succession of intra-cutaneous (*i.c.*) transduction of anti-*tyrosinase* (*Tyr*) pre-miRNA construct (50 µg) for 4 days (total 200 µg). The Tyr, a type-I membrane protein and copper-containing enzyme, catalyzes the critical and rate-limiting step of tyrosine hydroxylation in the biosynthesis of melanin (black pigment) in skins and hairs. Thirteen days after the treatment, the expression of melanin has been blocked in the miRNA transfections due to a significant loss of

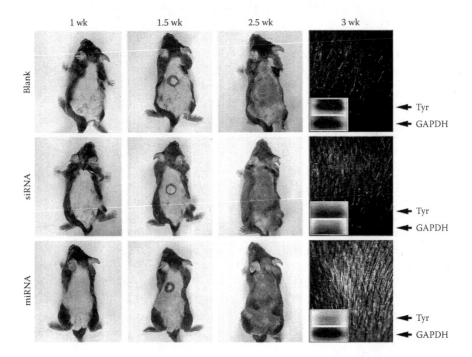

**FIGURE 15.4**    *In vivo* effects of anti-tyrosinase miRNA on the mouse pigment production of local skins. Transfection of the miRNA induced strong gene silencing of *tyrosinase* (*Tyr*) mRNA expression but not house-keeping *GAPDH*, whereas that of U6-directed siRNA triggered mild nonspecific RNA degradation of both *Tyr* and *GAPDH* gene transcripts. Since Tyr is an essential enzyme for black pigment melanin production, the success of gene silencing can be observed by a significant loss of the black color in mouse hairs. The red circles indicate the location of i.c. injections. Northern blot analysis of *Tyr* mRNA expression in local hair follicles confirmed the effectiveness and specificity of the miRNA-mediated gene silencing effect (small windows).

its intermediates resulted from the anti-*Tyr* miRNA-triggered gene silencing effect. Contrarily, the blank control and the Pol-III (U6)-directed siRNA transfections presented normal black skin color. Northern blot analysis using RNA–PCR-amplified mRNAs from hair follicles showed a $76.1\pm5.3\%$ reduction of *Tyr* expression 2 days after the miRNA transfection, which is consistent with the immunohistochemical staining obtained from the same skin area, whereas mild, nonspecific degradation of common gene transcripts was detected in the siRNA-transfected skins (smearing patterns of both house-keeping control *GAPDH* and targeted *Tyr* mRNAs). Given that Grimm et al. [47] have recently reported that high siRNA/shRNA concentrations generated by the Pol-III-directed RNAi systems could oversaturate the cellular miRNA pathway and cause global miRNA dysregulation, the siRNA pathway may be incompatible with the native miRNA pathway in some tissues of mammals. Therefore, these findings have shown that the utilization of intronic miRNA expression systems provided a powerful new approach for transgenic animal generation and in vivo gene therapy. It was noted that nontargeted skin hairs appear to be normal after miRNA transfection. This underscores the fact that the intronic miRNA is safe and effective in vivo. The results also indicated that the miRNA-mediated gene silencing effect is stable and efficient in knocking down the targeted gene expression over a relatively long period of time since the hair regrowth takes at least 10 days. Taken together, the intronic miRNA-mediated transgene approach may offer relatively safe, effective, and long-term gene manipulation in animals, preventing the nonspecific lethal effects of the conventional transgenic methods.

More recent advances of the utilization of intronic miRNA expression systems have been reported in mice. Chung et al. [17] have succeeded in expression of a cluster of polycistronic miRNAs using the Pol-II-mediated intronic miRNA expression system. A polycistronic miRNA cluster can be processed into multiple miRNAs via the cellular miRNA pathway. This new RNAi approach has a few advantages over the conventional Pol-III-mediated shRNA expression systems. First, Pol-II expression is tissue-specific, whereas Pol-III expression is not. Second, Pol-II expression is compatible with the native miRNA pathway, while Grimm et al. [47] have reported some incompatibility between the Pol-III-mediated shRNA and Pol-II-mediated native miRNA pathways. Third, excessive RNA accumulation and toxicity can be prevented by the NMD mechanism of a cellular Pol-II-mediated intron expression system, but not a Pol-III exon-like expression system [49]. Lastly, a Pol-II is able to express a large-size cluster (>10 kb) of polycistronic shRNAs, which can be further excised into multiple shRNAs via the native miRNA pathway, preventing the promoter conflict that often occurs in a multiple promoter vector system. For example, in many commercial U6-mediated shRNA expression systems, a self-inactivated vector promoter is often used to increase the U6 promoter activity.

### 15.3.1 First miRNA-Mediated Transgenic Animal Model—FXS Zebra Fish

Animal models of human development and disease are essential. Zebra fish (*Danio rerio*), a tropical fresh water fish, has set an impressive record as an in vivo viable model for studies of mechanisms involving embryogenesis, organogenesis, physiology, and behavior. One of the areas that stand to benefit most from the zebra fish model is developmental neuroscience. Advantages of using zebra fish include low cost, easy maintenance, rapid life cycle, small size, embryonic transparency, quick development (i.e., nervous system precursors present by 6–7 hour postfertilization; first neuron forms by 1 day postfertilization), large generation number (i.e., clutch sizes from a single mating pair range between 100 to 200 embryos), and the fact that phenotypes can be easily assessed in many high-throughput assays [50,51]. Ultimately, screening genetic suppressors is likely to add great value to the understanding of loss-of-gene-function phenotypes that are related to certain diseases, with the genes becoming logical drug target candidates. Also, screening for morphological or behavioral mutants is often more time- and cost-effective than equivalent assays in mouse. These advantages have provided great advances in understanding the detailed pathological mechanisms underlying brain disorders that may lead to functional and behavioral deficits. For example, zebra fish possess three *FMRP*-related genes, *fmrl*, *fxrl*, and *fxr2*; these genes are completely orthologous to the human *FMR1*, *FXR1*, and *FXR2* genes, respectively [52]. The expression patterns of these *FMRP*-related familial genes in zebra fish tissues are consistent with the expressions in mouse and human [52,53], suggesting that zebra fish is an excellent model for studying human *FMRP*-related disorders.

Fragile X syndrome (FXS) is one of the most common mental retardation and neuropsychiatric disorders in humans, affecting approximately one in 2000 males and one in 4000 females. The characteristic features of FXS in boys include a long face, prominent ears, large testes, delayed speech, hyperactivity, tactile defensiveness, gross motor delays, and autistic-like behaviors. Much less is known about girls with FXS. This disease is caused by a dynamic premutation (i.e., expansion of trinucleotide CGG repeats) at an inherited fragile site on the X chromosome, representing a typical X-linked disorder. Because this premutation is dynamic, it can change in length and hence in severity from generation to generation, from person to person, and even within a given person. Patients with the FXS usually have an increased number of CGG trinucleotide repeats—for example, r(CGG) >230 copies—in the 5′ untranslated region (5′ UTR) of a gene located in the long-arm of X chromosome, namely *FMR1*, and this CpG-rich portion of the *FMR1* gene is often methylated. Such excessive r(CGG) expansion and methylation leads to physical, neurocognitive, and emotional characteristics linked to the inactivation of the *FMR1* gene or the loss of its protein function. We have detected the expression of native miRNA-like molecules derived from the fish *fmrl* 5′ UTR r(CGG) expansion in zebra fish, as determined by fluorescent in situ hybridization (FISH) using locked nucleic acid (LNA) probes homologous to the nucleotide −25 to 45 region of the *fmrl* (accession number NM152963). However, no appropriate animal model is available for studying such miRNA-mediated FXS etiology because current *Drosophila* and mouse models are all based on the deletion of the *FMR1* gene, completely irrelevant to the 5′ UTR r(CGG) expansion mechanism.

To investigate the r(CGG)-derived miRNA-mediated *FMR1* methylation mechanism, we have developed a transgenic FXS model in zebra fish, of which *FMR1* is silenced by excess expression of the r(CGG) expansion isolated from the *FMR1* 5′ UTR [14]. Recent studies using this novel FXS model have shown that formation of synaptic connection was markedly reduced among the dendrites of the *FMR1*-deficient neurons, similar to the diseased hippocampal neurons in human FXS. Such *FMR1* deficiency often caused synapse deformity in the neurons essential for cognition and memory activities, damaging the activity-dependent synaptic neuron plasticity. Furthermore, the metabotropic glutamate receptor (mGluR)-activated long-term depression (LTD) was augmented after *FMR1* inactivation, suggesting that exaggerated LTD may be responsible for aspects of abnormal neuronal responses in FXS, such as autism. Thus, we have presented an appropriate animal model for mimicking the FXS etiology, which may provide insight into the pathological mechanism of the r(CGG) expansion in FXS-related disorders.

## 15.4   GENERATION OF INTRONIC miRNA-MEDIATED FXS ZEBRA FISH LINES

Following the concept of the FXS etiology proposed by Handa et al. [33] and Jin et al. [34], we were able to recreate the r(CGG) miRNA-mediated *FMR1* inactivation in transgenic zebra fish [14]. As shown in Figure 15.5, we have used a fish-compatible vector, *pICML1-rT16*, acquired from Huang et al. [54,55], to deliver an intronic miRNA transgene containing the *fmr1* 5' UTR r(CGG) expansion into eGFP-expressing Tg(UAS:gfp) zebra fish. We constructed the transgene based on the proof-of-principle design of the *SpRNAi-rGFP* transgene used in gene-knockout zebra fish as reported previously [13,15]. The transgene is directed against the nucleotide –25 to 45 region of the zebra fish *fmr1* 5' UTR methylation site (accession number NM152963). This target region containing 13 5' UTR (CGG) repeats was proposed to be the native anti-*FMR1* non-miRNA target site of the FXS mental retardation syndrome. This transgene was further cloned into the *pICML1-rT16* vector and transfected into 0-4 postfertilization (hpf)-stage zebra fish embryos as described previously [13–15]. Transgenic F0 zebra fish were then separated into four groups based on their levels of *fmr1* knockdown, as determined by Western blotting, including <50%, 50–75%, 75–90%, and >90% knockdown of *fmr1* expression (Figure 15.6). The zebra fish showing above 90% *fmr1* knockdown were too defective to be raised for a transgenic line, where we have succeeded in raising the fish with 75–90% *fmr1* knockdown to sexual maturity and mated them to generate the F1 founder line with a stable 75–85% *fmr1* knockdown rate. After genome typing and transgene sequencing analyses, the F1 and F2 transgenic lines were show to possess two copies of the transgene in a consistent genomic insertion site located in the chromosome 18 close to the 3'-proximity of the LOC565390 locus region, where encodes no gene. We have also identified that the >90% *fmr1* knockdown fishes possess on average 3–5 copies of the transgene located in two to three different genomic insertion locations.

The loss-of-*fmr1*-function zebra fish model and human FXS are based on the same pathological mechanism and molecular interaction between the r(CGG)-derived miRNA and the *FMR1* 5' UTR r(CGG) expansion. They are all triggered by intronic miRNA-mediated *FMR1* inactivation and result in similar physical defects in relation to FMR1 deficiency. By transgenically increasing the *fmr1* 5'

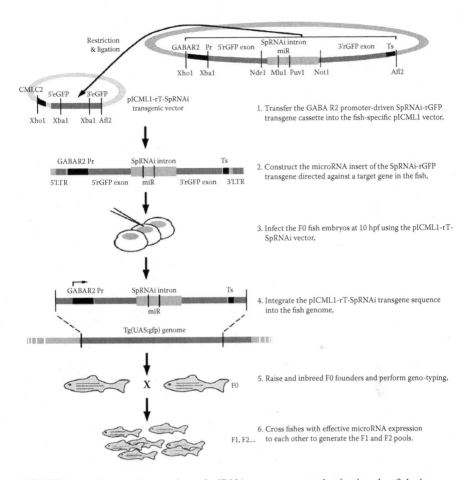

**FIGURE 15.5**   Schematic procedure of miRNA transgene transduction in zebra fish. A transgenic vector delivery approach is used to transfect a Pol-II (i.e., neuron-specific fish *GABA[A] receptor βZ2 [GABA R2]*) promoter-driven *SpRNAi-rGFP* transgene into the Tg(UAS:gfp) zebra fish for steady expression of a pre-designed miRNA directed against its target gene.

UTR r(CGG) RNA expression in zebra fish, we have shown to significantly elevate anti-*FMR1* miRNA concentrations over 4- to 6-fold in the transgenic zebra fish with 75–90% *fmr1* knockdown (Figure 15.6a). In comparison with the weak *FMR1* promoter (100–1000 copies of mRNA per cell), we used the fish *GABA R2* promoter (5000 to 15,000 copies of mRNA per cell) to boost the expression of the transgenic *fmr1* 5′ UTR r(CGG) expansion and successfully induce specific *fmr1* gene silencing in these transgenic zebra fish, as determined by Northern blot (Figure 15.6b) and Western blot (Figure 15.6c and 15.6d) analyses. Because we only select approximately 30% of the whole *fmr1* 5′ UTR r(CGG) expansion for transgenic insertion, it is estimated that each transgene, after *GABA R2* promoter-driven expression, will generate a 2- to 3-fold increase in anti-*fmr1* miRNA RNA production. As a result, the transgenic zebra fish with 75–90% *fmr1* knockdown actually possess 4- to 6-fold more miRNA concentration than do the wild-type zebra fish, resembling

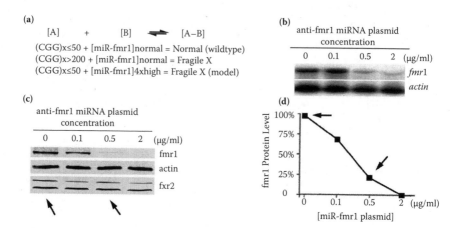

**FIGURE 15.6** Transgenic zebra fish with different levels of *fmr1* knockdown can be generated using the intronic miRNA-expressing *pICML1-rT16-SpRNAi* vector to transfectively express the *FMR1* 5' UTR r(CGG) expansion at different dosages. Strategy of this transgenic loss-of-*fmr1*-function zebra fish model was shown in (a). Instead of increasing r(CGG) expansion in *fmr1* 5' UTR, we directly overexpressed the r(CGG)-derived miRNA to trigger the *fmr1* inactivation. Knockdown levels of the *fmr1* mRNA and fmr1 protein were determined by Northern blot (b) and Western blot (c) analyses, respectively. The line chart (d) indicates the Western blotting results of (c). Arrows indicate the two compared groups: one is wild-type and the other is the transgenic zebra fish line with 75–90% *fmr1* knockdown, as shown in Figures 15.7–15.9.

the increased level of human FXS r(CGG) expansion (>200 copies) versus normal r(CGG) site (<50 copies). During early embryonic stages, both human and fish FXS models generate over a 4-fold increase of r(CGG)-derived miRNA expression, which is sufficient to trigger *FMR1* gene silencing via an RNAi pathway that is important for neuronal development and plasticity [14]. We have shown that both FXS models present similar abnormalities in synapse connectivity and neuron plasticity. To a less extent of vector transfection, the fish with 50–75% *fmr1* knockdown likely represent the case of FXTAS, which expresses a moderate increase of r(CGG) expansion (~120 copies) and often displays elevated *fmr1* mRNA levels but decreased fmr1 protein levels, as shown in Figure 15.6b and 15.6c, lanes 2, respectively.

In addition, we have used fluorescent in situ hybridization (FISH) assays to determine the different *fmr1* mRNA expression patterns in the wild-type and FXS zebra fish brains (Figure 15.7a). In wild-type zebra fish, abundant *fmr1* mRNA expression was shown in the junction (upper row, dot lines) of pallium and neocortical neurons and the dendritic projection of neocortical granule neurons adjacent to the granular layer (yellow arrows), where the Purkinje cells reside. In FXS zebra fish, no *fmr1* mRNA is detected in these areas (red + green = yellow; no blue), indicating that the *fmr1* gene is completely inactivated in the FXS neurons as suggested by Handa et al. [33] and Jin et al. [34]. To further show miRNA-mediated *fmr1* methylation, we have isolated a part of the *fmr1* 5' UTR r(CGG) expansion region from the nucleotide -1103 to 48 of the *fmr1* (accession number NM152963) for methylation-specific PCR analysis and digested the region with two -C CGG-cutting restriction enzymes, *HpaII*

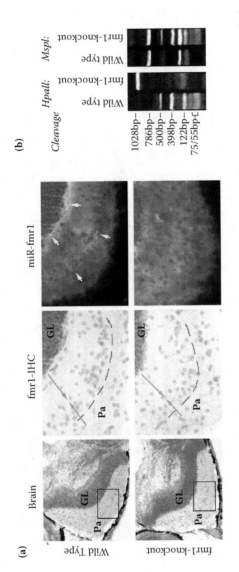

**FIGURE 15.7** *Color version of this figure follows page 238.* r(CGG) miRNA-mediated *fmr1* gene silencing at the (a) posttranscriptional and (b) transcriptional levels. (a) Fluorescent in situ hybridization (FISH) analysis of *FMR1* mRNA and anti-*FMR1* miRNA (miR-*FMR1*) expressions in day-7 transgenic zebra fish brains. Fluorescent labels are shown in neurons (green EGFP), anti-*fmr1* miRNA (red RGFP), and *fmr1* mRNA (blue DAPI). Gray arrows indicate the dendro-dendritic contacts between lateral pallium and neocortical neurons. (b) Methylation site cleavage assay with *HpaII* and *MspI* restriction enzymes shows the changes of CpG methylation patterns in the *fmr1* 5′ UTR r(CGG) expansion region of the wild-type versus loss-of-*fmr1*-function zebra fish brains.

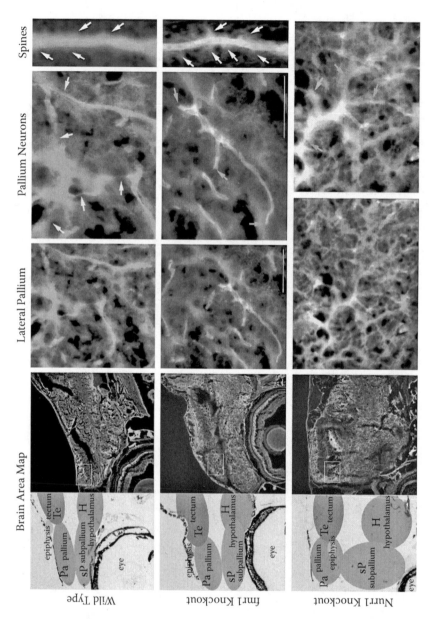

FIGURE 15.8

(methylation-insensitive) and *MspI* (methylation-sensitive), respectively. Figure 15.7b shows that at least six or seven CpG sties are methylated in this isolated *fmr1* 5′ UTR region from the FXS neurons, while only one or two methylated sites are in the wild-types. Therefore, the FXS r(CGG) miRNA-induced *fmr1* gene silencing can occur in both posttranscriptional and transcriptional levels. We have also noted that the r(CGG) RNA-mediated *fmr1* 5′ UTR methylation requires methyl-CpG binding protein (*MeCP2*) activity, a known candidate gene mechanism in both FXS-related autism and Rett syndrome [56].

## 15.5    SIMILAR miRNA-INDUCED ABNORMALITIES IN HUMAN AND FISH FXS

Despite some notable differences in sizes, the overall organization of major brain components in zebra fish is highly conserved with those of the human [57,58]. As in other vertebrates, zebra fish possess all of the classical sense modalities such as vision, hearing, olfaction, taste, tactile, balance, and their sensory pathways, sharing an overall homology with humans. We have compared the phenotypes between human and fish FXS in detail to provide an informative groundwork for the use of this novel r(CGG) miRNA-mediated animal model for FXS-related research and drug development. Figure 15.8 shows a fluorescent three-dimension (3D)-micrograph of abnormal neuron morphology and connectivity in the embryonic brains of the loss-of-*fmr1*-function fish transgenics, which are similar to those in human FXS. In fish lateral pallium (similar to human hippocampal-neocortical junction), wild-type neurons presented normal dendrite outline and a well connection to each other (yellow arrows), whereas the *fmr1*-knockdown transgenics exhibited thin, strip-shape neurons, reminiscent of the abnormal dendritic spine neurons in human FXS [59,60]. Synapse deformity frequently occurred in the FXS neurons (red arrows), indicating that *FMR1* play an important functional role in the synaptic neuron plasticity. Altered synaptic plasticity has been reported to be one of the major physiological damages in human FXS, particularly in the hippocampal stratum radiatum, layer IV/V cortex, and sometimes cerebellum of severe cases [60,61].

**FIGURE 15.8    (See facing page.)** Morphological changes of lateral pallium neurons in *fmr1*-knockdown zebra fish (middle row) and *Nurr1*-knockdown zebra fish (bottom row) in comparison to the wild-types (top row). Fluorescent 3D-micrograph showed distinct neuron morphology and connectivity between wild-type and *fmr1*-knockdown Tg(UAS:gfp) zebra fish. Because the entire Tg(UAS:gfp) zebra fish expresses an *actin* promoter-driven EGFP protein and the anti-*fmr1* miRNA transgene is marked with RGFP, we can easily observe the normal dendritic neurons versus the loss-of-*fmr1*-function neurons under a fluorescent microscope. This miRNA-mediated zebra fish FXS model is consistent with the hypothetic mechanism of human FXS disorders, in which the native anti-*FMR1* miRNA prevents synaptic strengthening and blocks local protein synthesis-dependent synaptic connections, a cascade of events for which FMR1 has been strongly implicated. As a comparison control, the *Nurr1* knockdowns have been shown to display a positive function in neuron connectivity. Abbreviations indicate: Pa, pallium; sP, subpallium; Te, tectum; H, hypothalamus.

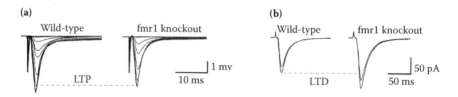

**FIGURE 15.9**   The responsiveness of hippocampal-neocortical LTP (a) is decrease while that of hippocampal LTD (b) is augmented in loss-of-*fmr1*-function zebra fish, which likely interferes with normal formation and maintenance of synapses required for particular learning and cognitive functions.

*FMR1* mRNA is present in dendritic spines and translated in response to activation of the type 1 metabotropic glutamate receptors (mGluR-1) in synaptoneurosomes [62,63]. The activation of mGluR1 stimulates a phosphorylation cascade, triggering rapid association of some mRNAs with translation machinery near synapses and leading to protein synthesis of the mRNAs [62]. FMR1 protein is, however, a translation inhibitor that binds with mRNA species involving in the regulation of microtubule-dependent synapse growth and function, including its own mRNA [30,64,65]. An increased density of long, immature dendritic spines in loss-of-*fmr1*-function neurons (Figure 15.8, rightmost panel), suggests that FMR1 has an important role in synaptic maturation and pruning, possibly through its regulation of gene products that are involved in synaptic development.

In male, 3-month-old FXS zebra fish, excitatory synapses in slices of the pallium-neocortical junction were found to exhibit diminished long-term potentiation (LTP), compared with wild-type controls (Figure 15.9a). LTP in hippocampus is a learning-related form of synaptic plasticity, highly involving in dendritic spine shape changes [66]. This result suggests that deficits in hippocampal-cortical LTP mechanisms may contribute to cognitive impairments in FXS disorders. On the other hand, postsynaptic stimulation of mGluR results in increased protein synthesis and subsequent internalization of α-amino-3-hydroxy-5-methyl-4-isoxazole propionic acid (AMPA) receptors, which is important in the expression of long-term depression (LTD). FMRP, which is also up-regulated by the activation of mGluR, serves to quench this process. The absence of FMR1 in FXS, therefore, results in overamplification of this LTD response. As shown in Figure 15.9b, the pallium neuron LTD could be augmented in the absence of *fmr1*, suggesting that exaggerated LTD may be responsible for aspects of abnormal neuronal responses in FXS, such as autism. This exaggerated LTD could be inhibited by treating the brain slices of the FXS zebra fish with mGluR-specific agonists, such as (1S,3R)-1-aminocyclopentane-1,3-dicarboxylic acid (1S,3R-trans-ACPD) and 3,5-dihydroxyphenylglycine (DHPG). These findings raise a possibility in FXS-associated autism, which is supported by the evidence that induction of mGluR1-dependent LTD is enhanced in pyramidal cells of the hippocampus in *FMR1*-deleted mice [59]. Thus, altered hippocampal LTP and LTD in both human and fish FXS may explain how and why *FMR1* inactivation interferes with normal formation and maintenance of synaptic connections, which are required for particular learning and cognitive functions. Based on this advanced FXS animal model, more detailed investigation in r(CGG)-derived miRNA-mediated neuronal damages is possible.

## 15.6  CONCLUSIONS

Increasing evidence of miRNA-induced gene silencing effects in zebra fish, chicken embryos, mouse stem cells, and human diseases demonstrates the preservation of an ancient intron-mediated gene regulation system in eukaryotes. In these animal models, the intron-derived miRNAs determine the activation of RNAi-like gene silencing pathways. We provide the first evidence for the biogenesis and function of intronic miRNAs in vivo. Given that natural evolution gives rise to more complexity and variety of introns in higher animal and plant species for coordinating their vast gene expression volumes and interactions, dysregulation of these miRNAs due to intronic expansion or deletion will likely cause genetic diseases, such as myotonic dystrophy and fragile X mental retardation. Thus, gene expression produces not only gene transcripts for its own protein synthesis but also intronic miRNAs that are capable of interfering with other gene expressions. By the same token, the expression of a gene results in concurrent gain-of-function of the gene and also loss-of-function of some other genes, which contain complementarity to the mature intronic miRNAs. An array of genes can swiftly and accurately coordinate their expression patterns with each other through the mediation of their intronic miRNAs, bypassing the time-consuming translation processes under quickly changed environments. Conceivably, intron-mediated gene regulation may be as important as the mechanisms by which transcription factors regulate the gene expression. It is likely that intronic miRNA is able to trigger cell transitions quickly in response to external stimuli without the tedious protein synthesis. Undesired gene products are reduced by both transcriptional inhibition and/or translational suppression via miRNA regulation. This process allows a rapid switch to new gene expression patterns without the need of producing various transcription factors. This regulatory property of miRNAs may serve as one of the most ancient gene modulation systems before the emergence of proteins. Because the variety of miRNAs and the complexity of genomic introns, thorough investigations of miRNA variants in the human genome will markedly improve the understanding of genetic diseases and the design of miRNA-based drugs. Understanding how to exploit such a novel gene regulation system in the future therapy will be a forthcoming challenge.

## ACKNOWLEDGMENTS

We thank the Zebrafish International Resource Center (supported by grant RR12546 from the NIH-NCRR) for the Tg(UAS:gfp) strain zebra fish and Dr. Huai-Jen Tsai, Institute of Molecular and Cellular Biology and Angiogenesis Research Center, National Taiwan University, Taipei, Taiwan, for the *pICML1-rT16* vector. Patents on all vectors, protocols, and transgenic animals invented in these studies were supported by NIH/NCI CA-85722 and assigned to the University of Southern California (USC Files #3443 and #3802) under the guidelines of the National Institute of Health.

## REFERENCES

1. Lee, R. C. and Ambros, V. (2001) An extensive class of small RNAs in *Caenorhabditis elegans. Science* 294: 862–864.
2. Ambros, V., Lee, R. C., Lavanway, A., Williams, P. T., and Jewell, D. (2003) MicroRNAs and other tiny endogenous RNAs in *C. elegans. Curr Biol* 13, 807–818.
3. Lin, S. L., Chang, D., Wu, D. Y., and Ying, S. Y. (2003) A novel RNA splicing-mediated gene silencing mechanism potential for genome evolution. *Biochem Biophys Res Commun* 310, 754–760.
4. Rodriguez, A., Griffiths-Jones, S., Ashurst, J. L., and Bradley, A. (2004) Identification of mammalian microRNA host genes and transcription units. *Genome Res* 14, 1902–1910.
5. Ying, S. Y. and Lin, S. L. (2004) Intron-derived microRNAs–fine tuning of gene functions. *Gene* 342, 25–28.
6. Zhang, G., Taneja, K. L., Singer, R. H., and Green, M. R. (1994) Localization of pre-mRNA splicing in mammalian nuclei. *Nature* 372, 809–812.
7. Lewis, B. P., Green, R. E., and Brenner, S. E. (2003) Evidence for the widespread coupling of alternative splicing and nonsense-mediated mRNA decay in humans. *Proc Natl Acad Sci USA* 100, 189–192.
8. Clement, J. Q., Qian, L., Kaplinsky, N., and Wilkinson, M. F. (1999) The stability and fate of a spliced intron from vertebrate cells. *RNA* 5, 206–220.
9. Lin, S. L. and Ying, S. Y. (2004) New drug design for gene therapy—Taking Advantage of Introns. *Lett Drug Design Discovery* 1, 256–262.
10. Lin, S. L. and Ying, S. Y. (2004) Novel RNAi therapy—Intron-derived microRNA drugs. *Drug Design Rev* 1, 247–255.
11. Lin, S. L. and Ying, S. Y. (2006) Gene silencing in vitro and in vivo using intronic microRNAs. *Methods Mol Biol* 342: 295–312.
12. Lin, S. L., Chang, D., and Ying, S. Y. (2006) Hyaluronan stimulates transformation of androgen-independent prostate cancer. *Carcinogenesis* 2006 July 24; Epub ahead of print.
13. Lin, S. L., Chang, D., and Ying, S. Y. (2005) Asymmetry of intronic pre-microRNA structures in functional RISC assembly. *Gene* 356: 32–38.
14. Lin, S. L., Chang, S. J. E., and Ying, S. Y. (2006) First in vivo evidence of microRNA-induced fragile X mental retardation syndrome. *Mol Psychiatry* 11: 616–617.
15. Lin, S. L., Chang, S. J. E., and Ying, S. Y. (2006) Transgene-like animal model using intronic microRNAs. *Methods Mol Biol* 342: 321–334.
16. Zhou, H. S. L., Xia, X. G., and Xu, Z. (2005) An RNA polymerase II construct synthesizes short hairpin RNA with a quantitative indicator and mediates high efficient RNAi. *Nucleic Acid Res* 33: e62.
17. Chung, K. H., Hart, C. C., Al-Bassam, S., Avery, A., Taylor, J., Patel, P. D., Vojtek, A. B., and Turner, D. L. (2006) Polycistronic RNA polymerase II expression vectors for RNA interference based on BIC/miR-155. *Nucleic Acid Res* 34: e53.
18. Lee, Y., Kim, M., Han, J., Yeom, K. H., Lee, S., Baek, S. H., and Kim, V. N. (2004) MicroRNA genes are transcribed by RNA polymerase II. *EMBO J.* 23, 4051–4060.
19. Lee, Y., Ahn, C., Han, J., Choi, H., Kim, J., Yim, J., Lee, J., Provost, P., Radmark, O., Kim, S., and Kim, V. N. (2003) The nuclear RNase III Drosha initiates microRNA processing. *Nature* 425, 415–419.
20. Denli, A. M., Tops, B. B., Plasterk, R. H., Ketting, R. F., and Hannon, G. J. (2004) Processing of primary microRNAs by the Microprocessor complex. *Nature* 432: 231–235.
21. Gregory, R. I., Yan, K. P., Amuthan, G., Chendrimada, T., Doratotaj, B., Cooch, N., and Shiekhattar, R. (2004) The Microprocessor complex mediates the genesis of microRNAs. *Nature* 432: 235–240.

22. Yi, R., Qin, Y., Macara, I. G., and Cullen, B. R. (2003) Exportin-5 mediates the nuclear export of pre-microRNAs and short hairpin RNAs. *Genes Dev* 17, 3011–3016.

23. Lund, E., Guttinger, S., Calado, A., Dahlberg, J. E., and Kutay, U. (2004) Nuclear export of microRNA precursors. *Science* 303, 95–98.

24. Schwarz, D. S., Hutvagner, G., Du, T., Xu, Z., Aronin, N., and Zamore, P. D. (2003). Asymmetry in the assembly of the RNAi enzyme complex. *Cell* 115, 199–208.

25. Khvorova, A., Reynolds, A., and Jayasena, S. D. (2003). Functional siRNAs and miRNAs exhibit strand bias. *Cell* 115, 209–216.

26. Lee, Y. S., Nakahara, K., Pham, J. W., Kim, K., He, Z., Sontheimer, E. J., and Carthew, R. W. (2004) Distinct roles for *Drosophila* Dicer-1 and Dicer-2 in the siRNA/miRNA silencing pathways. *Cell* 117, 69–81.

27. Liquori, C. L., Ricker, K., Moseley, M. L., Jacobsen, J. F., Kress, W., Naylor, S. L., Day, J. W., and Ranum, L. P. W. (2001) Myotinic dystrophy type 2 caused by a CCTG expansion in intron 1 of ZNF9. *Science* 293, 864–867.

28. Hagerman, R. J., Staley, L. W., O'Conner, R., Lugenbeel, K., Nelson, D., McLean, S. D., and Taylor, A. (1996) Learning-disabled males with a fragile X CGG expansion in the upper premutation size range. *Pediatrics* 97: 122–126.

29. Genc, B., Muller-Hartmann, H., Zeschnigk, M., Deissler, H., Schmitz, B., Majewski, F., von Gontard, A., and Doerfler, W. (2000) Methylation mosaicism of 5'-(CGG)(n)-3' repeats in fragile X, premutation and normal individuals. *Nucleic Acids Res* 28: 2141–2152.

30. Jin, P., Zarnescu, D. C., Zhang, F., Pearson, C. E., Lucchesi, J. C., Moses, K., and Warren, S. T. (2003) RNA-mediated neurodegeneration caused by the fragile X premutation rCGG repeats in *Drosophila. Neuron* 39: 739–747.

31. Jin, P., Zarnescu, D. C., Ceman, S., Nakamoto, M., Mowrey, J., Jongens, T. A., Nelson, D. L., Moses, K., and Warren, S. T. (2004) Biochemical and genetic interaction between the fragile X mental retardation protein and the microRNA pathway. *Nat Neurosci* 7: 113–117.

32. Willemsen, R., Hoogeveen-Westerveld, M., Reis, S., Holstege, J., Severijnen, L. A., Nieuwenhuizen, I. M., Schrier, M., van Unen, L., Tassone, F., Hoogeveen, A. T. et al. (2003) The FMR1 CGG repeat mouse displays ubiquitin-positive intranuclear neuronal inclusions; implications for the cerebellar tremor/ataxia syndrome. *Hum Mol Genet* 12: 949–959.

33. Handa, V., Saha, T., and Usdin, K. (2003) The fragile X syndrome repeats form RNA hairpins that do not activate the interferon-inducible protein kinase, PKR, but are cut by Dicer. *Nucleic Acids Res* 31: 6243–6248.

34. Jin, P., Alisch, R. S., and Warren, S. T. (2004) RNA and microRNAs in fragile X mental retardation. *Nat Cell Biol* 6: 1048–1053.

35. Eberhart, D. E., Malter, H. E., Feng, Y., and Warren, S. T. (1996) The fragile X mental retardation protein is a ribonucleoprotein containing both nuclear localization and nuclear export signals. *Hum Mol Genet* 5, 1083–1091.

36. Tamanini, F., Van Unen, L., Bakker, C., Sacchi, N., Galjaard, H., Oostra, B. A., and Hoogeveen, A. T. (1999) Oligomerization properties of fragile-X mental-retardation protein (FMRP) and the fragile-X-related proteins FXR1P and FXR2P. *Biochem J* 343 (Pt. 3), 517–523.

37. Tuschl, T. and Borkhardt, A. (2002) Small interfering RNAs: A revolutionary tool for the analysis of gene function and gene therapy. *Mol Interv* 2, 158–167.

38. Miyagishi, M. and Taira, K. (2002) U6 promoter-driven siRNAs with four uridine 3' overhangs efficiently suppress targeted gene expression in mammalian cells. *Nat Biotechnol* 20, 497–500.

39. Lee, N. S., Dohjima, T., Bauer, G., Li, H., Li, M. J., Ehsani, A., Salvaterra, P., and Rossi, J. (2002) Expression of small interfering RNAs targeted against HIV-1 rev transcripts in human cells. *Nat Biotechnol* 20: 500–505.

40. Paul, C. P., Good, P. D., Winer, I., and Engelke, D. R. (2002) Effective expression of small interfering RNA in human cells. *Nat Biotechnol* 20, 505–508.

41. Xia, H., Mao, Q., Paulson, H. L., and Davidson, B. L. (2002) siRNA-mediated gene silencing in vitro and in vivo. *Nat Biotechnol* 20, 1006–1010.

42. McCaffrey, A. P., Meuse, L., Pham, T. T., Conklin, D. S., Hannon, G. J., and Kay, M. A. (2002). RNA interference in adult mice. *Nature* 418, 38–39.

43. Gunnery, S., Ma, Y., and Mathews, M. B. (1999) Termination sequence requirements vary among genes transcribed by RNA polymerase III. *J Mol Biol* 286, 745–757.

44. Schramm, L. and Hernandez, N. (2002) Recruitment of RNA polymerase III to its target promoters. *Genes Dev* 16, 2593–2620.

45. Sledz, C. A., Holko, M., de Veer, M. J., Silverman, R. H., and Williams, B. R. (2003) Activation of the interferon system by short-interfering RNAs. *Nat Cell Biol* 5, 834–839.

46. Lin, S. L. and Ying, S. Y. (2004c) Combinational therapy for HIV-1 eradication and vaccination. *Int J Oncol* 24: 81–88.

47. Grimm, D., Streetz, K. L., Jopling, C. L., Storm, T. A., Pandey, K., Davis, C. R., Marion, P., Salazar, F., and Kay, M. A. (2006) Fatality in mice due to oversaturation of cellular microRNA/short hairpin RNA pathways. *Nature* 441: 537–541.

48. Xia, X. G., Zhou, H., Samper, E., Melov, S., and Xu, Z. (2006) Pol II-expressed shRNA knocks down Sod2 gene expression and causes phenotypes of the gene knockout in mice. *PLoS Genet* 2: e10.

49. Sheth, U. and Parker, R. (2006) Targeting of aberrant mRNAs to cytoplasmic processing bodies. *Cell* 125, 1095–1109.

50. Kimmel, C. B., Ballard, W. W., Kimmel, S. R., Ullmann, B., and Schilling, T. F. (1995) Stages of embryonic development of the zebrafish. *Dev Dyn* 203: 253–310.

51. Westerfield, M. (2003) The Zebrafish Book. Sprague, J., Clements, D., Conlin, T., Edwards, P., Frazer, K., Schaper, K., Segerdell, E., Song, P., Sprunger, B., and Westerfield, M. The Zebrafish Information Network (ZFIN): the zebrafish model organism database. *Nucleic Acids Res* 31: 241–243.

52. Tucker, B., Richards, R., and Lardelli, M. (2004) Expression of three zebrafish orthologs of human FMR1-related genes and their phylogenetic relationships. *Dev Genes Evol* 214: 567–574.

53. van 't Padje, S., Engels, B., Blonden, L., Severijnen, L. A., Verheijen, F., Oostra, B. A., and Willemsen, R. (2005) Characterisation of Fmrp in zebrafish: evolutionary dynamics of the fmr1 gene. *Dev Genes Evol* 215: 198–206.

54. Huang, C. J., Tu, C. T., Hsiao, C. D., Hsieh, F. J., and Tsai, H. J. (2003) Germ-line transmission of a myocardium-specific GFP transgene reveals critical regulatory elements in the cardiac myosin light chain 2 promoter of zebrafish. *Dev Dyn* 228: 30–40.

55. Huang, C. J., Jou, T. S., Ho, Y. L., Lee, W. H., Jeng, Y. T., Hsieh, F. J., and Tsai, H. J. (2005) Conditional expression of a myocardium-specific transgene in zebrafish transgenic lines. *Dev Dyn* 233: 1294–1303.

56. Lopez-Rangel, E. and Lewis, M. E. S. (2006) Loud and clear evidence for gene silencing by epigenetic mechanisms in autism spectrum and related neurodevelopmental disorders. *Clin Genet* 69: 21–25.

57. Wullimann, M. F. (1998) The central nerve system. In Evans, D.H. (Ed.) *The Physiology of Fishes*, 2nd ed. CRC press, New York, 245–281.

58. Tropepe, V. and Sive, H. L. (2003) Can zebrafish be used as a model to study the neurodevelopmental causes of autism? *Genes, Brain Behav* 2: 268–281.

59. Huber, K. M., Gallagher, S. M., Warren, S. T., and Bear, M. F. (2002) Altered synaptic plasticity in a mouse model of fragile X mental retardation. *Proc Natl Acad Sci USA* 99: 7746–7750.

60. Irwin, S. A., Christmon, C. A., Grossman, A. W., Galvez, R., Kim, S. H., DeGrush, B. J., Weiler, I. J, and Greenough, W. T. (2005) Fragile X mental retardation protein levels increase following complex environment exposure in rat brain regions undergoing active synaptogenesis. *Neurobiol Learn Mem* 83: 180–187.
61. Galvez, R., Gopal, A. R., and Greenough, W. T. (2003) Somatosensory cortical barrel dendritic abnormalities in a mouse model of the fragile X mental retardation syndrome. *Brain Res* 971: 83–89.
62. Weiler, I. J. and Greenough, W. T. (1999) Synaptic synthesis of the Fragile X protein: Possible involvement in synapse maturation and elimination. *Am J Med Genet* 83: 248–252.
63. Koekkoek, S. K., Yamaguchi, K., Milojkovic, B. A., Dortland, B. R., Ruigrok, T. J., Maex, R., De Graaf, W., Smit, A. E., VanderWerf, F., Bakker, C. E., Willemsen, R., Ikeda, T., Kakizawa, S., Onodera, K., Nelson, D. L., Mientjes, E., Joosten, M., De Schutter, E., Oostra, B. A., Ito, M., and De Zeeuw, C. I. (2005) Deletion of FMR1 in Purkinje cells enhances parallel fiber LTD, enlarges spines, and attenuates cerebellar eyelid conditioning in Fragile X syndrome. *Neuron* 47: 339–352.
64. Brown, V., Jin, P., Ceman, S., Darnell, J. C., O'Donnell, W. T., Tenenbaum, S. A., Jin, X., Feng, Y., Wilkinson, K. D., Keene, J. D., Darnell, R. B., and Warren, S. T. (2001) Microarray identification of FMRP-associated brain mRNAs and altered mRNA translational profiles in fragile X syndrome. *Cell* 107: 477–487.
65. Li, Z., Zhang, Y., Ku, L., Wilkinson, K. D., Warren, S. T., and Feng, Y. (2001) The fragile X mental retardation protein inhibits translation via interacting with mRNA. *Nucleic Acids Res* 29: 2276–2283.
66. Godfraind, J. M., Reyniers, E., De Boulle, K., D'Hooge, R., De Deyn, P. P., Bakker, C. E., Oostra, B. A., Kooy, R. F., and Willems, P. J. (1996) Long-term potentiation in the hippocampus of fragile X knockout mice. *Am J Med Genet* 64: 246–251.

# 16 MicroRNA-Based RNA Polymerase II Expression Vectors for RNA Interference in Mammalian Cells

*Anne B. Vojtek, Kwan-Ho Chung,*
*Paresh D. Patel, and David L. Turner*

## CONTENTS

## OVERVIEW

RNA interference (RNAi) is widely used to inhibit mammalian gene expression. It is mediated by cellular ribonucleoprotein complexes containing short-interfering RNAs (siRNAs) or microRNAs (miRNAs) that guide target recognition. Multiple vector-based strategies for RNAi have been developed, usually based on the synthesis of short hairpin RNAs (shRNAs) that are subsequently processed by cellular enzymes into siRNAs/miRNAs. Here, we review vector-based strategies for RNAi, focusing on vectors that utilize RNA polymerase II and modified versions of miRNA precursors to produce synthetic miRNAs/siRNAs in mammalian cells.

## 16.1  INTRODUCTION

RNA interference (RNAi) is an increasingly widely used technology for inhibiting the expression of specific genes (reviewed in Reference 1–4). In mammalian cells, RNAi initially was achieved by the introduction of in vitro synthesized 21 nucleotide (nt) synthetic short-interfering RNA (siRNA) duplexes into cells by transient transfections [5]. These siRNA duplexes, upon entering the cell, interact with the cytoplasmic RNA-induced silencing complex (RISC), and usually one strand of the siRNA duplex is preferentially incorporated into RISC. RISC, guided by the siRNA, cleaves mRNAs that contain complementary sequences, leading to the destruction of those target mRNAs (reviewed in References 6 and 7).

A complementary approach for RNAi in mammalian cells has been to use expression vector-based strategies that synthesize short RNAs in cells. Use of RNAi expression vectors provides prolonged or stable RNAi in cells, as well as new delivery options such as viral vectors. One popular method that has been adapted to a wide variety of vector types has been the use of RNA polymerase III to transcribe a short hairpin RNA (shRNA), in which two siRNA strands are joined at one end by a short loop to create a stem-loop or hairpin [8–13]. The shRNA is processed into siRNAs by the cellular Dicer enzyme [8,10]. Alternately, the two strands of an siRNA duplex can be transcribed as two independent short transcripts by RNA polymerase III, which can anneal in the cell to form the siRNA duplex [14,13]. A second vector-based method for RNAi has been to use RNA polymerase II to synthesize RNAs that contain synthetic microRNAs (miRNAs), based on modified versions of miRNA precursors [15, reviewed in 16]. Such synthetic miRNAs can function in a manner identical to siRNAs and mediate RNAi in mammalian cells. In addition, several RNA polymerase II-based vectors have been described that synthesize shRNAs that are not specifically miRNA based [17–19]. Here, we review the design and features of miRNA-based RNA polymerase II vectors for RNAi in mammalian cells, including miR-155-based vectors for RNAi that we recently described [20].

Endogenous miRNAs are abundant, small (~21–22 nt) noncoding regulatory RNAs derived from longer primary miRNA transcripts, or pri-miRNAs [reviewed by 6]. It appears that most pri-miRNAs are transcribed by RNA polymerase II [21–23], although recently it has been found that a subset is transcribed by RNA polymerase III [24,25]. The pri-miRNAs are processed by the nuclear endonuclease Drosha into ~70-nt hairpin intermediates, known as pre-miRNAs [26,27]. The pre-miRNAs are exported from the nucleus, a process that depends on Exportin-5 [28,29]. Once in the cytoplasm, pre-miRNAs are processed by Dicer into a short RNA duplex, and usually only one strand of the duplex then enters RISC or an RISC-like complex, while the other strand, referred to as the miRNA* strand, is degraded [30,31]. miRNAs that are perfectly or near-perfectly complementary to an mRNA function in a manner equivalent to synthetic siRNAs, by direct cleavage of matching mRNAs at a position opposite the middle of the miRNA [32,15]. However, in animals, most endogenous miRNAs are thought to interact with their target mRNAs via imperfect binding sites, often located in the 3′ untranslated region (UTR). These imperfect sites do not promote cleavage opposite the miRNA similar to an siRNA, but instead mediate

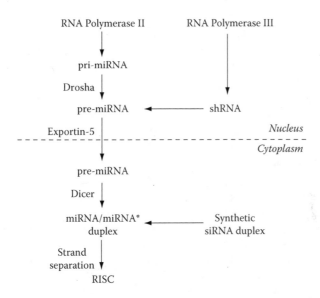

**FIGURE 16.1**  Entry points into the miRNA biogenesis pathway for different RNAi strategies. Synthetic miRNAs expressed from RNA polymerase II promoters (e.g., miR-155 or miR-30-based vectors) enter the cellular pathway at the level of the endogenous primary miRNA (pri-miRNA) transcripts. Synthetic shRNAs transcribed from RNA polymerase III promoters (e.g., vectors that use U6 or H1 promoters) bypass the Drosha processing step, but still require nuclear export and subsequent Dicer processing. Transfected synthetic siRNA duplexes bypass Drosha processing, nuclear export, and Dicer processing. The product of Dicer processing is the mature miRNA and its complementary strand, denoted as miRNA*. See text for additional information and references.

translational inhibition or RNA degradation by mechanisms that are not yet completely understood [33].

It has become clear that the various mammalian RNAi methods described earlier function in large part by utilizing cellular mechanisms for the synthesis and function of endogenous miRNAs. Different RNAi strategies enter the miRNA synthesis pathway at different steps (Figure 16.1). Transfected or transcribed siRNAs bypass both the Drosha- and Dicer-mediated processing steps, while RNA polymerase III-transcribed shRNAs mimic miRNA precursors and require Dicer processing, but bypass Drosha processing. In contrast, RNA polymerase II vectors expressing synthetic pri-miRNAs presumably enter the cellular pathways at the same point as an endogenous pri-miRNA transcript.

## 16.2  RNA POLYMERASE II miRNA-BASED RNAi EXPRESSION VECTORS

### 16.2.1  miR-30-Based Vectors

The first RNA polymerase II miRNA vectors for RNAi were described by Cullen and colleagues and were based on human miR-30 [15]. Their vector derivatives contained a minimal predicted stem-loop miR-30 precursor (~80 nt) under control of the human CMV immediate-early promoter. Vectors containing only the 21-nt miR-30 miRNA sequence under control of the CMV promoter failed to produce the mature miR-30 miRNA, indicating that expression of the miRNA stem-loop precursor was essential to produce the mature miRNA. Importantly, this work showed that a synthetic miRNA could function similar to an siRNA and trigger cleavage of an exactly matched target mRNA, opening up the possibility of using RNA polymerase II-based miRNA designs for RNAi (also see Reference 32). Note that while we refer to the products of miRNA-based RNAi vectors as synthetic miRNAs, they are functionally identical to siRNAs. Subsequent miR-30 derivatives lengthened the sequence of the minimal stem-loop precursor to increase the efficacy of miRNA processing [34,35]. Zhou et al. further optimized the miR-30 stem-loop for RNAi applications and also described miR-30-based vectors in which the miR-30 cassette is expressed from an intron. This design permits coexpression of the miRNA and a green fluorescent protein (GFP) marker protein encoded by an exon, using a single promoter [36].

A subset of endogenous miRNAs are synthesized from polycistronic transcripts [37–40,22], suggesting that miRNA-based vectors should be suitable for combinatorial RNAi using a single transcription unit. Cullen and colleagues demonstrated the expression of two miRNAs (one synthetic and one a copy of miR-30) from an expression vector using a single transcript, although this design was not used for RNAi [41]. However, Zhou et al. [36] found that two identical tandem copies of a synthetic miRNA based on miR-30 reduced the efficacy for RNAi relative to a single copy of the same synthetic miRNA. As described in the following text, we have developed polycistronic RNAi vectors based on a different miRNA, miR-155, in which tandem copies are efficiently processed. Others have expressed polycistronic clusters of cellular miRNAs using a single promoter [39,40], and recently, an miRNA cluster was adapted for polycistronic RNAi [42]. RNA polymerase II miRNA vectors with inducible or repressible miR-30 cassettes also have been developed. In pTRE-miR-30, expression of the siRNA/miRNA is regulated by a tetracycline-responsive promoter. The corresponding activator, expressed from a second transfected plasmid pTet-Off, is suppressed by doxycycline [16]. The pPRIME (potent RNA interference using microRNA expression) miR-30-based lentiviral vectors enable tetracycline-responsive knockdown of specific mRNAs at single copy [43].

## 16.2.2  MIR-155-BASED VECTORS

We developed a series of RNA polymerase II expression vectors for RNAi based on the miR-155 miRNA precursor [20] (Figures 16.2 and 16.3). The miR-155 precursor is contained within the evolutionarily conserved BIC (B-cell integration cluster) noncoding RNA (ncRNA) [44] and expression of BIC under a heterologous promoter leads to production of miR-155 [45,20]. Prior to the observation that BIC contained miR-155, the BIC ncRNA was characterized in multiple species, and it was shown that BIC could be expressed in heterologous cell types from an RNA polymerase II promoter in a retroviral vector [46–48]. These studies made BIC one of the better-characterized miRNA primary transcripts, suggesting that it would be suitable as the basis for an miRNA-based RNAi vector. We mapped the minimal region of

## a. UI4-GFP-SIBR

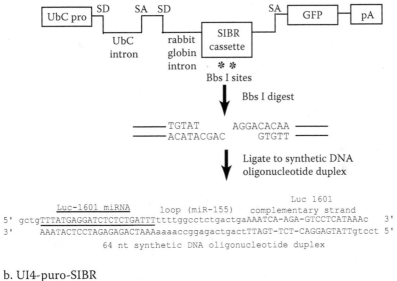

```
                                                          Luc 1601
          Luc-1601 miRNA        loop (miR-155)   complementary strand
5' gctgTTTATGAGGATCTCTCTGATTTttttggcctctgactgaAAATCA-AGA-GTCCTCATAAAc  3'
3'     AAATACTCCTAGAGAGACTAAAaaaaccggagactgactTTAGT-TCT-CAGGAGTATTgtcct 5'
              64 nt synthetic DNA oligonucleotide duplex
```

## b. UI4-puro-SIBR

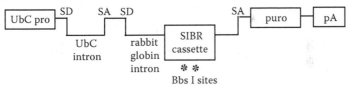

**FIGURE 16.2**  Schematic of the UI4-GFP-SIBR and UI4-puro-SIBR vectors. In these vectors, the human ubiquitin C promoter (UbC pro) drives the expression of a single transcript used for both synthetic miRNA expression (for RNAi) and protein expression (for visible/selectable marker expression). Intron 1 is a modified version of the first intron from human UbC. The SIBR cassette, which contains the synthetic miRNA, is inserted in the rabbit globin intron, located between exon 2 and exon 3. Exon 3 contains the coding region for either green fluorescent protein (GFP) (a) or puromycin-resistance protein (puro) (b). To construct SIBR vectors for RNAi, a synthetic 64-nt DNA duplex encoding the synthetic miRNA, the mouse miR-155 loop, and the complementary strand for the synthetic miRNA is inserted into the Bbs I-digested SIBR vector as shown in (a). The synthetic miRNA sequence shown (luc-1601) is complementary to luciferase [20]. Abbreviations: SA, splice acceptor; SD, splice donor; puro, puromycin-resistance gene; pA, SV40 polyA site.

the mouse BIC ncRNA necessary for miR-155 production, which contains the miR-155 stem-loop precursor and evolutionarily conserved flanking sequences [20]. As observed for other miRNAs, sequences adjacent to the precursor stem-loop structure are required for efficient production of miR-155, most likely by facilitating Drosha processing. Based on our mapping, we created a synthetic variant of a short region from BIC (nt 134–283) in which the 22-nt miR-155 sequence can be replaced by a synthetic 22-nt miRNA sequence complementary to a target gene (Figure 16.3). This short region functions as a synthetic miRNA expression cassette designated the

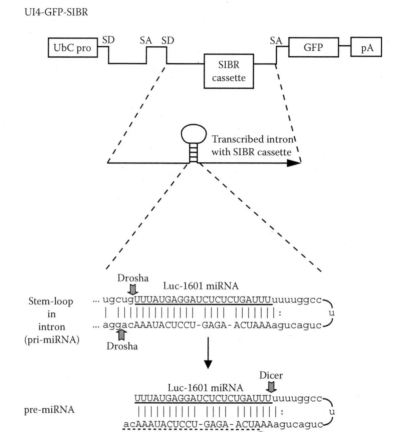

**FIGURE 16.3**  Generation of synthetic miRNAs from the UI4-GFP-SIBR Luc-1601 vector. The intronic SIBR cassette is transcribed and contains a stem-loop structure that can
be processed into a pre-miRNA and then a mature miRNA as indicated. The 22-nt miR-155
sequence has been replaced by a 22-nt synthetic miRNA sequence complementary to the
luciferase mRNA (underlined) [20]. The bases complementary to this synthetic miRNA have
also been adjusted to maintain the partial duplex (upper case) and two single nucleotide gaps
are indicated by dashes. The unmodified miR-155 loop and sequences derived from BIC are
shown in lowercase. The expected miRNA* strand is indicated by a dashed underline. See
Figure 16.2 legend for abbreviations.

SIBR (synthetic inhibitory BIC-derived RNA) cassette. Although miR-155 is located in an exon of the mouse BIC ncRNA [47], we find that the SIBR cassette can be expressed from either an exon or intron under the control of an RNA polymerase II promoter. In the current generation UI4-SIBR vectors (Figure 16.2 and 16.4), the SIBR cassette is expressed from the second intron of a protein-coding gene. As discussed in the following text, this permits efficient coexpression of a protein with the synthetic miRNA.

The SIBR vectors allow production of multiple synthetic miRNAs for effective RNAi from concatenated SIBR cassettes expressed as a single transcription unit from either introns or exons. The ability to express multiple SIBR cassettes can be used to enhance RNAi in two different ways. First, multiple miRNAs against different genes can be expressed from a single vector, allowing inhibition of multiple target mRNAs with a single RNAi vector. The SIBR cassette is flanked by unique but compatible restriction enzyme sites that permit the rapid and facile construction of vectors with two or more tandem SIBR cassettes. We have used SIBR vectors to simultaneously express synthetic miRNAs that target endogenous serine/threonine kinases B-Raf and c-Raf [20]. Either of the miRNAs expressed from two tandem SIBR cassettes was as effective at RNAi as the miRNAs expressed from individual SIBR cassettes. Second, multiple copies of a single SIBR cassette can be expressed from a single vector, providing enhanced inhibition of the target. Using this design, we expressed up to eight tandem copies of an SIBR cassette with increasing efficacy of target inhibition [20]. This contrasts with a previous report of decreased efficacy with two tandem copies in a miR-30-derived vector (Zhou et al., 2005). The reason for this apparent difference in the expression of tandem copies between the miR-155 and miR-30-based vectors is not clear. However, Xia et al. recently have described a system for expression of multiple miR-30-based synthetic miRNAs in which each miRNA is expressed from a separate intron [49], circumventing the need for tandem copies.

The use of RNA polymerase II-based RNAi vectors potentially enables the coregulated expression of a protein and one or more synthetic miRNA/siRNAs from a single transcript. Coexpression of a marker protein can be used to identify or select transfected cells, while expression of other proteins may be useful for gene replacement or functional studies. Three approaches to achieve coregulated expression of proteins and miRNAs have been employed. In one approach, the miRNA cassette is inserted in an untranslated region (UTR) of the marker message, most often the 3′ UTR [21]. This approach is flexible, since it places minimal constraints on vector design. However, the placement of the miRNA cassette in the 3′ UTR of an mRNA may create a situation where neither miRNA production nor marker translation is optimal. For efficient translation, mRNAs require a polyA tail, and they must be exported from the nucleus. Since Drosha cleaves transcripts in the nucleus to release the miRNA precursor, cleavage to release an miRNA precursor from the 3′ UTR is expected to separate the coding region of an mRNA from its polyA tail, leading to reduced translation (but also see Reference 21). In addition, mRNA export from the nucleus is enhanced by the presence of a polyA tail [50], suggesting that polyadenylation may reduce the fraction of the primary transcript that is processed by Drosha in the nucleus [20]. A different strategy for coregulated expression of an miRNA and a marker is to position the miRNA cassette in an intron [36,20,51], as is observed for

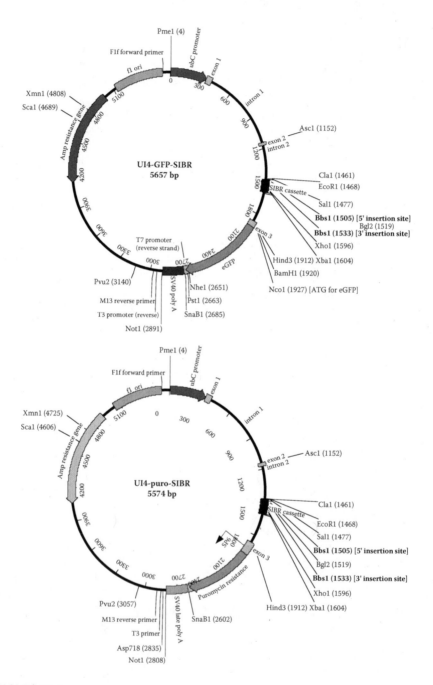

FIGURE 16.4

some naturally occurring miRNAs. The splicing of the intron, and subsequent processing of the miRNA precursor, does not appear to compete with production of an mRNA that can be translated. While allowing unfettered miRNA and marker expression, the use of introns is not generally feasible in lentiviral or retroviral vectors, since the intron (and miRNA) will be spliced out of the genomic transcript during vector production. A recently developed third approach, demonstrated with lentiviral vectors, is to use the transcriptional read-through of a weak poly(A) signal to generate coding and shRNA-containing RNAs from a single transcription unit [19].

We have used design rules developed for siRNAs [52] to select target sequences for synthetic miRNAs in the SIBR vectors (adapted to the 22-nt length of the miR-155-derived synthetic miRNAs). These rules were developed in part based on the analysis of strand bias among miRNAs and their precursors, and they are thought to identify sequences/structures that optimize entry into RISC and probably subsequent events in RNAi [36]. The miR-155 precursor, similar to many miRNA precursors, is an imperfect duplex, and we have maintained this partial duplex structure in the SIBR cassettes that we have designed for producing synthetic miRNAs [20]. Drosha processing is also required for miRNA precursor production from the pri-miRNA produced by miRNA-based vectors. Analyses of Drosha function indicate that the formation of a duplex by sequences flanking the miRNA precursor stem-loop, as well as the size of the stem-loop, influence Drosha function [26,35,53]. However, the sequence of the miRNA itself does not seem to play a major role in Drosha processing. Further, the use of the miR-155 loop as well as the 3′ overhang from the endogenous miR-155 miRNA in the SIBR vectors is likely to facilitate nuclear export and Dicer processing of the synthetic miRNA precursors [54,20].

## 16.3  COMPARING miRNA-BASED VECTORS AND shRNA VECTORS

As discussed earlier, one potential advantage for the SIBR vectors and other miRNA-based vectors when compared to conventional RNA polymerase III shRNA vectors is the ability to use a single transcription unit for both RNAi (expression of a synthetic miRNA) and protein expression (e.g., for a marker protein). Identification or selection of cells expressing an shRNA from an RNA polymerase III promoter requires expression of the marker protein from a separate RNA polymerase II promoter. Since the marker and the shRNA may not be precisely coregulated from the separate promoters, this may make the identification or selection of shRNA-expressing cells less accurate. In addition, the SIBR vectors have the ability to express multiple synthetic

**FIGURE 16.4** (See facing page.) Restriction maps of the UI4-GFP-SIBR and UI4-puro-SIBR vectors. All indicated restriction sites are present only once, except for Bbs1, which is present at two inverted sites that generate the incompatible 5′ overhangs available for the insertion of DNA duplexes (see Figure 16.2). The two Bbs1 sites and the intervening Bgl2 site are deleted in SIBR cassettes that contain a synthetic miRNA insert. Note that the GFP coding region in UI4-GFP-SIBR is flanked by unique restriction sites, facilitating the replacement of GFP with other protein-coding regions. The SIBR cassette in either UI4 vector is flanked by unique restriction sites that facilitate the creation of vectors with two or more tandem SIBR cassettes (see Reference 20 for details).

miRNAs from a single transcription unit. In each case, the potential advantage is the ability to use a single promoter to direct the expression of multiple products. Although it is clear that separate promoters can be used for the simultaneous expression of either multiple shRNAs or marker proteins and shRNAs [55,56], the use of a single promoter or transcription unit is likely to provide more tightly coordinated expression in many cases, and may make combinatorial RNAi more efficacious. The development of internal ribosome entry site (IRES) expression vectors provided an analogous advantage for the coordinated expression of two or more proteins from a single transcription unit in mammalian cells [57].

RNA polymerase III terminates transcription when it encounters four or more consecutive T's in the DNA template [58,59], providing a mechanism to transcribe shRNA molecules with a defined 3′ end. While this feature of RNA polymerase III was essential for the initial development of shRNA vectors, it prevents the use of shRNAs that contain internal stretches of four or more U's (e.g., to target mRNA sequences that contain four or more adjacent A's). RNA polymerase II-transcribed miRNA-based vectors are not subject to this restriction, allowing more flexibility in selecting target sequences for vector-based RNAi. For example, we have successfully expressed synthetic miRNA precursors with four consecutive U's in either the synthetic miRNA or its complementary strand (miRNA*) using the SIBR vectors [20]. In addition, the miR-155-derived loop used in all of the SIBR vectors includes four consecutive U's and would not be compatible with RNA polymerase III transcription (Figure 16.3).

RNA polymerase III promoters are expressed at high levels in transiently transfected cells, and they drive the accumulation of transcribed RNAs to levels as much as ten-fold higher than a strong RNA polymerase II promoter [60,61]. Nonetheless, we and others have observed effective RNAi with miRNA-based RNA polymerase II vectors, comparable to inhibition by shRNAs expressed from RNA polymerase III promoters in transient transfections [36,20]. Possibly entering the miRNA synthesis pathway at the level of Drosha processing facilitates miRNA maturation despite lower levels of initial synthesis. Such a possibility would be consistent with the accumulation of endogenous miRNAs to high levels per cell [62], even though many endogenous miRNAs are derived from primary transcripts synthesized by RNA polymerase II (discussed earlier in Section 16.1). Recently, it was reported that expression of a variety of different shRNAs in mouse liver in vivo can lead to toxicity and death, apparently at least in part because the shRNAs saturate the endogenous miRNA export pathway in the liver cells [63]. The shRNAs were expressed from an RNA polymerase III promoter on a multicopy, extrachromosomal viral vector (based on adenovirus-associated virus), and a high dose of virus was chosen to ensure infection of all liver cells, leading to high levels of shRNA expression. The authors conclude that by adjusting viral vector dose and/or shRNA selection it is possible to safely achieve in vivo RNAi in this model system. However, it would be interesting to determine if the use of RNA polymerase II miRNA-based vectors for RNAi, presumably expressing lower levels of miRNA precursor stem-loops than the corresponding shRNA vectors, might be less prone to toxic effects. A recent report indicates that the inclusion of pri-miRNA sequences in an RNA polymerase III-driven shRNA can help to reduce shRNA toxicity [64].

## 16.4 APPLICATIONS

The use of RNAi to inhibit the function of specific genes has enormous potential for scientific discovery, including identification of novel components of biochemical and biological pathways and target validation for drugs. Libraries of shRNAs or synthetic miRNAs for gene discovery in the human and mouse genomes have been described; see Chang et al. and Wiznerowicz et al. for excellent reviews on this topic [65,66].

Most RNA polymerase III promoters described to date are not tissue specific, although several modifications have been used to create regulated or inducible variants of constitutive RNA polymerase III promoters, some of which are suitable for tissue-specific expression [reviewed in Reference 66]. Nonetheless, an extensive literature exists on tissue-specific RNA polymerase II transcription, suggesting that the use of RNA polymerase II-based RNAi may facilitate tissue-specific expression in animals. Feasibility was recently demonstrated using a Rhox5 promoter to direct a non-miRNA RNAi expression vector to testes [18]. This paper also highlights that vector-based delivery of RNAi has great potential to facilitate rapid in vivo gene knockdown. In addition, RNA polymerase II-based synthetic miRNAs offer additional opportunities for conditional and reversible regulation of target genes [67,66]. Moreover, by allowing the creation of robust hypomorphs, transgenic RNAi also may facilitate the analysis of essential genes without the lethality associated with the targeted disruption of such genes. To this end, we have applied our miR-155-based SIBR vectors to knock-down expression of the mouse gene for neuronal tryptophan hydroxylase (TPH2), the rate-limiting enzyme for serotonin biosynthesis in brain. Transgenic mice containing a TPH2 RNAi construct similar to the UI4-GFP-SIBR vector [20] results in >80% reduction of mRNA for TPH2 in midbrain raphe nucleus, without an appreciable effect on raphe cell viability, as measured by the expression of other mRNAs typically expressed in raphe neurons (see Reference 68 and Patel et al., in preparation). Moreover, transgenic offspring expressing the SIBR vector are easily identified by fluorescent detection of the coexpressed GFP marker. A similar strategy was recently used for the global knockdown of superoxide dismutase 2, using an intronic miR-30-based RNAi vector [69]. The use of conventional RNA polymerase II transgenes for RNAi should facilitate a variety of experimental approaches, including rapid screening of candidate genes arising from genetic mapping experiments. In addition, the ability to simultaneously inhibit multiple genes by expressing two or more synthetic miRNAs from a single transgenic RNAi vector should facilitate future analyses of redundant gene functions.

## 16.5 CONCLUSIONS

Vector-based RNAi using RNA polymerase III-driven shRNAs has become a standard method for the functional analysis of genes. The use of RNA polymerase II to produce synthetic miRNAs for RNAi provides an alternative method that can facilitate coordinated protein expression and RNAi, as well as combinatorial inhibition of multiple target genes. Future application of this approach (e.g., in transgenic animals) holds considerable promise. The expanding range of vectors available for RNAi should facilitate research in many different areas of biology and medical science.

## ACKNOWLEDGMENTS

We thank Sarmad Al-Bassam for assistance with construction of the UI4-puro-SIBR vector and David Engelke for discussions on the relative expression levels of RNA polymerase II and III promoters. This work supported by NIMH MH073085 (ABV), NIMH MH063992 (PDP), NIH NS38698 (DLT), and the University of Michigan Center for Gene Therapy (DLT). Conflict of interest statement: The University of Michigan has patents pending on miR-155-based RNAi technology. The authors are potential recipients of royalties paid to the University of Michigan for licensed use.

## REFERENCES

1. Filipowicz, W. et al., Post-transcriptional gene silencing by siRNAs and miRNAs. *Curr Opin Struct Biol*, 2005. 15(3): 331–341.
2. Sandy, P. et al., Mammalian RNAi: A practical guide. *Biotechniques*, 2005. 39(2): 215–224.
3. Gaur, R.K. and J.J. Rossi, The diversity of RNAi and its applications. *Biotechniques*, 2006. April Suppl: S4–S5.
4. Sen, G.L. and H.M. Blau, A brief history of RNAi: The silence of the genes. *FASEB J*, 2006. 20(9): 1293–1299.
5. Elbashir, S.M. et al., Duplexes of 21-nucleotide RNAs mediate RNA interference in cultured mammalian cells. *Nature*, 2001. 411(6836): 494–498.
6. Murchison, E.P. and G.J. Hannon, miRNAs on the move: miRNA biogenesis and the RNAi machinery. *Curr Opin Cell Biol*, 2004. 16(3): 223–229.
7. Filipowicz, W., RNAi: The nuts and bolts of the RISC machine. *Cell*, 2005. 122(1): 17–20.
8. Brummelkamp, T.R. et al., A system for stable expression of short interfering RNAs in mammalian cells. *Science*, 2002. 296(5567): 550–553.
9. McManus, M.T. et al., Gene silencing using micro-RNA designed hairpins. *RNA*, 2002. 8(6): 842–850.
10. Paddison, P.J. et al., Short hairpin RNAs (shRNAs) induce sequence-specific silencing in mammalian cells. *Genes Dev*, 2002. 16(8): 948–958.
11. Paul, C.P. et al., Effective expression of small interfering RNA in human cells. *Nat Biotechnol*, 2002. 20(5): 505–508.
12. Sui, G. et al., A DNA vector-based RNAi technology to suppress gene expression in mammalian cells. *Proc Natl Acad Sci USA*, 2002. 99(8): 5515–5520.
13. Yu, J.Y. et al., RNA interference by expression of short-interfering RNAs and hairpin RNAs in mammalian cells. *Proc Natl Acad Sci USA*, 2002. 99(9): 6047–6052.
14. Lee, N.S. et al., Expression of small interfering RNAs targeted against HIV-1 rev transcripts in human cells. *Nat Biotechnol*, 2002. 20(5): 500–505.
15. Zeng, Y. et al., Both natural and designed micro RNAs can inhibit the expression of cognate mRNAs when expressed in human cells. *Mol Cell*, 2002. 9(6): 1327–1333.
16. Zeng, Y. et al., Use of RNA polymerase II to transcribe artificial microRNAs. *Methods Enzymol*, 2005. 392: 371–380.
17. Xia, H. et al., siRNA-mediated gene silencing in vitro and in vivo. *Nat Biotechnol*, 2002. 20(10): 1006–1110.

18. Rao, M.K. et al., Tissue-specific RNAi reveals that WT1 expression in nurse cells controls germ cell survival and spermatogenesis. *Genes Dev*, 2006. 20(2): 147–152.
19. Unwalla, H.J. et al., Novel Pol II fusion promoter directs human immunodeficiency virus type 1-inducible coexpression of a short hairpin RNA and protein. *J Virol*, 2006. 80(4): 1863–1873.
20. Chung, K.H. et al., Polycistronic RNA polymerase II expression vectors for RNA interference based on BIC/miR-155. *Nucleic Acids Res*, 2006. 34(7): e53.
21. Cai, X. et al., Human microRNAs are processed from capped, polyadenylated transcripts that can also function as mRNAs. *RNA*, 2004. 10(12): 1957–1966.
22. Cullen, B.R., Transcription and processing of human microRNA precursors. *Mol Cell*, 2004. 16(6): 861–865.
23. Lee, Y. et al., MicroRNA genes are transcribed by RNA polymerase II. *EMBO J*, 2004. 23(20): 4051–4060.
24. Pfeffer, S. et al., Identification of microRNAs of the herpesvirus family. *Nat Methods*, 2005. 2(4): 269–276.
25. Borchert, G.M. et al., RNA polymerase III transcribes human microRNAs. *Nat Struct Mol Biol*, 2006. 13: 1097–1101.
26. Lee, Y. et al., The nuclear RNase III Drosha initiates microRNA processing. *Nature*, 2003. 425(6956): 415–419.
27. Gregory, R.I. et al., The Microprocessor complex mediates the genesis of microRNAs. *Nature*, 2004. 432(7014): 235–240.
28. Yi, R. et al., Exportin-5 mediates the nuclear export of pre-microRNAs and short hairpin RNAs. *Genes Dev*, 2003. 17(24): 3011–3016.
29. Lund, E. et al., Nuclear export of microRNA precursors. *Science*, 2004. 303(5654): 95–98.
30. Hutvagner, G., Small RNA asymmetry in RNAi: Function in RISC assembly and gene regulation. *FEBS Lett*, 2005. 579(26): 5850–5857.
31. Preall, J.B. and E.J. Sontheimer, RNAi: RISC gets loaded. *Cell*, 2005. 123(4): 543–545.
32. Hutvagner, G. and P.D. Zamore, A microRNA in a multiple-turnover RNAi enzyme complex. *Science*, 2002. 1: 1.
33. Valencia-Sanchez, M.A. et al., Control of translation and mRNA degradation by miRNAs and siRNAs. *Genes Dev*, 2006. 20(5): 515–524.
34. Chen, C.Z. et al., MicroRNAs modulate hematopoietic lineage differentiation. *Science*, 2004. 303(5654): 83–86.
35. Zeng, Y. and B.R. Cullen, Efficient processing of primary microRNA hairpins by Drosha requires flanking nonstructured RNA sequences. *J Biol Chem*, 2005. 280(30): 27595–27603.
36. Zhou, H. et al., An RNA polymerase II construct synthesizes short-hairpin RNA with a quantitative indicator and mediates highly efficient RNAi. *Nucleic Acids Res*, 2005. 33(6): e62.
37. Houbaviy, H.B. et al., Embryonic stem cell-specific MicroRNAs. *Dev Cell*, 2003. 5(2): 351–358.
38. Rodriguez, A. et al., Identification of mammalian microRNA host genes and transcription units. *Genome Res*, 2004. 14(10A): 1902–1910.
39. Hayashita, Y. et al., A polycistronic microRNA cluster, miR-17-92, is overexpressed in human lung cancers and enhances cell proliferation. *Cancer Res*, 2005. 65(21): 9628–9632.
40. He, L. et al., A microRNA polycistron as a potential human oncogene. *Nature*, 2005. 435(7043): 828–833.
41. Zeng, Y. and B.R. Cullen, Sequence requirements for micro RNA processing and function in human cells. *RNA*, 2003. 9(1): 112–123.
42. Liu, Y.P. et al., Inhibition of HIV-1 by multiple siRNAs expressed from a single microRNA polycistron. *Nucleic Acids Res*, 2008. 36(9): 2811–2824.

43. Stegmeier, F. et al., A lentiviral microRNA-based system for single-copy polymerase II-regulated RNA interference in mammalian cells. *Proc Natl Acad Sci USA*, 2005. 102(37): 13212–13217.·

44. Lagos-Quintana, M. et al., Identification of tissue-specific microRNAs from mouse. *Curr Biol*, 2002. 12(9): 735–739.

45. Eis, P.S. et al., Accumulation of miR-155 and BIC RNA in human B cell lymphomas. *Proc Natl Acad Sci USA*, 2005. 102(10): 3627–3632.

46. Tam, W. et al., bic, a novel gene activated by proviral insertions in avian leukosis virus-induced lymphomas, is likely to function through its noncoding RNA. *Mol Cell Biol*, 1997. 17(3): 1490–1502.

47. Tam, W., Identification and characterization of human BIC, a gene on chromosome 21 that encodes a noncoding RNA. *Gene*, 2001. 274(1–2): 157–167.

48. Tam, W. et al., Avian bic, a gene isolated from a common retroviral site in avian leukosis virus-induced lymphomas that encodes a noncoding RNA, cooperates with c-myc in lymphomagenesis and erythroleukemogenesis. *J Virol*, 2002. 76(9): 4275–4286.

49. Xia, X.G. et al., Multiple shRNAs expressed by an inducible pol II promoter can knock down the expression of multiple target genes. *Biotechniques*, 2006. 41(1): 64–68.

50. Huang, Y. and G.C. Carmichael, Role of polyadenylation in nucleocytoplasmic transport of mRNA. *Mol Cell Biol*, 1996. 16(4): 1534–1542.

51. Du, G. et al., Design of expression vectors for RNA interference based on miRNAs and RNA splicing. *FEBS J*, 2006.

52. Khvorova, A. et al., Functional siRNAs and miRNAs exhibit strand bias. *Cell*, 2003. 115(2): 209–216.

53. Zeng, Y. et al., Recognition and cleavage of primary microRNA precursors by the nuclear processing enzyme Drosha. *EMBO J*, 2005. 24(1): 138–148.

54. Zeng, Y. and B.R. Cullen, Structural requirements for pre-microRNA binding and nuclear export by Exportin 5. *Nucleic Acids Res*, 2004. 32(16): 4776–4785.

55. Yu, J.Y. et al., Simultaneous inhibition of GSK3alpha and GSK3beta using hairpin siRNA expression vectors. *Mol Ther*, 2003. 7(2): 228–236.

56. Jazag, A. et al., Single small-interfering RNA expression vector for silencing multiple transforming growth factor-beta pathway components. *Nucleic Acids Res*, 2005. 33(15): e131.

57. Martinez-Salas, E., Internal ribosome entry site biology and its use in expression vectors. *Curr Opin Biotechnol*, 1999. 10(5): 458–464.

58. Tazi, J. et al., Mammalian U6 small nuclear RNA undergoes 3′ end modifications within the spliceosome. *Mol Cell Biol*, 1993. 13(3): 1641–1650.

59. Booth, B.L., Jr. and B.F. Pugh, Identification and characterization of a nuclease specific for the 3′ end of the U6 small nuclear RNA. *J Biol Chem*, 1997. 272(2): 984–991.

60. Bertrand, E. et al., The expression cassette determines the functional activity of ribozymes in mammalian cells by controlling their intracellular localization. *RNA*, 1997. 3(1): 75–88.

61. Good, P.D. et al., Expression of small, therapeutic RNAs in human cell nuclei. *Gene Ther*, 1997. 4(1): 45–54.

62. Lim, L.P. et al., The microRNAs of *Caenorhabditis elegans*. *Genes Dev*, 2003. 17(8): 991–1008.

63. Grimm, D. et al., Fatality in mice due to oversaturation of cellular microRNA/short hairpin RNA pathways. *Nature*, 2006. 441(7092): 537–541.

64. McBride, J.L. et al., Artificial miRNAs mitigate shRNA-mediated toxicity in the brain: Implications for the therapeutic development of RNAi. *Proc Natl Acad Sci USA*, 2008. 105(15): 5868–5873.

65. Chang, K. et al., Lessons from Nature: microRNA-based shRNA libraries. *Nat Methods*, 2006. 3(9): 707–714.

66. Wiznerowicz, M. et al., Tuning silence: Conditional systems for RNA interference. *Nat Methods*, 2006. 3(9): 682–688.
67. Szulc, J. et al., A versatile tool for conditional gene expression and knockdown. *Nat Methods*, 2006. 3(2): 109–116.
68. Mueller, H.M. et al., Characterization of a tryptophan hydroxylase 2 knockdown mouse: A hyposerotonergic animal model. *Soc Neurosci (Abs)*, 2006. 32(#290.7).
69. Xia, X.G. et al., Pol II-expressed shRNA knocks down Sod2 gene expression and causes phenotypes of the gene knockout in mice. *PLoS Genet*, 2006. 2(1): e10.

# 17 Transgenic RNAi
## A Fast and Low-Cost Approach to Reverse Genetics in Mammals

*Linghua Qiu and Zuoshang Xu*

## CONTENTS

## OVERVIEW

Reverse genetics in mammals has primarily relied on targeted gene mutations by homologous recombination. Although highly effective, its application in general biological studies has been hampered by its high cost, high complexity, lengthy process, and limitation to the mouse. Recently, transgenic RNAi has been demonstrated to silence genes and elicit phenotypes of gene dysfunction in vivo, indicating that it can be an alternative method for reverse genetics in mice and other mammalian species. Here, we review the progress and future challenges in improving and applying this approach.

## 17.1 INTRODUCTION

Reverse genetics studies gene functions by targeted gene mutations and observation of the functional consequences. It is an indispensable vehicle in building the wealth of knowledge on gene function in modern biology. Increasingly, it has been used to understand the mechanism of diseases and to generate animal models for diseases.

Thus, advancement in reverse genetics technology has directly impacted human health. Since the early 1980s, reverse genetics in mammals has primarily relied on gene knockout by homologous recombination technology in mice. Although the method is highly effective and has been instrumental in defining gene functions, several drawbacks have limited its use. First, only a few laboratories have access to this technology because of its technical complexity, high cost, and lengthy process. Even under optimum conditions, which assumes that the process is carried out by an expert laboratory and everything goes smoothly, the process of obtaining a gene knockout mouse takes 9 to 10 months [1]. Considering that problems do arise in experiments, the time required to complete a gene knockout is often much longer than this minimum. It is not uncommon for a laboratory to spend three to four years to complete a study using the knockout approach. Second, the knockout technology is limited to applications in mice, and therefore, cannot meet the demand for reverse genetics in other mammalian species. Third, gene knockout is not an ideal approach to generate disease models that require genetic hypomorphism, because standard gene knockout often results in embryonic lethality in the homozygote and a lack of phenotype in the heterozygote. To generate hypomorphic models, one has to knock in a mutant allele that has a reduced function. This approach has the same drawbacks of the standard gene knockout, and in addition, requires prior knowledge of the mutant.

RNA interference (RNAi) is a conserved gene silencing mechanism in most eukaryotes [2]. In mammalian cells, RNAi can be triggered by introducing into cells short double-stranded RNAs (dsRNAs) of 21–29 nucleotides in length (also known as small interfering RNA, or siRNA) or gene cassettes driven by Pol III or Pol II promoter that synthesizes short hairpin RNA (shRNA) or micro RNA (miRNA; see Figure 17.1) [3]. The dsRNA, shRNA, or miRNA are processed through several successive steps, whereby a single-stranded siRNA (called the *guide strand*) is incorporated into a protein complex to form RISC (RNA-induced silencing complex). The guide strand then recognizes the target RNA by Watson–Crick base pairing and directs the RISC to cleave the target RNA (Figure 17.1), resulting in gene silencing [4].

These mechanistic understandings and technological advances led to the idea of transgenic RNAi approach for reverse genetics in mammals. In this approach, a transgene that expresses an shRNA or an miRNA targeting a gene of interest is introduced into the mouse genome, making this transgene inheritable to the offspring. The pattern of the gene silencing is stable, and the effects of gene silencing can be studied in generations of the animals. Thus, this approach has the advantages of traditional transgene expression in terms of ease of the technology and quickness in obtaining transgenic founders, and many features of gene knockout can be achieved. In recent years, efforts by different research groups have improved the transgenic RNAi approach. As a result, the potential of this approach in accelerating reverse genetics in mammals and generation of models for human diseases is being realized. Similar to many experimental approaches, transgenic RNAi involves multiple steps. Each step faces multiple choices. In the following text, we review the methods that have been tried and the results of these trials. We discuss the pros and cons of these methods, and we hope that these discussions will help both the experienced and inexperienced investigators to apply this technology in their research.

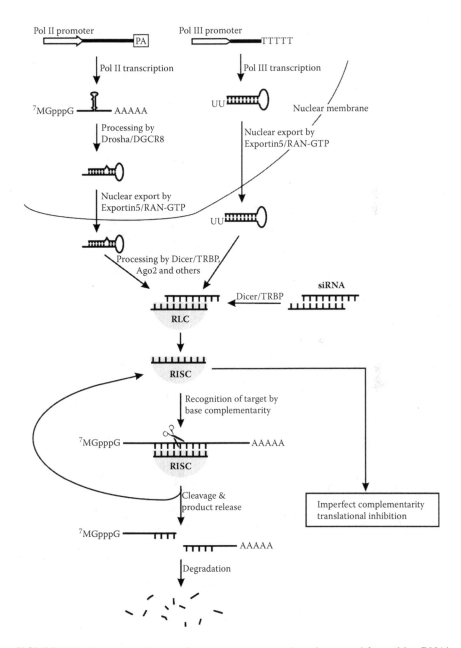

**FIGURE 17.1** Two general types of transgene constructs have been used for making RNAi transgenic mice. The Pol III-shRNA construct is composed of a Pol III promoter, a DNA segment encoding the shRNA and a string of 5 or more T's to signal transcription termination. The Pol II-miRNA construct is composed of a Pol II promoter, a DNA fragment encoding a pri-miRNA transcript carrying one or several pre-miRNAs. See text for the description of the processing steps.

## 17.2 THE CONSTITUTIVELY ACTIVE TRANSGENE BASED ON POL III PROMOTERS

The first step in generating transgenic RNAi animals is the same as that in constructing gene-overexpression animals [5], which is constructing a transgene that
synthesizes an shRNA. To accomplish this, one has to first select the target region
in an mRNA. There are Web-based algorithms that can help select the target region.
However, to obtain a highly efficient silencing target, several shRNAs have to be made
based on the algorithm's recommendations. DNA sequences encoding these shRNAs
are synthesized and cloned downstream of a Pol III promoter. The shRNA contains
a perfectly base-paired dsRNA stem of 19–29 nt, and the two strands of the stem are
connected with a commonly used 9-nt loop [6,7]. The stem sequence should be the
same as the sequence in the target mRNA. After synthesis by the Pol III, the shRNA
is exported by Exportin 5/RAN-GTP complex to the cytoplasm, where one strand of
the stem is incorporated into the RISC and mediates silencing (Figure 17.1).

The basic, constitutively active transgene construct is composed of a Pol III promoter followed by an shRNA-encoding sequence and a Pol III termination sequence
(Figure 17.2A). Early experiments demonstrated that shRNAs synthesized by constitutively active Pol III promoters such as U6 or H1 work well in silencing genes in

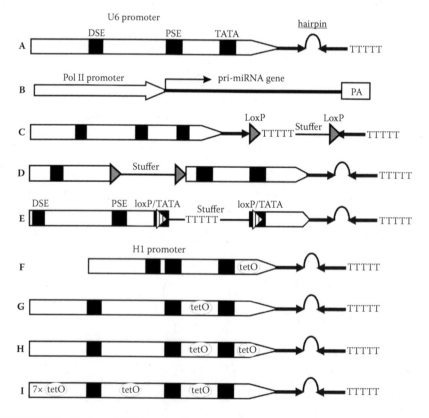

**FIGURE 17.2** Various shRNA and miRNA expression constructs.

cultured cells [3]. These constructs were quickly used for making transgenic mice. The initial experiments reported success in knocking down GFP in transgenic mice [8–10]. Despite the optimism raised by these reports, successful knockdown of mouse endogenous genes and expression of knockdown phenotypes in mice did not occur, despite using this approach for several years. Our experience, as well as that of others, suggests that the constitutive Pol III constructs are inefficient in making stable transgenic RNAi lines [7,11]. The main problems are that injection of these constructs into mouse embryos often yields few or no positive transgenic founders; of the few positives, the levels of shRNA expression are often too low to cause a significant target knockdown, resulting in no gene deficiency phenotype.

The reasons for this inefficiency are not known. One possibility is that the embryos that express high levels of shRNA suffer from a negative selection pressure. Constitutively active transgenes are expected to express the shRNA immediately following transgene injection into embryos. The shRNA could interfere with the miRNA functions in the embryo by competing for protein machinery that is involved in the miRNA processing and functioning. Because miRNAs play essential roles during embryonic development, compromising miRNA function can abort the developmental program, leading to the death of the embryo. This mechanism might have killed those embryos that express high levels of the shRNA. Silencing of the mouse endogenous target gene, especially if the target gene is essential in embryonic development, may be another negative selection factor contributing to the absence of high shRNA expression transgenic lines.

## 17.3   THE CONSTITUTIVELY ACTIVE TRANSGENE BASED ON POL II PROMOTERS

Because of the poor efficiency in generating transgenic mice using the Pol III promoter-based transgenes, several groups turned to using Pol II promoters. Pol II produces the majority of the endogenous miRNAs, which are first synthesized as Pol II transcripts that have 5′ cap and 3′ poly A, called primary micro RNAs (pri-miRNAs). Each pri-miRNA contains one or several precursor miRNAs (pre-miRNAs), which are similar to shRNAs but contain mismatches and/or bulges in their stems [12]. From the pri-miRNA, the pre-miRNA is excised by an RNase III enzyme complex composed of Drosha and DGCR8 [13–17]. Similar to the shRNA, the pre-miRNA is exported by Exportin 5/RAN-GTP to the cytoplasm, where it is further processed by another RNase complex composed of Dicer, TRBP, and Ago2 to generate siRNA, from which the guide strand, or the miRNA, is retained by the complex to form RISC [18,19]. This RISC can mediate translational suppression, if the target mRNA imperfectly complements with the miRNA, or RNAi, if the target mRNA complements with the miRNA perfectly (Figure 17.1) [20–22].

Zeng et al. initially demonstrated that human miR-30a pre-miRNA placed in a Pol II construct can express the miRNA and silence its target expression by either translational inhibition or RNAi [23]. Based on this work, several laboratories independently improved the design of Pol II-synthesized miRNA strategy and demonstrated that Pol II-synthesized miRNA can achieve similar silencing efficiency as the Pol III constructs in cultured cells [24–26]. Using this Pol II-miRNA strategy (Figure 17.2B), two groups reported success in knocking down endogenous mouse genes and in generating phenotypes from a lack of the target gene function [27,28]. Xia et al. targeted Mn superoxide dismutase (SOD2) mRNA. The pri-miRNA synthesis was driven by human ubiquitin C (UbC) promoter, a ubiquitously active promoter. Mice transgenic for this transgene expressed the miRNA widely, and SOD2 expression was knocked down by 60 to 90% in eight different tissues examined. This knockdown was correlated with a lowered SOD2 enzyme activity, an increased production of reactive oxygen species (ROS), an enhanced sensitivity to oxidative stress, and evidence of oxidative damage in heterozygous transgenic mice. Although these signs were consistent with SOD2 deficiency, these mice were viable and did not show the severe phenotypes characteristic of the SOD2 knockout. Xia et al. then crossed the two transgenic lines to obtain bigenic mice and succeeded in increasing the transgenic miRNA levels further. These bigenic mice expressed the severe phenotypes typical of SOD2 knockout mice, including retarded growth, fatty liver and muscle, dilated cardiomyopathy, and premature death [28]. These results demonstrate that RNAi can be an alternative to the gene knockout approach for reverse genetics in mice.

Rao et al. targeted the transcription factor Wilm's tumor 1 (WT1) [27]. To drive the expression of their miRNA, they used the proximal promoter from the mouse Pem (Rhox5) gene, which is active specifically in Sertoli cells in the testis. They observed reduced fertility in male transgenic mice due to increased apoptosis and other abnormalities in germ cells. These phenotypes were confirmed in another transgenic mouse line that expressed a dominant-negative mutant of WT1. These results demonstrate that Pol II-miRNA strategy can silence target gene and elicit phenotypes in a tissue- or cell-specific manner.

Why did the Pol II-miRNA constructs work, whereas Pol III-shRNA constructs failed? Transgenes made from Pol II promoters that are constitutively and ubiquitously active can also begin their synthesis of miRNA at the earliest stages of embryonic development, and therefore, could potentially interfere with the endogenous miRNA function and embryonic development. However, Pol II-miRNA transgenes might be expressed at lower levels than Pol III-shRNA transgenes because Pol III promoters are very strong and can synthesize very high levels of shRNA. Using the tissue-specific Pol II promoter circumvents the problem of interfering with miRNA function during early embryonic development before cell differentiation. In any case, the success of Pol II-miRNA transgenes in generating transgenic RNAi mice opens the door of RNAi to a wide array of Pol II promoters that have been proved successful in transgenic mice. These include ubiquitously active promoters, various tissue- or cell-specific promoters, developmentally regulated promoters, Cre-inducible promoters, and tetracycline-inducible promoters. The application of these approaches will undoubtedly increase the versatility of transgenic RNAi in a wide range of transgenic research situations.

## 17.4 THE INDUCIBLE TRANSGENES BASED ON POL III PROMOTERS

The constitutively active transgenes have several drawbacks. First, these genes might be activated immediately after their injection into the single-cell-stage embryos, resulting in high levels of shRNA that could interfere with the endogenous miRNA function in the early embryo. Second, if the target gene is essential, its silencing will cause embryonic lethality. As a consequence, no founder mice with good target silencing will be obtained. Third, these transgenes also do not provide the flexibility, spatially and temporally, to control the silencing. Using cell-specific promoters can partially circumvent these shortcomings. However, a powerful approach is to make RNAi transgenic mice by employing inducible strategies whereby the miRNA is not expressed initially, but can be induced to express at specific times or in specific cell types. This allows for the transgenic lines to be generated and the silencing effects to be observed upon induction in generations of the mouse progenies.

Different inducible Pol III promoters have been designed (Figure 17.2C–I). These can be divided into two classes. One class is based on the Cre-lox system, and the other is inducible systems that can be turned on and off by administration of small chemicals such as doxycycline (a tetracycline analog). Three designs of Cre-activated transgenes have been reported. The first places a stuffer sequence that is flanked by loxP sites and contains a Pol III termination signal (a string of four or more T's) in the loop region of the hairpin (Figure 17.2C) [29]. Before exposure to Cre, transcription terminates after the first loxP site when Pol III encounters the termination signal. Therefore, only the sense strand of the hairpin is synthesized. After exposure to Cre, the stuffer sequence along with the Pol III terminator is excised, and the shRNA is synthesized with the loxP as the loop. Although this construct has been shown to mediate RNAi upon Cre induction in cell culture, transgenic RNAi animals using this design have not been reported.

The second Cre-lox design places a stuffer sequence that is flanked by loxP sites between the DSE (distal sequence element) and PSE (proximal sequence element) of the U6 promoter (Figure 17.2D). Two groups reported successful knockdown of the target genes and phenotypes associated with the hypomorphic target gene function after the mice were crossed with a Cre-expressing transgenic mouse line [30,31]. Coumoul et al. targeted fibroblast growth factor receptor 2 (*fgfr2*) for silencing and observed embryonic abnormalities consistent with the lack of this receptor function following the ubiquitous or tissue-specific induction of the shRNA. Xia et al. targeted a mutant form of human SOD1 for silencing. This mutant SOD1 causes amyotrophic lateral sclerosis (ALS), a fatal neurodegenerative disease. After crossing the shRNA-expressing transgenic mice with mice that express this mutant SOD1, they observed a modest knockdown of SOD1 and alleviation of the disease. These two studies are largely consistent, except that Coumoul et al. reported no expression of the shRNA before crossing with Cre transgenic lines while Xia et al. reported a small leakage of shRNA expression before induction. The reason for this inconsistency is not clear but could be caused by different genomic integration sites or the length of the stuffer sequence. However, early studies on the U6 promoter have demonstrated that the DSE is not absolutely required for U6 transcription activity [32], which is consistent with the observations by Xia et al.

The third Cre-inducible design employs modified loxP sites that can function both as a Cre recombinase recognition site and a TATA box (Figure 17.2E) [33,34]. The stuffer sequence is inserted between the PSE and the TATA box. Because alterations in the distance between PSE and TATA box can eliminate the U6 promoter activity

[32], the insertion of the stuffer sequence completely blocked the shRNA expression. After removal of the stuffer by Cre, expression of shRNA and silencing of the target gene expression resumed [33,34]. Ventura et al. and Zhou et al. used this strategy and made transgenic RNAi mice [34,35]. They observed efficient removal of the loxP-flanked stuffer sequence after crossing with Cre-driver lines that express Cre in their germ line. However, whether these transgenic lines are useful in crosses with other cell-specific Cre driver lines remains unclear. One potential problem of this construct is that the modified loxP site reduces the efficiency of Cre-mediated recombination significantly [36]. This could compromise the recombination efficiency in vivo and generate high levels of mosaicism upon induction.

The small molecule inducible system can be a powerful method because, in theory, it can reversibly turn on and off gene expression. There are four differently designed tetracycline-regulatable (Figure 17.2F–I) and one ecdysone-regulatable Pol III promoters that have been reported in the literature [37–42]. All showed regulatable gene expression in cultured cells, although some results among these constructs are conflicting [7]. To date, only construct F (Figure 17.2) has been tested in transgenic mice [43]. Therefore, the in vivo utility of these constructs remains mostly unclear.

## 17.5   THE INDUCIBLE TRANSGENES BASED ON POL II PROMOTERS

Many inducible Pol II promoters have been well characterized and used in transgenic mice. Among these are the Cre-inducible and the tetracycline-inducible systems. The various combinations of these strategies to express genes in specific type of cells and at specific times have been reviewed in many excellent reviews [1,44,45]. All of these strategies can be adapted to express miRNA targeting specific genes for silencing. These strategies are particularly attractive because they can circumvent some critical shortcomings of the constitutively active transgenes. For example, by inducing transgene expression later, one can avoid the toxic effects of miRNA expression or target silencing in early embryonic development and establish transgenic lines that can be used for study in infinite number of generations. By induction of the miRNA expression in specific cells or at specific times, one can study the target gene function in specific cell types and at a specific age of the animal.

## 17.6   GENE TRANSFERRING METHODS

Once the transgene construct is made, the next step is to deliver the gene construct into the mouse genome so that it can be inherited by generations of mice, which can be used to study the phenotype and its mechanistic basis. There are different methods to transfer transgenes into mammalian embryos for generation of transgenic animals [5]. The most commonly used method for making transgenic mice is the pronuclear injection method. By this method, the transgene is injected into the fertilized eggs, which are then transplanted back into the pseudopregnant females to develop the embryos to term. After pups are born, they are genotyped. The animals that carry the transgene are propagated, and some animals are sacrificed to determine whether the transgene is expressed. The transgenic lines that do not show a desired tissue expression pattern are terminated, and the lines that do are further analyzed for target gene silencing and the consequent phenotypes. The advantages of this method are that it is simple, cheap, and fast. The disadvantages are that the transgenes are randomly incorporated into the animal's genome and the transgene expression is highly subjected to the influence by the genome location where the transgene is inserted. Thus, the expression pattern can be highly variable in different transgenic lines. For example, Xia et al. reported that in a line of transgenic mice constructed from a U6-shRNA construct, the shRNA was expressed highly in brain, spinal cord, heart and muscle, lower in ovary, and nearly undetectable in liver, stomach, and lung [46]. Similar variable knockdown patterns were observed in transgenic mice generated using an H1 promoter-driven shRNA construct [47]. The Pol II-miRNA constructs have worked more consistently thus far: both the UbC promoter and the Pem promoter worked as anticipated, giving rise to ubiquitous and tissue-specific expression of miRNA, respectively [27,28].

An alternative method consists of using lentivirus to deliver the transgene into the mouse genome [48,49]. In this approach, the transgene is inserted into the lentiviral genome. The packaged virus is then injected into the perivitelline space of the fertilized eggs. The virus infects the eggs, transferring the transgene into the eggs and inserting it into the mouse genome. Several groups reported success in generating transgenic mice that express genes encoding proteins or shRNA [9,48]. Several disadvantages are worth keeping in mind when using this approach. First, this approach requires expertise in lentiviral construction and production. Second, one egg is often infected by multiple lentiviral genomes, and each is inserted into the mouse genome independently and segregates independently in the offspring of the founder mouse. Consequently, each transgene-positive mouse descended from the founder could have different gene expression levels and patterns. This complicates the analysis of the transgenic lines. Third, there is variegation in transgene expression, although this can be overcome by breeding using transgenic males [50]. There are several advantages compared to the pronuclear injection method. First, this method is very efficient, and a large fraction of founders are positive for the transgene. Second, because lentivirus tends to insert itself into the active gene loci in the genome, a high proportion of the positive transgenic lines express the transgene. Third, this approach can work with the transgenes that are based on the constitutively active U6 promoter [50,51], although this is most likely limited to nonessential genes.

Another way of introducing transgenes into transgenic mice is by using mouse embryonic stem (ES) cells [5]. In this method, the transgene is first introduced into the ES cells in culture. The cells are then screened to obtain clones that express the transgene. The cloned cells are injected into the mouse blastocysts, which are transplanted back into pseudopregnant females to develop. This method is highly effective in studying gene functions in developing mouse embryos when combined with tetraploid complementation [52–54]. Postnatal gene silencing has also been reported using transgenes with the constitutively active U6 promoter [9,11,55].

A variation of the ES cell method is to target the Pol III-driven transgene into defined loci in the mouse genome, where the transgene will not be silenced. This could overcome the limitation of the pronuclear injection method, with which the transgene expression pattern cannot be controlled. One group tested this approach by targeting Pol III-driven shRNA expression cassettes into the Rosa 26 locus and observed widespread knockdown of luciferase, Lac Z, and leptin receptor expression [56]. The degree of target gene knockdown varied in different tissues, however, ranging from ~40% to >95%. Another two groups targeted their Cre-inducible Pol III-shRNA construct to the HPRT [57] and Rosa 26 locus [58], respectively. After crossing with Cre-expressing mice, gene silencing was observed in various embryonic tissues. Overall, transgenic RNAi using ES cells has potential and does have some advantages. However, this approach also shares some of the drawbacks of the gene knockout approach owing to its requirement of ES cells. For example, it will not be practical to apply this approach to nonmurine mammalian species.

## 17.7   FUTURE CHALLENGES AND APPLICATIONS

The application of transgenic RNAi technology is similar to gene knockout. Both can be used to determine the function of a gene at the whole animal level. However, in choosing which method to use, it is important to keep in mind the pros and cons of each method. In addition to the contrasts between the two methods in complexity, time, and cost as discussed earlier, one also needs to consider several technical issues. If designed properly, the knockout approach can eliminate the target gene expression in all tissues or in specific cell groups. If complete elimination of the gene function is desired, the knockout approach should be the choice. RNAi, on the other hand, cannot eliminate gene expression, although in some cases, it can knock down gene expression to undetectable levels. There is also tissue variation due to the different levels of the transgenic miRNA and the target mRNA in different tissues. However, for studies on embryonic development, RNAi might be equivalent to gene knockout [52–54]. This is possibly due to less tissue-specific regulation of gene expression in embryos than in postnatal animals; and universal expression of miRNA and target gene knockdown is more readily accomplishable.

If models for hypomorphic gene function are desired, then RNAi will be the choice. Because gene knockout often generate animals with ~50% of gene expression in the heterozygotes and no expression in homozygotes, the degree of variation of gene expression is limited. Frequently, the heterozygotes have no phenotype, while the homozygotes are embryonic lethal. This makes studies on the gene function in postnatal animals and in aging difficult. RNAi, on the other hand, can produce animals with various degrees of target gene knockdown. This is because when the transgene is injected into the fertilized eggs, each egg will incorporate the transgene into different genome loci and different transgene copies. Both variations can influence the transgene (miRNA) expression levels. Therefore, when transgenic lines are established, each transgenic line will express different levels of miRNA, thus knocking down the target gene to different degrees. By analyzing these various transgenic lines, one may find the dosing effect of the target gene expression.

An additional means to generate graded levels of knockdown is to use shRNAs or miRNAs with different knockdown efficacy. Perhaps the most sophisticated way to generate graded knockdown is to use the tetracycline- or other regulatable systems to express miRNA. For the tetracycline-regulatable systems, it may be possible to induce miRNA expression at different levels by administering different doses of doxycycline (a tetracycline analog that is commonly used), thus knocking down the target gene to different degrees. Since the tetracycline-inducible gene expression has been successfully applied in generating transgenic mice [1,44], this strategy may be used to express miRNA.

Further potential of transgenic RNAi may be realized by the recently reported Pol II-based multiple miRNAs expression systems that are capable of silencing multiple gene targets simultaneously [59–61]. This approach can greatly accelerate studies on the effects of compound hypomorphism of multiple genes in mammalian systems.

A key application of the transgenic RNAi technology will be to generate gene-deficiency models in nonmurine mammals. These models can have many uses, including modeling diseases for mechanistic experimentation, drug tests, and commercial production of genetically modified animals. For example, one recent experiment has explored generation of prion-deficient goat and bovine [62]. If successful, this approach can generate livestock that are resistant to prion diseases [63]. Nonmurine transgenic animals may model human diseases better than mice in some cases. A precedent for this is the human inflammatory spondyloarthropathies, which is associated with the activities of human HLA27 and $\beta$2-microglobulin genes. Expression of HLA27 and human $\beta$2-microglobulin caused no phenotype in mice, suggesting no role of these two molecules in this disease [64]. However, expression of the same two molecules in rats triggered spontaneous spondyloarthropathies [65]. The rat model has been instrumental in understanding the mechanism of this disease [66]. Other disease examples include hypertension and HIV infection, where the rat models also have advantages over the mouse models [67,68]. Because many diseases are caused by loss of gene functions in humans and the gene knockout mice fail to recapitulate the disease phenotypes, better disease models may be established in nonmurine mammalian species using RNAi.

## 17.8   CONCLUSIONS

In summary, much progress has been made in transgenic RNAi technology. Success has been achieved using Pol III-shRNA gene constructs to investigate gene functions in embryonic development and in adult mice. Various regulatable Pol III-shRNA constructs have been developed, and some have been used successfully to silence genes in transgenic mice. In addition, Pol II-miRNA constructs have been created and used to generate mouse models for hypomorphic gene function. The regulatable Pol II-miRNA constructs offer flexibility and controllability in vivo and have the potential to become a mainstream method for reverse genetics in mammals. These new approaches can be used as an alternative method of gene knockout to determine the gene function in vivo, establish animal models for diseases, and test the potential drug targets.

## ACKNOWLEDGMENTS

This work is supported by grants from the ALS Association, NIH/NINDS (RO1NS048145), NIH/NIA (R21AG028528), and The Robert Pachard Center for ALS Research at Johns Hopkins to Zuoshang Xu.

## REFERENCES

1. Beglopoulos, V. and Shen, J. (2004). Gene-targeting technologies for the study of neurological disorders. *Neuromolecular Med 6*, 13–30.
2. Mello, C. C. and Conte, D. (2004). Revealing the world of RNA interference. *Nature 431*, 338–342.
3. Shi, Y. (2003). Mammalian RNAi for the masses. *Trends Genet 19*, 9–12.
4. Tomari, Y. and Zamore, P. D. (2005). Perspective: machines for RNAi. *Genes Dev 19*, 517–529.
5. Deng, H.-X. and Siddique, T. (2000). Transgenic mouse models and human neurodegenerative disorders. *Arch Neurol 57*, 1695–1702.
6. Siolas, D., Lerner, C., Burchard, J. et al. (2005). Synthetic shRNAs as potent RNAi triggers. *Nat Biotechnol 23*, 227–231.
7. Xia, X. G., Zhou, H., and Xu, Z. (2005). Promises and challenges in developing RNAi as a research tool and therapy for neurodegenerative diseases. *Neurodegener Dis 2*, 220–231.
8. Hasuwa, H., Kaseda, K., Einarsdottir, T. et al. (2002). Small interfering RNA and gene silencing in transgenic mice and rats. *FEBS Lett 532*, 227–230.
9. Rubinson, D. A., Dillon, C. P., Kwiatkowski, A. V. et al. (2003). A lentivirus-based system to functionally silence genes in primary mammalian cells, stem cells and transgenic mice by RNA interference. *Nat Genet 33*, 401–406.
10. Tiscornia, G., Singer, O., Ikawa, M. et al. (2003). A general method for gene knockdown in mice by using lentiviral vectors expressing small interfering RNA. *Proc Natl Acad Sci USA 100*, 1844–1848.
11. Carmell, M. A., Zhang, L., Conklin, D. S. et al. (2003). Germline transmission of RNAi in mice. *Nat Struct Biol 10*, 91–92.
12. Bartel, D. P. (2004). MicroRNAs: Genomics, biogenesis, mechanism, and function. *Cell 116*, 281–297.
13. Lee, Y., Ahn, C., Han, J. et al. (2003). The nuclear RNase III Drosha initiates microRNA processing. *Nature 425*, 415–419.
14. Han, J., Lee, Y., Yeom, K.-H. et al. (2004). The Drosha-DGCR8 complex in primary microRNA processing. *Genes Dev 18*, 3016–3027.
15. Gregory, R. I., Yan, K.-P., Amuthan, G. et al. (2004). The Microprocessor complex mediates the genesis of microRNAs. *Nature 432*, 235–240.
16. Landthaler, M., Yalcin, A., and Tuschl, T. (2004). The human DiGeorge syndrome critical region gene 8 and Its *D. melanogaster* homolog are required for miRNA biogenesis. *Curr Biol 14*, 2162–2167.
17. Denli, A. M., Tops, B. B. J., Plasterk, R. H. A. et al. (2004). Processing of primary microRNAs by the Microprocessor complex. *Nature 432*, 231–235.
18. Chendrimada, T. P., Gregory, R. I., Kumaraswamy, E. et al. (2005). TRBP recruits the Dicer complex to Ago2 for microRNA processing and gene silencing. *Nature 436*, 740–744.
19. Gregory, R. I., Chendrimada, T. P., Cooch, N. et al. (2005). Human RISC couples microRNA biogenesis and posttranscriptional gene silencing. *Cell 123*, 631–640.
20. Du, T. and Zamore, P. D. (2005). microPrimer: The biogenesis and function of microRNA. *Development 132*, 4645–4652.
21. Cullen, B. R. (2004). Transcription and processing of human microRNA precursors. *Mol Cell 16*, 861–865.
22. Kim, V. N. (2005). MicroRNA biogenesis: Coordinated cropping and dicing. *Nat Rev Mol Cell Biol 6*, 376–385.
23. Zeng, Y., Wagner, E. J., and Cullen, B. R. (2002). Both natural and designed micro RNAs can inhibit the expression of cognate mRNAs when expressed in human cells. *Mol Cell 9*, 1327–1333.

24. Zhou, H., Xia, X. G., and Xu, Z. (2005). An RNA polymerase II construct synthesizes short-hairpin RNA with a quantitative indicator and mediates highly efficient RNAi. *Nucleic Acids Res 33*, e62.

25. Dickins, R. A., Hemann, M. T., Zilfou, J. T. et al. (2005). Probing tumor phenotypes using stable and regulated synthetic microRNA precursors. *Nat Genet 37*, 1289–95. *Epub* October 2, 2005.

26. Silva, J. M., Li, M. Z., Chang, K. et al. (2005). Second-generation shRNA libraries covering the mouse and human genomes. *Nat Genet 37*, 1281–8. *Epub* October 2, 2005.

27. Rao, M. K., Pham, J., Imam, J. S. et al. (2006). Tissue-specific RNAi reveals that WT1 expression in nurse cells controls germ cell survival and spermatogenesis. *Genes Dev 20*, 147–152.

28. Xia, X. G., Zhou, H., Samper, E. et al. (2006). Pol II-expressed shRNA knocks down Sod2 gene expression and causes phenotypes of the gene knockout in mice. *PLoS Genet 2*, e10.

29. Kasim, V., Miyagishi, M., and Taira, K. (2004). Control of siRNA expression using the Cre-loxP recombination system. *Nucleic Acids Res 32*, e66.

30. Coumoul, X., Shukla, V., Li, C. et al. (2005). Conditional knockdown of Fgfr2 in mice using Cre-LoxP induced RNA interference. *Nucleic Acids Res 33*, e102.

31. Xia, X. G., Zhou, H., and Xu, Z. (2006). Transgenic RNAi: Accelerating and expanding reverse genetics in mammals. *Transgenic Res 15*, 271–275.

32. Paule, M. R. and White, R. J. (2000). Survey and summary transcription by RNA polymerases I and III. *Nucleic Acids Res 28*, 1283–1298.

33. Tiscornia, G., Tergaonkar, V., Galimi, F. et al. (2004). From The Cover: CRE recombinase-inducible RNA interference mediated by lentiviral vectors. *Proc Natl Acad Sci USA 101*, 7347–7351.

34. Ventura, A., Meissner, A., Dillon, C. P. et al. (2004). Cre-lox-regulated conditional RNA interference from transgenes. *Proc Natl Acad Sci USA 101*, 10380–10385.

35. Zhou, H., Falkenburger, B. H., Schulz, J. B. et al. (2007). Silencing of the Pink1 gene expression by conditional RNAi does not induce dopaminergic neuron death in mice. *Int J Biol Sci 3*, 242–250.

36. Lee, G. and Saito, I. (1998). Role of nucleotide sequences of loxP spacer region in Cre-mediated recombination. *Gene 216*, 55–65.

37. van de Wetering, M., Oving, I., Muncan, V. et al. (2003). Specific inhibition of gene expression using a stably integrated, inducible small-interfering-RNA vector. *EMBO Rep 4*, 609–615.

38. Kuninger, D., Stauffer, D., Eftekhari, S. et al. (2004). Gene disruption by regulated short interfering RNA expression, using a two-adenovirus system. *Hum Gene Ther 15*, 1287–1292.

39. Matsukura, S., Jones, P. A., and Takai, D. (2003). Establishment of conditional vectors for hairpin siRNA knockdowns. *Nucleic Acids Res 31*, e77.

40. Lin, X., Yang, J., Chen, J. et al. (2004). Development of a tightly regulated U6 promoter for shRNA expression. *FEBS Lett 577*, 376–380.

41. Chen, Y., Stamatoyannopoulos, G., and Song, C. Z. (2003). Down-regulation of CXCR4 by inducible small interfering RNA inhibits breast cancer cell invasion in vitro. *Cancer Res 63*, 4801–4804.

42. Gupta, S., Schoer, R. A., Egan, J. E. et al. (2004). From the Cover: Inducible, reversible, and stable RNA interference in mammalian cells. *Proc Natl Acad Sci USA 101*, 1927–1932.

43. Seibler, J., Kleinridders, A., Kuter-Luks, B. et al. (2007). Reversible gene knockdown in mice using a tight, inducible shRNA expression system. *Nucleic Acids Res 35*, e54.

44. Lewandoski, M. (2001). Conditional control of gene expression in the mouse. *Nat Rev Genet 2*, 743–755.

45. Ristevski, S. (2005). Making better transgenic models: conditional, temporal, and spatial approaches. *Mol Biotechnol 29,* 153–163.

46. Xia, X., Zhou, H., Huang, Y. et al. (2006). Allele-specific RNAi selectively silences mutant SOD1 and achieves significant therapeutic benefit in vivo. *Neurobiol Dis 23,* 578–586.

47. Peng, S., York, J. P., and Zhang, P. (2006). A transgenic approach for RNA interference-based genetic screening in mice. *Proc Natl Acad Sci USA 103,* 2252–2256.

48. Lois, C., Hong, E. J., Pease, S. et al. (2002). Germline transmission and tissue-specific expression of transgenes delivered by lentiviral vectors. *Science 295,* 868–872.

49. van den Brandt, J., Wang, D., Kwon, S.-H. , Heinkelein, M., and Reichardt, H.M. (2004). Lentivirally generated eGFP-transgenic rats allow efficient cell tracking in vivo. *Genesis 39,* 94–99.

50. Chen, Z., Stockton, J., Mathis, D. et al. (2006). Modeling CTLA4-linked autoimmunity with RNA interference in mice. *Proc Natl Acad Sci USA 103,* 16400–16405.

51. Hou, J., Shan, Q., Wang, T. et al. (2007). Transgenic RNAi depletion of claudin-16 and the renal handling of magnesium. *J Biol Chem 282,* 17114–17122.

52. Kunath, T., Gish, G., Lickert, H. et al. (2003). Transgenic RNA interference in ES cell-derived embryos recapitulates a genetic null phenotype. *Nat Biotechnol 21,* 559–561.

53. Lickert, H., Cox, B., Wehrle, C. et al. (2005). Dissecting Wnt/{beta}-catenin signaling during gastrulation using RNA interference in mouse embryos. *Development 132,* 2599–2609.

54. Lickert, H., Takeuchi, J. K., Von Both, I. et al. (2004). Baf60c is essential for function of BAF chromatin remodeling complexes in heart development. *Nature 432,* 107–112.

55. Saito, Y., Yokota, T., Mitani, T. et al. (2005). Transgenic small interfering RNA halts amyotrophic lateral sclerosis in a mouse model. *J Biol Chem 280,* 42826–42830.

56. Seibler, J., Kuter-Luks, B., Kern, H. et al. (2005). Single copy shRNA configuration for ubiquitous gene knockdown in mice. *Nucleic Acids Res 33,* e67.

57. Oberdoerffer, P., Kanellopoulou, C., Heissmeyer, V. et al. (2005). Efficiency of RNA interference in the mouse hematopoietic system varies between cell types and developmental stages. *Mol Cell Biol 25,* 3896–3905.

58. Yu, J. and McMahon, A. P. (2006). Reproducible and inducible knockdown of gene expression in mice. *Genesis 44,* 252–261.

59. Chung, K. H., Hart, C. C., Al-Bassam, S. et al. (2006). Polycistronic RNA polymerase II expression vectors for RNA interference based on BIC/miR-155. *Nucleic Acids Res 34,* e53.

60. Xia, X. G., Zhou, H., and Xu, Z. (2006). Multiple shRNAs expressed by an inducible pol II promoter can knock down the expression of multiple target genes. *Biotechniques 41,* 64–68.

61. Sun, D., Melegari, M., Sridhar, S. et al. (2006). Multi-miRNA hairpin method that improves gene knockdown efficiency and provides linked multi-gene knockdown. *Biotechniques 41,* 59–63.

62. Golding, M. C., Long, C. R., Carmell, M. A. et al. (2006). Suppression of prion protein in livestock by RNA interference. *Proc Natl Acad Sci USA 103,* 5285–5290.

63. Mabbott, N. A. and MacPherson, G. G. (2006). Prions and their lethal journey to the brain. *Nat Rev Microbiol 4,* 201–211.

64. Taurog, J. D., Hammer, R. E., Maika, S. D. et al. (1990). HLA-B27 transgenic mice as potential models of human disease. In *Transgenic Mice and Mutants in MHC Research* (I. K. Egorov and C. S. David, Eds.), 268–275. Springer-Verlag, Berlin.

65. Hammer, R. E., Maika, S. D., Richardson, J. A. et al. (1990). Spontaneous inflammatory disease in transgenic rats expressing HLA-B27 and human beta 2m: an animal model of HLA-B27-associated human disorders. *Cell 63,* 1099–1112.

66. Taurog, J. D. and Hammer, R. E. (1996). Experimental spondyloarthropathy in HLA-B27 transgenic rats. *Clin Rheumatol 15 Suppl 1*, 22–27.
67. Mullins, J. J., Peters, J., and Ganten, D. (1990). Fulminant hypertension in transgenic rats harbouring the mouse Ren-2 gene. *Nature 344*, 541–544.
68. Keppler, O. T., Welte, F. J., Ngo, T. A. et al. (2002). Progress toward a human CD4/CCR5 transgenic rat model for de novo infection by human immunodeficiency virus type 1. *J Exp Med 195*, 719–736.

# 18 Symphony of AIDS
## A microRNA-Based Therapy

*Yoichi R. Fujii*

## CONTENTS

> Our remedies oft in ourselves do lie, which we ascribe to heaven. The fated sky gives
> us free scope; only doth backward pull our slow designs when we ourselves are dull.
>
> **William Shakespeare, *All's Well That Ends Well***

## OVERVIEW

MicroRNA (miRNA) is a small and single-stranded RNA. miRNA inhibits expression of the target protein. miRNA precursor from the stem-loop structure is diced and can be considered a double-stranded RNA (dsRNA), which seems to closely resemble some short interfering RNAs (siRNAs). miRNA dominantly blocks protein translation based on its complementarity to the 3′ untranslated region (UTR) of mRNA, whereas in some cases, miRNA triggers the digestion of the mRNA transcript through RNA silencing machinery similar to siRNAs. Therefore, viral miRNAs can control gene expression of RNA viruses without causing the viral genomic RNA to be degraded by the latent infection. Although from in vivo prophylactic and therapeutic experiments with adult animals and from a phase I trial in humans, a small amount of trigger dsRNAs is sufficient to inactivate target viral RNAs without toxicity; the signal for miRNAs might imply spreading out to untreated cells and tissues as RNA waves, and then the information may be inherited by subsequent generations. To interfere with the viral replication cycle by suppression of expression of the latent viral genome, miRNA can be a workhorse

333

for development of nucleic acid drugs (NAD) and probably for an oral vaccine for HIV-1 infection.

## 18.1   INTRODUCTION

Previously, the central dogma of molecular biology was a one-way ticket from replication to transcription and translation. The pathway starts from the transcription of DNA genetic codes, which contain only 1–2% of information of genomic sequences, and subsequently, proteins are translated from mRNA of complementary DNA codes, then the translated protein can regulate cellular functions. The outside process of this dogma that is discovered as RNA interference (RNAi) is cosuppression of gene expression by small RNAs from ~95% of the junk DNA codes of the human genome. The junk contains noncoding DNA. RNA retroviruses and transgenes present in the noncoding region of human genome could have acted as transposon-like genes, designated as *retrotransposon*. From stem-looped RNA of these noncoding genes including the retrotransposon, biogenesis of miRNA, single-stranded RNA, 18- to 25-nt long is involved, and miRNA is responsible for posttranscriptional gene silencing (PTGS) and transcriptional gene silencing (TGS). The transcription, splicing, and maturation of mRNA can be aided by an intronic miRNA derived from inserting transposons into an intron. Further, some inherited diseases are caused by the dysfunction of the miRNA gene. Therefore, information of miRNAs may be inherited. The PTGS by miRNAs from transposon may not be an antiviral system that can completely eradicate virus genes from eukaryotic organisms. Here, I hypothesize that not only can RNA silencing with miRNAs control the expression of transposable elements but also the miRNA gene may be necessary to incorporate the transposable elements—a "nonselfish" gene—into the mammalian genome as a "selfish" gene via evolution of its own genome (Figure 18.1). The RNA silencing process has been presented as the oldest and most ubiquitous silencing system.

## 18.2   THE RNA WAVES HYPOTHESIS

PTGS against tobacco ringspot viruses in tobacco plants is known [1]. It is initiated by the production of small single-stranded RNAs. miRNA can induce PTGS via translational silencing as well as TGS, including CpG methylation and methylation of histone. Viroid RNAs, which form double-stranded and rod-like RNA structures, would be an ideal substrate for the Dicer. As well as for HIV-1, precedents for mobile RNAs include viroids, that is, small non-protein-coding RNA pathogens, and miRNAs from the transcript of RNA viruses target a homeobox protein mRNA [2]. The origin of miRNAs from RNA viruses was found in the viroid case. To induce the production of miRNA, Dicer ribonuclease processes dsRNA to approximately 21–23-nt duplexes from an endogenous long dsRNA, which is derived from hairpin-like RNA.

The stem-loop-like construct (pri-miRNA) from the DNA of the host genome induces approximately 70-nt pre-miRNA of fold-back dsRNA precursor by Drosha ribonuclease III with DiGeorge syndrome critical region 8 (DGCR8), and

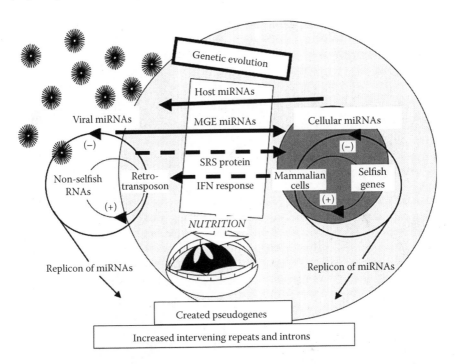

Genetic evolution

Host miRNAs

Viral miRNAs    MGE miRNAs    Cellular miRNAs

(−)    SRS protein    (−)

Non-selfish    Retro-    IFN response    Mammalian    Selfish
RNAs    transposon    cells    genes

(+)    (+)

*NUTRITION*

Replicon of miRNAs    Replicon of miRNAs

Created pseudogenes

Increased intervening repeats and introns

**FIGURE 18.1** Model of genetic evolution based on miRNAs as the MGE: One of the most fascinating questions in molecular biology is the evolutionary origin of intervening repeats [75]. The panels show that mammalian genome containing selfish genes [76] may be increased by intervening repeats and introns on its genome size by miRNAs from extracellular transposon as nonselfish genes via oral ingesting of nutrition-containing miRNAs included in plant, animal, and/or viral transposons. The pseudogenes may also be created by miRNAs. Mammalian genomes would encode several hundred miRNAs. Further, virally encoded miRNAs have been discovered. Vaccinia and influenza viral proteins E3L and NS1 suppress RNA interference machinery, called suppression of the RNA interference system (SRS). The longer RNAs are induced to produce nonspecific IFN into mammalian cells. Although one possible pathway upon miRNA replication in mammalian cells contains amplification with pol II, RT of human endogenous retrotransposon is also a candidate of a cellular RdRP [77–79]. The miRNA genes as the MGE contained in the inserted intervening sequences in food may be ingested and amplified by RNA wave, followed by miRNA-induced RNA silencing. Finally, they could be converted into DNA by reverse transcription with RTs and then integrated into nonhomologous sequences within selfish genes. The DNA codes from the converted miRNAs and miRNA itself may be inherited in this model. Evolution of social behavior [80] may finally be involved in the RNA waves hypothesis [81].

then pre-miRNA is exported from the nucleus to the cytoplasm by Exportin 5. By another ribonuclease III Dicer, pre-miRNA is processed from pre-miRNA to the asymmetric miRNA strand. Both siRNAs and miRNAs initiate degradation by a ribonuclease-containing RNA-induced silencing complex (RISC) [3]. Argonaute 2 (AGO2), TAR-binding protein (TRBP), and a protein activator of PKR (PACT) as components of the RISC execute target mRNA cleavage. In some cases, RNA-directed RNA polymerase (RdRP) synthesizes dsRNA from single-stranded RNA (ssRNA) templates to initiate or amplify the RNA silencing effects of *Caenorhabditis elegans*, *Arabidopsis thaliana*, *Neurospora crassa* [4,5], and RdRP activities of endogenous RT and HIV-1 RT have been proposed as an RdRP candidate of mammalian cells to amplify miRNA [6]. The RdRP activity could provide amplification of miRNA similar to a PCR, on various routes in a primer-independent manner. Although miRNA dominantly blocks the protein translation based on its complementarity to the 3′ UTR of mRNA, in some cases miRNA triggers the digestion of the mRNA transcript through RNAi machinery similar to siRNAs. However, HIV-1 Tat protein recruits TRBP and disturbs construction of RISC [7]; finally, it can block the digestion of HIV-1 RNA. Therefore, viral miRNAs can control gene expression of RNA viruses without causing the viral genomic RNA to be degraded that may be related to the latent infection. Further, sometimes miRNA might serve as the primer and add two or three more nucleotides at the 5′ end, which may correspond to 3′ overhanged nt because the siRNA-primed RdRP reaction converts target mRNA into dsRNA [8]. The second miRNA would dice and induce the second PTGS. The PCR of RdRP with miRNA ([n] miRNA) and the target mRNA is repeated. The RNA PCR wave was transmitted and simultaneously the PTGS wave may be shifted to the 5′ upstream and finally could completely inhibit the initiation of translation (Figure 18.2), which could show effectiveness similar to HIV-1 *nef*-specific long hairpin RNAs [9]. Our study showed that several 21–25-nt bands were detected in the region of nef/3′ LTR on the enhanced Northern blotting system [10] with an isolated clone of miRNA, *miR-N367*, suggesting that PTGS wave possibly occurred and the duration time of the reaction may have been short, and repeat associated siRNA (ra-siRNA)-like bands may be produced as miRNA-amplified nt, which have been observed in the afore-mentioned HIV-1 transcripts on Northern blotting.

Mammalian genomes encode more than 300 miRNAs and full-length transcripts that form stem structures [11]. Expression of human genomic miRNAs may downregulate the expression of HIV-1 genes [12], and the liver-specific miRNA 122 (miR-122) may facilitate replication of the hepatitis C virus [13]. Further, viral miRNAs are encoded in Epstein–Barr virus (EBV) [14], the Kaposi sarcoma-associated virus (KSHV), the mouse gammaherpesvirus (MHV), the human cytomegalovirus (HCMV) [15], and human immunodeficiency virus type 1 (HIV-1) [10,16]. miRNAs from the herpesvirus family have no notable sequence similarity with known host cell mRNAs; however, a new computational method has revealed that the sequence homology of miRNA for *nef* gene of HIV-1 directs to 2′–5′ oligoadenylate synthetase, major histocompatibility complex class II and IkB kinase-b (IKK-beta) mRNAs [16], suggesting that HIV-1 may have counterdefense strategies of *cis*-action using RNA silencing machinery to thwart the host immunodefense system. In viroid infection, pathogen-derived si/miRNAs could target endogenous genes [17]. It has been reported that vaccinia and influenza viral proteins E3L and NS1, named *SRS proteins*, which are both essential for inhibiting the activation of PKR with their RNA-binding motifs [18], may suppress RNA silencing machinery in mammalian cells [2]. HIV-1 also has the evading protein Tat of *trans*-action, which inhibits the RNA-binding ability of RISC and finally blocks host RNA silencing machinery [7]. PACT may be compensatory to be involved

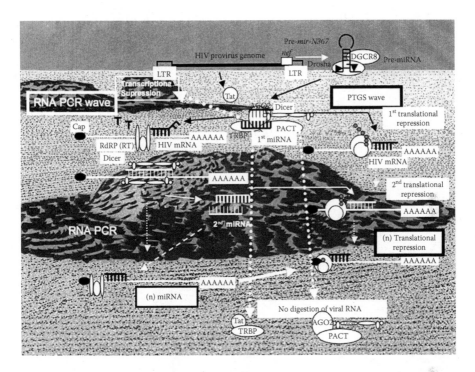

**FIGURE 18.2** The RNA waves hypothesis: The panels represent RdRP (RT)-dependent synthetic pathways leading to generation and amplification of dsRNA and the subsequent rendering of PTGS wave. The fold-back pre-miRNA, pre-*mir-N367* from HIV-1 *nef*/LTR region of provirus DNA is processed by Drosha and then Dicer. The *mir-N367* as a trigger dsRNA of PCR is amplified by RT and the cycle is repeated with new miRNA to (n) miRNA extending to 3′ Cap of HIV-1 RNA as a guide way. Finally, miRNA may inhibit Cap formation of viral mRNAs. During this cycle of dsRNA generation, Tat protein binds to TRBP as a part of RISC and locks the classical RNA silencing degradative pathway. The first, second, …, or (n) translational repression wave is initiated by the first, the second, … continues to (n) miRNA, respectively (as one might see symbolized in the stone garden at the famous Ryuan temple in Japan).

in miRNA accumulation. These instances may show an intriguing possibility that the movement of miRNA as a nonselfish gene plays some role in the regulation of endogenously selfish genes (Figure 18.1) under the RNA waves hypothesis. Thus, it is deduced from external evidence that the miRNA gene may be a mobile genetic element (MGE) through biological inheritance.

## 18.3 USAGE OF FUNCTIONAL miRNAs FOR HIV-1 THERAPY

### 18.3.1 Application of miRNAs for Oral/Edible Vaccine

siRNA has been delivered by liposome and nanoparticle-based deliveries via nostril, trachea, muscle, vagina, vein, and peritoneum (Table 18.1). siRNAs are now synthesized at low cost in industry compared to protein or antibody production.

**TABLE 18.1**
**Summary of Preclinical Experiments for Infectious Diseases on RNA Silencing Effects In Vivo**

| Viruses | Target Genes | Delivery Methods | Animals and Insects | Results | Toxicity | References |
|---|---|---|---|---|---|---|
| HSV-2 | UL27, UL29 | siRNA, liposome, vaginally | Mouse | Effective, $p < 0.001$ (prophylactic); $p < 0.04$ (therapeutic) | No | 61 |
| FMDV | 1D, 3D | shRNA, adenovirus vector, intramuscularly | Guinea pig and swine | Effective, $p < 0.012$ (prophylactic) | No | 62 |
| EBOV | L | siRNA, liposome, i.p. | Guinea pig | Effective, $p < 0.020$ (therapeutic) | No | 63 |
| DENV-2 | Mnp | shRNA, plasmid, intrathoracically | Mosquito | Effective (prophylactic) | No | 64 |
| JEV | C, M, NS3 | shRNA, plasmid, i.p. | Mouse | Effective, $p < 0.05$ (codelivery) | No | 65 |
| SARS-CoV | Spike, Orfb1 | siRNA, intratracheally | Mouse and rhesus macaque | Effective, $p < 0.05$ (prophylactic); $p < 0.05$ (codelivery); $p < 0.05$ (therapeutic) | No | 66 |
| RSV, PIV | P | Synthesized siRNA, i.n. | Mouse | Effective, $p < 0.002$ (prophylactic) | No | 67 |
| RSV | NS1 | shRNA, plasmid with a nanochitosan, i.n. | Mouse | Effective, $p < 0.05$ (prophylactic); $p < 0.05$ (therapeutic) | IFN production (2-fold) | 68 |
| HIV-1 | nef | miRNA, foamy virus vector, i.v. | Mouse | Effective, $p < 0.003$ (prophylactic) | No | 10 |

| Virus | Target | Delivery | Model | Effect | | Ref. |
|---|---|---|---|---|---|---|
| Flu | PA, NP, PB1 | Synthesized siRNA, i.v.; Lentivector, i.n. | Mouse | Effective, p < 0.0003 (prophylactic; i.v.); p < 0.004 (therapeutic) (i.n.) | No | 69 |
| Flu | PA, NP | Synthesized siRNA, HT, i.v., i.n. | Mouse | Effective, p < 0.020 (prophylactic; i.v.); p < 0.0001 (codelivery) (i.n.) | No | 70 |
| HBV | S | siRNA, HT, i.v. | Mouse | Effective, (codelivery) | No | 71 |
| HBV | S, Core | shRNA, plasmid, HT, i.v. | Mouse | Effective, p < 0.0001 (therapeutic) | No | 72 |
| HBV | S, Core | siRNA, HT, i.v. | Mouse | Effective, (codelivery) | No | 73 |
| HCV | NS5B | Synthetic siRNA, plasmid, HT, i.v. | Mouse | Effective, p < 0.0115 (codelivery) | No | 74 |

*Note:* HSV-2, Herpes simplex virus type 2; JEV, Japanese encephalitis virus; SARS-CoV, respiratory syndrome associated coronavirus; RSV, respiratory syncytial virus; PIV, parainfluenza virus; HIV-1, human immunodeficiency virus type 1; Flu, influenza virus; HBV, hepatitis B virus; HCV, hepatitis C virus; FMDV, foot-and-mouth disease virus; DENV, dengue virus; EBOV, ebola virus; i.v., intravenously; i.p., intraperitonial; i.n., intranasally; HT, hydrodynamic transfection; Shaded row, miRNA-based therapy.

siRNAs, therefore, focus on pharmaceutical properties and can be delivered to a wide range of organs such as the brain. However, their instability in blood is one problem, and another is specifically targeting the organ, both challenging to the development of RNAi agents. The former may be improved with the modification of the backbone and nucleotides of RNA, such as locked nucleic acid (LNA) [19] and boranophosphate-modified siRNAs [20]. The latter problem may be alleviated with specifically targeting particles. The cholesterol-conjugated siRNA can target the low-density lipoprotein (LDL) receptor to cholesterol on liver cells and decrease the ApoB protein synthesis in vivo [21]. Action against virally infectious diseases was taken for about 12 viruses (see Table 18.1) and significant results in vivo against both DNA and RNA viruses were obtained. Although several approaches of delivery of si/miRNAs have been used, a recent report has shown that oral administration of *Escherichia coli*-encoded short hairpin RNA (shRNA) induces target gene silencing in the intestinal epithelium in mice [22]. In *C. elegans*, ingested *E. coli* expressing dsRNAs can confer specific RNAi [23]. These results of oral administration have important implications for HIV-1 therapy with RNA silencing system because simian immunodeficiency virus (SIV), of which RNA was detected in feces [24] and HIV-1 is also transmitted by sex [25], indicating that lentiviruses can be blocked on epithelial cells in situ.

Oral NADs could be clinically feasible for systemic application and effective against HIV infection in intestine in situ because of the protection they give against infection of mucosal cells [26]. The incorporation of ingested phage DNA via intestinal mucosa further transports to spleen and liver of mice [27]. Therefore, feeding transgenic fruits expressing sequence-specific NADs targeting viral genes can protect against the invasion of real pathogens. The concept of oral NADs is shown in Figure 18.3. We have reported that the model plant, *A. thaliana,* may be able to produce miRNAs similar to those produced by mammalian cells [6]. Off-target effects by siRNAs occurred at the protein synthesis level; however, the recombinant plant could filter the danger of severe off-target effects because of the absence of callus growth in plants. Further, the application of these kinds of plant-based systems could avoid many potential issues such as poor resources and contamination. Though there is no effective AIDS vaccine, considerable evidence suggests that development of NADs technology may be able to overcome the present difficulties [28]. Analogous to micRNA in the bacterial system, miRNA oral agents could also avoid opportunistic and infectious pathogens during AIDS pathogenesis (Figure 18.3). Creating a banana miRNA vaccine might benefit prophylaxis and therapy of AIDS, but biological efficacy data are needed to test this concept.

The NADs treatment may have potential to be extended beyond mice to humans. Several clinical trials including siRNAs targeting VEGFR [29] that used shRNA carried on lentiviral vector for ex vivo treatment of AIDS are also encouraging because these trials showed considerable success without side effects (see Chapter 20) [30]. Thus, these novel technologies using NADs may some day resolve the AIDS pandemic [31]. Simultaneously, there is the intriguing possibility that transfer of RNA between food and species may occur in natural interaction, which may bring about the evolution of species following the Darwinism model (Figure 18.1) [32].

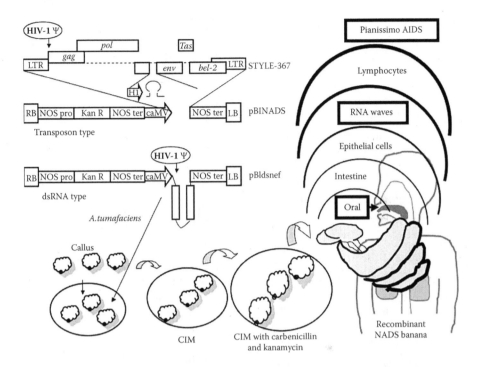

**FIGURE 18.3**  Pianissimo AIDS with NADs banana: The concept of our research lay upon transgenic fruits expressing sequence-specific miRNAs targeting HIV-1 *nef* genes. The hairpin loop of HIV-1 SF2 *nef* fragments and the STYLE-367 full length DNA into the pBI121 plasmids were made as pBIdsnef and pBINADS. The plasmids contain the HIV-1 ψ structure in upstream of the inserted *nef* fragment and *gag* region of the STYLE-367. The constructed plasmids were used for transformation of *Agrobacter tumafaciens* callus of banana cells by incubation with the transformed *A. tumafaciens* cells. The transformed aggregates were selected with carbenicillin and kanamycin in CIM medium. The grown recombinant NADs banana could be delivered as nutrition via oral. The dsRNA-type NADs expressed miRNA in the banana; therefore, ingested miRNA could directly affect epithelial macrophages and T lymphocytes, and then PTGS wave may be transmitted to PBMCs in blood and lymphoid tissues for silencing the HIV-1 expression. In the case of transposon-type NADs, oral ingested miRNA plasmid would be transposed to epithelial cells in the intestine, and then expressed *mir-N367* induced PTGS of HIV-1 genes in situ. The plasmid was integrated into the genomic DNA of epithelial cells and it could permanently express mir-N367. Both plasmids can act as pseudo-HIV-1 genes because of bearing HIV-1 ψ. Occasionally, readily infected HIV-1 capsid would misincorporate the miRNA plasmids. Even if secondary infection of HIV-1 happened, the miRNA expressing pseudo-HIV-1 particles would decrease the infectivity of virulent HIV-1.

## 18.3.2  AN OVERVIEW OF USAGE OF STYLE

The aim of an AIDS intervention project is to suppress the HIV-1 transposon expression rendered in both *cis*- and *trans*-action of HIV-1 genes corresponding to the new dogma of gene regulation. Further, it has been assumed that the dynamics of cell infection with HIV-1 and HIV-1 virion production are represented by

$$dT^*/dt = kVT - \delta T^* \tag{18.1}$$

and

$$dV/dt = N\delta T^* - cV \tag{18.2}$$

where

$T$ = target cells

$T^*$ = infected cells

$k$ = an infectious rate constant

$V$ = the concentration of viral particles in plasma

$\delta$ = the rate of loss of viral-producing cells

$N$ = the number of new virions produced per HIV-1-infected cell during its lifetime

$c$ = the rate constant for virion clearance [33]

Ritonavir does not affect the survival or rate of virion production from HIV-1-infected cells, but the repression of expression of HIV-1 genes with miRNAs $V$ and $N$ can go down to zero, whereupon treatment with si/miRNAs, from Equations 18.1 and 18.2, $dT^*/dt = -\delta T^*$. By allowing the outcome of $T^*$ to increase but the dynamics of cell infection to become negative, HIV-1-infected cells can be decreased.

Although HIV-1 *nef* can be expressed as the mRNA and the protein, the *cis*-acting element of the defective *nef* mRNA might maintain persistent infection and the *trans*-acting soluble Nef protein might induce memory T cell loss [34,35], severe fatty acid loss in T cells [36], weight loss [10], class switch suppression in B-cells [37], and the complex severe pulmonary vascular lesions [38] via Nef receptor on the cell surface [35]. As a target gene of HIV-1 therapy, the HIV-1 *nef* may be suitable for RNA silencing machinery because the *nef* has immense and significant influence on the development of clinical symptoms of AIDS as mentioned earlier.

Next, we need a vector for the delivery of miRNA, whose vector can Tat-independently be induced and work. Further, the vector is required for its characterization of a broad range of hosts, large packaging capacity, stable virions that can be concentrated, prolonged survival of the preintegration complex in quiescent cells, effectiveness for transduced hematopoietic stem cells, little influence on integration site selection, and no pathogenicity in their infectious host, especially no association with the development of malignancies. Since we have previously shown that FFV Tas specifically can *trans*-activate the LTR of FFV but not that of HIV-1, the *trans*-activation is Tat-independent [10] and FFV can be a safe vector for HIV-1 therapy. Therefore, we used FFV for the construction of a vector. To construct the FFV-based vectors for delivery of miRNAs, the *env* gene portion of pSKY3.0 was replaced with the shRNA expression cassette, as described earlier, under the control of the H1 promoter to produce STYLE, which was then transfected into the FFV envelope-expressing packaging cells, CRFKsugi. We observed that the STYLE could substantially inhibit not only the transcription of *nef* expression but also replication of the live vector. To the best of our knowledge and also in light of results described earlier, the concrete piece of evidence to show that significant inhibition in the expression of a target gene in HIV-1 can be induced by using virally-mediated expression of miRNA in mammalian cells. To test the impact of our STYLE vaccine system for use against HIV infections in mice, we used nested PCR at day 14 to evaluate the degree of *nef* mRNA expression in the liver tissues and hematopoietic cells of some of our model animals in the three groups. We found that *nef* mRNA was decreasingly expressed significantly in the tissues and cells of the live-vector-infected animals. However, this was suppressed among the animals that were vaccinated in the groups. These results suggest that miRNA could strongly suppress both virion production (*N* in Equation 18.2) and survival rate of the virion (*V*). Since *miR-N367* could inhibit HIV-1 replication in human T cells in vitro, this was significant since it tacitly implies that not only miRNA itself but also the same vector-mediated delivery, such as lentivector [39], adeno-associated vector, as well as foamy vector [10,41], etc., could be used in prophylactic and therapeutic strategies for HIV-1 and possibly other viruses infecting human beings and animals. According to the author's prediction, HIV-1 packaging signal bearing STYLE would be misincorporated into the HIV-1 virion instead of its own genome. Actually, a plasmid containing HIV-1 ψ and *nef* stem-loop was effective in the prevention of HIV-1 proliferation in vitro [9]. In vitro, off-target effects, leading escape mutants, and unknown toxicity such as rendering interferon were reported [42,43]; however, in adult animals, both prophylactic and therapeutic experiments for virally infectious diseases evidenced that all reports show effective results and lack of potential risk for side effects (Table 18.1). Further, although the production of escape mutant by treatment with siRNA is a problematic issue, it can be overcome by the use of the long-hairpin RNAs of *nef*, which inhibited HIV-1 production without the emergence of viral escape mutants [9]. Multiple siRNAs [44] and miRNA-based approach [10,45] would also elucidate the aforementioned issues. Agendas of clinical trials with RNA silencing should be set aside for the sake of public health in resource-poor countries, and simultaneously, we have to promise present and future generations that we will never use RNA silencing technology as a biological weapon. All good technologies having promise come with merits and demerits, and the merits of RNA silencing technology are larger than life; we should take advantage of this revolutionary technology.

## 18.4   ROSETTA STONE: AN HIV-1 miRNAs CODE

First of all, single-stranded and small RNA would be there as the origin of genome, similar to miRNAs. As an internal template, the enzyme telomerase repeatedly recruits and copies part of the miRNA into genomic DNA to extend chromosome ends [46]. In *Drosophila melanogaster*, *HeT-A* and *TART*, which are two telomerase-specific retrotransposable elements, are present in multiple copies on normal telomeres. The *HeT-A* has *gag* open reading frame (Orf) and, 5′ and 3′ UTR. *TART* also codes additional *pol* gene. The Gag protein and sense-stranded RNA complex can be transported into the nucleus and the template RNA is delivered specifically to its target at the telomere. From analogy with the transposon of *D. melanogaster* to retroviruses, it is suggested that if HIV-1 *nef* small RNA (single and sense-stranded RNA) is presented, although HIV-1 codes *gag* and *nef* are located in the 3′ end of HIV-1 genome, similar to 3′ UTR, Gag-*nef* /LTR miRNA complex may be able to target the 5′ LTR and the small RNA could regulate HIV-1 transcription, further the process may be related with the latent infection. In addition, the Gag complex includes reverse transcriptase (RT), and the *nef* miRNA gene could be reverse-transcribed into DNA; therefore, host small RNA as a nonselfish gene may also present and might target the *nef* gene as a selfish DNA [47].

Previously, the mutated *nef* gene quasispecies have been reported to be present in the HIV-infected patient, and it has been hypothesized that this may be related to the long-term nonprogression (LTNP) in some HIV patients [48]. The truncated *nef* gene usually does not code Nef protein; therefore, if there are some mechanisms to introduce the attenuation of HIV-1 virulence, it may be ideal as a therapeutic intervention strategy. A single-stranded sense or antisense *nef* plasmid was not so effective; however, at least, it might be an unreasonable mixture of both sense and antisense *nef* plasmids workhorse. Therefore, we hypothesized that it may be due to RNAs from HIV-1's own gene. Previously, we have shown that the defective variants of *nef* dsRNA from the LTNPs containing the U3 region inhibited HIV-1 replication, which is not caused by interferon (IFN) response [28]. The observation might implicate that the *cis*-acting element of the U3 region interfere with HIV-1 replication. But in mammalian cells, since the longer RNAs induce to produce nonspecific IFN in mammalian cells, the RNA silencing investigation has never been allowed to meet the challenge of RNA silencing therapy as a gene-specific approach.

Later, several investigations were made that revealed 21-nt duplexes of siRNA induce an effective antiviral response in cultured cells [49–56]. siRNAs have been made artificially in the aforementioned experiments; the mechanisms of RNAi were that exogenous siRNAs, which are double stranded and perfectly complementary to the target gene, destroy the RNAs complementary to the targeted gene transcripts. But, until now, there has been no report about endogenous siRNA in mammalian cells. It is expected that the dominant *nef* RNAs of total HIV-1 transcripts may contain small naked RNAs. Supporting this argument, miRNAs were detected in Southern blotting with small DNAs reverse-transcribed from small RNAs by DNA probes at first, and then followed up by using microarray assays [36]. This confirmed the evidence, which was further validated by Northern blotting with RNA probes and RT-PCR from HIV-1 corresponding to *nef* gene transcripts [57]. siRNA from HIV-1 *env* gene has been reported [7,16]. In our system, miRNAs could make a *nef* miRNA complex, and the complex may enter the nucleus similar to the aforementioned *HeT-A* and *TART*, and then the *nef* miRNA may regulate the integrated LTR by TGS, including histone deacetylation/methylation and/or DNA methylation [58–60]. Thus, *nef* miRNA may target provirally genomic DNA and induce abrogation of *nef* sequences and/or DNA genome mutation, which may be followed to the latent state of infection as well as LTNP [6].

## 18.5 CONCLUSIONS

The miRNA gene may be an origin of the MGE. Considering its small-sized information, miRNA has an important role for expression of inherited exquisite phenotypes in cells and individuals. While miRNA information flow could work effectively, the amplification of miRNA is necessary for success. Given these assumptions, it is predicted that one small effect of RNA silencing could expand to an organ- and/ or a naturewide concern. It is RNA waves. Thus, designing natural miRNA genes may be suitable for and can contribute to the development of NADs, and probably an edible vaccine for viral infections to prevent viral proliferation through silencing of expression of the latent viral genes.

## ACKNOWLEDGMENTS

I profoundly thank Dr. N. K. Saksena, Head, Retroviral Genetics Division, Westmead Millennium Institute, Sydney, Australia, and Dr. E. A. Brisibe in Molecular Bio-Sciences Ltd., Calabar, Nigeria, for employing my "RNA waves" hypothesis to extend my design of miRNA vaccine against HIV-1 infection.

## REFERENCES

1. Wingard, S.A. (1928). Hosts and symptoms of ring spot, a virus disease of plants. *J Agric Res 37*, 127–153.
2. Ding, S.-W., Li, H., Lu, R., Li, F., and Li, W.-X (2004). RNA silencing: A conserved antiviral immunity of plants and animals. *Virus Res 102*, 109–115.
3. Hutvagner, G. (2005). Small RNA asymmetry in RNAi: Function in RISC assembly and gene regulation. *FEBS Lett 579*, 5850–5857.
4. Sijen, T., Fleenor, J., Simmer, F., Thijssen, K.L., Parrish, S., and Fire, A. (2001). On the role of RNA amplification in dsRNA-triggered gene silencing. *Cell 107*, 465–476.
5. Matzke, M.A. and Birchler, J.A. (2005). RNAi-mediated pathways in the nucleus. *Nat Rev 6*, 24–35.
6. Fujii, Y.R. (2006). Lost in translation: Regulation of HIV-1 by microRNAs and a key enzyme of RNA-directed RNA polymerase. In *microRNAs: From Basic Science to Disease Biology*, K. Appasani, Ed. (Cambridge UK: Cambridge University Press), pp. 424–440.
7. Bennasser, Y., Le, S.Y., Benkirane, M., and Jeang, K.T. (2005). Evidence that HIV-1 encodes an siRNA and a suppressor of RNA silencing. *Immunity 22*, 607–619.
8. Lipardi, C., Wei, Q., and Paterson, B.M. (2001). RNAi as random degradative PCR: siRNA primers convert mRNA into dsRNAs that are degraded to generate new siRNAs. *Cell 107*, 297–307.
9. Konstantinova, P., de Vries, W., Haasnoot, J., ter Brake, O., de Haan, P., and Berkhout, B. (2006). Inhibition of human immunodeficiency virus type 1 by RNA interference using long-hairpin RNA. *Gene Ther 18*, 1–11.
10. Omoto, S., Ito, M., Tsutsumi, Y., Ichikawa, Y., Okuyama, H., Brisibe, E. A., Saksena, N.K., and Fujii, Y.R. (2004). HIV-1 nef suppression by virally encoded microRNA. *Retrovirology 1*, 44.
11. John, B., Sander, C., and Marks, D.S. (2006). Prediction of human microRNA targets. In *MicroRNA Protocols*, S.-Y. Ying Ed. (Totowa, Humana Press), pp. 101–113.
12. Yeung, M.L., Bennasser, Y., Myers, T.G., Jiang, G., Benkirane, M., and Jeang, K.-T. (2006). Changes in microRNA expression profiles in HIV-1-transfected human cells. *Retrovirology 2*, 81.
13. Jopling, C.L., Yi, M., Lancaster, A.M., Lemon, S.M., and Sarnow, P. (2005). Modulation of hepatitis C virus RNA abundance by a liver-specific microRNA. *Science 309*, 1577–1581.

14. Pfeffer, S., Zavolan, M., Grasser, F.A., Chien, M., Russo, J.J., Ju, J., John, B., Enright, A.J., Marks, D., Sander, C., and Tuschl, T. (2004). Identification of virus-encoded microRNAs. *Science 304*, 734–736.

15. Pfeffer, S., Sewer, A., Lagos-Quintana, M., Sheridan, R., Sander, C., Grasser, F.A., van Dyk, L.F., Ho, C.K., Shuman, S., Chien, M., Russo, J., Ju, J., Randall, G., Lindenbach, B.D., Rice, C.M., Simon, V., Ho, D.D., Zavolan, M., and Tuschl, T. (2005). Identification of microRNAs of the herpesvirus family. *Nat Methods 2*, 269–276.

16. Bennasser, Y., Le, S.Y., Yeung, M.L., and Jeang, K.T. (2004). HIV-1 encoded candidate micro-RNAs and their cellular targets. *Retrovirology 1*, 43.

17. Wang, M.-B., Bian, X.Y., Wu, L.M., Smith, N.A., Isenegger, D., Wu, R.M., Masuta, C., Vance, V.B., Watson, J.M., Rezaian, A., Dennis, E.S., and Waterhouse, P.M. (2004). On the role of RNA silencing in the pathogenicity and evolution of viroids and viral satellites. *Proc Natl Acad Sci USA 101*, 3275–3280.

18. Garcia-Sastre, A. (2002). Mechanisms of inhibition of the host interferon alpha/beta-mediated antiviral responses by viruses. *Microbes Infect 4*, 647–655.

19. Braasch, D.A., Jensen, S., Liu, Y., Kaur, K., Arar, K., White, M.A., and Corey, D.R. (2003). RNA interference in mammalian cells by chemically-modified RNA. *Biochemistry 42*, 7967–7975.

20. Hall, A.H.S., Wan, J., Spesock, A., Sergueeva, Z., Shaw, B.R., and Alexander, K.A. (2006). High potency silencing by single-stranded boranophosphate siRNA. *Nucleic Acids Res 34*, 2773–2781.

21. Soutschek, J., Akinc, A., Bramlage, B., Charisse, K., Constien, R., Donoghue, M., Elbashir, S., Geick, A., Hadwiger, P., Harborth, J., John, M., Kesavan, V., Lavine, G., Pandey, R.K., Racie, T., Rajeev, K.G., Röhl, I., Toudjarska, I., Wang, G., Wuschko, S., Bumcrot, D., Koteliansky, V., Limmer, S., Manoharan, M., and Vornlocher, H.-P. (2004). Therapeutic silencing of an endogenous gene by systemic administration of modified siRNAs. *Nature 432*, 173–178.

22. Xiang, S., Fruehauf, J., and Li, C.J. (2006). Short hairpin RNA-expressing bacteria elicit RNA interference in mammals. *Nat Biotechnol 24*, 697–702.

23. Timmons, L. and Fire, A. (1998). Specific interference by ingested dsRNA. *Nature 395*, 854.

24. Worobey, M., Santiago, M.L., Keele, B.F., Ndjango, J.-B.N., Joy, J.B., Labama, B.L., Rambaut, A., Sharp, P.M., Shawand, G.M., Hahn, B.H. (2004). Origin of AIDS: Contaminated polio vaccine theory refuted. *Nature 428*, 820.

25. Schwärtlander, B., Pisani, E., and Mertens, T. (1998). The global epidemiology of HIV. In *Human Immunodeficiency Virus*, N.K. Saksena Ed. (Genoa, Italy, Medical Systems S.P.A.), pp. 35–56.

26. Aquaro, S., Calio, R., Balzarini, J., Bellocchi, M.C., Garaci, E., and Perno, C.F. (2002). Macrophages and HIV infection: Therapeutical approaches toward this strategic virus reservoir. *Antiviral Res 55*, 209–225.

27. Schubbert, R., Renz, D., Schmitz, B., and Doerfler, W. (1997). Foreign (M13) DNA ingested by mice reaches peripheral leukocytes, spleen, and liver via the intestinal wall mucosa and can be covalently linked to mouse DNA. *Proc Natl Acad Sci USA 94*, 961–966.

28. Yamamoto, T., Omoto, S., Mizuguchi, M., Mizukami, H., Okuyama, H., Okada, N., Saksena, N.K., Brisibe, E.A., Otake, K., and Fujii, Y.R. (2002). Double-stranded *nef* RNA interferes with human immunodeficiency virus type 1 replication. *Microbiol Immunol 46*, 809–817.

29. Whelan, J. (2005). First clinical data on RNAi. *Drug Discov Today 10*, 1014–1015.

30. Rossi, J.J. (2008). Mammalian transcriptional gene silencing in small RNAs, in *Regulation of Gene Expression by Small RNAs*, Taylor & Francis, Boca Raton, FL, 2009, ch. 20.

31. von Beethoven, L. (1824). Symphony no. 9 in D-minor Op., 125.

32. Ruse, M. (2003). *Darwin and Design*. (New York, Harvard University Press).

33. Perelson, A.S., Neumann, A.U., Markowitz, M., Leonard, J.M., and Ho, D.D. (1996). HIV-1 dynamics in vivo: Virion clearance rate, infected cell life-span, and viral generation time. *Science 271*, 1582–1585.

34. Fujii, Y., Otake, K., Tashiro, M., and Adachi, A. (1996). Soluble Nef antigen of HIV-1 is cytotoxic for human CD4+ T cells. *FEBS Lett 393*, 93–96.

35. Otake, K., Ohta, M., Minowada, J., Hatama, S., Takahashi, A., Ikemoto, A., Okuyama, H., and Fujii, Y.R. (2000). Extracellular Nef of HIV-1 can target CD4 memory T population. *AIDS 14*, 1662–1664.

36. Otake, K., Omoto, S., Yamamoto, T., Okuyama, H., Okada, H., Okada, N., Kawai, M., Saksena, N., and Fujii, Y.R. (2004). HIV-1 Nef in the nucleus influences adipogenesis as well as viral transcription through the peroxisome proliferators-activated receptors. *AIDS 18*, 189–198.

37. Qiao, X., He, B., Chiu, A., Knowles, D., Chadburn, A., and Cerutti, A. (2006). Human immunodeficiency virus 1 Nef suppresses CD40-dependent immunoglobulin class switching in bystander B cells. *Nat Immunol 7*, 302–310.

38. Marecki, J.C., Cool, C.D., Parr, J.E., Beckey, V.E., Luciw, P.A., Tarantal, A.F., Carville, A., Shannon, R.P., Cota-Gomez, A., Tuder, R.M., Voelkel, N.F., and Flores, S.C. (2006). HIV-1 Nef is associated with complex pulmonary vascular lesions in SHIV-*nef*-infected Macaques. *Am Respir Crit Care Med 174*, 437–445.

39. Morris, K.V. and Rossi, J.J. (2006). Lentivirus-mediated RNA interference therapy for human immunodeficiency virus type 1 infection. *Human Gene Ther 17*, 479–486.

40. Wang, Z., Zhu, T., Qiao, C., Zhou, L., Wang, B., Zhang, J., Chen, C., Li, J., and Xiao, X. (2005). Adeno-associated virus serotype 8 efficiently delivers genes to muscle and heart. *Nat Biotechnol 23*, 321–336.

41. Park, J., Nadeau, P., Zucali, J.R., Johnson, C.M., and Mergia, A. (2005). Inhibition of simian immunodeficiency virus by foamy virus vectors expressing siRNAs. *Virology 343*, 275–282.

42. Kim, D.-H., Longo, M., Han, Y., Lundberg, P., Cantin, E., and Rossi, J.J. (2004). Interferon induction by siRNAs and ssRNAs synthesized by phage polymerase. *Nat Biotechnol 22*, 321–325.

43. Snøve Jr., O. and Holen, T. (2004). Many commonly used siRNAs risk off-target activity. *Biochem Biophys Res Commun 319*, 256–263.

44. Ter Brake, O. and Berkhout, B. (2005). A novel approach for inhibition of HIV-1 by RNA interference: Counteracting viral escape with a second generation of siRNAs. *J RNAi Gene Silencing 1*, 56–65.

45. Yeung, M.L., Bennasser, Y., Myers, T.G., Jiang, G., Benkirane, M., and Jeang, K.-T. (2006). Changes in microRNA expression profiles in HIV-1-transfected human cells. *Retrovirology 2*, 81.

46. Pardue, M.-L., Rashkova, S., Casacuberta, E., DeBaryshe, P.G., George, J.A., and Traverse, K.L. (2005). Two retrotransposons maintain telomeres in *Drosophila*. *Chromosome Res 13*, 443–453.

47. Hariharan, M., Scaria, V., Pillai, B., and Brahmachari, S.K. (2005). Targets for human encoded microRNAs in HIV genes. *Biochem Biophys Res Commun 337*, 1214–1218.

48. Deacon, N.J., Tsykin, A., Solomon, A., Smith, K., Ludford-Menting, M., Hooker, D.J., Mcphee, D.A., Greenway, A.L., Ellett, A., Chatfield, C., Lawson, V.A., Crowe, S., Mearz, A., Sonza, S., Learmont, J., Sullivan, J.S., Cunningham, A., Dwyer, D., Dowton, D., and Mills, J. (1995). Genomic structure of an attenuated quasi species of HIV-1 from a blood transfusion donor and recipients. *Science 270*, 988–991.

49. Hu, W.Y., Myers, C.P., Kilzer, J.M., Pfaff, S.L., and Bushman, E.D. (2002). Inhibition of retroviral pathogenesis by RNA interference. *Curr Biol 12*, 1301–1311.

50. Jacque, J.M., Triques, K., and Stevenson, M. (2002). Modulation of HIV-1 replication by RNA interference. *Nature 418*, 435–438.
51. Paddison, P.J., Caudy, A.A., and Hannon, G.J. (2002). Stable suppression of gene expression by RNAi in mammalian cells. *Proc Natl Acad Sci USA 99*, 1443–1448.
52. Lee, N.S., Dohjima, T., Bauer, G., Li, H., Li, M.J., Ehsani, A., Salvaterra, P., and Rossi, J. (2002). Expression of small interfering RNAs targeted against HIV-1 rev transcripts in human cells. *Nat Biotech 20*, 500–505.
53. Coburn, G.A. and Cullen, B.R. (2002). Potent and specific inhibition of human immunodeficiency virus type 1 replication by RNA interference. *J Virol 76*, 9225–9231.
54. Novina, C.D., Murray, M.F., Dykxfoorn, D.M., Beresford, P.J., Riess, J., Lee, S.K., Collman, R.G., Lieberman, J., Shankar, P., and Sharp, P.A. (2002). siRNA-directed inhibition of HIV-1 infection. *Nat Med 8*, 681–686.
55. Xia, H., Mao, Q., Paulson, H.L., and Davidson, B.L. (2002). siRNA-mediated gene silencing in vitro and in vivo. *Nat Biotechnol 20*, 1006–1010.
56. Capodici, J., Kariko, K., and Weissman, D. (2002). Inhibition of HIV-1 infection by small interfering RNA-mediated RNA interference. *J Immunol 169*, 5196–5201.
57. Omoto, S. and Fujii, Y.R. (2005). Regulation of human immunodeficiency virus 1 transcription by *nef* microRNA. *J Gen Virol 86*, 751–755.
58. Morris, K.V., Chan, S.W., Jacobsen, S.E., and Looney, D.J. (2004). Small interfering RNA-induced transcriptional gene silencing in human cells. *Science 305*, 1289–1292.
59. Castanotto, D., Tomamasi, S., Li, M., Yanow, S., Pfeifer, G.P., and Rossi, J.J. (2005). Short hairpin RNA-directed cytosine (CpG) methylation of the RASSF1A gene promoter in HeLa cells. *Mol Therapy 12*, 179–183.
60. Suzuki, K., Shijuuku, T., Fukamachi, T., Zaunders, J., Guillemin, G., Cooper, D., and Kelleher, A. (2005). Prolonged transcriptional silencing and CpG methylation induced by siRNAs targeted to the HIV-1 promoter region. *J RNAi Gene Silencing 1*, 66–68.
61. Palliser, D., Chowdhury, D., Wang, Q.-Y., Lee, S.J., Bronson, R.T., Knipe, D.M., and Lieberman, J. (2006). An siRNA-based microbicide protects mice from lethal herpes simplex virus 2 infection. *Nature 439*, 89–94.
62. Chen, W., Liu, M., Jiao, Y., Yan, W., Wei, X., Chen, J., Fei, L., Liu, Y., Zuo, X., Yang, F., Lu, Y., and Zheng, Z. (2006). Adenovirus-mediated RNA interference against foot-and-mouth disease virus infection both in vitro and in vivo. *J Virol 80*, 3559–3566.
63. Geisbert, T.W., Hensley, L.E., Kagan, E., Yu, E.Z., Geisbert, J.B., Daddario-DiCaprio, K., Fritz, E.A., Jahrling, P.B., McClintock, K., Phelps, J.R., Lee, A.C.H., Jeffs, L.B., and MacLachlan, I. (2006). Postexposure protection of guinea pigs against a lethal Ebola virus challenge is conferred by RNA interference. *J Infec Disease 193*, 1650–1657.
64. Franz, A.W.E., Sanchez-Vargas, I., Adelman, Z.N., Blair, C.D., Beaty, B.J., James, A.A., and Olson, K.E. (2006). Engineering RNA interference-based resistance to dengue virus type 2 in genetically modified *Aedes aegypti*. *Proc Natl Acad Sci USA 103*, 4198–4203.
65. Murakami, M., Ota, T., Nukuzuma, S., and Takegami, T. (2005). Inhibitory effect of RNAi on Japanese encephalitis virus replication in vitro and in vivo. *Microbiol Immunol 49*, 1047–1056.
66. Li, B.-J., Tang, Q., Cheng, D., Qin, C., Xie, F.Y., Wei, Q., Xu, J., Liu, Y., Zheng, B.-J., Woodle, M.C., Zhong, N., and Lu, P.Y. (2005). Using siRNA in prophylactic and therapeutic regimens against SARS coronavirus in Rhesus macaque. *Nat Med 9*, 944–951.
67. Bitko, V., Musiyenko, A., Shulyayeva, O., and Barik, S. (2005). Inhibition of respiratory viruses by nasally administered siRNA. *Nat Med 11*, 50–55.
68. Zhang, W., Yang, H., Kong, X., Mohapatra, S., Juan-Vergara, H.S., Hellemann, G., Behera, S., Singam, R., Lockey, R.F., and Mohapatra, S. (2005). Inhibition of respiratory syncytial virus infection with intranasal siRNA nanoparticles targeting the viral NS1 gene. *Nat Med 11*, 56–62.

69. Ge, Q., Filip, L., Bai, A., Nguyen, T., Eisen, H.N., and Chen, J. (2004). Inhibition of influenza virus production in virus-infected mice by RNA interference. *Proc Natl Acad Sci USA 101*, 8676–8681.

70. Tompkins, S.M., Lo, C.-Y., Tumpey, T.M., and Epstein, S.L. (2004). Protection against lethal influenza virus challenge by RNA interference in vivo. *Proc Natl Acad Sci USA 101*, 8682–8686.

71. Giladi, H., Ketzinel-Gilad, M., Rivkin, L., Felig, Y., Nussbaum, O., and Galun, E. (2003). Small interfering RNA inhibits hepatitis B virus replication in mice. *Mol Therapy 8*, 769–776.

72. McCaffrey, A.P., Nakai, H., Pandey, K., Huang, Z., Salazar, F.H., Xu, H., Wieland, S.F., Marion, P.L., and Kay, M.A. (2003). Inhibition of hepatitis B virus in mice by RNA interference. *Nat Biotechnol 21*, 639–644.

73. Klein, C., Bock, C.T., Wedemeyer, H., Wustefeld, T., Locarnini, S., Dienes, H.P., Kubicka, S., Manns, M.P., and Trautwein, C. (2003). Inhibition of hepatitis B virus replication in vivo by nucleoside analogues and siRNA. *Gastroenterology 125*, 9–18.

74. McCaffrey, A.P., Meuse, L., Pham, T.-T., Conklin, D.S., Hannon, G.J., and Kay, M.A. (2002). RNA interference in adult mice. *Nature 418*, 38, 39.

75. Sharp, P.A. (1985). On the origin of RNA splicing and introns. *Cell 42*, 397–400.

76. Dawkins, R. (1976). *The Selfish Gene* (New York, Oxford University Press).

77. Martienssen, R.A., (2003). Maintenance of heterochromatin by RNA interference of tandem repeats. *Nat Gene 35*, 213–214.

78. Nishikawa, K. (2001). A short primer on RNAi: RNA-directed RNA polymerase acts as a key catalyst. *Cell 107*, 415–418.

79. Ahlquist, P. (2002). RNA-dependent RNA polymerases, viruses, and RNA silencing. *Science 296*, 1270–1273.

80. Hamilton, W.D. (1964). The genetic evolution of social behaviour I. *J Theo Biol 7*, 1–16.

81. Fujii, Y.R. (2008). Formulation of new algorithmics for miRNAs. *Open Virol J 2*, 37–43.

# 19 MicroRNAs and Cancer
## Connecting the Dots*

Sumedha D. Jayasena

## CONTENTS

---

\* This chapter reflects the thoughts of the author; they should not be attributed to Amgen Inc.

## OVERVIEW

microRNAs (miRNAs, miRs) represent a super family of small noncoding RNA sequences that regulate the expression of a significant fraction of genes in metazoans. The human genome may transcribe as many as ~1000 miRNA sequences, and close to 500 miRNA molecules have already been identified. Together, these miRNAs have the capacity to regulate about 30% of the entire transcriptome, qualifying them to become a family of master gene regulators. Consequently, deregulated miRNAs could induce altered phenotypes, including human cancer. Understanding of deregulated miRNA expression profiles in different tumor types may provide a new approach to cancer diagnosis, as well as novel molecular targets for therapeutic intervention. The current knowledge of underlying mechanisms by which miRNAs contribute to human cancer is limiting, but expanding at a rapid pace. This chapter reviews recent discoveries, pointing to how miRNA molecules are potentially linked to human cancer.

## 19.1   INTRODUCTION

Gene regulation is vital to every aspect of a cell. Precise and coordinated expression of genes in a spatial and temporal manner leads to the proper development of a multicellular organism from a single cell progenitor, the fertilized egg. Processes such as proliferation, cell-type specification, cell migration, and programmed cell death (apoptosis) are vital to the natural development. Deregulation of any one of these processes could lead to a disastrous outcome. In cancer, the natural process of cellular development goes awry, leading to uncontrolled cell growth with little or no differentiation, proliferative advantage, and resistance to apoptosis and invasion of normal biological tissues. The process of deregulated gene expression lies at the heart of oncogenesis. At the most fundamental level, the gene regulatory circuits or the networks that precisely control the biological processes of normal development are altered in cancer.

Conceptually, gene regulation occurs at two major levels: transcriptional and posttranscriptional. While epigenetic control at the chromatin level is still important, gene regulatory circuits consisting of transcription factors (TFs) that constitute both activators and repressors dominate the regulation at the transcriptional level. At the posttranscriptional level, mechanisms that regulate the translational process as well as the life span of an mRNA are very important. A recently discovered class of noncoding RNA referred to as microRNAs has already exhibited a significant role in the posttranscriptional regulation of genes in metazoans. As the new player in gene regulatory networks, deregulated miRNAs are now becoming important contributors to cancer, similar to TFs and the factors involved in chromatin remodeling, which have already established their roles in this complex disease.

## 19.2   miRNAs: TINY RNAs THAT REGULATE GENE EXPRESSION

### 19.2.1   Biogenesis

miRNAs are a family of 20–25-nucleotide small RNAs that are derived from long transcripts resulting from RNA polymerase II activity on *miRNA* genes [1–4]. These initial transcripts are called primary miRNAs (pri-miRNAs) [3,5], from which fold-back stem-loop structures containing precursor miRNAs (pre-miRNAs) are derived. A significant number of human miRNAs (~80%) are derived from introns [6]. About 25% of such intronic miRNA are embedded within protein coding genes [7]. Most of these intron-embedded miRNAs are in the same orientation of coding sequences, indicating that such miRNAs are coexpressed and coregulated with their host genes. miRNAs in their primary transcripts can exist as either single units (monocystronic) or arranged in clusters (multicistronic). In humans, most miRNAs are expressed in clusters with an average of 2–3 miRNAs per cluster [8].

Upon transcription of a pri-miRNA, the Ribonuclease III (RNAse III) enzyme Drosha chops off the pre-miRNA containing the stem-loop structure [9–12] (Figure 19.1). The Drosha cleavage defines one end of the miRNA [13] in either the 3′ or the 5′ arm of the stem-loop [14]. For further processing, pre-miRNAs are exported into the cytoplasm by a nuclear-cytoplasmic transporter, Exportin 5, with help from a cofactor, Ran, a GTPase [15–17]. In the cytoplasm, HIV-1 TAR RNA binding protein (TRBP) helps a second RNase III enzyme, Dicer, to cleave the pre-miRNA to generate the mature miRNA duplex and mark the other end of the miRNA sequence.

To perform gene silencing, miRNAs enter into the RNA interference (RNAi) pathway by enlisting in an effector protein assembly called RNA-induced silencing complex (RISC). Other small RNA molecules of different origin, for example small interfering RNA (siRNA), could also be incorporated into similar effector complexes to participate in RNAi. To help distinguish from other effector complexes, miRNA-embedded RISC is referred to as miRISC. The duplex miRNA contains the miRNA strand (the "guide") and its complement (the "passenger"), also referred to as miRNA*. In almost all cases, only the guide miRNA strand gets incorporated into the miRISC. The relative thermodynamic stability of the two termini of a miRNA duplex dictates which strand gets incorporated into the miRISC [18,19]. Most miRNA duplexes have evolved to acquire differential stability at the two ends such that only the miRNA strand gets preferentially incorporated into the miRISC [19]. However, there are a few exceptional cases in which both miRNA and miRNA* strands get incorporated into the miRISC.

## 19.2.2   MOLECULAR MECHANISMS OF GENE SILENCING

RNAi is a natural, yet sophisticated, mechanism that has been evolved to capture endogenous miRNAs into the miRISC, and then uses them as guides to identify and downregulate specific target messages. Gene silencing by miRNAs is mediated by two ways: either by inhibiting the translation process or by reducing the life span of target mRNAs or both. The immediate fate of the target mRNA is dictated by the degree of base-pairing between the miRNA and its recognition site. Generally, miRNA recognition sites are located within the 3′ untranslated region (3′ UTR) of target messages. In animals, pairing between miRNA and their target mRNAs is not usually perfect, although there are a few exceptional cases where perfect or near-perfect recognition exists. If near-perfect base pairing between the miRNA and its target exists, then the cleavage of the target mRNA can be mediated by the

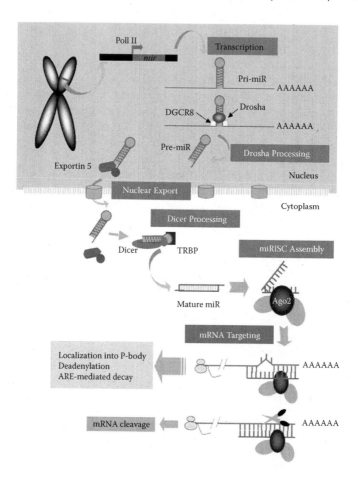

**FIGURE 19.1**  *Color version of this figure follows page 238.* microRNA biogenesis. The major steps in the biogenesis pathway are highlighted in blue boxes. A microRNA gene is transcribed by RNA polymerase II to generate a long primary miRNA transcript (pri-miR) that contains an miRNA sequence embedded in an ~70-nucleotide stem-loop structure. The RNAse III enzyme Drosha, with the help of its RNA-binding partner, DGCR8, cleaves at the base of the stem-loop structure to liberate the precursor miRNA (pre-miR), which is then exported into the cytoplasm by Exportin 5 nuclear transporter. In the cytoplasm, the second RNAse III enzyme, Dicer, coordinates with TRBP (HIV-1 TAR RNA Binding Protein) to process pre-miR to generate mature miRNA upon cleaving off the loop. The Drosha and the Dicer cleavages specify the two ends of the mature miRNA. The miRNA strand in the mature miRNA duplex gets incorporated into a multiprotein complex referred to as miRISC. The miRNA strand guides the miRISC to identify its targets through complementary base pairing between the miRNA and the 3′ UTR of target mRNAs. The 5′ end (nucleotides 2–7) defines the "seed" region of an miRNA that is critical to target recognition. The fate of the mRNA targeted by an miRNA is dictated by the degree of base pairing between the two. Near-perfect base pairing between miRNA and mRNA leads to message cleavage by the nuclease "slicer" associated with hAgo-2 protein within the miRISC and subsequently degraded rapidly. Imperfect base pairing within these two entities guides the target message for translation inhibition through one of several mechanisms, as indicated.

endonuclease activity (slicer activity) in the RISC [20–23]. A stretch of nucleotides at the 5′ end of a miRNA (2–7 nucleotides from the 5′ end) is commonly referred to as the "seed region," and is critical to the recognition of the target mRNA, as evident from the perfect complementarity within the seed region in pairing with target mRNA sequences. In the majority of cases, imperfect base pairing between miRNAs and the target messages results in translational inhibition through a mechanism that is not fully understood in detail. Translational repression requires the recruitment of human Argonaute-2 (hAgo-2), also known as EIF2C2, a major constituent of the RISC, to the 3′ UTR of the target message. In fact, hAgo-2 is the endonuclease of the RISC [24]. It has been demonstrated that translational repression by hAGO-2 recruited to the targeted message is dependent on the presence of the 5′ cap structure, as messages derived from IRESs (internal ribosome entry sites) are generally insensitive to silencing [25,26]. Messages with recruited hAgo-2 are localized to subcellular foci called *processing bodies* (P-bodies), where the majority of the mRNAs get degraded in the cell via decapping and 5′-3′ exonuclease activity (reviewed in Reference 27). Accumulation within P-bodies that lack ribosomes may be the key mechanism by which translational repression occurs. Further, the P-bodies may represent a temporary storage depot of miRNA-targeted messages, which may become available for translation in the future, as the need arises.

In addition to translational repression by sequestering messages within P-bodies, there are other mechanisms that miRNA targeting utilizes for gene silencing. Short-lived mRNAs often contain AU-rich elements (AREs) in their 3′ UTRs that are known to mediate rapid decay. Recent evidence suggests that miR-16 could target AREs through complementary binding and recruiting factors that mediate rapid decay, thereby silencing genes containing AREs [28]. Yet another mechanism by which specific microRNAs control gene expression is the induction of deadenylation of targeted messages, as observed with the rapid clearance of maternal transcripts with the expression of miR-430 during maternal to zygotic transition [29]. Deadenylated mRNAs are unstable and degrade rapidly.

With the development of novel approaches for discovering miRNAs, the predicted number of miRNAs in humans is estimated to be ~800–1000 [30–33], far exceeding the original estimates [34]. Studies employing microarray analysis have shown that a single miRNA has the capacity to regulate more than 100 genes [35]. Hence, the entire miRNA pool is expected to regulate a significant portion of transcripts, about one in every three transcripts. By doing so, miRNAs regulate genes that control a large number of biological processes which are critical to the homeostasis of a cell. Deregulation of some of these processes due to the misappropriate expression of master regulatory miRNA molecules could lead to tumor genesis.

### 19.2.3  COMBINATORIAL GENE REGULATION

The challenge of precise regulation of gene expression in a temporal and spatial manner in multicellular organisms has been worked out by creating master regulators that act in concert in a combinatorial fashion. In general, at the mechanistic level, TF and chromatin modulators function as "on-off" switches in regulating gene expression. In contrast, miRNA-mediated gene regulation grants a range of gene expression levels, akin to the function of a "dimmer" switch. This is accomplished by the targeting of miRNAs to multiple sites separated by appropriate distances within a given 3' UTR or even within coding regions of regulated transcripts. These target sites may have different affinities toward miRNAs as dictated by different base-pairing schemes. The presence of multiple sites with different binding strengths grants the modulation of translational efficacy over a broad range. The recruitment of different types of miRNAs to these multiple sites allows a significant control over this process, as different miRNAs can be expressed under different conditions; for example, in response to various cellular stimuli or during different developmental stages or both.

More than one binding site for a given miRNA or multiple miRNAs have been identified or predicted within 3' UTRs of genes under miRNA regulation [20,36–42]. Even a single miRNA binding site can give rise to a varying degree of in vivo translational efficacy of a reporter gene, depending on the position of a mismatch pairing within the 5' seed [43]. However, it is unlikely that natural targets would use only one target site for a given miRNA, since the combinatorial approach would be more advantageous. Currently, very little is known about the rules of combinatorial regulation of transcripts by miRNAs (Figure 19.2).

## 19.3   ABERRANT EXPRESSION OF miRNAs AND THE DEVELOPMENT OF CANCER

Upon deregulation, many molecular pathways that control normal embryonic development can cause cancer. To date, a substantial volume of experimental evidence supports the existence of a prominent role for miRNAs in the control of cellular differentiation process in animals, especially during embryonic development. Interestingly, the founding members of miRNAs were discovered owing to this very feature; that is, defects in lin-4 and let-7 miRNAs, leading to the reiteration of larval stages of *C. elegans*. These mutant cells continued to divide without differentiation into the next stage of the life cycle. Losing the ability to differentiate and acquiring limitless replicative potential are among the key characteristics of cancer. The initial discoveries of lin-4 and let-7 identified by forward genetic screens in the worm are, in fact, serendipitous, as the elimination of many miRNAs may give very subtle or no phenotypic differences owing to functional and sequence redundancies [44].

In *Drosophila*, the *bantam* locus was identified in a gain-of-function screen directed to identify genes that control tissue growth [45], and the *bantam* gene encodes an miRNA (bantam-miR) that stimulates cell proliferation and inhibits apoptosis, characteristics that are intimately connected with a malignant transformation. Bantam-miR inhibits apoptosis by suppressing the proapoptotic gene *hid* [42]. However, a target gene that functions as a negative regulator of cell growth and is inhibited by bantam-miR has not yet been identified. In flies, miR-14 is involved in apoptosis. An enhanced apoptosis triggered by the upregulated *Drice*, an apoptotic effecter caspase, has been observed in flies that lack miR-14. They also show semilethality, stress-induced death, and reduced life span [46]. The inhibition of miR-2, -6, -11, -13, and -308 led to widespread apoptosis in the fly embryos because of the upregulation of proapoptotic genes *grim, hid, reaper,* and *sickle* [47]. In humans, miR-21 plays an antiapoptotic role in cell lines derived from glioblastoma, as well

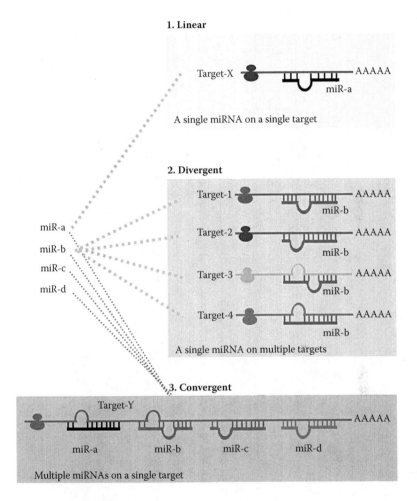

**FIGURE 19.2** Three principal methods of miRNA-mediated target recognition. 1. Linear recognition. A single miRNA could bind to a single mRNA. This mode of recognition is extremely rare in nature, and is expected to have a near-perfect base pairing or a very strong 5′-seed region between the target and miRNA to become effective in translational inhibition. 2. Divergent recognition. A single miRNA binds to multiple targets employing various base-pair recognition schemes; each one may be unique to the given target. This is one of the common mechanisms for miRNA-mediated target recognition. It is likely that these targets are also recognized by other miRNAs. 3. Convergent recognition. This is the most common mode of target recognition, in which each target is recognized by multiple miRNAs.

as in glioblastoma patient tumor tissues. In these samples, miR-21 shows marked upregulation, and the inhibition of miR-21 resulted in cell death through caspase activation [48].

These examples indicate that the well-being of an organism is dependent on the coordinated and precise expression of miRNAs that control genes critical to cellular homeostasis. Changes to this process could be detrimental to the cell. In most cases, cells lose a survival advantage and die, and in some cases cells adopt aberrant

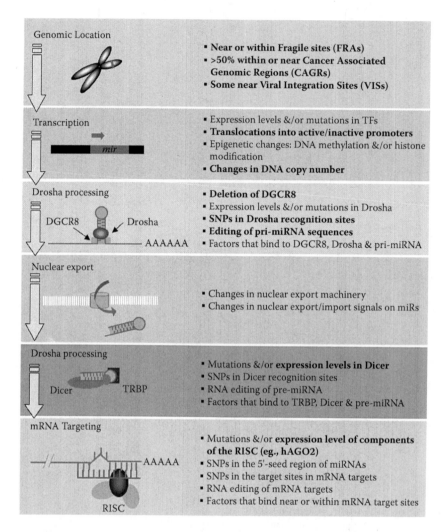

**FIGURE 19.3**  Key stages of miRNA biogenesis that could contribute to the deregulation of miRNAs. Different steps/stages in the miRNA biogenesis pathway (left) could be subjected to changes by many processes (right). Examples already exist to support the processes highlighted in bold.

phenotypes, contributing to tumor formation. As summarized in Figure 19.3, various stages during miRNA biogenesis could contribute to deregulation of their expression. Intriguingly, there are examples of human phenotypes or tumors that correlate with defects at each of these critical steps involved in miRNA biogenesis.

### 19.3.1  ABERRATIONS AT THE CHROMOSOMAL LEVEL

The most common chromosomal abnormality in chronic lymphocytic leukemia (CLL), the prevalent form of adult leukemia, gave the first hint as to the possible association of specific miRNAs to a given human cancer [49]. More than half of CLL patients carry heterozygous and/or homozygous deletions at 13q14. This aberration is also found in other malignancies: >50% in Mantle cell lymphoma, 16–40% in multiple myeloma, and 60% in prostate tumors. In prostate cancer, LOH (loss of heterozygozity) in 13q14 is frequently associated with high-grade tumors with poor prognosis [50]. This led to the belief that 13q14 deletion is likely to carry a tumor suppressor (TS), but efforts by several groups failed to identify a candidate gene in this 30-kb region with any certainty (summarized in Reference 49). Interestingly, two miRNA genes, *mir-15a* and *mir-16-1*, which are clustered within 0.5 kb, lie in this region, suggesting a possible TS role for these two sequences. Downregulation of both miR-15a and miR-16-1 was found in 68% of CLL patient samples as well as in prostate cancer cell lines. Thirty-eight percent of CLL patients showed pri-miR-15 but not the mature form, suggesting that the biogenesis of miR-15 is affected in certain cases where deletion of the gene was absent. Compared to the other chromosomal abnormalities found in CLL, patients with the deletion of 13q14 locus show a more favorable prognosis. The molecular basis for this clinical observation is not clear. However, it could be speculated that the deletion of miR-15 and miR-16 may contribute to the genesis of CLL. The deletion of 13q14 does not completely eliminate the level of these two miRNAs, owing to the presence of a second homologous locus for these two miRNAs, *mir-16-2* and *mir-15b*, on chromosome 3 [51]. The deletion of *mir-15a* and *mir-16-1* could increase the expression of their target genes, whose activities may allow the accumulation of additional genetic lesions, a so-far-untested hypothesis. This may be important for an ubiquitously expressed miRNA such as miR-16. The deletion of *miR-16* that regulate ARE-directed mRNA decay [28] may contribute to the oncogenic process by allowing the expression of a high-turnover gene set for a long period. Among the genes that are downregulated by miR-16 and miR-15 is BCL2, a key apoptotic inhibitor. The overall reduced level of the negative regulators could lead to prevention of apoptosis, giving an opportunity to the tumor-initiating cell population to accumulate additional genetic defects in this class of tumor.

On the basis of common or similar seed sequences, miRNAs are grouped into families. A significant number of miRNAs have their family members transcribed from different chromosomal locations, suggesting that their biological roles may be quite critical to the cell. The expression levels of such miRNA families can be controlled from different genomic locations and allows the cell to build reinforced mechanisms to ensure tight regulation of their targets. Furthermore, some miRNAs have homologues that are transcribed from different genetic locations, and miR-15 and miR-16 belong to this category of miRNAs. In addition to the deregulation of miR-15 and -16 in CLL, there are a limited number of cases where miRNA deregulation has been observed in human cancer, as indicated in the following text.

Patient tissues from precancerous adenomatous polyps and adenocarcinoma showed markedly reduced levels of miR-143 and miR-145 compared to the matched normal tissues [52]. These two miRNAs are colocated on chromosome 5 (5q32-33), the same genetic region associated with prostate cancer and myelodysplastic syndrome [8]. Chromosomal translocations can bring oncogenes (or other genes that potentiate oncogenesis) under the direction of inappropriate promoters that could lead to tumor formation. An example has been noted in a cancer patient with insertion of *mir-125b-1* into a rearranged immunoglobulin heavy chain locus in a leukemic recurrence as ovarian tumor seven years after allogeneic bone marrow transplantation for acute B-cell leukemia. This transposition of *mir-125b-1* at the 11q24 locus was found in both secondary ovarian tumor as well as in the primary ALL [53], suggesting a common ancestral origin for both tumors, implying that miRNAs can be used to track tumor histories in patients. In breast tumors, miR-125a, miR-125b-1, and miR-125b-2 are all downregulated [54]. Interestingly, *C. elegan's lin-4* is an ortholog of *mir-125* family in humans that is involved in determining cellular fate, and the mutation of which caused defects in lineage progression in the worm. In humans, miRNAs of this family appear to be involved in both hematopoietic and solid tumors, preserving their function during evolution.

These observations, which point to the possible involvement of specific miRNAs in different human tumors, were further supported by the results obtained from a global search for the distribution of *mir* genes among human chromosomes [8]. The global search, carried out using 186 *mir* genes (the total number available in the miRNA registry [55] at the time of the study), uncovered interesting correlations:

- A significant number of *mir*'s (19%) are located in or within 3 Mb of known chromosomal fragile sites (FRAs). FRAs appear as gaps in karyotype analysis and represent dynamic regions with ongoing activities that allow shuffling of genetic materials such as sister chromatid exchange, viral integration, activation of mobile genetic elements, or gene copy number amplification or deletion. Deregulation of such activities is common among human tumors and the presence of miRNA genes in or near these sites may contribute to tumor formation through expression changes of a large number of target genes.
- About 50% of *mir*'s are located within Cancer Associated Genomic Regions (CAGRs). A CAGR is characterized by a minimal genetic element showing one or more following features:
  a.  A loss of heterozygocity (LOH), suggestive of a tumor suppressor gene (TSG)
  b.  An amplified region, suggestive of an oncogene (OG)
  c.  A common breakpoint region in or near TSG or OG

About 40% of *mir* genes are located within CAGRs that satisfy conditions a and b (see preceding list) and have been described in association with both liquid and solid tumors. Some CAGRs are known to bear either TSG or OG attributing to the molecular cause for the associated cancer type. However, candidate TSGs or OGs have not been identified within most CAGRs. It is possible that miRNAs present within or near CAGRs that lack known candidate genes may in fact be involved in the respective tumor. Such studies would also reveal whether miRNAs fulfill the roles of classical OGs or TSGs or bear OG-like or TSG-like functions, with which these types of miRNAs have recently been identified [56,57].

The chromosomal integration of the human papilloma virus (HPV) is a common risk factor in cervical cancer. Frequent integration sites of HPV occur close to *mir* genes, suggesting that HPV integration could change the expression levels of such miRNAs. Further investigations await the confirmation of this hypothesis. Viral integration into human chromosomes could activate genetic elements close to the integration sites, leading to direct activation or inactivation of genes whose activities subsequently contribute to cancer. The expression of miRNA genes could be affected in a similar manner. In fact, 7% of miRNAs are located within 2.5 Mb of 45% of known integration sites of HPV16 [49], a known causative agent of cervical cancer.

A high level of DNA copy number alterations of *miR* genes have been discovered through array-based comparative genomic hybridization of 227 specimens (primary tumors and cell lines) of epithelial cancers (ovarian, breast, and melanoma) probed with 283 miRNA sequences [58]. These changes reflected both amplifications and deletions of *miR* genes. While alterations in 41 different miRNAs (14%) were common to the three tumor types, individual tumors carried changes in unique miRNAs as well: 37% in ovarian, 72% in breast, and 85% in melanoma. In these tumors, miRNA expression levels were correlated with their changes in gene copy numbers.

## 19.3.2 DEFECTS IN MICRORNA BIOGENESIS PATHWAY

The deregulated expression of certain miRNAs can be rationalized based on their genomic locations, as discussed previously. Further, factors that control the post-transcriptional processing of miRNA transcripts can also affect levels of miRNAs, which are described in the following text.

### 19.3.2.1 Drosha Processing

In humans, the microprocessor complex that regulates the processing of pri-miRNAs contains Drosha, the RNAse III nuclease, and DGCR8, the cofactor for Drosha [10,59]. *DGCR8* is among a number of genes within the genetic locus 22q11.2, whose monoallelic deletion causes >90% of three syndromes (DGS-DiGeorge syndrome, VCFS-velocardiofacial syndrome, and CAFS-conotrunal anomaly face syndrome) that constitute a common human genetic syndrome with a frequency of 1 in 4000 live births [60,61]. These syndromes present complex clinical manifestations: learning disabilities, psychiatric disorders, craniofacial abnormalities, and immune, hormone, and cardiovascular defects. The causative gene(s) of this complex disease has not been completely characterized. However, a correlation between the disease manifestation and the tissue expression profile of DGCR8 in mouse embryo has been noted [60]. It was hypothesized that the global defect in miRNA biogenesis upon deletion of DGCR8 could lead to developmental defects observed in the foregoing syndromes [62]. It is possible that the deletion of DGCR8 could lead to reduced levels of miRNAs rather than the complete elimination of the miRNA pool, as the latter is detrimental to the developing embryo. However, deletion of DGCR8 by gene targeting in mouse ES cells demonstrated a global loss of miRNA pool, and did not produce viable mice owing to embryonic lethality [63].

It is also possible that the processing of pri-miRNA transcripts can be regulated by Drosha itself, independent of DGRC8, although the function of DGCR8 is essential for Drosha activity. The activity of RNA-editing enzyme ADAR deaminase, which converts adenosines to inosines on pri-miRs, has been shown to inhibit Drosha processing [64], presumably by changing secondary structure or the primary sequence of the substrate of Drosha. Human tumors of multiple tissue origins show a global change of miRNAs, compared to those in normal adjacent tissues [65]. This change is very prominent for the expression of mature miRNAs in tumors and not for pri-miRNA sequences [66], indicating that Drosha activity, but not *miR* transcription activity, is regulated in tumors. The exact biochemical nature of the differential Drosha activity in tumors is unclear, but factors that interact with either Drosha or pri-miR transcripts, or both, could contribute to this phenomenon.

### 19.3.2.2   Dicer Processing

The activity of the Dicer enzyme generates mature miRNA duplexes from pre-miRNAs, and represents a key step in the biogenesis process of all miRNAs. The microRNA pool is critical to the developing vertebrate embryo, as demonstrated by the embryonic lethality at E 7.5 in *dicer-1* knockout mice [67]. In zebra fish, embryos show growth arrest at days 8–10, and die at day 14 upon elimination of Dicer [68]. When the full-length wild-type Dicer was replaced by a truncated Dicer that lacks the exons 1 and 2 (Dicer hypomorph) in mouse embryo, the embryonic development was arrested between days 12.5 and 14.5 in gestation. These embryos showed markedly reduced embryonic angiogenesis with altered expression of key angiogenic factors, VEGF, FLT-1, KDR, and TIE-1 [69].

The conditional knockouts of Dicer in different tissues have unique defects in mice; the absence of proper morphology and enhanced cell death in the developing limb [70], markedly reduced alveolar branching and enhanced cell death in developing lungs [71], enhanced cell death in T lymphocytes [72,73], evaginated hair germs with destabilized epidermal organization, and postnatal death within 4–5 days on account of dehydration are the hallmarks of Dicer elimination in epithelial progenitor cells [74,75]. Furthermore, embryonic stem (ES) cells derived from the *dicer*-null mouse embryos do not differentiate into embryonic bodies (EBs) with distinct cell types representing endo-, meso-, and ectoderm. Interestingly, these *dicer*-null ES cells do not form tumors (teratomas) in nude mice. All these evidence point to the critical role of the miRNA pool derived from Dicer activity on the proper development of tissues, especially during early development.

The expression levels of Dicer have been shown to be different in two different tumor types of epithelial origin. In lung cancer, overall Dicer level is reduced, and the patients with the lowest level of Dicer had worse postoperative survival outcome compared to the subpopulation with a relatively high Dicer level [76]. This suggests that in lung tumors, low level of the miRNA pool resulting from reduced Dicer levels could lead to the activation of a number of genes that contribute to tumorigenesis with OG-like function. In prostate cancer, Dicer appears to play the opposite role; there is an upregulation of Dicer expression as monitored by immunohistochemistry studies (IHC) that correlates with clinical and lymph node stage and Gleason score [77]. Dicer is well expressed in the basal layer of the normal prostate and not in the luminal cells. With the transformation into prostate intraepithelial neoplasia (PIN), the luminal cells start to express Dicer, and are markedly enhanced in the prostate adenocarcinoma. In this case, the high level of Dicer activity is expected to produce a high level of miRNA pool, as corroborated by independent analysis of miRNA expression [58] that will suppress a selected set of genes with TS-like activity. Interestingly, amplification of Dicer at the DNA copy number level has been noticed in ovarian tumors as well [78].

It is likely that certain miRNAs whose expression is restricted to specific tissues are under tissue-specific promoters [79,80]. Additionally, the processing of certain microRNA precursors can be regulated in a tissue-specific manner by Drosha, Dicer, and their partner proteins [66,81]. There is some evidence to support the existence of specific cellular factors that inhibit Dicer processing of specific pre-miR sequences that are expressed in a tissue-specific manner [66]. Nucleotide mutations in miRNA precursor sequences that affect the processing have been identified in CLL patients, indicating that single-nucleotide polymorphisms (SNPs) within microRNA genes could predispose individuals to cancer. Likewise, SNPs within the 3′ UTRs of the target mRNAs that are being specifically recognized by the seed regions of miRNAs may equally contribute to deregulation of gene expression by microRNAs. However, experimental evidence to support the latter has not so far been forthcoming.

### 19.3.2.3   miRNA Targeting

Among children, Wilms' tumor, or nephroblastoma, is the most common type of kidney tumor with genetic heterogeneity. A subset of patients shows a deletion on chromosome 11 (11q13) carrying a tumor suppressor *WT-1* gene, a transcription factor involved in normal kidney development. LOH analysis has revealed the possible existence of other TSGs in 11q15, 16p, and 1p, but TSGs have not been adequately characterized [82]. One such chromosomal deletion associated with Wilms' tumor includes a region that harbors hAgo2, the endonuclease of the RISC, on chromosome 1 (1p34-35) [83]. Human Ago-2 belongs to the Argonauts family proteins, which are intimately involved in the RNAi pathway [84,85]. The high expression of hAgo-2 has been noted in normal embryonic kidney and lung, the deletion of which may affect normal kidney development. In ovarian tumors hAgo-2 also shows a gain in DNA copy number [78], suggesting that the miRNA activity is likely to be elevated in these tumors.

Argonaute proteins are evolutionary conserved and contain PAZ and PIWI domains. The piwi family proteins, a subfamily of Argonaute proteins is essential to RNAi, stem cell self-renewal, and gametogenesis [83]. Hiwi, the human homologue of Piwi gene family, has an association with germ line testicular seminomas and is not found in somatic testicular tumors [86]. The genomic location of Hiwi, 12q24.33, has a genetic linkage to the development of testicular germ cell tumors. This evidence collectively supports the view that atypical changes in molecules that regulate miRNA biogenesis could lead to developmental defects and cancer in humans.

## 19.4   UNIQUE miRNA PROFILES IN SPECIFIC CANCER TYPES

Northern blot analysis, a classical molecular biology technique, has been used to preview the expression changes of a handful of miRNAs at a time in specific tumor types [8,49,52,87]. Intriguing results obtained from these studies have inspired the development of approaches that facilitate parallel analysis of global expression patterns of miRNAs. To that effect, parallel analysis of most, if not all, known miRNAs has been achieved by oligonucleotide chip arrays or bead-based liquid arrays.

Chronic lymphocytic leukemia (CLL) comes in two common forms: an aggressive form, which requires the treatment within a short period of time of disease diagnosis, and the indolent type, which progresses slowly into a stage that requires therapy. While the aggressive form is characterized by few or no mutations in the immunoglobulin heavy-chain variable-region (IgVH) or enhanced expression of the 70-kDa zeta-associated protein (ZAP-70), the inverse correlation of these two features is associated with the indolent form. A chip-based miRNA expression profiling on samples of CLL patients revealed a 13-miRNA signature that can be correlated with the expression level of ZAP-70 protein [88]. A subgroup of 9 miRNAs out of the 13 miRNAs in the prognostic signature was able to differentiate between patients with a shorter time (~40 months) to initial therapy from those with a longer time (~88 months) to therapy from the initial diagnosis. This is a powerful diagnostic tool when it comes to managing patients and assessing possible treatment options, perhaps, much earlier in the patients who will need treatment.

microRNA expression profiles of individual tumor types employing a relatively small number of patient-derived tissue samples have been described for lung [89], papillary thyroid carcinoma (PTC) [90], follicular thyroid carcinoma (FTC) [91], colorectal cancer (CRC) [92], and prostate and breast cancers [93]. These studies have revealed unique profiles in each tumor type with both up- and downregulated miRNA sequences. While these initial results are quite encouraging, analysis of large sample cohorts of tumors of different types would be very informative in identifying unique miRNA signatures for each tumor type.

A bead-based liquid array with the capacity to probe 217 miRNAs in parallel was used in a systematic analysis of diverse human tissues and tumor types (332 samples) [65]. This study provided several insights into cancer stratification, diagnosis, and progression, all by monitoring the changes in a couple of hundreds of miRNAs (a modest number) as opposed to several thousands of transcripts in mRNA profiling using chip-based oligonucleotide arrays.

- Tumors of different tissue origin (colon, kidney, prostate, uterus, lung, and breast) showed globally reduced miRNA expression in comparison to normal tissues, indicating that miRNA expression profiles could distinguish tumors from normal tissues. About 50% of miRNAs were downregulated in tumors, compared to normal tissues, reflecting the poor cellular differentiation in all cancer types. In spite of the global reduction of miRNAs in tumor tissues, it is not uncommon to find uniquely upregulated miRNAs in almost all cancer types. For example, miR-192, miR-194, and miR-215 are highly expressed in GI tumors.
- Further, miRNA expression profiles revealed the developmental origin of cancer types. For example, all samples with hematopoietic malignancies were clustered away from the cancers with epithelial origin, resulting in two hierarchical clusters.
- Within each hierarchical cluster, subclusters encoded with specific information emerged. Reflecting the common endodermic origin, tumors from

colon, liver, pancreas, and stomach clustered together under the epithelial cluster. Interestingly, the sampling of 16,000 mRNA transcripts has failed to cluster GI tumors, whereas miRNA profiling succeeded, suggesting that changes in a limited number of miRNAs is more informative than changes observed in a much larger number of transcripts. This is more likely to be due to the noise associated with high-density mRNA profiles.

- Within the hierarchical cluster of the hematopoietic origin emerged three sublineages with telltale signs of the mechanistic origin of tumors: first, tumors with BCR/ABL translocation (t9;22) and TEL/AML1 (t12;21); second, samples with T-cell Acute Lymphoblastic Leukemia; and third, samples with MLL rearrangements.

- miRNA expression profiling helped classify tumors of unknown origin with uncertain diagnosis. A classifier developed using a training set based on the miRNA profiles of 68 well-differentiated tumors was challenged with the expression profiles of 17 less defined tumor samples. This approach correctly predicted 12 out of 17 poorly differentiated tumors, a remarkable improvement in comparison to an analogous approach based on mRNA expression profiles.

These initial results are highly encouraging in terms of identifying specific miRNA profiles characteristic of individual tumor types that will help improve clinical diagnosis and choice of treatment options for cancer patients. These observations remain to be corroborated by independent investigations using an even wider selection of tumors and patient samples.

In a separate study that employed a chip-based microarray tool and a larger number of tumor samples (540) representing six solid tumors of the epithelial origin, a set of deregulated miRNAs common to three or more tumor types has been uncovered [58]. In contrast to the results obtained with bead-based profiling, this study did not reveal global downregulation of miRNAs in the six solid tumor types analyzed. The discrepancy between these observations may be due to the differences in the number of tumor samples used, and the technique employed, or both. In the future, it may be quite possible to identify unique miRNA profiles for a given tumor type carrying crucial information on the origin, stage of development, and the mechanism of oncogenic transformation of a specific tumor to help dictate the best possible treatment options. It would also provide opportunities for early diagnosis of cancer, and facilitate the monitoring of disease progression and the patient's response to treatment. In combination with nanotechnology, expression analysis of miRNAs from a few circulating tumor cells derived from a patient's blood would become a reality in the future.

## 19.5   ONCOMiRs: PLAYING A DIRECT ROLE IN ONCOGENESIS

Could deregulated expression of miRNAs cause cancer? As discussed earlier, evidence has accumulated to favor this hypothesis. However, at the time of this article, very little evidence exists to support that deregulated miRNA expression alone is sufficient to cause cancer. In most cases, it is very likely that deregulated miRNAs cooperate with another genetic alteration to either initiate oncogenesis, or to facilitate cancer progression into metastasis, or both.

The very first direct experimental evidence that the RNAi pathway could contribute to oncogenesis came from an experiment performed in the *Eμ-myc* transgenic mouse model carrying the *C-Myc* oncogene under the immunoglobulin heavy-chain enhancer, *Eμ* [94]. These mice, already harboring a single mutation for oncogene activation, develop B-cell lymphoma after a long latency period of 4–6 months of age [95], indicating the requirement for additional mutations for oncogenesis. The latency period get shortened upon the establishment of a second mutation within an OG or a TSG that cooperates with C-Myc [96]. Reconstitution of bone marrow in lethally irradiated mice with ex vivo modified hematopoietic stem cells (HSCs) using a retroviral vector carrying investigative genetic elements provides a venue to study hematopoietic malignancies in vivo. In this approach, the introduction of small hairpin RNA (shRNA) designed to downregulate a TSG, p53, in *Eμ-myc* transgenic mice led to rapid B-cell lymphomagenesis within 9–14 weeks, depending on the efficacy of p53 knockdown by individual shRNA [94]. This indicated that an artificially designed small hairpin RNA that enters into the RNAi pathway that is shared by endogenous miRNAs could actively contribute to oncogenesis through posttranscriptional gene silencing of a specific and well-characterized TSG. This experimental evidence suggests the possibility of the involvement of naturally occurring miRNAs in oncogenesis upon their misappropriate expression. Figure 19.4 illustrates different mechanisms by which miRNA deregulation could lead to tumor formation through changing the overall expression of a number of genes involved in multiple biological processes that are intimately involved in establishing and maintaining neoplastic transformation.

## 19.5.1 MiR-17-92 MULTICISTRON

A high level of chromosomal amplification at 13q31-q32 is present in both hematological and solid tumors, including follicular lymphoma, mantle cell lymphoma, diffuse large B-cell lymphoma (DLBCL), primary cutaneous B-cell lymphoma, glioma, non-small-cell lung carcinoma, head and neck cancer, bladder cancer, and alveolar rhabdomyosarcoma [97,98]. In contrast, 13q31-q32 locus is deleted in hepatocellular carcinoma [99]. A gene called *c13orf25* (chromosome 13 open reading frame 25), found within a common region of 13q31-q32 amplification, produces two transcripts: one with a short open reading frame for a short peptide, and the other carrying the pri-miRNA of the miR-17-92 multicistron [98]. microRNAs in miR-17-92 multicistron are upregulated in several B-cell lymphoma cell lines [98] [57]. The miR-17-92 cluster arose through the duplication and the loss of individual microRNAs. While miR-92 shows homology to invertebrate miRNAs (of both fly and worm), miR-17 and miR-19 are restricted to vertebrates, and must have evolved later [100]. A truncated multicistron (miR-17-19b) consisting of only the vertebrate-specific miRNAs derived from miR-17-92 cluster was used in vivo to probe its tumorgenic potential by cooperating with a known oncogene [57]. Upon introduction of the miR-17-19b multicistron into HSC using retroviral delivery, followed by engraftment into lethally irradiated *Eμ-myc* transgenic mice, lymphomagenesis was found to be accelerated to

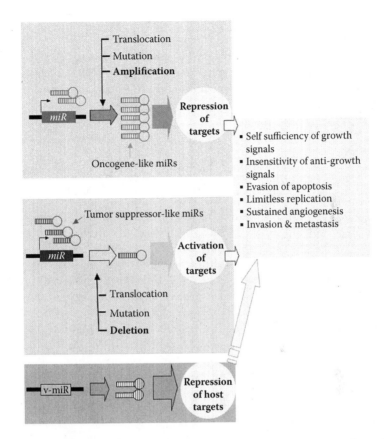

**FIGURE 19.4** Different approaches by which miRNAs contribute to tumor formation. A class of *miRNA* genes that are generally expressed at a low level can get upregulated through multiple mechanisms (top box). These miRNAs are likely to have oncogene-like behavior and may downregulate genes with tumor suppressor-like activities to establish multiple hallmarks in cancer, as shown in the right-hand box. A group of *miRNA* genes that are generally expressed at a high level may have tumor suppressor-like property and downregulate a class of target genes that otherwise may act as oncogenes (middle box). Under pathologic conditions, these miRNAs can get downregulated, leading to overexpression of their oncogenic targets, which eventually facilitate tumor formation. In addition to these cellular endogenous miRNAs, viral infections can introduce virally encoded miRNAs (v-mIRs), which may repress host genes, leading to oncogenesis as well (bottom box).

an average period of 7 weeks, in comparison to 3–6 months of latency in the animals that received the control virus. Furthermore, the lymphoma induced by the enforced expression of miR-17-92 cluster showed resistance to apoptosis. The observed acceleration of lymphomagenesis in this mouse model suggested that the miR-17-92 multicistron harbors an oncogenic activity. Hence, these miRNAs were called oncoMiRs. It is important to note that oncoMiRs are normal miRNAs that undergo deregulated expression by an unknown mechanism that contributes to oncogenesis, most likely through cooperating with other factors such as activated oncogene or deactivated or deleted tumor suppressor. Alternatively, oncomiRs can also represent miRNAs with mutations, predominantly in the 5′-seed region that could lead to the regulation of unintended targets.

Amplification of miR-17-92 cluster and the overexpression of individual miRNAs have been noticed in lung cancer, as well [101]. Two miRNAs in this cluster, miR-19a and miR-92-1, are upregulated in CLL patient samples and certain other members of this cluster are differentially expressed between the two CLL clusters [88].

## 19.5.2  MiR-155

The proto-oncogene c-Myc plays an important role in lymphomagenesis in birds and mammals. However, the activation of c-Myc alone is not sufficient for the tumorigenesis, as the development of lymphomas in Eμ-Myc transgenic mice requires a long latency [95], allowing the accumulation of additional genetic alterations that cooperate with c-myc. Several such combinations of c-myc and other candidate genes have been characterized (reviewed in Reference 96). Avian and mouse models have been useful in delineating the molecular mechanisms of oncogenesis [102,103]. In an avian lymphoma model, a rapid lymphomagenesis occurs with the infection of avian leucosis virus (ALV) where the viral integration occurs at a common chromosomal integration site that activates a specific locus called bic [104]. Activation of the bic gene by promoter insertion of ALV led to the transcription of Bic with no known ORF. Intriguingly, the bic noncoding RNA transcript cooperates with c-Myc in lymphomagenesis in an avian model through an unknown mechanism [105], suggesting an oncogene-like role for the Bic transcript. The normal expression of bic is tissue specific (found in the bursa, thymus, and spleen), and is developmentally regulated as its expression is very low in early embryonic bursa but is elevated in the hatching chicks and in mature animals [106], suggesting that the deregulation of bic could be pathogenic in specific tissues during their development. However, in humans, Bic is upregulated in activated T- and B-lymphocytes [107]. High level of Bic expression was found in several types of B-cell lymphomas including diffuse large B-cell (DLBCL) [108], Hodgkins lymphoma and Burkitt lymphoma in children [109,110].

Analysis of the Bic transcript revealed a phylogenetically conserved region forming a stem-loop [111] with sequence homology to miR-155 isolated from mouse tissue [51]. The proposed oncogene-like activity associated with Bic is in miR-155. Bic and miR-155 show parallel expression in several lymphoma-derived cell lines and patient samples [108,109]. DLBCL is an aggressive malignancy of mature B-cells that accounts for ~40% of cases of non-Hodgkin's lymphoma with a highly variable clinical outcome; patients with germinal center-type (GC) have a significantly better overall survival compared to those with activated B-cell type (ABC) (references in Reference 112). A microarray analysis identified expression clusters representing genes that are expressed during B-cell differentiation and activation that distinguish GC and ABC phenotypes [112]. Interestingly, higher levels of Bic/MiR-155 have been observed in cell lines and patient samples of the ABC phenotype than those derived from the GC-type [108], which reflects the developmental history of these two subgroups and is also in accordance with the elevated expression of Bic in activated lymphocytes [107].

In order to investigate the potential role of miR-155 in B-cell malignancy, a transgenic mouse was recently generated by expressing miR-155 under the $V_H$ promoter-Ig heavy chain Eµ enhancer that gets activated at the late pro-B cell stage of B-cell development [113]. These Eµ–miR-155 transgenic mice showed atypical lymphoid populations, and by 6 months, showed a great increase in lymphoid blasts in both bone marrow and spleen. Flow cytometry analysis revealed similar cell surface marker profile characteristics of pre-B cell population found in acute lymphoblastic leukemia or lymphoblastic lymphoma in humans. Six-month-old animals exhibited splenomegaly due to the expansion of B-cell blasts. Microarray analysis of Eµ–miR-155 transgenics had revealed that ~200 genes were upregulated while 50 genes were downregulated. One of the upregulated genes is VpreB1, which is expected to rise with pre-B cell proliferation. Among other targets, PU.1 and C-CEBP transcription factors have been predicted as targets for miR-155 [40,114], but so far have not been experimentally validated. Overall, the transgenic Eµ–miR-155 mice showed lymphoproliferative malignancies observed in an avian model as well as in humans, suggesting that the enhanced expression of a single miRNA could, in fact, dictate the biology of a disease. Understanding of how the network of these genes and other miRNAs interact would be a step forward in elucidating the underlying molecular mechanism of lymphoproliferative disorders.

### 19.5.3   MiR-21

The overexpression of miR-21 was initially reported in Glioblastoma patient samples and cell lines derived from this tumor type [48]. Subsequently, and recently, overexpression of miR-21 was observed in breast tumors as well [54,115]. The inhibition of miR-21 in cell lines derived from both these tumor types resulted in cell death through enhanced apoptosis. MCF-7 breast tumor cell line transfected with a chemically modified oligonucleotide complementary to miR-21 resulted in retarded tumor growth in a xenograft animal model study. The tumors generated from the cells transfected with an anti-miR-21 oligonucleotide were smaller in size and lower in weight compared to tumors resulting from the same cell line transfected with a control oligonucleotide. Further, these tumors expressed a low level of Ki-67, a marker for proliferation. It has been reported that the antiapoptotic function of the miR-21 is due to its ability to downregulate the proapoptotic Bcl-2 protein [115]. These observations suggest that miR-21 with an antiapoptotic function could potentiate malignant transformation.

### 19.6   ONCOGENIC PATHWAYS CONTROLLED BY DEREGULATED MICRORNAS

At any given stage of cellular differentiation, the deregulation of *mir* expression by any one or combination of mechanisms outlined earlier changes the dosage of gene expression, which includes those of OGs and TSGs. Furthermore, any specific mutations within these proteins could contribute to any altered phenotypes as well. Alternatively, any alterations to the expression levels by deletion, amplification, and promoter activities could also contribute to tumor formation. microRNA deregulation could contribute to the foregoing processes by changing the threshold levels of

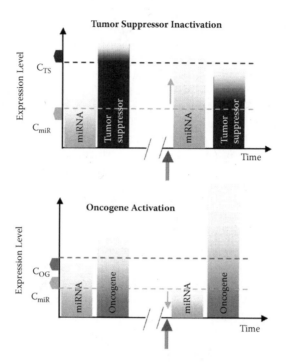

**FIGURE 19.5** *Color version of this figure follows page 238.* A general model for miRNA-mediated oncogenesis. In this model, it is assumed that human tumors could originate in the absence of activated or inactivated mutations within genes that could function as oncogenes or tumor suppressors, respectively. It is proposed that the tumor formation is initiated or maintained by changing the dosage of gene expression of certain genes through changes in microRNA expression. Those genes that changed their expression levels in response to deregulated miRNA levels may exert oncogenic and tumor suppressor activities. However, other genetic changes could work in tandem with the miRNA-mediated gene dosage changes to contribute to oncogenesis. The upper panel illustrates how the overexpression of certain miRNAs beyond a certain threshold value decreases the expression of genes with tumor suppressor activity. Tumor formation occurs when their expression level decreases beyond a threshold value. The lower panel shows the opposite effect, in which the expression of certain miRNAs falls below a threshold level. This change causes an upregulation above a threshold level of a number of genes with oncogenic activity. The red arrow denotes an event that contributes to the change in miRNA expression.

TSGs and OGs, as illustrated in Figure 19.5. So far, a few examples exist in which changes to known oncogenic pathways by deregulated *mir* expression have already been discovered; they are discussed in detail in the following text. With more investigations in the future, the total number of oncogenic pathways that is influenced by deregulated miRNAs is expected to rise.

## 19.6.1 MIRNAS AND PROTO-ONCOGENE C-MYC

The proto-oncogene, *c-myc,* encodes a master transcription factor that regulates key cellular functions, including cell cycle regulation, apoptosis, metabolism, cell adhesion, and cellular differentiation [116] by regulating nearly 10% of human genes [117] through interaction with the E-box element within the promoter regions of target genes. Approximately 30% of human tumors carry either a mutated or constitutively active *c-Myc* [118,119].

Interestingly, it has been observed that the induction of the miR-17-92 cluster expression by c-Myc was discovered in a cell line capable of inducible expression of *c-myc* [120]. The upregulation of the miR-17-92 cluster was also found when normal primary human fibroblasts grown under serum deprivation were subjected to serum stimulation that transiently induces *c-myc* expression. The chromatin immunoprecipitation (ChIP) studies revealed that c-Myc directly binds to the noncanonical E-Box upstream element of the miR17-92 cluster and upregulates the latter. Several members of the E2 Factor (E2F) family transcription factors have been shown to be targeted by miRNAs of the miR-17-92 cluster [114,120–122]. There is evidence to support that miR-17-5p regulates E2F1 and miR-20a regulates E2F1, E2F2, and E2F3.

The E2F family transcription factors represent a transcriptional network consisting of both activators and repressors that controls a range of functions from G1 to S phase cell cycle control, DNA repair and recombination, apoptosis, and cellular differentiation (reviewed in Reference 123). Among the nine E2F family members, E2F1, E2F2, and E2F3a represent transcriptional activators, and E2F3b and E2F8 are transcriptional repressors. Each member appears to have specific cellular roles. For example, E2F1 controls apoptosis in cells, and mice lacking E2F1 exhibit a propensity for spontaneous tumor formation [124,125]. E2F1 and c-Myc are directly transactivated by each other, placing them in a tight feedback control loop of their own expressions [126,127]. In addition to this direct feedback control, the miR-17-92 cluster offers an indirect route for c-Myc to control its own expression. Furthermore, E2F1, E2F3a can activate this cluster by directly binding to an upstream promoter region (Sylvestre, Y. 2006 e-pub, JBC; Woods, K. et al., 2006, e-pub, JBC). Taken together, the activation of the miR-17-92 cluster will inhibit apoptosis by damping E2F1 and promoting E2F3-induced cell proliferation (Figure 19.6). Further, miR-17-92 cluster exhibits a proangiogenic role by downregulating anti-angiogenic targets in a solid tumor model in an activated c-Myc background complemented with mutated RAS [128]. Under these conditions, angiogenesis is favored owing to the downregulation of thrombospondin-1 and connective tissue growth factor by miR-19 and miR-18, respectively.

The enforced expression of the individual miRNAs in the miR-17-92 cluster did not result in accelerated lymphomagenesis in the *Eµ-myc* transgenic mouse model [57], indicating that the cooperation between two or more miRNAs are necessary for tumor formation. In an independent study, a lung cancer cell line with the enforced expression of the entire miR-17-92 cluster, but not individual miRNAs of the cluster, showed enhanced cell proliferation [101].

The tumor suppressor, PTEN, is the predicted and validated target for miR-19a [114]. PTEN is the main negative regulator of the PI3K/AKT pathway, whose activation leads to cancer, and PTEN is a common target for mutation in sporadic cancer. It is the most frequently mutated gene in prostate and endometrial cancer. Downregulation of PTEN by activation of the miR-17-92 cluster could in turn activate the PI3K/AKT pathway, leading to tumor genesis in the absence of any

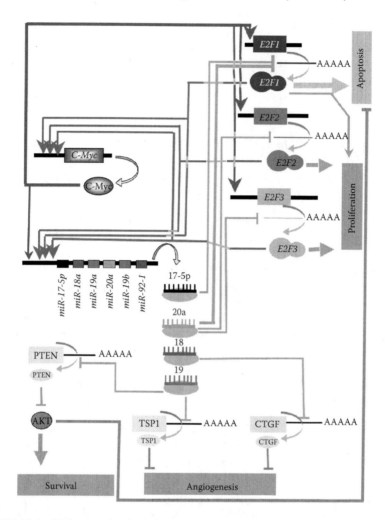

**FIGURE 19.6**  Molecular circuitry illustrating the interaction between the miRNAs within miR-17-92 cluster and their biological targets to generate a cellular environment conducive to tumor formation. Three transcription factors, E2F1, E2F2, and E2F3, activate c-Myc transcription upon binding to its upstream promoter region. In turn, c-Myc binds to the upstream promoter region of the miR-17-92 cluster and activates the expression of miRNAs. Two miRNAs, miR-17-5p and miR-20a, target 3′ UTRs of E2F1-3 and downregulate their expression. Compared to E2F-2 and E2F-3, which promote cell proliferation, E2F-1, which induces cellular apoptosis is significantly downregulated by both miRNAs. This creates a cellular environment that is resistant to apoptosis and favors cell growth. The expression of c-Myc is under tight regulation through a negative regulatory loop in which the miRNAs that were induced by c-Myc inhibit E2Fs, the inducers of c-Myc. In addition to the regulation of the c-Myc/E2F loop, two additional miRNAs in this cluster target additional genes; miR-18 downregulates connective tissue growth factor (CTGF), thrombospondin-1 (TSP1), and the tumor suppressor PTEN. The downregulation of PTEN leads to the activation of the AKT pathway with significant oncogenic activity: increased resistance to apoptosis, inhibition of cell cycle arrest and DNA repair, enhanced cell survival, and growth promotion.

deactivating (in PTEN) or activating (in PI3K/AKT) mutations. In a nutshell, the miR-17-92 cluster can activate the oncogene *c-myc* and downregulate the tumor suppressor PTEN and a family of transcription factors closely involved in cell cycle control and apoptosis. This network would become even more complex upon validation of other candidate targets in the future. The existing body of evidence indicates that the individual miRNAs in the miR-17-92 cluster control multiple targets involved in different cellular processes, as summarized in Figure 19.6. This may be the modus operandi of how miRNAs exert their biological effects both in normal and pathological states.

## 19.6.2 miRNAs in the Oncogenic RAS Pathway

Lung cancer is an aggressive form of cancer, with less than 10% overall five-year survival. RAS mutations are found in 15–20% of non-small-cell lung carcinoma (NSCLC) and in 30–50% adenocarcinoma. Mutations in Ras are necessary for the induction and maintenance of lung cancer [129], and are associated with a poor prognostic outcome for overall survival in NSCLC patients, as revealed by a recent retrospective study [130].

About 40% of lung tumor tissues express significantly low levels, as low as 80%, of let-7 miRNA [56,131]. This is also true for 60% of lung cancer cell lines. A better survival outcome for patients with tumors expressing high levels of let-7 expression compared to those expressing low levels of let-7 was noted in both NSCLC and adenocarcinoma patients, suggesting a tumor suppressor role for let-7 miRNA [56]. Further, in this patient population, the let-7 level emerged as an independent and significant prognostic factor, next to the disease stage. In addition to lung cancer, genomic loci that encode different members of let-7 family are also mapped to CAGRs of various tumors of different tissues, including lung, breast, cervix, and ovary [8].

The possible molecular connection between the let-7 expression and oncogenesis was made in a separate study, performed first in *C. elegans* and subsequently in human cells [131]. A cellular phenotype with uncontrolled cell division in the absence of differentiation, a characteristic of cancer, was initially observed in let-7 mutants in worms [132]. A computer search identified let-60, the *C. elegans* ortholog of the human oncogene RAS, as the target for let-7. Interestingly, both let-7 and its target RAS are conserved between *C. elegans* and human. Binding sites for let-7 family members have been identified in the 3′ UTRs of H-RAS, K-RAS, and N-RAS, the three human RAS genes identified [133]. Further, Let-7 regulated RAS expression was experimentally validated in HeLa cells and HepG2 cells [131]. In light of this evidence, it is likely that the reduction in let-7 levels in lung tumors could lead to the upregulation of the Ras protein. This has been demonstrated in human lung tumors, in which the expression of the Ras protein was inversely correlated with the let-7 levels. It is important to note that Ras is just one target out of several hundreds of predicted targets for the let-7 miRNA family. Their involvement in the oncogenesis process awaits further investigations.

It is likely that under normal conditions, RAS expression is tightly controlled and kept at a relatively low level through regulated expression of let-7 miRNA. When let-7 level is deregulated, as in lung tumors, the overall expression of RAS can get upregulated to a level that activates a cascade of oncogenic pathways. The initiation of RAS-driven oncogenesis, under this scenario, may take place in the absence of any activating mutations in RAS. RAS proteins downregulate anti-angiogenic proteins such as Thrombospondin-1 and upregulate the activities of VEGF and matrix metalloproteases required for endothelial cell migration, contributing to overall tumor growth and metastatic potential [134].

## 19.7   miRNAs IN STEM CELLS AND EXTENSION TO CANCER STEM CELLS

Embryonic stem cells (ES) are able to differentiate into various kinds of somatic tissues (pluripotent) and recapitulate the early embryonic development process. Since they have self-renewing capacity in an environment in which other cells do not divide, ES provide an unlimited source of material for studying the developmental processes of advanced animals, including humans. Since miRNAs are often expressed in a tissue-specific and developmental stage-specific manner, ES are likely to be affected by miRNA regulation. Dicer1-null animals are embryonic lethal and show depletion of ES, suggesting that the miRNAs are regulating the ability of ES for self-renewal in vertebrates [67,68]. The overall effect of miRNAs on adult germ-line stem cell (GSC) development was studied using a conditional gene knockout approach in *Drosophila* [135]. The reduction of Dicer1 activity in female flies resulted in a marked depletion of developing egg chambers, owing to the reduction of cell division of GSCs with increase in Cyclin E, the gatekeeper of G1/S cell cycle checkpoint. The Dicer1 mutant male flies also exhibited higher number of GSCs with elevated Cyclin E, demonstrating that the impairment of miRNA biogenesis affects cell divisions in GSCs in general. Further, miRNAs help GSC to maintain the stem cell character of self-renewal by facilitating the G1/S checkpoint transition upon inhibiting cyclin-dependent kinase (CDK) inhibitor, Dacapo, a fly homologue of p21/p27 family CDK inhibitors.

Certain functional members in the RISC, such as Ago-2, belong to the Argonaute family proteins. Other members, *piwi* in fly and *prg-1* and *prg-2* in worms, are crucial in maintaining stem cell characteristics [136] within germ-line stem cells (reviewed in Reference 83). In humans, *hiwi*, the *piwi* homologue, is specifically expressed in the germ-line cells [86] and CD34$^+$ hematopoietic progenitors [137]. Enhanced expression of hiwi has been found in 12 out of 19 testicular seminomas, tumors originating from embryonic germ cells and retaining the germ cell phenotype, and not in nonseminomas, tumors that lack germ cell characteristics but originate from the same precursor cells [86]. Taken together, these observations suggest that the RNAi pathway, and most likely the miRNAs, play a crucial role in maintaining stem cell biology.

The limited number of miRNAs isolated so far from ES indicates that a few are conserved between human and mouse, and some are specific to either human or mouse ES, suggesting that some of the regulatory pathways between human and mouse may not be conserved. Interestingly, many of the ES-specific miRNAs of both human and mouse are highly homologous and form miRNA families, and are expressed in a multicistronic fashion, an organization with distinct advantages: it ensures robust regulation of common targets and also allows simultaneous regulation of the expression of more than one miRNA during the differentiation of ES. Interestingly, miR-302a, -302b, -302c, and -302d that are encoded within this multicistron are members of an miRNA family whose other members include miR-17-5p and miR-20. MiR-17-5p and miR-20 are in the oncogenic multicistron miR-17-92, upregulation of which accelerates B-cell lymphoma in a mouse model [57], and their overexpression has been noted in human breast tumors [54,101].

Multicistronic miRNAs specific to ES get downregulated upon differentiation into embryoid bodies (EB) in both mouse and humans. In humans, miR-371-373 multicistron in chromosome 19 was downregulated rapidly upon differentiation into EB, and the miRNAs from this particular multicistron were absent in EC (embryonic carcinoma). On the other hand, the second multicistron miR-302b-367 was expressed in ES, EB, and EC, but not in HeLa cells, implying that EC is a poorly differentiated cell line that retains stem cell character. These results suggest that deregulated expression of miRNAs that are normally expressed during embryonic development could lead to tumor formation.

To study the spatiotemporal expression of miRNAs during mammalian development, a technique used earlier in *Drosophila* [42] was recently adapted to the mouse [23]. In a transient transgenic mouse model, a sensor reporter gene (β-galactosidase) responsive to a given miRNA was engineered to express ubiquitously in developing embryo. The silencing of the reporter gene reflected the expression of the specific miRNA in a given tissue at a given time. Using this approach, two let-7 family members showed restricted expression in developing limb buds. Let-7c expression revealed anterior posterior gradient, akin to the expression of the *HOX* genes that are regulated in this fashion during embryogenesis.

The *HOX* genes constitute a subset of homeobox genes that represent the vertebrate counterparts of the bithorax and antennapedia genes in *Drosophila* that specify positional identity along the anterior-posterior axis. In vertebrates, the *HOX* genes encode transcriptional regulatory proteins containing a helix-turn-helix DNA binding domain and are involved in positional identity along the anterior-posterior axis. Cells that are actively proliferating as well as those that are undifferentiated tend to express the *HOX* genes. Deregulation of *HOX* gene expression is associated with different types of cancer [138,139]. The *HOX* genes are organized in clusters that are distributed among different chromosomes. Mammals have four HOX clusters (A through D) containing 39 genes distributed into 13 paralogous subgroups (Figure 19.7). Interestingly, several *miR* genes, arranged in the same direction of HOX transcription, are located within the *HOX* gene clusters; five individual *mir* genes belonging to two families are located within the *HOX* gene clusters with at least one *mir* gene per cluster. The *HOX* genes show collinear expression with respect to their chromosomal location (genes near the 5' end are expressed more posteriorly, and vice versa). Using the sensor transient transgenic mouse model, Mansfield et al. identified that MiR-196 is expressed in the posterior trunk of the embryo [23]. This collinear expression of *mir* gene is similar to the expression pattern of the *HOX* genes, suggesting similar regulatory elements that control the *HOX* gene expression may also regulate miR-196, and probably other miRNAs in the HOX clusters.

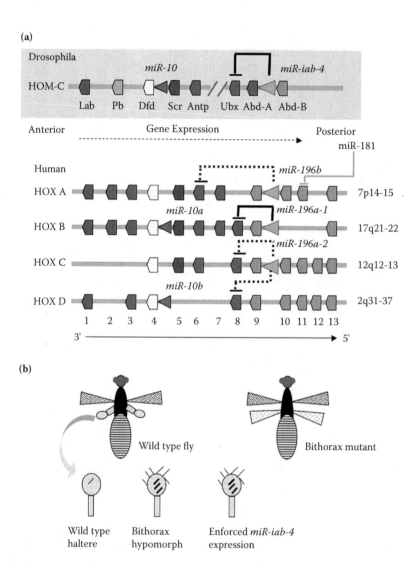

**FIGURE 19.7**

MiR-196 downregulates several *HOX* genes. HOXB8, by direct cleavage of a single perfect complementary site within the 3′ UTR [22,23]; and HOXA7, HOXC8, and HOXD8, by translational repression as they carry multiple target sites of imperfect complementarity within their 3′ UTRs [22]. Interestingly, the posterior limit of HOXB8 expression coincides with the anterior limit of miR-196 expression, indicating how an miRNA dictates the spatial control of a *HOX* gene expression, which in turn controls embryonic development. MiR-196 is a member of the let-7 family, and one would expect to find common targets that are being shared by this family. It is quite likely that the *HOX* genes may be regulated by many other miRNAs that are not physically located within HOX genes. An example is miR-181, which downregulates HOX A11 during the differentiation of mammalian myoblasts [140]. miR-181 is expressed at very low levels in terminally differentiated muscle, but upregulated in regenerating muscle fibers as well as in undifferentiated myoblasts. In this context, the target of miR-181, HOX A11, is involved in limb muscle patterning, and it inhibits MyoD expression. The depletion of miR-181 affected muscle regeneration and myoblast differentiation by upregulating HOX A11 (a repressor of the differentiation process), which in turn downregulates the transcription factor, MyoD.

Upon deregulation of their expression, *HOX* genes that regulate embryonic development have been involved in multiple types of cancer [138]. In addition, *HOX* genes are differentially expressed during hematopoiesis, and deregulation of this coordinated process leads to hematopoietic malignancies. A well-known

**FIGURE 19.7** (See facing page.) miRNAs in HOX gene regulation. a. Homeotic (HOX) genes encode a family of transcription factors that dictate segment identity along the anterior-posterior axis of metazoans. In *Drosophila melanogaster*, HOX genes are organized in two clusters, Antennapedia complex and Bithorax, separated by an ~7.5 Mb region in the homeotic gene complex (HOM-C). Antennapedia complex: lab (labial), Pb (proboscipedia), Dfb (Deformed), Scr (Sex combs reduced), and Antp (Antennapedia); Bithorax: Ubx (Ultrabithorax), Abd-A (Abdominal-A), and Abd-B (Abdominal-B). During evolution, these primodial HOX genes underwent duplication and divergence to generate 39 HOX genes in humans (color-coded to indicate their evolutionary history), which are arranged in four clusters. Each cluster has 9 to 11 members and 13 paralog groups, as color-coded. HOX genes are expressed sequentially from 3′ to 5′ along the anterior-posterior axis with a low-number gene (HOXA-1) being expressed earlier and more anteriorly than a higher-number gene (HOXA-8). In humans, 5 *miR*s (indicated in black and pink arrowheads) from two families are located within the 4 HOX clusters. The orthologs of the same two miRNA families are found in *Drosophila* HOM-C, showing their conservation during evolution. Among these two miRNA families, the miR-196 family targets several HOX genes in humans, as indicated by either solid lines with existing biochemical validation or dotted lines with bioinformatic predictions. HOX genes can also be regulated by miRNAs located outside the HOX clusters, as reported for miR-181, that target HOXA-11. In *Drosophila*, miR-10 is located within the antennapedia complex, whereas miR-iab-4, which is analogous to miR-196 in vertebrates, is located in the Bithorax cluster. The posterior miR-iab-4 controls the expression of anterior Ubx. The expression pattern of Ubx shows inverse correlation to that of miR-iab-4. While bithorax mutant shows a fully developed second pair of wings in place of haltere (haltere-to-wing homeotic transformation), bithorax hypomorphs develop triple rows of sensory bristles characteristic of wings. A similar phenotype has been observed upon the ectopic expression of miR-iab-4, which led to the reduced levels of Ubx protein.

example is the rearrangement of the MLL gene, which occurs in a range of aggressive leukemias, including ALL and AML. The normal function of MLL is to regulate a number of genes, including several genes in HOX clusters, HOXA9 and HOXC8. In the oncogenic form, the N-terminal domain MLL that contains the DNA binding elements combines with a variety of partners to form chimeric proteins leading to deregulation of its normal function. On this note, it is also conceivable that the deregulation of miRNAs that control *HOX* gene expression could contribute to hematopoietic malignancies, as evident by miR-181 in directing B-cell lineage [141].

## 19.8   microRNAs AND HEMATOPOIESIS

Hematopoiesis is a life-long activity in which a pluripotent self-renewing hematopoietic stem cell (HSC) gives rise to all blood cell lineages through a highly controlled multistage process. HSC is one of the most extensively studied systems. A great deal of knowledge has been derived from the studies on differentiation of hematopoietic progenitor cells into specific cell lineages, and the development of malignant tumors as a result of specific chromosomal translocations. Some of the fusion proteins that originate from translocations have been characterized to be the causative agent of specific malignancies. Two specific examples include 8;14 translocation in Burkitt lymphoma, in which a portion of Myc in chromosome 8 is in juxtaposition with immunoglobulin heavy chain (IGH), which is actively transcribed in B cells; and 9;22 reciprocal translocation in CML, in which the ABL gene on chromosome 9 is fused with the promoter of the BCR gene on chromosome 22. The products of these aberrant fusions are often involved as transcription factors that affect the differentiation process, leading to hematopoietic malignancies. Concerted activities of growth factors and unique cocktails of transcription factors (TF) dictate stage-specific differentiation during hematopoiesis (reviewed in Reference 142), and drastic changes to such TF milieu could have profound effects on the developmental process.

Given the intimate role played by miRNA-mediated gene regulation, along with some of their genomic locations within the FRAs and CAGRs, it is possible that deregulated miRNA expression could impact hematopoiesis as well, leading to hematopoietic malignancies. Some experimental evidence exists to support this notion. Hematopoietic lineage differentiation is affected by the expression of specific miRNAs, as demonstrated by bone marrow transplantation in mice [141]. miRNAs cloned from mouse bone marrow revealed that some of the miRNAs were differentially expressed in hematopoietic organs, as well as in hematopoietic lineages [141]. MiR-223 is selectively expressed in bone marrow that contains a mixed population of cells undergoing hematopoiesis, especially in CD34⁻ cell population, which consists mostly of lineage-committed precursors [143]. MiR-142 is highly expressed only in hematopoietic tissues. MiR-181, strongly expressed in the thymus, bone marrow, and spleen, is present in lineage negative progenitors (lin⁻) and upregulated only in differentiated B-lymphocytes. MiR-181 showed a significant biological effect when it was ectopically expressed in lin⁻ hematopoietic progenitor cells from mouse bone marrow in a 10-day in vitro culture on a stromal cell layer supplemented with a cocktail of cytokines. Overexpression of miR-181 resulted in the doubling of cells in

B-cell lineage without noticeable effect on the T-cell lineage. Environmental cues that trigger lineage commitment and further differentiation found under in vitro experimental conditions may be different from those in the in vivo microenvironment. Interestingly, when miR-181 was ectopically expressed in Lin⁻ progenitors that were reconstituted in vivo by bone marrow engraftment, the B-cell lineage was increased by ~50%. However, under in vivo conditions, miR-181 acted on a T-cell lineage as well, by reducing the T-cell population, specifically CD-8+ T-helper cells, by more than 80%, indicating that miR-181 acts simultaneously on two different lineages.

A genome-wide miRNA expression analysis in different hematopoietic lineages derived from human cord blood identified miR-221 and miR-222 with a markedly low level of expression in erythropoietic culture [144]. Additional biochemical studies revealed a potential link with the enhanced expression of Kit receptor, which may be the direct target of these two miRNAs. Further, it was also shown that the Kit receptor is necessary for the expansion of erythroblasts in culture, suggesting how miRNA expression contributes to direct different lineages during hematopoiesis.

The role of miR-223 in directing myeloid progenitor lineage and the possible involvement in acute promyelocytic leukemia (APL) is an elegant example of how a single microRNA could direct a hematopoietic lineage in humans [143]. A clonal expansion of hematopoietic progenitors blocked at various stages in the differentiation of myeloid lineage is found in acute myeloid leukemia (AML). In APL, a subtype of AML, the differentiation of hematopoietic precursors is blocked at the promyelocytic stage owing to specific chromosomal translocations involving the retinoic-acid receptor–α (RAR-α) and generating RAR-α fusion genes. The resulting gene fusions promote the hematopoietic malignancy by repressing RAR-α directed gene expression and deregulated chromatin modification at the chromosomal level. Retinoic acid (RA) treatment can induce terminal differentiation of APL blasts both in vitro and in vivo. A gene family of the CCAAT enhancer proteins (C/EBPs) that are known to regulate cell growth and differentiation are among the genes targeted by the RAR-α fusion proteins, and can be induced by RA treatment. Upon expression in the early myeloid precursors, C/EBPα directs them to the granulocytic lineage [145]. Further, the loss of C/EBPα has been linked to leukemogenesis [146], indicating its prominent role as a lineage differentiation factor combined with a tumor suppressor role. C/EBPα is regulated by miR-223 by binding to its upstream promoter region. One of the validated targets for miR-223 is the transcription factor NF1-A, a CCAAT-related binding protein involved in controlling cell growth [143]. Interestingly, NF1-A competes with C/EBPα at the upstream of miR-223 promoter, placing miR-223 expression in a regulatory loop (Figure 19.8). It appears that the involvement in a negative feedback loop is a common feature among miRNAs whose expressions need to be tightly regulated (see above for EIF and miR 17-92 cluster, and Figure 19.7). However, miR-223 is in a regulatory circuit in which it downregulates its own repressor, NF1-A, to maintain a steady level of its expression (Figure 19.8). Upon RA treatment, the expression of miR-223 is upregulated within 72 hr in both peripheral blood cells isolated from untreated APL patients and APL cell lines, but not in AML cell lines of non-APL origin [143]. Ectopic expression of miR-223 in an APL cell line led to the differentiation of cells into the granulocytic lineage and not along the monocytic lineage, similar to the outcome of the RA treatment. Interestingly, the NF1-A binding site overlaps with one of

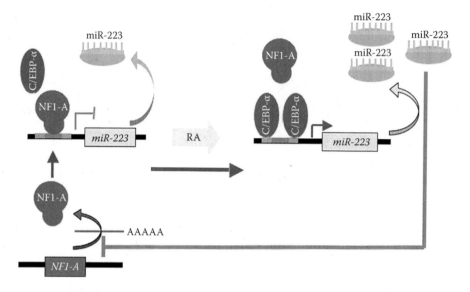

**FIGURE 19.8**   The involvement of transcription factors in the regulation of miR-223 expression in granulopoiesis in humans. The upstream promoter element of miR-223 has binding sites for C/EBP-α as well as for NF1-A. In acute promyelocytic leukemia (APL), the level of miR-223 expression has been detected to be unusually low, possibly owing to suppression of its expression through the binding of NF1-A to the promoter. Pharmacological treatment of retinoic acid (RA) displaces NF1-A and allows CEBP-α binding to the promoter, thereby upregulating miR-223 expression. The high level of miR-223 expression triggered by the RA treatment also leads to the downregulation of NF1-A through the interaction of miR-223 with the NF1-A message. The latter process maintains miR-223 expression level at a high steady state level after a RA treatment to allow miR-223-mediated downstream events that will lead to differentiation of APL cells into granulocytic lineage.

the two C/EBPα binding sites in the miR-223 promoter, and the occupation of NF1-A decreases miR-223 expression. Upon RA treatment, C/EBPα displaces NF1-A and activates the expression of miR-223 by only about 3- to 5-fold. This small change in the miRNA level is sufficient to trigger differentiation of these cells into the granulocyte lineage. These data show how a single microRNA could direct the lineage fate during myelopoesis, by changing its expression even to a modest level.

Hematopoietic stem cells and multipotential progenitors are known to display mixed lineage patterns of gene expression [147,148] and, during lineage-specific differentiation, the suppression and the activation of specific genes pertinent to a specific lineage is required. It is possible that specific miRNAs such as miR-181, -221, -222, and -223 play a pivotal role in activating and suppressing specific messages during hematopoiesis (Figure 19.9). While suppression of gene expression is a direct interaction between the miRNAs and their targets, an activation of gene expression can be achieved by inhibiting the expression of specific repressors that block the transcription of specific genes.

Deregulation of miRNAs could take place at virtually any stage of the cell cycle, during and after differentiation. Given stem cells' capacity to repopulate various cell

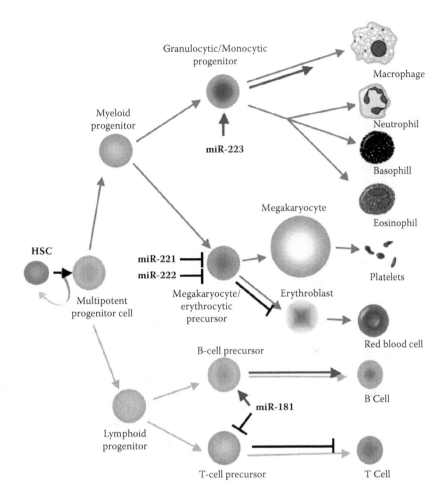

**FIGURE 19.9** miRNAs that regulate hematopoiesis. A number of hematopoietic cell lineages originate from a hematopoietic stem cell (HSC) that divides into two daughter cells, one of which becomes a multipotent progenitor cell that gives rise to myeloid and lymphoid lineages. So far, a limited number of miRNAs that control the hematopoiesis have been identified. The expression of miR-223 directs the myeloid lineage toward differentiation into macrophages (monocytic lineage). On the other hand, the expression of miR-221 and miR-222 inhibits erythropoiesis and directs the generation of platelets. The fate of the lymphoid lineage could be dictated by the level of miR-181 expression that inhibits the T-cell differentiation while promoting B-cell production.

lineages and tissues, in addition to their self-renewal, deregulation of miRNAs in a stem cell population would be a formidable challenge in controlling cancer.

## 19.9 VIRAL miRNAs

Plants and insects use RNA silencing as a defense mechanism to fend off viruses (reviewed in Reference 149). This evolutionary primitive mechanism of defense is still conserved in humans as well, as shown by a recent report in which a cellular miRNA protects human cells from a retrovirus [150]. It is also conceivable that viruses would have evolved a similar mechanism to exploit their hosts. To that effect, several pathogenic viruses, including Epstein-Barr (EBV) [151], herpesvirus [152–154], HIV [155], and Simian Virus-40 (SV-40) [156], have been shown to harbor specific miRNAs of their own. The number of virally encoded miRNA repertoires is expected to rise with the exploration of more and more viral genomes.

So far, 10 viral *mir* genes (v-miRs) specific to Kaposi's sarcoma-associated herpesvirus (KSHV) have been identified from cell lines infected with the virus, and all of them, arranged in one direction, are confined to an ~4-kb region within a coding gap in latency-associated ORF73 of previously unknown function [152,153]. Nine out of the 10 miRNAs are located within a single intron of ORF73. Taken together, it is likely that these 9 miRNAs are coexpressed. Yet, TPA (12-O-tetradecanoylphorbol-13-acetate) induction resulted in the upregulation of only 2 miRNAs [152], raising the possibility of posttranscriptional regulation of this cluster. Further, most of these v-miRs are likely to maintain the latency stage of the viral activities. In EBV, two miRNA clusters have been identified, and miRNAs in one cluster (BHRF1-1, -2, and -3) were differentially expressed in three different cell lines characteristics of latency stages I, II, and III of the virus, again suggesting posttranscriptional regulation of individual miRNAs derived from a single transcript. On the other hand, the expression of two miRNAs from the second cluster was markedly elevated in a cell line representative of lytic stage from the levels in latency stage cell lines. An miRNA encoded within the nef-gene of HIV suppresses HIV transcription in infected human T cells, helping the virus to commit to its latency stage [157,158].

In SV-40, a single miRNA with unique characteristics has been identified within the late pre-mRNA [156]. A 62-nucleotide transcript with an unknown function referred to as SAS (SV40-associated small RNA), discovered earlier, has recently been identified to encode a miRNA. Both arms of this miRNA (SAS-miR) get incorporated into the RISC and are functional in cleaving the SV-40 early mRNA at two complementary sites, indicating the resourceful nature of the virus in harnessing information coded on both strands to regulate gene expression. Furthermore, SAS-miR downregulates the T-antigen (a target of cytotoxic T lymphocyte response) to evade the immune response of the host.

Overall, viral miRNAs have been evolved to effectively control their life cycles by regulating the expression of viral genes. However, the most intriguing quest is to uncover the host genes regulated by viral miRNAs. A number of human genes regulating cell proliferation, apoptosis, survival, signal transduction, and immune response have been predicted to be candidates for viral miRNAs, but have not been proved experimentally. However, in light of the existing evidence, it is very likely that certain viruses known to cause cancer may do so by manipulating the miRNA pathways either using their own miRNAs or the miRNAs of the host to potentiate oncogenesis. In an interesting investigation, Lu and Cullen have reported that ~160-nt-long VA1 noncoding RNA of adenovirus can inhibit miRNA biogenesis by competing with pre-miRNA exporter Exportin 5 and Dicer [159]. Analogous factors derived from viruses are likely to exist for regulation of the miRNA pathway globally as well as specifically, contributing to pathogenesis in the host, including cancer.

## 19.10   CONCLUSIONS AND FUTURE PERSPECTIVES

The initial discovery of microRNAs in *C. elegans*, and later in other species, has rapidly opened up a whole new area of research that will pave the way for an understanding of the biology of the human genome. Ever since its inception, the exploration of miRNA biology has already made significant inroads into cancer research. In a more pragmatic sense, future discoveries will focus on the rapid diagnosis of human cancer and novel treatment options for patients, depending on an understanding of miRNA biology. Most of the major human cancer types have subtypes dictated by, among other factors, the molecular mechanisms of tumor formation. The correct diagnosis is key to identifying the right treatment options for patients with a given tumor subtype. For example, breast cancer collectively represents tumors that arise in the breast. These tumors come in multiple forms, such as those with differential estrogen, progesterone, ERBB2, ERBB3 receptor expression, ductal or lobular, in situ or invasive, and differential involvement of lymph nodes. In the future, each subtype of a human cancer may be rapidly diagnosed with microRNA expression profiles derived from a small number of tumor cells from a patient's needle biopsy of the tumor or from circulating tumor cells in blood.

The discovery of unique miRNA expression profiles associated with each tumor type will help identify a unique set of targets that are being regulated by these miRNAs. These will become the next wave of molecular targets for the development of novel therapeutics for the pharmaceutical industry. Biomarkers associated with each tumor type will help design clinical trials for the right patient population, and also facilitate the interpretation of therapeutic efficacy rapidly. The unique miRNA profiles in each tumor type could also become a biomarker signature to facilitate drug development and to monitor therapeutic response. An understanding of miRNA profiles and their potential targets in each tumor type will enable the elucidation of novel molecular circuitries associated with oncogenesis. Nucleic-acid-based therapeutic approaches have been developed to directly target specific miRNAs when they are abnormally overexpressed using complementary or antisense oligonucleotide sequences [160–164]. Alternatively, downregulated miRNAs can be supplemented with synthetic pre-miRs or mature miRNA duplexes. This can also be achieved by using viral vectors engineered to transcribe hairpin miRNA precursors. While specificity is the clear advantage of this approach, along with the ease of use as a research tool, it is still faced with the hurdle of delivering relatively large polyanionic molecules into the tissues of whole animals, especially into tumors.

The most challenging aspect of microRNA biology is the assignment of a specific biological function to each miRNA. Screens that are designed to inhibit specific miRNAs would enable the identification of miRNAs with a common biological role, involving perhaps a single molecular pathway. Computational algorithms based on seed pairing and evolutionary conservation of miRNA target sites across multiple species could facilitate target prediction. However, the biological validation of predicted targets is extremely important. This is further complicated by the combinatorial nature of target regulation predominantly associated with miRNAs. One approach is to undertake a systematic knockout of miRNA genes in animal models to identify their function. As the elimination of a single *miR* at a time may give rise to subtle or no phenotypes [44], elimination of either miRNA families altogether, or multicistronic miRNAs or both in model organisms may accelerate the discovery of their biological functions [47].

## ACKNOWLEDGMENTS

The author is thankful to Vineetha Jayasena and Glenn Begley for the comments and critical reading of the chapter and Rick Kendall for the encouragement.

## REFERENCES

1. Lee, Y. et al. MicroRNA genes are transcribed by RNA polymerase II. *Embo J* **23**, 4051-4060 (2004).
2. Bracht, J., Hunter, S., Eachus, R., Weeks, P. and Pasquinelli, A.E. Trans-splicing and polyadenylation of let-7 microRNA primary transcripts. *Rna* **10**, 1586-1594 (2004).
3. Cai, X., Hagedorn, C.H. and Cullen, B.R. Human microRNAs are processed from capped, polyadenylated transcripts that can also function as mRNAs. *Rna* **10**, 1957-1966 (2004).
4. Houbaviy, H.B., Dennis, L., Jaenisch, R. and Sharp, P.A. Characterization of a highly variable eutherian microRNA gene. *Rna* **11**, 1245 - 1257(2005).
5. Ambros, V. and Lee, R.C. Identification of microRNAs and other tiny noncoding RNAs by cDNA cloning. *Methods Mol Biol* **265**, 131-158 (2004).
6. Kim, Y.K. and Kim, V.N. Processing of intronic microRNAs. *EMBO Journal* **26**, 775-783 (2007).
7. Ying, S.Y. and Lin, S.L. Intron-derived microRNAs--fine tuning of gene functions. *Gene* **342**, 25-28 (2004).
8. Calin, G.A. et al. Human microRNA genes are frequently located at fragile sites and genomic regions involved in cancers. *Proc Natl Acad Sci U S A* **101**, 2999-3004 (2004).
9. Denli, A.M., Tops, B.B., Plasterk, R.H., Ketting, R.F. and Hannon, G.J. Processing of primary microRNAs by the Microprocessor complex. *Nature* **432**, 231-235 (2004).
10. Gregory, R.I. et al. The Microprocessor complex mediates the genesis of microRNAs. *Nature* **432**, 235-240 (2004).
11. Allen, E. et al. Evolution of microRNA genes by inverted duplication of target gene sequences in Arabidopsis thaliana. *Nat Genet* **36**, 1282-1290 (2004).
12. Landthaler, M., Yalcin, A. and Tuschl, T. The human DiGeorge syndrome critical region gene 8 and Its D. melanogaster homolog are required for miRNA biogenesis. *Curr Biol* **14**, 2162-2167 (2004).
13. Ambros, V., Lee, R.C., Lavanway, A., Williams, P.T. and Jewell, D. MicroRNAs and other tiny endogenous RNAs in *C. elegans*. *Curr Biol* **13**, 807-818 (2003).
14. Lee, Y., Jeon, K., Lee, J.T., Kim, S. and Kim, V.N. MicroRNA maturation: stepwise processing and subcellular localization. *Embo J* **21**, 4663-4670 (2002).
15. Gwizdek, C. et al. Exportin-5 mediates nuclear export of minihelix-containing RNAs. *J Biol Chem* **278**, 5505-5508 (2003).
16. Lund, E., Guttinger, S., Calado, A., Dahlberg, J.E. and Kutay, U. Nuclear export of microRNA precursors. *Science* **303**, 95-98 (2004).
17. Bohnsack, M.T., Czaplinski, K. and Gorlich, D. Exportin 5 is a RanGTP-dependent dsRNA-binding protein that mediates nuclear export of pre-miRNAs. *Rna* **10**, 185-191 (2004).
18. Schwarz, D.S. et al. Asymmetry in the assembly of the RNAi enzyme complex. *Cell* **115**, 199-208 (2003).
19. Khvorova, A., Reynolds, A. and Jayasena, S.D. Functional siRNAs and miRNAs exhibit strand bias. *Cell* **115**, 209-216 (2003).

20. Doench, J.G., Petersen, C.P. and Sharp, P.A. siRNAs can function as miRNAs. *Genes Dev* **17**, 438-442 (2003).
21. Zeng, Y., Wagner, E.J. and Cullen, B.R. Both natural and designed micro RNAs can inhibit the expression of cognate mRNAs when expressed in human cells. *Mol Cell* **9**, 1327-1333 (2002).
22. Yekta, S., Shih, I.H. and Bartel, D.P. MicroRNA-directed cleavage of HOXB8 mRNA. *Science* **304**, 594-596 (2004).
23. Mansfield, J.H. et al. MicroRNA-responsive 'sensor' transgenes uncover Hox-like and other developmentally regulated patterns of vertebrate microRNA expression. *Nat Genet* **36**, 1079-1083 (2004).
24. Meister, G. et al. Human Argonaute2 mediates RNA cleavage targeted by miRNAs and siRNAs. *Mol Cell* **15**, 185-197 (2004).
25. Humphreys, D.T., Westman, B.J., Martin, D.I. and Preiss, T. MicroRNAs control translation initiation by inhibiting eukaryotic initiation factor 4E/cap and poly(A) tail function. *Proceedings of the National Academy of Sciences of the United States of America* **102**, 16961-16966 (2005).
26. Pillai, R.S. et al. Inhibition of translational initiation by Let-7 MicroRNA in human cells. *Science* **309**, 1573-1576 (2005).
27. Parker, R. and Song, H. The enzymes and control of eukaryotic mRNA turnover. *Nature Structural and Molecular Biology* **11**, 121-127 (2004).
28. Jing, Q. et al. Involvement of microRNA in AU-rich element-mediated mRNA instability. *Cell* **120**, 623-634 (2005).
29. Giraldez, A.J. et al. Zebrafish MiR-430 promotes deadenylation and clearance of maternal mRNAs. *Science* **312**, 75-79 (2006).
30. Berezikov, E. et al. Phylogenetic shadowing and computational identification of human microRNA genes. *Cell* **120**, 21-24 (2005).
31. Xie, X. et al. Systematic discovery of regulatory motifs in human promoters and 3' UTRs by comparison of several mammals. *Nature* **434**, 338-345 (2005).
32. Bentwich, I. et al. Identification of hundreds of conserved and nonconserved human microRNAs. *Nat Genet* **37**, 766-770 (2005).
33. Cummins, J.M. et al. The colorectal microRNAome. *Proceedings of the National Academy of Sciences of the United States of America* **103**, 3687-3692 (2006).
34. Lim, L.P., Glasner, M.E., Yekta, S., Burge, C.B. and Bartel, D.P. Vertebrate microRNA genes. *Science* **299**, 1540 (2003).
35. Lim, L.P. et al. Microarray analysis shows that some microRNAs downregulate large numbers of target mRNAs. *Nature* **433**, 769-773 (2005).
36. Lee, R.C., Feinbaum, R.L. and Ambros, V. The C. elegans Heterochronic Gene *lin-4* Encodes Small RNAs with Antisense Complementarity to *lin-14*. *Cell* **75**, 843-854 (1993).
37. Ha, I., Whitma, B. and Ruvkun, G. A bulged l*in-4/lin-14* RNA duplex is sufficient for *Caenorhabditis elegans* lin-14 temporal gradient formation. *Genes and Dev.* **10**, 3041-3050 (1996).
38. Slack, F.J. et al. The lin-41 RBCC gene acts in the C. elegans heterochromic pathway between the let-7 regulatory RNA and the LIN-29 transcription factor. *Mol Cell* **5**, 659-669 (2000).
39. Krek, A. et al. Combinatorial microRNA target predictions. *Nat Genet* **37**, 495-500 (2005).
40. John, B. et al. Human MicroRNA targets. *PLoS Biol* **2**, e363 (2004).
41. Banerjee, D. and Slack, F. Control of developmental timing by small temporal RNAs: a paradigm for RNA-mediated regulation of gene expression. *Bioessays* **24**, 119-129 (2002).
42. Brennecke, J., Hipfner, D.R., Stark, A., Russell, R.B. and Cohen, S.M. bantam encodes a developmentally regulated microRNA that controls cell proliferation and regulates the proapoptotic gene hid in *Drosophila*. *Cell* **113**, 25-36 (2003).

43. Brennecke, J., Stark, A., Russell, R.B. and Cohen, S.M. Principles of microRNA-target recognition. *PLoS Biol* **3**, e85 (2005).

44. Lai, E.C. microRNAs: runts of the genome assert themselves. *Curr Biol* **13**, R925-936 (2003).

45. Hipfner, D.R., Weigmann, K. and Cohen, S.M. The *bantam* gene regulates *Drosophila* growth *Genetics* **161**, 1527-1537 (2002). ·

46. Xu, P., Vernooy, S.Y., Guo, M. and Hay, B.A. The *Drosophila* microRNA Mir-14 suppresses cell death and is required for normal fat metabolism. *Curr Biol* **13**, 790-795 (2003).

47. Leaman, D. et al. Antisense-Mediated Depletion Reveals Essential and Specific Functions of MicroRNAs in Drosophila Development. *Cell* **121**, 1097-1108 (2005).

48. Chan, J.A., Krichevsky, A.M. and Kosik, K.S. MicroRNA-21 Is an Antiapoptotic Factor in Human Glioblastoma Cells. *Cancer Res* **65**, 6029-6033 (2005).

49. Calin, G.A. et al. Frequent deletions and down-regulation of micro- RNA genes miR15 and miR16 at 13q14 in chronic lymphocytic leukemia. *Proc Natl Acad Sci U S A* **99**, 15524-15529 (2002).

50. Dong, J.T., Boyd, J.C. and Frierson, H.F., Jr. Loss of heterozygosity at 13q14 and 13q21 in high grade, high stage prostate cancer. *Prostate* **49**, 166-171 (2001).

51. Lagos-Quintana, M. et al. Identification of tissue-specific microRNAs from mouse. *Curr Biol* **12**, 735-739 (2002).

52. Michael, M.Z., SM, O.C., van Holst Pellekaan, N.G., Young, G.P. and James, R.J. Reduced accumulation of specific microRNAs in colorectal neoplasia. *Mol Cancer Res* **1**, 882-891 (2003).

53. Sonoki, T., Iwanaga, E., Mitsuya, H. and Asou, N. Insertion of microRNA-125b-1, a human homologue of lin-4, into a rearranged immunoglobulin heavy chain gene locus in a patient with precursor B-cell acute lymphoblastic leukemia. *Leukemia* **19**, 2009-2010 (2005).

54. Iorio, M.V. et al. MicroRNA gene expression deregulation in human breast cancer. *Cancer Research* **65**, 7065-7070 (2005).

55. Griffiths-Jones, S. The microRNA Registry. *Nucleic Acids Res* **32**, D109-111 (2004).

56. Takamizawa, J. et al. Reduced expression of the let-7 microRNAs in human lung cancers in association with shortened postoperative survival. *Cancer Res* **64**, 3753-3756 (2004).

57. He, L. et al. A microRNA polycistron as a potential human oncogene. *Nature* **435**, 828-833 (2005).

58. Volinia, S. et al. A microRNA expression signature of human solid tumors defines cancer gene targets. *Proc Nat Acad Sci USA* **103**, 2257-2261 (2006).

59. Han, J. et al. The Drosha-DGCR8 complex in primary microRNA processing. *Genes Dev* **18**, 3016-3027 (2004).

60. Shiohama, A., Sasaki, T., Noda, S., Minoshima, S. and Shimizu, N. Molecular cloning and expression analysis of a novel gene DGCR8 located in the DiGeorge syndrome chromosomal region. *Biochem Biophys Res Commun* **304**, 184-190 (2003).

61. Lindsay, E.A. Chromosomal microdeletions: dissecting del22q11 syndrome. *Nat Rev Genet* **2**, 858-868 (2001).

62. Gregory, R.I. and Shiekhattar, R. MicroRNA biogenesis and cancer. *Cancer Res* **65**, 3509-3512 (2005).

63. Wang, Y., Medvid, R., Melton, C., Jaenisch, R. and Blelloch, R. DGCR8 is essential for microRNA biogenesis and silencing of embryonic stem cell self-renewal. *Nat Genet* **39**, 380-385 (2007).

64. Yang, W. et al. Modulation of microRNA processing and expression through RNA editing by ADAR deaminases.[see comment]. *Nature Structural and Molecular Biology* **13**, 13-21 (2006).

65. Lu, J. et al. MicroRNA expression profiles classify human cancers. *Nature* **435**, 834-838 (2005).

66. Thomson, J.M. et al. Extensive post-transcriptional regulation of microRNAs and its implications for cancer. *Genes and Development* **20**, 2202-2207 (2006).

67. Bernstein, E. et al. Dicer is essential for mouse development. *Nat Genet* **35**, 215-217 (2003).

68. Wienholds, E., Koudijs, M.J., van Eeden, F.J., Cuppen, E. and Plasterk, R.H. The microRNA-producing enzyme Dicer1 is essential for zebrafish development. *Nat Genet* **35**, 217-218 (2003).

69. Yang, W.J. et al. Dicer is required for embryonic angiogenesis during mouse development. *J Biol Chem* **280**, 9330-9335 (2005).

70. Harfe, B.D., McManus, M.T., Mansfield, J.H., Hornstein, E. and Tabin, C.J. The RNaseIII enzyme Dicer is required for morphogenesis but not patterning of the vertebrate limb. *Proceedings of the National Academy of Sciences of the United States of America* **102**, 10898-10903 (2005).

71. Harris, K.S., Zhang, Z., McManus, M.T., Harfe, B.D. and Sun, X. Dicer function is essential for lung epithelium morphogenesis. *Proceedings of the National Academy of Sciences of the United States of America* **103**, 2208-2213 (2006).

72. Muljo, S.A. et al. Aberrant T cell differentiation in the absence of Dicer. *J Exp Med* **202**, 261-269(2005).

73. Cobb, B.S. et al. T cell lineage choice and differentiation in the absence of the RNase III enzyme Dicer. *J Exp Med* **201**, 1367-1373 (2005).

74. Yi, R. et al. Morphogenesis in skin is governed by discrete sets of differentially expressed microRNAs.[see comment]. *Nature Genetics* **38**, 356-362 (2006).

75. Andl, T. et al. The miRNA-processing enzyme dicer is essential for the morphogenesis and maintenance of hair follicles. *Current Biology* **16**, 1041-1049 (2006).

76. Karube, Y. et al. Reduced expression of Dicer associated with poor prognosis in lung cancer patients. *Cancer Sci* **96**, 111-115 (2005).

77. Chiosea, S. et al. Up-regulation of dicer, a component of the MicroRNA machinery, in prostate adenocarcinoma. *American Journal of Pathology* **169**, 1812-1820 (2006).

78. Zhang, L. et al. microRNAs exhibit high frequency genomic alterations in human cancer. *Proceedings of the National Academy of Sciences of the United States of America* **103**, 9136-9141 (2006).

79. Chen, J.F. et al. The role of microRNA-1 and microRNA-133 in skeletal muscle proliferation and differentiation. *Nature Genetics* **38**, 228-233 (2006).

80. Zhao, Y., Samal, E. and Srivastava, D. Serum response factor regulates a muscle-specific microRNA that targets Hand2 during cardiogenesis. *Nature* **436**, 214-220 (2005).

81. Obernosterer, G., Leuschner, P.J., Alenius, M. and Martinez, J. Post-transcriptional regulation of microRNA expression. *Rna-A Publication of the Rna Society* **12**, 1161-1167 (2006).

82. Ruteshouser, E.C. et al. Genome-wide loss of heterozygosity analysis of WT1-wild-type and WT1-mutant Wilms tumors. *Genes Chromosomes Cancer* **43**, 172-180 (2005).

83. Carmell, M.A., Xuan, Z., Zhang, M.Q. and Hannon, G.J. The Argonaute family: tentacles that reach into RNAi, developmental control, stem cell maintenance, and tumorigenesis. *Genes Dev* **16**, 2733-2742 (2002).

84. Grishok, A. et al. Genes and mechanisms related to RNA interference regulate expression of the small temporal RNAs that control C. elegans developmental timing. *Cell* **106**, 23-34 (2001).

85. Mourelatos, Z. et al. miRNPs: a novel class of ribonucleoproteins containing numerous microRNAs. *Genes Dev* **16**, 720-728 (2002).

86. Qiao, D., Zeeman, A.M., Deng, W., Looijenga, L.H. and Lin, H. Molecular characterization of hiwi, a human member of the piwi gene family whose overexpression is correlated to seminomas. *Oncogene* **21**, 3988-3999 (2002).

87. Ramkissoon, S.H. et al. Hematopoietic-specific microRNA expression in human cells. [see comment]. *Leukemia Research* **30**, 643-647 (2006).
88. Calin, G.A. et al. MicroRNA profiling reveals distinct signatures in B cell chronic lymphocytic leukemias. *Proc Natl Acad Sci U S A* **101**, 11755-11760 (2004).
89. Yanaihara, N. et al. Unique microRNA molecular profiles in lung cancer diagnosis and prognosis. *Cancer Cell* **9**, 189-198 (2006).
90. He, H. et al. The role of microRNA genes in papillary thyroid carcinoma. *Proceedings of the National Academy of Sciences of the United States of America* **102**, 19075-19080 (2005).
91. Weber, F., Teresi, R.E., Broelsch, C.E., Frilling, A. and Eng, C. A limited set of human MicroRNA is deregulated in follicular thyroid carcinoma. *Journal of Clinical Endocrinology and Metabolism* **91**, 3584-3591 (2006).
92. Bandres, E. et al. Identification by Real-time PCR of 13 mature microRNAs differentially expressed in colorectal cancer and non-tumoral tissues. *Molecular Cancer* **5**, 29 (2006).
93. Mattie, M.D. et al. Optimized high-throughput microRNA expression profiling provides novel biomarker assessment of clinical prostate and breast cancer biopsies. *Molecular Cancer* **5**, 24 (2006).
94. Hemann, M.T. et al. An epi-allelic series of p53 hypomorphs created by stable RNAi produces distinct tumor phenotypes in vivo. *Nat Genet* **33**, 396-400 (2003).
95. Adams, J.M. et al. The c-myc oncogene driven by immunoglobulin enhancers induces lymphoid malignancy in transgenic mice. *Nature* **318**, 533-538 (1985).
96. Adams, J.M. and Cory, S. Oncogene co-operation in leukaemogenesis. *Cancer Surv* **15**, 119-141 (1992).
97. Knuutila, S. et al. DNA copy number amplifications in human neoplasms: review of comparative genomic hybridization studies. *Am J Pathol* **152**, 1107-1123 (1998).
98. Ota, A. et al. Identification and characterization of a novel gene, C13orf25, as a target for 13q31-q32 amplification in malignant lymphoma. *Cancer Res* **64**, 3087-3095 (2004).
99. Lin, Y.W. et al. Loss of heterozygosity at chromosome 13q in hepatocellular carcinoma: identification of three independent regions. *Eur J Cancer* **35**, 1730-1734 (1999).
100. Tanzer, A. and Stadler, P.F. Molecular evolution of a microRNA cluster. *J Mol Biol* **339**, 327-335 (2004).
101. Hayashita, Y. et al. A polycistronic microRNA cluster, miR-17-92, is overexpressed in human lung cancers and enhances cell proliferation. *Cancer Research* **65**, 9628-9632 (2005).
102. Hayward, W.S., Neel, B.G. and Astrin, S.M. Activation of a cellular onc gene by promoter insertion in ALV-induced lymphoid leukosis. *Nature* **290**, 475-480 (1981).
103. Cory, S. and Adams, J.M. Transgenic mice and oncogenesis. *Annu Rev Immunol* **6**, 25-48 (1988).
104. Clurman, B.E. and Hayward, W.S. Multiple proto-oncogene activations in avian leukosis virus-induced lymphomas: evidence for stage-specific events. *Mol Cell Biol* **9**, 2657-2664 (1989).
105. Tam, W., Hughes, S.H., Hayward, W.S. and Besmer, P. Avian bic, a gene isolated from a common retroviral site in avian leukosis virus-induced lymphomas that encodes a noncoding RNA, cooperates with c-myc in lymphomagenesis and erythroleukemogenesis. *J Virol* **76**, 4275-4286 (2002).
106. Tam, W., Ben-Yehuda, D. and Hayward, W.S. bic, a novel gene activated by proviral insertions in avian leukosis virus-induced lymphomas, is likely to function through its noncoding RNA. *Mol Cell Biol* **17**, 1490-1502 (1997).
107. Haasch, D. et al. T cell activation induces a noncoding RNA transcript sensitive to inhibition by immunosuppressant drugs and encoded by the proto-oncogene, BIC. *Cell Immunol* **217**, 78-86 (2002).

108. Eis, P.S. et al. Accumulation of miR-155 and BIC RNA in human B cell lymphomas. *Proc Natl Acad Sci U S A* **102**, 3627-3632 (2005).

109. Metzler, M., Wilda, M., Busch, K., Viehmann, S. and Borkhardt, A. High expression of precursor microRNA-155/BIC RNA in children with Burkitt lymphoma. *Genes Chromosomes Cancer* **39**, 167-169 (2004).

110. van den Berg, A. et al. High expression of B-cell receptor inducible gene BIC in all subtypes of Hodgkin lymphoma. *Genes Chromosomes Cancer* **37**, 20-28 (2003).

111. Tam, W. Identification and characterization of human BIC, a gene on chromosome 21 that encodes a noncoding RNA. *Gene* **274**, 157-167 (2001).

112. Alizadeh, A.A. et al. Distinct types of diffuse large B-cell lymphoma identified by gene expression profiling. *Nature* **403**, 503-511 (2000).

113. Costinean, S. et al. Pre-B cell proliferation and lymphoblastic leukemia/high-grade lymphoma in E(mu)-miR155 transgenic mice. *Proceedings of the National Academy of Sciences of the United States of America* **103**, 7024-7029 (2006).

114. Boffelli, D. et al. Phylogenetic shadowing of primate sequences to find functional regions of the human genome. *Science* **299**, 1391-1394 (2003).

115. Si, M.-L. et al. miR-21-mediated tumor growth. *Oncogene* **25**, 1-5 (2006).

116. Boxer, L.M. and Dang, C.V. Translocations involving c-myc and c-myc function. *Oncogene* **20**, 5595-5610 (2001).

117. Fernandez, P.C. et al. Genomic targets of the human c-Myc protein. *Genes Dev* **17**, 1115-1129 (2003).

118. Cole, M.D. and McMahon, S.B. The Myc oncoprotein: a critical evaluation of transactivation and target gene regulation. *Oncogene* **18**, 2916-2924 (1999).

119. Janz, S. Uncovering MYC's full oncogenic potential in the hematopoietic system. *Oncogene* **24**, 3541-3543 (2005).

120. O'Donnell, K.A., Wentzel, E.A., Zeller, K.I., Dang, C.V. and Mendell, J.T. c-Myc-regulated microRNAs modulate E2F1 expression. *Nature* **435**, 839-843 (2005).

121. Woods, K., Thomson, M.J. and Hammond, S.M. Direct Regulation of an Oncogenic Micro-RNA Cluster by E2F Transcription Factors. *J. Biol. Chem.* **282**, 2130 - 2134 (2007).

122. Sylvestre, Y. et al. An E2F/miR-20a Autoregulatory Feedback Loop. *J. Biol. Chem.* **282**, 2135-2143 (2007).

123. Bracken, A.P., Ciro, M., Cocito, A. and Helin, K. E2F target genes: unraveling the biology. *Trends Biochem Sci* **29**, 409-417 (2004).

124. Field, S.J. et al. E2F-1 functions in mice to promote apoptosis and suppress proliferation. *Cell* **85**, 549-561 (1996).

125. Yamasaki, L. et al. Tumor induction and tissue atrophy in mice lacking E2F-1. *Cell* **85**, 537-548 (1996).

126. Matsumura, I., Tanaka, H. and Kanakura, Y. E2F1 and c-Myc in cell growth and death. *Cell Cycle* **2**, 333-338 (2003).

127. Thalmeier, K., Synovzik, H., Mertz, R., Winnacker, E.L. and Lipp, M. Nuclear factor E2F mediates basic transcription and trans-activation by E1a of the human MYC promoter. *Genes and Development* **3**, 527-536 (1989).

128. Dews, M. et al. Augmentation of tumor angiogenesis by a Myc-activated microRNA cluster. *Nature Genetics* **38**, 1060-1065 (2006).

129. Fisher, G.H. et al. Induction and apoptotic regression of lung adenocarcinomas by regulation of a K-Ras transgene in the presence and absence of tumor suppressor genes. *Genes Dev* **15**, 3249-3262 (2001).

130. Mascaux, C. et al. The role of RAS oncogene in survival of patients with lung cancer: a systematic review of the literature with meta-analysis. *Br J Cancer* **92**, 131-139 (2005).

131. Johnson, S.M. et al. RAS is regulated by the let-7 microRNA family. *Cell* **120**, 635-647 (2005).

132. Reinhart, B.J. et al. The 21-nucleotide let-7 RNA regulates developmental timing in Caenorhabditis elegans. *Nature* **403**, 901-906 (2000).

133. Rodenhuis, S. and Slebos, R.J. The ras oncogenes in human lung cancer. *Am Rev Respir Dis* **142**, S27-30 (1990).

134. Rak, J., Yu, J.L., Kerbel, R.S. and Coomber, B.L. What do oncogenic mutations have to do with angiogenesis/vascular dependence of tumors? *Cancer Research* **62**, 1931-1934 (2002).

135. Hatfield, S.D. et al. Stem cell division is regulated by the microRNA pathway. *Nature* **435**, 974-978 (2005).

136. Cox, D.N. et al. A novel class of evolutionarily conserved genes defined by piwi are essential for stem cell self-renewal. *Genes Dev* **12**, 3715-3727 (1998).

137. Sharma, A.K. et al. Human CD34(+) stem cells express the hiwi gene, a human homologue of the Drosophila gene piwi. *Blood* **97**, 426-434 (2001).

138. Abate-Shen, C. Deregulated Homeobox Gene Expression in Cancer: Cause or Consequence? . *Nature Reviews Cancer* **2**, 777-785 (2002).

139. Cillo, C., Faiella, A., Cantile, M. and Boncinelli, E. Homeobox genes and cancer. *Experimental Cell Research* **248**, 1-9 (1999).

140. Naguibneva, I. et al. The microRNA miR-181 targets the homeobox protein Hox-A11 during mammalian myoblast differentiation. *Nature Cell Biology* **8**, 278-284 (2006).

141. Chen, C.Z., Li, L., Lodish, H.F. and Bartel, D.P. MicroRNAs modulate hematopoietic lineage differentiation. *Science* **303**, 83-86 (2004).

142. Matthias, P. and Rolink, A.G. Transcriptional networks in developing and mature B cells. *Nature Reviews. Immunology* **5**, 497-508 (2005).

143. Fazi, F. et al. A minicircuitry comprised of microRNA-223 and transcription factors NFI-A and C/EBPalpha regulates human granulopoiesis. *Cell* **123**, 819-831 (2005).

144. Felli, N. et al. MicroRNAs 221 and 222 inhibit normal erythropoiesis and erythroleukemic cell growth via kit receptor down-modulation. *Proceedings of the National Academy of Sciences of the United States of America* **102**, 18081-18086 (2005).

145. Tenen, D.G. Disruption of differentiation in human cancer: AML shows the way. *Nature Reviews. Cancer* **3**, 89-101 (2003).

146. Pabst, T. et al. AML1-ETO downregulates the granulocytic differentiation factor C/EBPalpha in t(8;21) myeloid leukemia.[see comment]. *Nature Medicine* **7**, 444-451 (2001).

147. Miyamoto, T. and Akashi, K. Lineage promiscuous expression of transcription factors in normal hematopoiesis. *International Journal of Hematology* **81**, 361-367 (2005).

148. Miyamoto, T. et al. Myeloid or lymphoid promiscuity as a critical step in hematopoietic lineage commitment. *Developmental Cell* **3**, 137-147 (2002).

149. Voinnet, O. Induction and suppression of RNA silencing: insights from viral infections. *Nat Rev Genet* **6**, 206-220 (2005).

150. Lecellier, C.H. et al. A cellular microRNA mediates antiviral defense in human cells. *Science* **308**, 557-560 (2005).

151. Pfeffer, S. et al. Identification of virus-encoded microRNAs. *Science* **304**, 734-736 (2004).

152. Cai, X. et al. Kaposi's sarcoma-associated herpesvirus expresses an array of viral microRNAs in latently infected cells. *Proc Natl Acad Sci U S A* **102**, 5570-5575 (2005).

153. Samols, M.A., Hu, J., Skalsky, R.L. and Renne, R. Cloning and identification of a microRNA cluster within the latency-associated region of Kaposi's sarcoma-associated herpesvirus. *J Virol* **79**, 9301-9305 (2005).

154. Pfeffer, S. et al. Identification of microRNAs of the herpesvirus family. *Nat Methods* **2**, 269-276 (2005).

155. Bennasser, Y., Le, S.Y., Yeung, M.L. and Jeang, K.T. HIV-1 encoded candidate micro-RNAs and their cellular targets. *Retrovirology* **1**, 43 (2004).
156. Sullivan, C.S., Grundhoff, A.T., Tevethia, S., Pipas, J.M. and Ganem, D. SV40-encoded microRNAs regulate viral gene expression and reduce susceptibility to cytotoxic T cells. *Nature* **435**, 682-686 (2005).
157. Omoto, S. and Fujii, Y.R. Regulation of human immunodeficiency virus 1 transcription by nef microRNA. *J Gen Virol* **86**, 751-755 (2005).
158. Omoto, S. et al. HIV-1 nef suppression by virally encoded microRNA. *Retrovirology* **1**, 44 (2004).
159. Lu, S. and Cullen, B.R. Adenovirus VA1 noncoding RNA can inhibit small interfering RNA and MicroRNA biogenesis. *J Virol* **78**, 12868-12876 (2004).
160. Cheng, A.M., Byrom, M.W., Shelton, J. and Ford, L.P. Antisense inhibition of human miRNAs and indications for an involvement of miRNA in cell growth and apoptosis. *Nucleic Acids Res* **33**, 1290-1297 (2005).
161. Meister, G., Landthaler, M., Dorsett, Y. and Tuschl, T. Sequence-specific inhibition of microRNA- and siRNA-induced RNA silencing. *Rna* **10**, 544-550 (2004).
162. Orom, U.A., Kauppinen, S. and Lund, A.H. LNA-modified oligonucleotides mediate specific inhibition of microRNA function. *Gene* **372**, 137-141 (2006).
163. Hutvagner, G., Simard, M.J., Mello, C.C. and Zamore, P.D. Sequence-specific inhibition of small RNA function. *PLoS Biol* **2**, E98 (2004).
164. Krutzfeldt, J. et al. Silencing of microRNAs in vivo with 'antagomirs'. *Nature* **438**, 685-689 (2005).

# 20 Mammalian Transcriptional Gene Silencing by Small RNAs

*Daniel H. Kim and John J. Rossi*

## CONTENTS

## OVERVIEW

Small RNAs regulate gene expression at the transcriptional and posttranscriptional levels through various RNA silencing pathways, including the well-characterized mechanism of RNA interference (RNAi). Classes of small RNAs include small interfering RNAs (siRNAs) and microRNAs (miRNAs), which associate with Argonaute (AGO) protein family members to mediate sequence-specific gene silencing of their targets. In mammalian cells, siRNAs that exhibit perfect Watson–Crick base pairing to gene promoters direct transcriptional gene silencing (TGS) through epigenetic modifications involving histone and/or DNA methylation events. While a role for miRNAs in TGS has not been explored, evidence in plants suggests that miRNAs may also play a role in directing heritable, epigenetic gene silencing pathways. Here, we provide an overview of the mounting evidence that small RNAs regulate transcriptional silencing in mammals, along with recently emerging findings of their therapeutic potential in treating human diseases.

## 20.1   INTRODUCTION

Distinct pathways of RNA silencing exist in the nucleus and cytoplasm of eukaryotic cells.[1] Posttranscriptional gene silencing (PTGS) occurs mainly in the cytoplasm through mRNA degradation,[2] while TGS causes epigenetic modifications to histone tails and DNA bases in the nucleus.[3] The phenomenon of PTGS in canonical RNAi is well described as taking place predominantly in the cytoplasm of eukaryotes, including mammals. While TGS has been described extensively in model eukaryotes such as plants, fission yeast, worms, and flies, the evidence for mammalian TGS has not been overwhelming to date. Recent findings have begun to unravel mechanistic aspects of this nuclear RNA silencing pathway, and both parallels and differences exist between mammalian TGS and analogous pathways in other eukaryotes.

## 20.2   CLASSES OF SMALL RNAs

### 20.2.1   SMALL INTERFERING RNAs

siRNAs are double-stranded RNA (dsRNA) molecules ~21 base pairs (bp) in length with two nucleotide (nt) 3′ overhangs, a processing characteristic of the RNase III enzyme Dicer.[4] These small RNAs serve as effector molecules for the RNAi machinery by providing sequence-specificity to the process of target recognition and degradation. In PTGS, siRNAs are loaded into a silencing complex known as the RNA-induced silencing complex (RISC).[5] Only one strand of the siRNA, known as the antisense or guide strand, is required for targeting a specific mRNA. The sense strand, referred to as the passenger strand, is cleaved by the RNase H-like PIWI domain of the endonuclease Argonaute 2 (AGO2), facilitating the subsequent loading of the guide strand into RISC.[6,7] AGO2 is the effector protein component of RISC that elicits RNAi by "slicing" target mRNAs.[8] Along with AGO2, Dicer and its dsRNA-binding proteins, HIV-1 TAR RNA-binding protein (TRBP)[9] and protein activator of protein kinase PKR (PACT),[10] comprise additional key components of RISC in mammalian cells.

### 20.2.2   MIRNAs

Endogenous miRNAs generally exhibit only partial sequence complementarity to their target mRNA 3′ untranslated regions (3′ UTRs).[11] In mammalian cells, long transcripts known as pri-miRNAs are transcribed initially in the nucleus by RNA polymerase II[12] or III,[13] and processed by an RNase III enzyme known as Drosha into ~70-nt stem-loop structures called pre-miRNAs.[14] Similar to Dicer, Drosha acts in concert with the dsRNA-binding protein DGCR8 to generate pre-miRNAs within the nucleus.[15] Exportin 5, another dsRNA-binding protein, then transports pre-miRNAs into the cytoplasm,[16] where a complex containing Dicer, TRBP, and PACT processes pre-miRNAs into miRNA duplexes, which are then loaded into RISC.[17] The miRNA-loading pathway into RISC, however, does not involve cleavage of the passenger strand, and somehow bypasses this mechanism instead.[17] Additionally, located at the 5′ end of a mature miRNA is the seed sequence comprised of the first 2–8 nt, and extensive sequence complementarity between the 5′ end of the miRNA with its target mRNA 3′ UTR is necessary for target recognition.[11]

## 20.3 TRANSCRIPTIONAL GENE SILENCING

### 20.3.1 GENE PROMOTERS

RNA silencing in the nucleus of mammalian cells regulates gene expression at the level of DNA and chromatin. TGS is induced by siRNAs with sequence complementarity to gene promoters, directing epigenetic modifications to the targeted promoter regions.[2] Exogenous siRNAs designed against promoters in human cell lines have targeted promoter sequences relatively close to known transcription start sites (TSS), within 200 bp (Table 20.1),[18–26] although the requirement for TSS proximity has not been thoroughly investigated. siRNA sequences have also been delivered into mammalian cells as short hairpin RNAs (shRNAs), small RNAs with a stem of 19–29 bp and a loop of 4–10 nt that are processed by Dicer into siRNAs.[2] Expressed from vectors, shRNAs have been shown to induce TGS of RASSF1A[20] and TGFbRII.[25] Stable HeLa cells lines expressing shRNA against RASSF1A knocked down gene expression by ~70% at the mRNA level, while shRNAs against TGFbRII were cloned into lentiviral vectors and used to transduce rat hepatic stellate cells (SBC10), leading to ~50% knockdown of TGFbRII mRNA levels.

**TABLE 20.1**

**Gene Promoters Amenable to siRNA-Induced TGS in Mammalian Cells**

| Gene Promoter | Distance from TSS | Cell Line |
|---|---|---|
| EF1A | −106 to −86 | 293FT |
| RASSF1A | −49 to −28 | HeLa |
| CDH1 | −181 to −161 | HCT116 and MCF-7 |
| PR | −26 to −7 | T47D |
| MVP | −9 to +10 | SW1573/2R120 |
| AR | −8 to +11 | T47D |
| COX-2 | −9 to +10 | T47D |
| CCR5 | −183 to −163 | 293T |
| HTT | −9 to +10 | T47D |
| TGFbRII | −161 to −141 | SBC10 |
| uPA | −131 to −111 | PC3 and DU145 |

## 20.3.2   DNA Methylation

Initial observations of siRNA-mediated TGS in human cells have shown that DNA methylation of a targeted promoter region occurs, but subsequent observations have shown that transcriptional silencing can also occur in the absence of promoter DNA methylation. In the first example of TGS in mammalian cells,[18] siRNAs were targeted against the elongation factor 1 alpha (EF1A) promoter, driving expression of a green fluorescent protein (GFP) reporter gene stably integrated into 293FT cells. At 48 hours posttransfection of siRNAs, mRNA levels were detectable at less than 1% relative to control siRNA-transfected samples. Treatment of cells with 5-azacytidine (5-azaC), an inhibitor of DNA methyltransferase (DNMT) activity, reversed the silencing effect, implicating a role for DNMTs in the mechanism of silencing. In mammalian cells, DNMT3a and DNMT3b are characterized as de novo DNMTs,[27] suggesting that one or both enzymes could play a role in the de novo methylation of siRNA-targeted promoter sequences.

Studies of TGS in plants have demonstrated a link between small RNAs and de novo DNA methylation, and the mechanism of silencing in plants indicates that small RNAs serve as sequence-specific guides to epigenetic modifications at the DNA level known as RNA-directed DNA methylation (RdDM).[3] siRNAs targeted against the endogenous EF1A promoter indicated de novo DNA methylation when assayed by DNA methylation-sensitive enzyme cleavage. Two subsequent studies published soon thereafter targeted the endogenous RASSF1A and CDH1[19] promoters, respectively, and using both methylation-specific PCR and the more sensitive bisulfite sequencing analysis, both findings indicated that small RNAs directed against endogenous promoters yielded little to no de novo DNA methylation of their targeted promoters. For HeLa cells stably expressing shRNAs against RASSF1A, 3 out of 16 stable clones indicated de novo DNA methylation, for a relatively low frequency of ~19%. Of the HeLa clones showing DNA methylation, 15 out of 16 CpGs were methylated, suggesting that while the frequency of RdDM in mammalian cells may be relatively low, the robustness may be relatively high when it does indeed occur. Additionally, TGS studies of the PR and AR promoters in T47D cells using siRNAs also indicated a lack of RdDM when assessed using bisulfite sequencing,[21] suggesting that DNA methylation is not a requirement for mammalian TGS to occur.

Other DNA methylation analyses of CDH1 in HCT116 and MCF-7 cells were performed using siRNAs transfected at 24-hour intervals over a span of 7 days.[19] Bisulfite sequencing results indicated that there was no change in DNA methylation levels at the siRNA-targeted CDH1 promoter in both HCT116 and MCF-7 cells. HCT116 cells with double knockouts (DKO) for maintenance DNMT1 and de novo DNMT3b were also used in these TGS studies of CDH1, and TGS occurred robustly in the DNMT1 and DNMT3b DKO cells, suggesting that neither DNMT plays a role in mediating mammalian TGS. To investigate a role for de novo DNMT3a, followup studies of the EF1A promoter revealed that DNMT3a interacted with siRNAs targeted against EF1A, and in particular, only the EF1A-targeting siRNA antisense or guide strand.[22] The siRNAs targeting the EF1A promoter did not interact with DNMT1, indicating that DNMT3a was specifically associated with the promoter-targeting siRNAs. In vitro studies have also indicated that siRNAs bind to purified recombinant DNMT3a but not to DNMT1,[28] indicating that DNMT3a may contain a small RNA-binding domain.

More recent studies also provide evidence for RdDM in mammalian cells. SBC10 cells transduced with lentiviral vectors, expressing shRNAs against TGFbRII, revealed de novo promoter DNA methylation using bisulfite sequencing analysis.[25] Two different shRNA sequences were both shown to induce RdDM extensively in the TGFbRII-promoter region, based on 12 different analyzed clones, when compared to 5 different control clones. Treatment of transduced SBC10 cells with 5-azaC resulted in the loss of TGFbRII mRNA knockdown and TGS. In another study, the PC3 prostate cancer cell line was transfected with a pool of four siRNAs targeting the promoter of the urokinase plasminogen activator (uPA) gene.[26] This induced robust DNA methylation of the promoter region that was detected using methylation-specific PCR and bisulfite sequencing analysis of seven different clones, relative to ten different control clones.

Taken together, these various studies indicate that while RdDM can occur in mammalian cells at some targeted promoters, DNA methylation does not seem to play an important role in TGS of various other mammalian gene promoters targeted by siRNAs. Although some level of controversy continues to exist in the field regarding whether or not siRNAs induce DNA methylation of promoter regions, the topic of RdDM in mammalian cells continues to be an area of active investigation.

## 20.3.3　Histone Methylation

Evidence for small RNA-directed histone methylation is supported by studies in the fission yeast *Schizosaccharomyces pombe*, where siRNAs direct H3K9me2 to regions of heterochromatin nucleation.[29] In fission yeast, cryptic loci regulator 4 (Clr4) is the histone methyltransferase responsible for dimethylation of histone H3 lysine 9 (H3K9me2) mediated by siRNAs. This histone mark in turn recruits Swi6, which contains a chromodomain and is a homologue of heterochromatin protein HP1. Regions enriched for H3K9me2 and Swi6 in the fission yeast genome are found at centromeres, the mating-type locus, and telomeres, which contain *dg* and *dh* repeats that serve to generate heterochromatic siRNAs.

In mammalian cells, targeting a promoter region with sequence-specific siRNAs also leads to epigenetic modifications that silence gene expression at the level of chromatin. Histone marks characteristic of facultative heterochromatin,[30] namely H3K9me2 and trimethylation of histone H3 lysine 27 (H3K27me3), have been shown to correlate with siRNA-targeted promoter regions. In particular, the CDH1 and CCR5[23] promoters exhibited increased levels of H3K9me2 when targeted with siRNAs; H3K9me2 and H3K27me3 were both present at the siRNA-targeted EF1A promoter, and H3K27me3 was found to be enriched at the shRNA-targeted RASSF1A promoter. Furthermore, inhibition of histone deacetylases using trichostatin A (TSA) resulted in the loss of TGS at several siRNA-targeted promoters. These findings indicate that siRNAs appear to regulate the "histone code"[31] at their sequence-complementary promoters in mammalian cells by recruiting histone deacetylases and histone methyltransferases that mediate the characteristic H3K9me2 and H3K27me3 epigenetic silencing marks.

## 20.4 POLYCOMB GROUP AND RNAi COMPONENTS

### 20.4.1 POLYCOMB GROUP SILENCING

The epigenetic histone marks H3K9me2 and H3K27me3 that are characteristic of siRNA-mediated TGS in human cells are known to be found in genomic regions of facultative heterochromatin, such as the inactive X chromosome.[30] In X inactivation, a 19-kb-long noncoding RNA known as *Xist* acts in *cis* to initiate silencing of the X chromosome. This leads to the loss of H3K9 acetylation and H3K4 methylation, which are both histone marks of actively transcribed euchromatin. Silencing histone marks H3K9me2, H3K27me3, and H4K20me subsequently appear on the inactive X chromosome, generating a facultative heterochromatin domain. In euchromatic regions, the H3K9me2 histone mark can be mediated by the histone methyltransferase G9a, among other enzymes, although it is not clear whether G9a is involved in X inactivation. The H3K27me3 mark is catalyzed by a specific histone methyltransferase, EZH2, which is a Polycomb group (PcG) component found in Polycomb repressive complex 2 (PRC2) that also contains the Eed and Suz12 proteins. The *Xist* RNA appears to recruit EZH2 to silence the X chromosome, although the mechanism by which this occurs is not entirely clear. Once the X chromosome has been heterochromatinized, DNA methylation also occurs and leads to permanent and heritable silencing of the inactive X chromosome.

Evidence in human cells suggests that promoter-specific siRNAs may also direct histone methylation and silence transcription by recruiting PcG components, such as EZH2, to their target promoters.[23] The recruitment of PRC2 to an siRNA-targeted promoter would potentially result in the formation of a facultative heterochromatin domain. Additionally, EZH2 is also known to associate with DNA methyltransferases DNMT3a, DNMT3b, and DNMT1,[32] possibly providing a connection between siRNA-mediated TGS and DNA methylation of targeted promoter regions. Since DNMT3a has also been shown to bind with promoter-specific siRNAs[22] and also siRNAs in general in vitro,[28] this association may be facilitated by the recruitment of DNMT3a by EZH2.

### 20.4.2 ARGONAUTE PROTEINS

AGO proteins play important roles in the nuclear RNA-silencing pathways of eukaryotes, including fission yeast, plants, flies, and worms.[1] The role of AGO proteins in TGS has been best characterized in fission yeast,[29] which utilizes the RNA-induced transcriptional silencing (RITS) complex comprised of Ago1, Chp1, which contains a chromodomain, and Tas3, whose function has not yet been identified. RITS is initially loaded with siRNAs derived from heterochromatic repeat regions and is required for H3K9me2 mediated by Clr4 recruitment. A second complex known as the RNA-directed RNA polymerase complex (RDRC) is also recruited by RITS and contains RNA-directed RNA polymerase 1 (Rdp1), Cid12, a poly(A) polymerase, and Hrr1, an RNA helicase. The "slicer" function of fission yeast Ago1 is required for the recruitment of RDRC and the spreading of siRNA-directed heterochromatin.[1] RITS binds to a nascent RNAPII transcript and "slices" this transcript, which is used as a template by RDRC to generate a double-stranded RNA substrate for Dicer (Dcr1) to process into siRNAs. While Rdp1 is conserved in plants and worms, among other eukaryotes, flies and mammals do not possess an Rdp1 homologue. AGO proteins, however, are conserved in most eukaryotes, and a function for Ago4 in directing TGS and epigenetic modifications in plants has been well defined, where Ago4 is required for RdDM to occur.[3] Ago1 in flies also appears to play a role in transcriptional regulatory pathways, as Ago1 associates with PcG proteins in distinct nuclear bodies, along with the RNAi component proteins Dicer-2 and PIWI.[33] AGO proteins thus appear to play a conserved role in diverse eukaryotic TGS pathways.

Since the effector molecules of TGS in mammalian cells are siRNAs targeting promoter regions, nuclear RNA silencing effector complexes are also presumed to exist into which these siRNAs would be loaded. AGO proteins are at the core of cytoplasmic RNAi effector complexes in mammals and other eukaryotes, suggesting that a putative, nuclear-localized silencing complex also would contain an AGO protein family member in mammals. In human cells, AGO1 is involved in directing histone methylation and siRNA-mediated TGS at targeted promoter regions.[23] Specifically, AGO1 localizes to the siRNA-targeted CCR5 promoter and shRNA-targeted RASSF1A promoter. The induction of H3K9me2 at the targeted CCR5 promoter is dependent on the presence of AGO1, and AGO1 colocalizes with TRBP, EZH2, and H3K27me3 at the targeted RASSF1A promoter. TGS also appears to require the presence of TRBP, but the role of TRBP as a double-stranded RNA-binding protein in TGS is not clear. One possible role is that TRBP initially binds duplex siRNAs and serves to facilitate delivery of the proper guide strand to AGO1. Another possibility is that TRBP is part of an effector complex along with AGO1 and other unknown proteins, similar to cytoplasmic RISC being comprised of TRBP and AGO2 in part. This putative nuclear effector complex may also bind with RNA polymerase II (RNAPII), since AGO1 has been shown to associate with RNAPII in human cells. In fission yeast, Ago1 appears to associate with RNAPII as well, and evidence in both fission yeast and mammalian cells suggests that proper RNAPII function is required for TGS to occur.[1]

The involvement of an AGO protein further suggests that an RNA:RNA interaction is required for TGS to occur, as opposed to an RNA:DNA interaction where the siRNA guide strand would bind directly to the promoter DNA. Furthermore, the transcriptional activity of RNAPII is necessary for siRNA-mediated TGS in human cells.[22] A low-copy, nascent promoter-associated RNA that is transcribed through promoter regions by RNAPII is involved in the silencing mechanism,[34] further supporting the RNA:RNA model of siRNA-mediated TGS. The antisense guide strand of the promoter-targeting siRNA directs TGS, leading to the formation of an RNA:RNA duplex between the siRNA guide strand and the nascent promoter-associated RNA, which in turn results in silencing of the targeted promoter through recruitment of histone methyltransferase activity, including EZH2. Furthermore, because AGO1 associates with RNAPII, this suggests that AGO1 is tethered to RNAPII as it transcribes through promoter regions, possibly allowing for an efficient scanning mechanism to identify potential siRNA target sequences in promoters. The promoter-associated RNA serves as a scaffold for the siRNA guide strand and AGO1, and the promoter-associated RNA is only detectable in the sense orientation with respect to the downstream transcribed gene. Furthermore, the promoter-associated RNA also associates with the histone marks H3K9me2 and H3K27me3, providing additional support for its role as a scaffolding for the RNA-based transcriptional silencing machinery.

Blocking or destroying the guide strand recognition sequence on the promoter-associated RNA inhibits transcriptional silencing from occurring. The promoter-associated RNA for EF1A in particular is transcribed from ~230 bp upstream of the known, major transcription start site (TSS) of EF1A, indicating the presence of a previously undetected, minor TSS upstream. This observation also correlates with the observation that siRNAs that induce TGS in mammalian cells have thus far been targeted only within 200 bp of TSS. Depending on the length of the promoter-associated RNA, siRNA-targeting of a promoter region that lies upstream of a promoter-associated RNA TSS would seemingly be unable to induce TGS. Given the widespread levels of transcription in the mammalian genome,[35] the existence of promoter-associated RNAs may be a broad phenomenon and an important component of transcriptional regulatory processes. Indeed, promoter-associated RNAs have been detected in the promoter regions of EF1A, CCR5, RASSF1A, p16, NF-kB, and Cyclin D1 in various mammalian cell types.

A role for AGO2 has also been shown in mammalian TGS,[24] but the mechanism of silencing is unclear for promoter regions that are enriched for AGO2 when targeted with siRNAs. AGO1 and AGO2 both associate with the siRNA-targeted PR promoter, but the mechanism of transcriptional silencing does not seem to involve histone modifications. The use of TSA has no effect on the TGS effect observed, suggesting that histone deacetylation is not involved in silencing the PR promoter. Additionally, H3K9me2 is not induced at the PR promoter when targeted with siRNAs. While this observation does not exclude a role for H3K27me3 in mediating the observed transcriptional silencing at the PR promoter, the involvement of AGO2 may be promoter-specific, as AGO2 does not appear to be involved in the transcriptional silencing of the CCR5 promoter. DNA methylation also does not appear to be involved in TGS of the PR promoter,[21] suggesting that a different mechanism of transcriptional silencing is occurring at the AGO2-associated PR promoter. Since AGO2 has a catalytic PIWI domain that enables it to bind a single strand of RNA in RNA-RNA duplexes, AGO1 does not contain the critical PIWI-domain DDH motif (AGO1 contains a DDR motif) and lacks the RNA cleavage activity in mammalian cells[36]; one possibility is that siRNA-loaded AGO2 is cleaving a promoter-associated RNA at the PR promoter, leading to subsequent degradation of the PR gene transcript through other mechanisms, such as an unknown exonuclease activity. What does seem clear, however, is that AGO proteins are directly involved in mediating mammalian TGS pathways (Figure 20.1).

## 20.5  THERAPEUTIC APPLICATIONS OF TGS

### 20.5.1  CANCER

The use of a TGS approach for therapeutic strategies against human diseases would be relevant for disease states in which long-term, heritable epigenetic

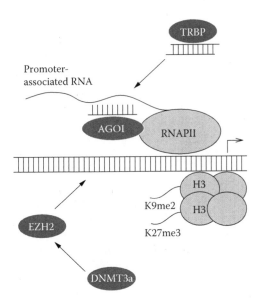

**FIGURE 20.1** A putative model for mammalian TGS by small RNAs. Small double-stranded RNAs may initially be bound by the double-stranded RNA-binding domains of TRBP. TRBP may then "hand-off" a single-stranded guide strand to AGO1, which mediates recognition of the target sequence in the promoter-associated RNA. Once the guide strand target sequence is recognized, AGO1 may stall the transcribing RNAPII complex, which could in turn lead to the recruitment of histone methyltransferases, including EZH2, leading to the induction of the silencing histone marks H3K9me2 and H3K27me3, and the formation of a facultative heterochromatin domain. EZH2 may then recruit DNMT3a to mediate de novo DNA methylation of the small RNA-targeted promoter region, leading to permanent and heritable epigenetic gene silencing.

silencing of a particular gene is beneficial. In human prostate cancer, the uPA gene is overexpressed and causes an increase in metastasis.[26] The mechanism by which this occurs involves DNA demethylation of the uPA promoter region. To test whether siRNAs that reinduce DNA methylation of the uPA promoter in prostate cancer cells could inhibit tumor growth in vivo, siRNA-transfected PC3 cells were introduced into immunodeficient mice. When compared to mice containing control siRNAs, mice implanted with TGS-inducing siRNAs in their PC3 cells showed tumor volumes that were significantly smaller at day 40 postimplantation. Additionally, uPA expression was silenced long-term by TGS-inducing siRNAs when assessed at day 40. There was also no detectable tumor metastasis in mice receiving TGS-inducing siRNAs, whereas mice with control siRNAs exhibited tumor metastasis on the lung surface. When tumor morphology was compared between the two sample groups, mice with TGS-inducing siRNAs displayed normal histology, while control mice showed significant vascularity in their tumors. There was also noticeable suppression of tumor angiogenesis in mice receiving TGS-inducing siRNAs, indicating the therapeutic potential in vivo of a TGS-based approach for long-term silencing of a disease-related gene. Evidence in worms also indicates that a single dose of siRNAs can lead to prolonged epigenetic gene silencing over 80 generations,[37] suggesting that TGS is a heritable, epigenetic process.

### 20.5.2  HIV

Therapeutic approaches using TGS are potentially amenable for targeting chronic diseases such as human immunodeficiency virus (HIV) infection and consequent acquired immune deficiency syndrome (AIDS).[38] The long-terminal repeat (LTR) of HIV-1 has been targeted by siRNAs for TGS of viral gene expression.[22] Analogous to endogenous promoters, the HIV-1 LTR showed an induction of H3K27me3 when targeted by siRNAs in the U3 region of the LTR. Additionally, the chemokine (C-C motif) receptor 5 (CCR5), which serves as a co-receptor for HIV-1, has been targeted by siRNAs and shown to be amenable to TGS.[23] From a therapeutic perspective, knockdown of CCR5 would potentially prevent R5-tropic HIV-1 from gaining entry into its target cells. Studies of CCR5 TGS are still preliminary, however, and require testing in primary cells such as monocytes and macrophages that express CCR5 and are the natural targets of HIV-1 within the host.

### 20.6  CONCLUSIONS

Small RNAs regulate gene expression at the level of transcription in eukaryotic cells, including mammalian cells. The process of siRNA-mediated TGS in mammalian cells is beginning to be understood in more detail based on several recent studies, but an endogenous role for small RNA-mediated TGS has yet to be uncovered. One possibility for the observation of mammalian TGS is that exogenous, promoter-targeting siRNAs are capitalizing on an endogenous mechanism of epigenetic silencing involving small RNAs and PcG proteins. Identification of additional components of the RNAi and PcG silencing pathways will provide further insights into this mechanism of RNA silencing in the nucleus. A more in-depth understanding of this RNA-silencing pathway may allow investigators to harness the capacity of TGS for gene regulation in therapeutically relevant settings and preclinical models for the treatment of human diseases.

### ACKNOWLEDGMENTS

We thank S. Jacobsen, R. Martienssen, and K. Morris for insights on TGS, and R. Lin, R. Natarajan, G. Pfeifer, and members of our laboratory for discussions and support. This work was supported by grants from the NIH to J.J.R.

### REFERENCES

1. Zaratiegui, M., Irvine, D. V., and Martienssen, R. A., Noncoding RNAs and gene silencing, *Cell* 128 (4), 763–776, 2007.
2. Kim, D. H. and Rossi, J. J., Strategies for silencing human disease using RNA interference, *Nat Rev Genet* 8 (3), 173–184, 2007.
3. Matzke, M. A. and Birchler, J. A., RNAi-mediated pathways in the nucleus, *Nat Rev Genet* 6 (1), 24–35, 2005.

4. Bernstein, E., Caudy, A. A., Hammond, S. M., and Hannon, G. J., Role for a bidentate ribonuclease in the initiation step of RNA interference, *Nature* 409 (6818), 363–366, 2001.
5. Martinez, J., Patkaniowska, A., Urlaub, H., Luhrmann, R., and Tuschl, T., Single-stranded antisense siRNAs guide target RNA cleavage in RNAi, *Cell* 110 (5), 563–574, 2002.
6. Matranga, C., Tomari, Y., Shin, C., Bartel, D. P., and Zamore, P. D., Passenger-strand cleavage facilitates assembly of siRNA into Ago2-containing RNAi enzyme complexes, *Cell* 123 (4), 607–620, 2005.
7. Rand, T. A., Petersen, S., Du, F., and Wang, X., Argonaute2 cleaves the anti-guide strand of siRNA during RISC activation, *Cell* 123 (4), 621–629, 2005.
8. Liu, J., Carmell, M. A., Rivas, F. V., Marsden, C. G., Thomson, J. M., Song, J. J., Hammond, S. M., Joshua-Tor, L., and Hannon, G. J., Argonaute2 is the catalytic engine of mammalian RNAi, *Science* 305 (5689), 1437–1441, 2004.
9. Chendrimada, T. P., Gregory, R. I., Kumaraswamy, E., Norman, J., Cooch, N., Nishikura, K., and Shiekhattar, R., TRBP recruits the Dicer complex to Ago2 for microRNA processing and gene silencing, *Nature* 436 (7051), 740–744, 2005.
10. Lee, Y., Hur, I., Park, S. Y., Kim, Y. K., Suh, M. R., and Kim, V. N., The role of PACT in the RNA silencing pathway, *EMBO J* 25 (3), 522–532, 2006.
11. Bartel, D. P., MicroRNAs: genomics, biogenesis, mechanism, and function, *Cell* 116 (2), 281–297, 2004.
12. Lee, Y., Kim, M., Han, J., Yeom, K. H., Lee, S., Baek, S. H., and Kim, V. N., MicroRNA genes are transcribed by RNA polymerase II, *EMBO J* 23 (20), 4051–4060, 2004.
13. Borchert, G. M., Lanier, W., and Davidson, B. L., RNA polymerase III transcribes human microRNAs, *Nat Struct Mol Biol* 13 (12), 1097–1101, 2006.
14. Lee, Y., Ahn, C., Han, J., Choi, H., Kim, J., Yim, J., Lee, J., Provost, P., Radmark, O., Kim, S., and Kim, V. N., The nuclear RNase III Drosha initiates microRNA processing, *Nature* 425 (6956), 415–419, 2003.
15. Gregory, R. I., Yan, K. P., Amuthan, G., Chendrimada, T., Doratotaj, B., Cooch, N., and Shiekhattar, R., The Microprocessor complex mediates the genesis of microRNAs, *Nature* 432 (7014), 235–240, 2004.
16. Yi, R., Qin, Y., Macara, I. G., and Cullen, B. R., Exportin-5 mediates the nuclear export of pre-microRNAs and short hairpin RNAs, *Genes Dev* 17 (24), 3011–3016, 2003.
17. Preall, J. B. and Sontheimer, E. J., RNAi: RISC gets loaded, *Cell* 123 (4), 543–545, 2005.
18. Morris, K. V., Chan, S. W., Jacobsen, S. E., and Looney, D. J., Small interfering RNA-induced transcriptional gene silencing in human cells, *Science* 305 (5688), 1289–1292, 2004.
19. Ting, A. H., Schuebel, K. E., Herman, J. G., and Baylin, S. B., Short double-stranded RNA induces transcriptional gene silencing in human cancer cells in the absence of DNA methylation, *Nat Genet* 37 (8), 906–910, 2005.
20. Castanotto, D., Tommasi, S., Li, M., Li, H., Yanow, S., Pfeifer, G. P., and Rossi, J. J., Short hairpin RNA-directed cytosine (CpG) methylation of the RASSF1A gene promoter in HeLa cells, *Mol Ther* 12 (1), 179–183, 2005.
21. Janowski, B. A., Huffman, K. E., Schwartz, J. C., Ram, R., Hardy, D., Shames, D. S., Minna, J. D., and Corey, D. R., Inhibiting gene expression at transcription start sites in chromosomal DNA with antigene RNAs, *Nat Chem Biol* 1 (4), 216–222, 2005.
22. Weinberg, M. S., Villeneuve, L. M., Ehsani, A., Amarzguioui, M., Aagaard, L., Chen, Z. X., Riggs, A. D., Rossi, J. J., and Morris, K. V., The antisense strand of small interfering RNAs directs histone methylation and transcriptional gene silencing in human cells, *RNA* 12 (2), 256–262, 2006.
23. Kim, D. H., Villeneuve, L. M., Morris, K. V., and Rossi, J. J., Argonaute-1 directs siRNA-mediated transcriptional gene silencing in human cells, *Nat Struct Mol Biol* 13 (9), 793–797, 2006.

24. Janowski, B. A., Huffman, K. E., Schwartz, J. C., Ram, R., Nordsell, R., Shames, D. S., Minna, J. D., and Corey, D. R., Involvement of AGO1 and AGO2 in mammalian transcriptional silencing, *Nat Struct Mol Biol* 13 (9), 787–792, 2006.

25. Kim, J. W., Zhang, Y. H., Zern, M. A., Rossi, J. J., and Wu, J., Short hairpin RNA causes the methylation of transforming growth factor-beta receptor II promoter and silencing of the target gene in rat hepatic stellate cells, *Biochem Biophys Res Commun* 359 (2), 292–297, 2007.

26. Pulukuri, S. M. and Rao, J. S., Small interfering RNA directed reversal of urokinase plasminogen activator demethylation inhibits prostate tumor growth and metastasis, *Cancer Res* 67 (14), 6637–6646, 2007.

27. Okano, M., Bell, D. W., Haber, D. A., and Li, E., DNA methyltransferases Dnmt3a and Dnmt3b are essential for *de novo* methylation and mammalian development, *Cell* 99 (3), 247–257, 1999.

28. Jeffery, L. and Nakielny, S., Components of the DNA methylation system of chromatin control are RNA-binding proteins, *J Biol Chem* 279 (47), 49479–49487, 2004.

29. Grewal, S. I. and Elgin, S. C., Transcription and RNA interference in the formation of heterochromatin, *Nature* 447 (7143), 399–406, 2007.

30. Heard, E. and Disteche, C. M., Dosage compensation in mammals: fine-tuning the expression of the X chromosome, *Genes Dev* 20 (14), 1848–1867, 2006.

31. Jenuwein, T. and Allis, C. D., Translating the histone code, *Science* 293 (5532), 1074–1080, 2001.

32. Vire, E., Brenner, C., Deplus, R., Blanchon, L., Fraga, M., Didelot, C., Morey, L., Van Eynde, A., Bernard, D., Vanderwinden, J. M., Bollen, M., Esteller, M., Di Croce, L., de Launoit, Y., and Fuks, F., The Polycomb group protein EZH2 directly controls DNA methylation, *Nature*, 2005.

33. Grimaud, C., Bantignies, F., Pal-Bhadra, M., Ghana, P., Bhadra, U., and Cavalli, G., RNAi components are required for nuclear clustering of Polycomb group response elements, *Cell* 124 (5), 957–971, 2006.

34. Han, J., Kim, D., and Morris, K. V., Promoter-associated RNA is required for RNA-directed transcriptional gene silencing in human cells, *Proc Natl Acad Sci USA*, 2007.

35. Kapranov, P., Cheng, J., Dike, S., Nix, D. A., Duttagupta, R., Willingham, A. T., Stadler, P. F., Hertel, J., Hackermuller, J., Hofacker, I. L., Bell, I., Cheung, E., Drenkow, J., Dumais, E., Patel, S., Helt, G., Ganesh, M., Ghosh, S., Piccolboni, A., Sementchenko, V., Tammana, H., and Gingeras, T. R., RNA maps reveal new RNA classes and a possible function for pervasive transcription, *Science* 316 (5830), 1484–1488, 2007.

36. Peters, L. and Meister, G., Argonaute proteins: mediators of RNA silencing, *Mol Cell* 26 (5), 611–623, 2007.

37. Vastenhouw, N. L., Brunschwig, K., Okihara, K. L., Muller, F., Tijsterman, M., and Plasterk, R. H., Gene expression: long-term gene silencing by RNAi, *Nature* 442 (7105), 882, 2006.

38. Rossi, J. J., RNAi as a treatment for HIV-1 infection, *Biotechniques* Suppl, 25–29, 2006.

# 21 Regulation of Gene Expression by RNA-Mediated Transcriptional Gene Silencing

*Kevin V. Morris*

## CONTENTS

## OVERVIEW

RNA interference, specifically, the use of small interfering RNAs (siRNAs), represents a new paradigm in gene knockout technology. siRNAs can be used to knock down the expression of a targeted transcript in what has been termed posttranscriptional gene silencing (PTGS). While there are a plethora of reports applying siRNA-mediated PTGS, there are apparent limitations such as the duration of the effect and saturation of the RNA-induced silencing complex. There is, however, recent data indicating that an alternative pathway is operative in human cells where siRNAs have been shown, similar to plants, *Drosophila, Caenorhabditis elegans,* and *Shizosaccharomyces pombe,* to mediate transcriptional gene silencing (TGS). Transcriptional gene silencing is operative by siRNA, or small antisense RNAs, targeting of the gene promoter regions, which results in epigenetic modifications that lead to silent-state chromatin marks and heterochromatization of the targeted gene. The observation that small RNA-directed TGS is operative via epigenetic modifications suggests that, similar to plants and *S. pombe*, human genes may also be able to be silenced more permanently or for longer periods following a single treatment. Undoubtedly, the potential to employ small RNA technology is broader than once envisioned. The potential to utilize small RNAs to direct epigenetic changes in local chromatin structure offers a new therapeutic avenue that could prove robust and of immeasurable therapeutic value in the directed control of gene expression.

## 21.1 INTRODUCTION

RNA interference (RNAi) is the process in which double-stranded small interfering RNAs (siRNAs) modulate gene expression. RNAi, first described in plants and termed *cosuppression* (reviewed in Reference 1), can suppress gene expression via two distinct pathways involving small interfering RNAs (siRNAs): transcriptional (TGS) and posttranscriptional gene silencing (PTGS) [2,3]. Small interfering RNAs are generated by the action of the ribonuclease (Rnase) III-type enzyme Dicer [4]. The Dicer-processed siRNAs then pair with the complementary target mRNA, where cleavage of the mRNA is instigated by the action of Argonaute 2 (Ago-2) [5]. The Ago-2 protein interacts specifically with the 3′ end of one of the siRNA strands via the Ago-2- conserved PAZ domain. PTGS involves siRNA targeting of mRNA and in human cells operates in both the cytoplasm and nucleus [6,7].

Double-stranded RNAs can also produce TGS in *Arabidopsis*, *S. pombe*, *Drosophila*, and mammalian cells (reviewed in Reference 8). TGS was first observed when doubly transformed tobacco plants exhibited a suppressed phenotype of a transgene that was essentially the result of directed methylation [9]. The observation of TGS in plants turned out to be mediated by dsRNAs as was substantiated in viriod-infected plants [10] and was shown to be the result of RNA-dependent DNA methylation (RdDM). The action of RdDM requires a dsRNA, which is subsequently processed to yield short RNAs [10,11]. Interestingly, it was these short double-stranded RNAs in the doubly transformed tobacco plant that happened to include sequences that were identical to genomic promoter regions and ultimately led to TGS via methylation of the homologous promoter. In general transcriptional gene silencing in plants is carried out by a larger-size class of siRNAs, 24–26 nt in length [12,13].

Recently, members of the Argonaute protein family in *Arabidopsis* have been shown to play a vital role in RdDM of promoter DNA and transposon silencing [14]. Specifically, Ago4 is known to direct siRNA-mediated silencing, and Ago4 mutants display reactivation of silent *SUP* alleles, along with a corresponding decrease in both CpNpG DNA and H3K9 methylation [13]. Consequently, in plants, siRNAs that include sequences with homology to genomic promoter regions are capable of directing the methylation of the homologous promoter and subsequent transcriptional gene silencing.

## 21.2 TRANSCRIPTIONAL GENE SILENCING (TGS) IN *S. POMBE*

Similar to plants, the fission yeast *S. pombe* also employ TGS via siRNAs to silence heterochromatic regions. However, mechanistically, *S. pombe* lacks the epigenetic mechanism of DNA methylation. Instead, *S. pombe* utilizes Argonaute 1 (Ago1) to direct histone methylation, and heterochromatin formation [14]. Typically, euchromatin (less condensed and relatively transcriptionally active) is associated with acetylated histones and with histone H3 di-methylation on lysine 4 (H3mLys-4), whereas heterochromatin (condensed and relatively transcriptionally inactive) is associated with histone H3 di-methylation on lysine 9 (H3mLys-9) [15]. The acetylation of histone tails by histone acetyltransferasese (HAT) results in relaxing the chromatin and a disruption of histone-DNA interactions and gene activation, while the deacetylation of histones by histone deacetylases (HDACs) result in condensation of the chromatin and transcriptional repression (reviewed in Reference 16). Histone H3 Lys-9 methylation directly recruits Swi6/HP1 (mammalian Heterochromatin Protein 1), and this recruitment coincides with spreading in H3 Lys-9 methylation in *cis* [17].

RNAi-mediated TGS in *S. pombe* operates specifically through Histone 3 Lysine-9 methylation (H3K9) [18]. In *S. pombe* mutants in *dcrl* (Dicer homologue) and the only known Argonaute (Ago-1) were shown to be reduced in centromeric repeat H3K9 methylation, which is necessary for centromere function [18]. These data suggested a link between RNAi and directed targeting of specific histone modifications to genomic sequences that, upon modification, recruit or interact with Swi6, resulting in regulation of the heterochromatic state [18]. Additional investigation demonstrated that the *dcrl* processed double-stranded RNAs, which correspond to centromeric repeats in *S. pombe*, interact with Ago-1, Chip1 (chromodomain protein), and Tas1 (previously uncharacterized protein) to form the RITS complex [19] (Figure 21.1). The presence of siRNAs in the RITS complex was shown to require Rdp1, Hrr1 (*helicase required* for *RNA*–heterochromatin assembly *1*), and Cid12 (a 38-kDa protein involved in mRNA polyadenylation) [20]. The siRNA containing RITS complex then associates with chromatin-binding factors Swi6 and Clr4 (Suv39H6 human homologue) to silence targeted genomic regions [19] in an RNA polymerase II-dependent fashion [21], which ultimately results in an Ago-1-mediated slicing of the centromeric-expressed RNAs [22] and silencing through histone methylation of the corresponding centromeric region (Figure 21.1).

## 21.3   TGS IN *C. ELEGANS, DROSOPHILA,* AND *NEUROSPORA*

RNAi-mediated TGS is pervasive throughout biological systems. While the majority of work has been performed on *S. pombe* and plants, other model organisms have demonstrated siRNA-mediated TGS with slight variations in the underlying mechanism. In the fungi *Neurospora crassa*, the silencing of homologous sequences has been termed quelling [23]. Interestingly, in *Neurospora*, the PTGS and TGS pathways both utilize histone 3 lysine 9 di-methylation and moreover appear to be distinct from one another [24], whereas in *Drosophila* the two pathways of RNAi, PTGS, and TGS, respectively, appear to be connected via the *piwi* protein [2]. The piwi family of proteins has several homologues (piwi/sign/elF2C/rde1/Argonaute) and is conserved from plants to animals [25–27]. It is worth noting that in the nematode *C. elegans*, the PAZ-PIWI like protein Rde-1 has been shown to play an essential role in RNA-mediated silencing in the soma of *C. elegans*. Whereas other RNA-mediated silencing mechanisms operative in the germ line do not appear to require Rde-1 [28,29].

More recently, RNA-mediated transcriptional silencing of somatic transgenes in *C. elegans* was shown to be the result of ADAR-encoding genes, adr-1 and adr-2 and

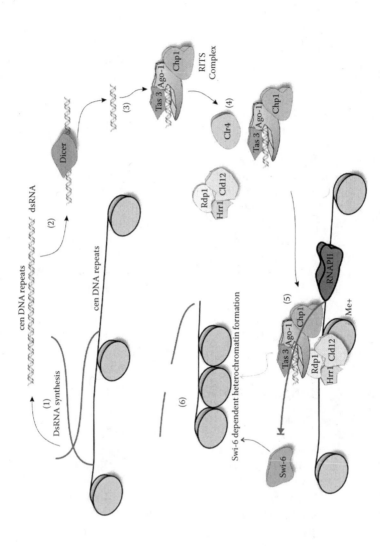

**FIGURE 21.1** *Color version of this figure follows page 238.* TGS in *S. pombe.* DsRNAs are generated from the transcription of centromeric DNA repeats (cen DNA repeats, (1)), which are then processed by Dicer (2) to 21-22 bp siRNAs (3). Next, the dicer processed siRNAs are loaded into the RITS complex (4). The RITS complex then interacts with the RNA dependent RNA polymerase complex (RDRC, (5)) and the histone methyltransferase Clr4, which can then lead to H3K9 methylation (5) and silencing of the siRNA targeted cen DNA repeat regions and/or swi-6 dependent heterochromatin formation (6). The siRNA targeted cen DNA repeat RNAs are also targeted by the action of Ago-1 and sliced (6).

is dependent on Rde-1 [30]. The observed silencing and requirements for both adr-1 and 2 as well as Rde-1 was the result of siRNA targeting of pre-mRNAs and corresponded with a decrease in both RNA polymerase II and acetylated histones at the targeted genomic region [30]. Interestingly, when an RNAi screen was carried out in *C. elegans* to identify components of the RNAi-induced TGS, genes encoding RNA-binding, Polycomb, and chromodomain proteins as well as histone methyltransferases were detected [30]. These data were essentially recapitulated in *C. elegans* in an interesting set of experiments that demonstrated long-term transcriptional gene silencing by RNAi. In essence, one dose of siRNAs was capable of inducing gene silencing that was inherited indefinitely in the absence of the original siRNA trigger [31]. This long-term inheritance of siRNA-mediated TGS appeared to require had-4 (a class II histone deacetylase), K03D10.3 (a histone acetyltransferase of the MYST family), isw-1 (a homologue of the yeast chromatin-remodeling ATPase ISW1), and mrg-1 (a chromodomain protein) [31]. Taken together, these data strongly suggest that RNA-mediated transcriptional gene silencing in *C. elegans* contains a convergence of pathways that include epigenetic modifying factors as well as RNA and that, indeed, RNA is more intricately involved in the regulation of gene expression than has previously been envisioned.

## 21.4   TGS IN HUMAN CELLS

A few commonalities can be discerned from the work performed in plants, *S. pombe*, *C. elegans*, and *Drosophila*: (1) siRNAs can direct transcriptional gene silencing in a specific and directed manner, (2) epigenetic modifications such as histone methylation are observed at the siRNA targeted chromatin, and (3) PIWI-related proteins appear to be involved. As such, and based on the theory of evolution and natural selection, one would also expect to observe similarities between plants, *C. elegans*, *S. pombe*, *Drosophila*, and humans.

In human cells, as well as other organisms, DNA is packaged by being essentially wrapped around histone proteins forming nucleosomes, and is also known as chromatin. Chromatin can exist in various states such as euchromatin or heterochromatin. Euchromatin is typically less condensed and comparatively transcriptionally active and tends to be associated with acetylated histones as well as with histone H3 di-methylation on lysine 4 (H3mLys-4), whereas heterochromatin typically associated with histone H3 di-methylation on lysine 9 (H3mLys-9) and tends to be more condensed and relatively transcriptionally inactive [15]. The acetylation of histone tails by histone acetyltransferasese (HAT) results in relaxing of the chromatin and a disruption of histone-DNA interactions and gene activation. Conversely, the deacetylation of histones by histone deacetylases (HDACs) results in condensation of the chromatin and transcriptional repression (reviewed in Reference 16). Currently, there are six well-defined gene regulatory histone deacetylase complexes (HDAC) that appear to be involved in the suppression of gene expression (reviewed in Reference 32).

Observations of siRNA-mediated TGS in mammalian cells has lagged behind the work done in other model organisms such as *Arabidopsis* and *S. pombe* [33]. However, recent studies by our group and others has revealed that siRNA-mediated TGS in mammalian cells does occur and appears to be the result of the siRNA-directed

H3K9, H3K27 methylation at the siRNA-targeted promoter [33–37]. While some DNA methylation has also been observed at the siRNA-targeted promoters, the role which DNA methylation plays in the observed silencing is debatable [36,38–40]. Indeed, the ability for siRNAs to direct-target DNA methylation could result in a much greater duration of suppression, as DNA methylation tends to correlate better with long-term suppressed genes than does histone methylation. However, to date, little is known regarding how long siRNA-directed TGS of RNA polymerase II promoter (RNAPII) can persist in human cells. The majority of siRNA-directed TGS experiments in human cells have been carried out so far with synthetic siRNAs targeted to RNAPII promoters [33,34,36,37,40–44] or constitutive-expressing lentiviral-transduced cell lines [34]. Nonetheless, that observation that siRNA-directed TGS is operative via epigenetic modifications argues favorably for a longer-term effect relative to siRNA-directed PTGS, provided the siRNAs are efficiently delivered to the nucleus. However, experimental evidence supporting this claim is still lacking.

While the duration of siRNA-directed effect remains to be determined, what has become more clear recently is that siRNAs directed to RNAPII-promoter regions can mediate transcriptional silencing in human cells and that a repressive histone methyl mark is observed at the targeted promoter [33,35,40,43]. Furthermore, only the antisense strand of the siRNA is required to mediate histone methylation and silencing of the targeted RNAPII promoter [43]. Moreover, RNAPII appears required for siRNA-mediated TGS and DNA methyltransferase 3A (DNMT3A) co-immunoprecipitates (co-IP) with biotin-linked siRNAs at the H3K27me3-targeted promoter [43]. Recently, Ago-1 and possibly also Ago-2 have been shown to be involved in siRNA-mediated TGS in human cells [41,42]. Ago-1 was shown to co-IP with RNAPII, and an enrichment of EZH2 and the TAR-binding protein (TRBP) have also been observed at Ago-1 containing siRNA-targeted promoters [42]. Clearly, in human cells, similar to observations in other organisms such as *S. pombe* and *Arabidopsis*, Argonaute proteins and H3K9 methylation are required for transcriptional silencing, linking the RNAi and chromatin silencing machinery.

## 21.5 MODEL OF TGS IN HUMAN CELLS

A model for the mechanism of how siRNA-directed TGS is directed and initiated in human cells has begun to emerge and appears to contain some similarities, as well as distinct differences with the previously established models for TGS in *S. pombe* and plants. Similar to *S. pombe,* siRNA-mediated TGS in human cells involves the siRNAs, particularly the antisense strand, RNAPII, and histone methylation [41,42,45,46] (Figures 21.2a and 21.2b). However, in human cells DNMT3A has been shown to coimmunoprecipitate along with the antisense strand of the promoter-specific siRNA at the siRNA-targeted promoter [43]. Interestingly, DNMT3A has also been shown to bind siRNAs in vitro [47]. Similar to observations in *S. pombe* [21], TGS in human cells requires RNAPII [43], possibly suggesting that the promoter region is transcribed, and either an RNAPII-expressed transcript covers the targeted gene/chromatin (Figure 21.2a) or that during transcription, RNAPII unwinds the targeted gene and subsequently allows access of the antisense strand of the siRNA (Figure 21.2b) to the targeted gene. Either mechanism would explain

(a)

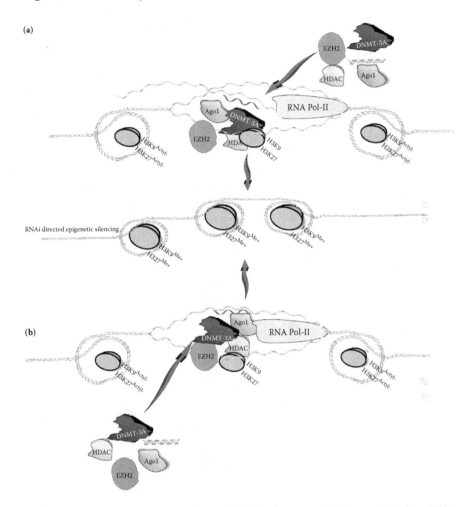

**FIGURE 21.2** Proposed mechanisms for TGS in human cells. Two models for siRNA-mediated TGS have been proposed: either an RNA/RNA or a RNA/DNA-mediated mode of silencing. (a) The RNA/RNA model might operate via the siRNAs interacting with a low-copy, possibly noncoding RNA transcript that spans the chromatin of the targeted promoter region. One notion is that the siRNAs are unwound, possibly by the action of an Ago-1-containing complex (though there is currently no data supporting this function in Ago-1). The unwinding of the siRNA might then allow interactions with a complex also containing DNMT3A. Thus, the antisense strand of the siRNA, Ago-1 [42] and DNMT3A [43,47], as well as factors previously shown to bind DNMT3A—such as HDAC-1, and Suv39H1 [58,59], and possibly EZH2 [60]—might all localize to the targeted promoter region, possibly by interacting with a low-copy RNA being produced by RNAPII during transcription of the siRNA targeted gene. The result of targeting these factors to this region would be the methylation of the local histones, H3K9me2 and H3K27me3, ultimately resulting in targeted TGS. (b) Alternatively, siRNA-mediated TGS might function through an RNA/DNA intermediate. The RNA/DNA model would be expected to function by the antisense strand of the siRNA, gaining access to the targeted DNA by the effects of RNAPII unwinding the targeted genomic region, which would then allow the Ago-1, DNMT3a [58], HDAC-1, Suv39H1 [61], and possibly EZH2 chromatin remodeling factors to gain access, ultimately leading to gene silencing of the siRNA-targeted promoter.

how the antisense strand of the siRNA can localize to the particular targeted gene-promoter region and fundamentally recruit or direct chromatin modifications that ultimately lead to TGS.

Preliminary evidence from siRNA-mediated TGS of the EF1 alpha and HIV-1 promoters has demonstrated a paradigm for the underlying mechanism in which the antisense strand of the siRNA by itself can direct TGS via the induction of a corresponding silent histone methyl-mark at the targeted promoter [43]. Interestingly, this siRNA and/or antisense RNA-directed histone methylation is capable of spreading distal in a 5′–>3′ direction of the targeted region corresponding to the homologous siRNA. This spreading of histone methylation might be due to an interaction with Ago-1 and RNAPII [42]; specifically, histone methylation H3K9me2, and the observed H3K27me3 methyl mark at the siRNA-targeted promoter, is reminiscent of those phenotypes observed in the X-inactivation-like pathway. One cannot help but wonder if siRNA-mediated RNAi-like responses in the cell (PTGS and TGS) are simply a remnant of a more ancient pathway utilized by the cell to deal with competing genes, such as is observed in mammalian X-inactivation.

The observation that RNAPII is involved in siRNA-mediated TGS in both *S. pombe* and more recently in human cells [42,43] suggests two potential models. The first model, the RNA/RNA model, proposes that there is an RNAPII-expressed noncoding transcript that is homologous to the targeted promoter/gene, which somehow remains associated with the local chromatin corresponding to the targeted gene (Figure 21.2A). This noncoding RNA may somehow remain affiliated with the nucleosome(s), and as such would permit the RNAi machinery to direct chromatin modifications of the targeted genomic region ultimately leading to TGS. This putative non-coding RNA could be envisioned to act as a local "address" to allow chromatin and RNAi modification complexes, guided by siRNAs, access to the targeted gene (Figure 21.2A). Supporting the RNA/RNA model is the observation that heterchromatin formation in mouse cells involves HP1 proteins and treatment with RNase causes a dispersion of HP1 proteins from pericentromeric foci [48], and higher-order chromatin structures appear to contain an uncharacterized RNA component that might function as a scaffolding in chromatin remodeling [49]. Whether such a noncoding RNA exists, if the siRNAs target this noncoding RNA during RNAPII transcription, or if the noncoding RNA somehow associates with local chromatin remains to be determined.

An alternative model, the RNA/DNA model, could also be envisioned to operate in an RNAPII-dependent manner, where RNAPII essentially unwinds the targeted DNA and permits the intercalation of the antisense strand of the siRNA, as well as RNAi and chromatin remodeling machinery, to access to the targeted promoter region (Figure 21.2B). Supporting this model is the observation that RNAPII has been shown to associate with complete unfolding of 1.85 out of 3 nucleosomes upstream of the transcription start site for the PH05 promoter [50]; the siRNA EF52 target site, shown to initiate TGS in human cells is ~1 nucleosome upstream of the TATAA transcriptional start site [33]. Moreover, RNA Pol-II is associated with nearly 60 subunits and a mass in excess of 3 million Daltons [50]. Overall, these reports suggest, at least in human cells, that a good, albeit speculative, hypothesis for the role of RNAPII in siRNA-mediated TGS might be to essentially unwind the targeted promoter region to allow the promoter-directed siRNAs access to their respective target.

## 21.6   ENDOGENOUS SMALL RNAs INVOLVED IN GENE REGULATION

While it is becoming apparent in human cells that there is a mechanism in place by which siRNAs or antisense RNAs can transcriptionally modulate gene expression, it is less clear what the endogenous signal utilizing this pathway might be, that is, siRNA, small antisense RNAs, micro RNAs, etc. In *C. elegans,* deep sequencing revealed a class of 21 U-RNAs that consists of 21-nt-long RNAs, which contain a uridine 5′-monophophate and 3′ terminal ribose [51]. These 21 U-RNAs were found to correlate with 5700 genomic loci and were essentially dispersed between protein-coding regions [51]. Overall, these data suggest that noncoding regions of the genome, while not necessarily coding for a protein, do appear to express RNAs that might possibly be involved in gene regulation. As many of these small 21 U-RNAs matched noncoding regions, it is possible they are involved in transcriptional modulation of *C. elegans* gene expression [51]. In human cells, however, far less is known, and the majority of inferences have been surmised from computational approaches. One recent observation involved the discovery and characterization of a vast array of small (21 to 26 nt), noncoding RNAs which fundamentally suggested that there is an RNA component, possibly involved in gene regulation, that is, siRNA-mediated TGS, within the basic fabric of the human cells [52]. Another interesting observation was noted when the intergenic and intronic regions of the human genome were assessed: a subset of ~127,998 patterns that were essentially ~22 nucleotides in length were discerned and termed "pyknons" [53]. Many of these pyknons were overlapping genes involved in cell communication, transcription, regulation of transcription, and cell signaling [53]. Overall, these data suggest that another layer of complexity is operative in the genome of many organisms and supports a paradigm in which RNA is actively involved in the regulation of DNA. One cannot help but contemplate the concept that DNA and genomes are simply repositories of information that is actively managed by RNA, which oddly enough is encoded from the DNA/genes it is actively managing.

## 21.7   CONCLUSIONS

The observation in human cells that siRNAs, particularly the antisense strand alone, can direct TGS, and that this event involves DNMT3a, histone methylation, and RNAPII, strongly suggests that siRNAs can be used to specifically direct epigenetic modifications in human cells. Indeed the modification of histone tails, such as methylation, results in a "histone code." The histone code hypothesis argues that the local histone environment (specifically in the nucleosomes) can have an effect on the expression profile of the corresponding local gene [54]. These "marked" histone tails are then capable of dictating the recruitment of various specialized chromatin remodeling factors [55,56]. To date, the histone code is best exemplified by the sheer multiplicity of modifications that can occur to histones (reviewed in Reference 57). Interestingly, the fundamental underlying mechanism responsible for governing the histone code is not yet understood. One potential mechanism to regulating the histone code could be mediated by siRNAs and, in particular, small antisense

RNAs. Interestingly, the recent discovery and characterization of a vast array of small (21 to 26 nt), noncoding RNAs suggests that there is an RNA component possibly involved in gene regulation—that is, siRNA-mediated TGS woven into the basic fabric of the cell—and this has been, to date, overlooked [52]. While it is becoming apparent that RNAi goes beyond the confines of the cytoplasm, the observations with siRNA-mediated TGS are evocative of an antisense-related phenomenon that is deeply seeded in the fabric of the cell. Indeed, one day it may be possible to harness RNA to direct permanent epigenetic modifications, resulting in superlative control of the human genome.

## ACKNOWLEDGMENTS

I would like to thank Rob Martienssen and John J. Rossi for their comments, criticisms, and valued conversations on siRNA-mediated TGS. This work was supported by NIH HLB R01 HL83473 to KVM.

## REFERENCES

1. Tijsterman, M., R.F. Ketting, and R.H. Plasterk, The genetics of RNA silencing. *Annu Rev Genet*, 2002. **36**: 489–519.
2. Pal-Bhadra, M., U. Bhadra, and J.A. Birchler, RNAi related mechanisms affect both transcriptional and posttranscriptional transgene silencing in *Drosophila*. *Mol Cell*, 2002. **9**: 315–327.
3. Sijen, T., I. Vign, A. Rebocho, R. Blokland, D. Roelofs, J. Mol, and J. Kooter, Transcriptional and posttranscriptional gene silencing are mechansitically related. *Curr Biol*, 2001. **11**: 436–440.
4. Bernstein, E. et al., Role for a bidentate ribonuclease in the initiation step of RNA interference. *Nature*, 2001. **409**(6818): 363–366.
5. Liu, J. et al., Argonaute2 Is the Catalytic Engine of Mammalian RNAi. *Science*, 2004. **305**(5689): 1437–1441.
6. Langlois, M.A. et al., Cytoplasmic and nuclear retained DMPK mRNAs are targets for RNA interference in myotonic dystrophy cells. *J Biol Chem*, 2005. **280**(17): p. 16949–16954.
7. Robb, G.B. et al., Specific and potent RNAi in the nucleus of human cells. *Nat Struct Mol Biol*, 2005. **12**(2): 133–137.
8. Matzke, M.A. and J.A. Birchler, RNAi-mediated pathways in the nucleus. *Nat Rev Genet*, 2005. **6**(1): 24–35.
9. Matzke, M.A., M. Primig, J. Trnovsky, and A.J.M. Matzke, Reversible methylation and inactivation of marker genes in sequentially transformed tobacco plants. *EMBO J*, 1989. **8**: 643–649.
10. Wassenegger, M., M.W. Graham, and M.D. Wang, RNA-directed de novo methylation of genomic sequences in plants. *Cell*, 1994. **76**: 567–576.
11. Mette, M.F., W. Aufsatz, J. Van der Winden, A.J.M. Matzke, and M.A. Matzke, Transcriptional silencing and promoter methylation triggered by double-stranded RNA. *EMBO J*, 2000. **19**(19): 5194–5201.
12. Hamilton, A. et al., Two classes of short interfering RNA in RNA silencing. *EMBO J*, 2002. **21**(17): 4671–4679.
13. Zilberman, D., X. Cao, and S.E. Jacobsen, ARGONAUTE4 control of locus-specific siRNA accumulation and DNA and histone methylation. *Science*, 2003. **299**(5607): 716–719.

14. Lippman, Z. et al., Distinct Mechanisms Determine Transposon Inheritance and Methylation via Small Interfering RNA and Histone Modification. *PLoS Biol*, 2003. **1**(3): p. E67.

15. Lippman, Z. et al., Role of transposable elements in heterochromatin and epigenetic control. *Nature*, 2004. **430**(6998): 471–476.

16. Lusser, A., Acetylated, methylated, remodeled: chromatin states for gene regulation. *Curr Opin Plant Biol*, 2002. **5**(5): 437–443.

17. Hall, I.M., G.D. Shankaranarayana, K. Noma, N. Ayoub, A. Cohen, and S.I.S. Grewal, Establishment and Maintenance of a Heterochromatin Domain. *Science*, 2002. **297**: 2232–2237.

18. Volpe, T.A., C. Kidner, I.M. Hall, G. Teng, S.I.S. Grewal, and R.A. Martienssen, Regulation of Heterchromatic Silencing and Histone H3 Lysine-9 Methylation by RNAi. *Science*, 2002. **297**(September 13, 2002): 1833–1837.

19. Verdel, A. et al., RNAi-mediated targeting of heterochromatin by the RITS complex. *Science*, 2004. **303**(5658): 672–676.

20. Motamedi, M.R. et al., Two RNAi complexes, RITS and RDRC, physically interact and localize to noncoding centromeric RNAs. *Cell*, 2004. **119**(6): 789–802.

21. Kato, H. et al., RNA polymerase II is required for RNAi-dependent heterochromatin assembly. *Science*, 2005. **309**(5733): 467–469.

22. Irvine, D.V. et al., Argonaute slicing is required for heterochromatic silencing and spreading. *Science*, 2006. **313**(5790): 1134–1137.

23. Romano, N. and G. Macino, Quelling: transient inactivation of gene expression in Neurospora crassa by transformation with homologous sequences. *Mol Microbiol*, 1992. **6**(22): 3343–3353.

24. Chicas, A. et al., Small interfering RNAs that trigger posttranscriptional gene silencing are not required for the histone H3 Lys9 methylation necessary for transgenic tandem repeat stabilization in Neurospora crassa. *Mol Cell Biol*, 2005. **25**(9): 3793–3801.

25. Fagard, M. et al., AGO1, QDE-2, and RDE-1 are related proteins required for post-transcriptional gene silencing in plants, quelling in fungi, and RNA interference in animals. *Proc Natl Acad Sci USA*, 2000. **97**(21): 11650–11654.

26. Tabara, H. et al., The rde-1 gene, RNA interference, and transposon silencing in *C. elegans*. *Cell*, 1999. **99**(2): 123–132.

27. Grishok, A. et al., Genes and mechanisms related to RNA interference regulate expression of the small temporal RNAs that control C. elegans developmental timing. *Cell*, 2001. **106**(1): 23–34.

28. Dernburg, A.F. et al., Transgene-mediated cosuppression in the C. elegans germ line. *Genes Dev*, 2000. **14**(13): 1578–1583.

29. Ketting, R.F. and R.H. Plasterk, A genetic link between co-suppression and RNA interference in C. elegans. *Nature*, 2000. **404**(6775): 296–298.

30. Grishok, A., J.L. Sinskey, and P.A. Sharp, Transcriptional silencing of a transgene by RNAi in the soma of C. elegans. *Genes Dev*, 2005.

31. Vastenhouw, N.L. et al., Gene expression: long-term gene silencing by RNAi. *Nature*, 2006. **442**(7105): 882.

32. Dobosy, J.R. and E.U. Selker, Emerging connections between DNA methylation and histone acetylation. *Cell Mol Life Sci*, 2001. **58**(5–6): 721–727.

33. Morris, K.V. et al., Small interfering RNA-induced transcriptional gene silencing in human cells. *Science*, 2004. **305**(5688): 1289–1292.

34. Castanotto, D. et al., Short hairpin RNA-directed cytosine (CpG) methylation of the RASSF1A gene promoter in HeLa cells. *Mol Ther*, 2005. **12**(1): 179–183.

35. Buhler, M. et al., Transcriptional silencing of nonsense codon-containing immunoglobulin minigenes. *Mol Cell*, 2005. **18**(3): 307–317.

36. Janowski, B.A., K.E. Huffman, J.C. Schwartz, R. Ram, D. Hardy, D.S. Shames, J.D. Minna, and D.R. Corey, Inhibiting gene expression at transcription start sites in chromosomal DNA with antigene RNAs. *Nat Chem Biol*, 2005. **1**: 210–215.

37. Suzuki, K., T. Shijuuku, T. Fukamachi, J. Zaunders, G. Guillemin, D. Cooper, and A. Kelleher, Prolonged transcriptional silencing and CpG methylation induced by siRNAs targeted to the HIV-1 promoter region. *J RNAi Gene Silencing*, 2005. **1**(2): 66–78.

38. Park, C.W. et al., Double-stranded siRNA targeted to the huntingtin gene does not induce DNA methylation. *Biochem Biophys Res Commun*, 2004. **323**(1): 275–280.

39. Svoboda, P. et al., Lack of homologous sequence-specific DNA methylation in response to stable dsRNA expression in mouse oocytes. *Nucleic Acids Res*, 2004. **32**(12): 3601–3606.

40. Ting, A.H. et al., Short double-stranded RNA induces transcriptional gene silencing in human cancer cells in the absence of DNA methylation. *Nat Genet*, 2005. **37**(8): 906–910.

41. Janowski, B.A. et al., Involvement of AGO1 and AGO2 in mammalian transcriptional silencing. *Nat Struct Mol Biol*, 2006.

42. Kim, D.H. et al., Argonaute-1 directs siRNA-mediated transcriptional gene silencing in human cells. *Nat Struct Mol Biol*, 2006.

43. Weinberg, M.S., L.M. Villeneuve, A. Ehsani, M. Amarzguioui, L. Aagaard, Z. Chen, A.D. Riggs, J.J. Rossi, and K.V. Morris, The antisense strand of small interfering RNAs directs histone methylation and transcriptional gene silencing in human cells. *RNA*, 2005. **12**(2).

44. Zhang, M., H. Ou, Y.H. Shen, J. Wang, J. Wang, J. Coselli, and X.L. Wang, Regulation of endothelial nitric oxide synthase by small RNA. *Proc Natl Acad Sci USA*, 2005. **102**(47): 16967–16972.

45. Morris, K.V., siRNA-mediated transcriptional gene silencing: the potential mechanism and a possible role in the histone code. *Cell Mol Life Sci*, 2005. **62**(24): 3057–3066.

46. Morris, K.V., Therapeutic potential of siRNA-mediated transcriptional gene silencing. *Biotechniques*, 2006. Suppl: 7–13.

47. Jeffery, L. and S. Nakielny, Components of the DNA methylation system of chromatin control are RNA-binding proteins. *J Biol Chem*, 2004. **279**(47): 49479–49487.

48. Muchardt, C. et al., Coordinated methyl and RNA binding is required for heterochromatin localization of mammalian HP1alpha. *EMBO Rep*, 2002. **3**(10): 975–981.

49. Maison, C. et al., Higher-order structure in pericentric heterochromatin involves a distinct pattern of histone modification and an RNA component. *Nat Genet*, 2002. **19**: 19.

50. Boeger, H. et al., Structural basis of eukaryotic gene transcription. *FEBS Lett*, 2005. **579**(4): 899–903.

51. Ruby, J.G., C. Jan, C. Player, M.J. Axtell, W. Lee, C. Nusbaum, H. Ge, and D.P. Bartel, Large-Scale Sequencing Reveals 21U-RNAs and Additional MicroRNAs and Endogenous siRNAs in *C. elegans*. *Cell*, 2006. **127**(December 15): 1193–1207.

52. Katayama, S. et al., Antisense transcription in the mammalian transcriptome. *Science*, 2005. **309**(5740): 1564–1566.

53. Rigoutsos, I. et al., Short blocks from the noncoding parts of the human genome have instances within nearly all known genes and relate to biological processes. *Proc Natl Acad Sci USA*, 2006. **103**(17): 6605–6610.

54. Jenuwein, T. and C.D. Allis, Translating the histone code. *Science*, 2001. **293**(5532): 1074–1080.

55. Strahl, B.D. et al., Methylation of histone H3 at lysine 4 is highly conserved and correlates with transcriptionally active nuclei in tetrahymena [in process citation]. *Proc Natl Acad Sci USA*, 1999. **96**: 14967–14972.

56. Turner, B.M., Histone acetylation and an epigenetic code. *Bioessays*, 2000. **22**(9): 836–845.

57. Berger, S.L., Histone modifications in transcriptional regulation. *Curr Opin Genet Dev*, 2002. **12**(2): 142–148.
58. Fuks, F. et al., Dnmt3a binds deacetylases and is recruited by a sequence-specific repressor to silence transcription. *EMBO J*, 2001. **20**(10): 2536–2544.
59. Fuks, F. et al., DNA methyltransferase Dnmt1 associates with histone deacetylase activity. *Nat Genet*, 2000. **24**(1): 88–91.
60. Vire, E. et al., The Polycomb group protein EZH2 directly controls DNA methylation. *Nature*, 2005.
61. Fuks, F. et al., The DNA methyltransferases associate with HP1 and the SUV39H1 histone methyltransferase. *Nucleic Acids Res*, 2003. **31**(9): 2305–2312.

# Index